Study Guide & Solutions Manual

Volume 2

to accompany

# ORGANIC
# CHEMISTRY

Third Edition

# Study Guide & Solutions Manual

# Volume 2

to accompany

# ORGANIC CHEMISTRY

Third Edition

G. Marc Loudon
Joseph G. Stowell
*Purdue University*

**The Benjamin/Cummings Publishing Company, Inc.**

Menlo Park, California  •  Reading, Massachusetts
New York • Don Mills, Ontario • Wokingham, U.K.  •  Amsterdam
Bonn • Paris • Milan • Madrid • Sydney • Singapore • Tokyo
Seoul • Taipei • Mexico City • San Juan, Puerto Rico

Executive Editor: Sally Elliot
Sponsoring Editor: Anne Scanlan-Rohrer
Associate Editor: Leslie With
Editorial Assistant: Sharon Sforza
Accuracy Reviewers: Morton Golub; Steven Hardinger, California State University
        at Fullerton; Martin Newcomb, Wayne State University
Production Editor: Larry Olsen
Manufacturing Supervisor: Merry Free Osborn
Cover Designer: Yvo Riezebos
Cover Illustration: Joseph Maas

ISBN 0-8053-6664-4

5 6 7 8 9 10—CRS— 99

The Benjamin/Cummings Publishing Company, Inc.
2725 Sand Hill Road
Menlo Park, CA 94025

# Preface

The purpose of this preface is to give you an overview of the organization and features of this Study Guide and Solutions Manual so that you might use it more effectively.

## Organization

Each chapter corresponds to the chapter of the text with the same number and title. The following sections are found within each chapter.

1. **Terms**

   This is a list of new or important terms encountered within the corresponding chapter of the text along with the location of each term. The lists of terms in this manual comprise the vocabulary of organic chemistry that you will need to master the subject. These lists will probably be most useful if you can *write* a definition of each term *in your own words* and give an example, if appropriate. Don't forget that if you need a more general index of terms, the text itself has a very detailed index that can be used to locate the definition of any term of interest.

2. **Concepts**

   This is a concise summary of the concepts within the chapter in outline form. In some cases, these are presented from a somewhat different perspective, and organized somewhat differently, than they are in the text. These outlines can be used in various ways. For example, they can provide a quick review after you have read the chapter; they can be used for ready reference while you are working problems; or they can serve as a refresher prior to class periods.

3. **Reactions**

   This summary consolidates the reactions of the chapter in one place and presents not only the reactions themselves but also the essential features of their mechanisms and stereochemistry for cases in which these issues are discussed in the text. Studying and learning reactions is a skill that should be cultivated, and Study Guide Link 5.1 on page 88, Vol. 1, of this manual will help you to use these reaction summaries to best advantage.

4. Study Guide Links

These are short extensions of text material that are called out by margin icons within the text. Study Guide Links are of two types. A Study Guide Link of the first type, flagged with a checked (✓) icon, provides additional explanations of topics that typically cause difficulty. For example, Study Guide Link 5.1, "How to Study Organic Reactions," (mentioned in item 3 above) is flagged with a check at a point in the text at which some students begin to have difficulty learning reactions. This Study Guide Link provides detailed hints on how to study and learn organic reactions efficiently. A Study Guide Link of the second type, called out in the text by an unchecked icon, provides more in-depth information about the particular topic or a different way of looking at the same topic. Although it may at first seem that the material in the Study Guide Links should have been included in the text, doing so would have made the text unacceptably long, and, in some cases, would have interrupted the logical flow of material.

5. Solutions

We have endeavored not only to provide answers to problems but also to show how these answers are deduced. Solutions to the problems within the text come first, and these are followed by solutions to the additional problems. Only solutions to the *asterisked* problems are provided. (See the discussion of paired problems, below.) The best use of solutions is to try to work problems with the solutions closed or covered, and consult a solution only when you have answered the problem or have given it a reasonable effort.

# Features

This manual contains certain features that are designed to help you use it more effectively.

1. Typesetting Conventions

The text and typesetting conventions are the same ones used in the text.

2. Index Tabs

You will find on the first right-hand page of each Solutions section a black rectangle at the edge of the page. (For example, the first of these in this volume appears on page 471.) The edges of these rectangles can be seen when the manual is viewed end-on and the pages are bent back slightly. These serve as index tabs that can be used to locate the solutions sections rapidly. In Volume 1, they are arranged in two rows of nine, and in Volume 2, they are arranged in one row of nine. Thus, to locate the solutions for Chapter 20, bend back the pages of Volume 2 and open to the second index mark from the top.

3. Icon Comments

Within the solutions to the problems, you will occasionally find comments marked with icons of two different types.

 This icon indicates a comment that provides additional information about the topic covered in the problem, an alternate correct answer, or another way of looking at the problem.

 This icon indicates a special caution about some aspect of the problem or the material on which the problem is based—something you should be careful about to avoid confusion.

4. **Paired Problems**

Many of the problems within the text are *paired.* This means that some problems (or problem parts) marked with an asterisk are followed by problems (or problems parts) of a similar type, in most cases of equal or lesser difficulty, which are not asterisked. Only the answers to the asterisked problems appear in the Solutions sections of this manual. If you understand the solution to an asterisked problem (or part), you should be able to work the unasterisked problem (or part) that follows using similar reasoning. (In one or two cases, an unasterisked problem follows a worked-out Study Problem in the text rather than an asterisked problem.) Note that not every problem is paired in this way; some asterisked problems have no unasterisked partner. You may want to use the unasterisked problems as a "test bank" that can be used for a review prior to examinations.

5. **Extensive Cross-Referencing**

The further into the text you go, the more the new material depends on earlier material. To help you review earlier material, we have provided in the solutions sections of this manual a large number of cross-references both to the text and to preceding parts of this manual. Take advantage of this cross-referencing to review and reinforce principles that you have learned earlier. Remember that continued reinforcement is a key to effective learning.

This manual has undergone three separate accuracy checks by the authors and a critical reading by three accuracy checkers. We have tried hard to make this manual as free of errors as possible. Nevertheless, if errors are found, we would like to know about them so that we can correct them on reprint if possible. You can send these or any comments to us by electronic mail at *loudonm@sage.cc.purdue.edu,* by FAX at 317-494-7880, or by mail at RHPH Building, Purdue University, West Lafayette IN 47907-1330.

We would like to gratefully acknowledge the many comments of the users of the previous edition of this manual, especially Mark Cushman, John Schwab, Ron Magid, and Phil Fuchs, as well as their teaching assistants and students. We particularly appreciate the conscientious work of the accuracy checkers, Professor Martin E. Newcomb, Professor Steve Hardinger, and Dr. Morton Golub.

We sincerely hope that you find this manual useful in your study of organic chemistry.

November 9, 1995
G. Marc Loudon and Joseph G. Stowell

# Contents

# Chemistry of Carboxylic Acid Derivatives    521

# Chemistry of Enolate Ions, Enols, and α,β-Unsaturated Carbonyl Compounds    565

# Chemistry of Amines    615

# Chemistry of Naphthalene and the Aromatic Heterocycles    653

# Pericyclic Reactions    691

# 19

# Chemistry of Aldehydes and Ketones; Carbonyl-Addition Reactions

## Terms

# Concepts

## I. Introduction to Aldehydes and Ketones

### A. GENERAL

1. Carbonyl compounds contain the carbonyl group (C=O) and include
   a. aldehydes
   b. ketones
   c. carboxylic acids and most carboxylic acid derivatives:
      i. esters
      ii. amides
      iii. anhydrides
      iv. acid chlorides
2. In an aldehyde, at least one of the groups at the carbonyl carbon atom is a hydrogen, and the other may be alkyl, aryl, or a second hydrogen.
3. In a ketone, the groups bound to the carbonyl carbon are alkyl or aryl groups.

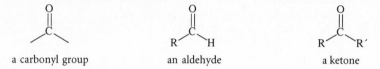

| a carbonyl group | an aldehyde | a ketone |

4. The carbonyl carbon of a typical aldehyde or ketone is $sp^2$-hybridized with bond angles approximating 120°.
5. A carbon-oxygen double bond:
   a. consists of a $\sigma$ bond and a $\pi$ bond.
   b. is shorter than a carbon-carbon double bond.

### B. COMMON NOMENCLATURE

1. In the common nomenclature of aldehydes, the suffix *aldehyde* is added to a prefix that indicates the chain length of the group attached to the carbonyl group.

| form | H— | isobutyr | $(CH_3)_2CH_2$— |
| acet | $CH_3$— | valer | $CH_3CH_2CH_2CH_2$— |
| propion | $CH_3CH_2$— | isovaler | $(CH_3)_2CHCH_2$— |
| butyr | $CH_3CH_2CH_2$— | benz | Ph— |

   a. Common names are almost always used for the simplest aldehydes.
   b. Benzaldehyde is the simplest aromatic aldehyde.

isovaleraldehyde                          benzaldehyde

2. The common name of some ketones are constructed by citing the two groups on the carbonyl carbon followed by the word *ketone*.

dicyclopentyl ketone

3. Certain aromatic ketones are named by attaching the suffix *ophenone* to the appropriate prefix.

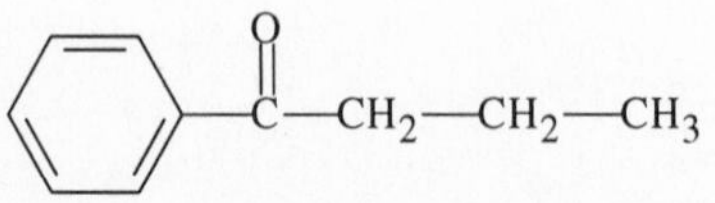

butyrophenone

4. Simple substituted aldehydes and ketones can be named in the common system by designating the position of substituents with Greek letters, beginning at the position adjacent to the carbonyl group.

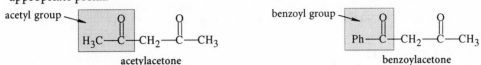

$\gamma$-chlorovaleraldehyde

   a. A carbon adjacent to the carbonyl group is termed the $\alpha$-carbon.
   b. The hydrogens on the $\alpha$-carbon are termed $\alpha$-hydrogens.
5. Many common carbonyl-containing substituent groups are named by adding *yl* or *oyl* to the appropriate prefix.

acetyl group

acetylacetone

benzoyl group

benzoylacetone

   a. Such groups are called in general acyl groups.
   b. To be named as an acyl group, a substituent group must be connected to the remainder of the molecule at its carbonyl carbon.

## C. SUBSTITUTIVE NOMENCLATURE

1. The substitutive name of an aldehyde is constructed from a prefix indicating the length of the carbon chain followed by the suffix *al*; the prefix is the name of the corresponding hydrocarbon without the final *e*.
   a. The carbonyl carbon is designated as carbon-1 in numbering the carbon chain of an aldehyde.
   b. The final *e* is not dropped when the carbon chain has more than one aldehyde group.
   c. When an aldehyde group is attached to a ring, the suffix *carbaldehyde* is appended to the name of the ring.
       *i.* In aldehydes of this type, carbon-1 is not the carbonyl carbon, but rather the ring carbon attached to the carbonyl group.
       *ii.* In older nomenclature, the suffix carboxaldehyde was used.

3-methylbutanal

propanedial

3-methylcyclopentanecarbaldehyde

   d. The name benzaldehyde is used in both common and substitutive nomenclature.
2. A ketone is named by giving the hydrocarbon name of the longest carbon chain containing the carbonyl group, dropping the final *e*, and adding the suffix *one*.
   a. The position of the carbonyl group is given the lowest possible number.
   b. The final *e* of the hydrocarbon name is not dropped in the nomenclature of diones, triones, etc.
   c. When a ketone carbonyl group is treated as a substituent, its position is designated by the term *oxo*.

1,2-cyclobutanedione

4-(1-oxoethyl)benzaldehyde

3. Aldehyde and ketone carbonyl groups receive higher priority than —OH or —SH groups for citation as principal groups. (A complete list of group priorities is given in Appendix I, text page A-1.)

—CH=O (aldehyde) > $\overset{\backslash}{\underset{/}{C}}$=O (ketone) > —OH > —SH

6-mercapto-4,4-dimethyl-5-oxo-2-hexynal

### D. Physical Properties of Aldehydes and Ketones

1. Most simple aldehydes and ketones are liquids.
   a. Formaldehyde is a gas.
   b. Acetaldehyde has a boiling point very near room temperature.
   c. Aldehydes and ketones have higher boiling points than alkenes or alkanes with similar molecular weights and shapes because of their polarity, which is result of the C=O bond dipole.
   d. Aldehydes and ketones are not hydrogen-bond donors and thus have boiling points that are considerably lower than those of the corresponding alcohols.
2. Aldehydes and ketones with four or fewer carbons have considerable solubilities in water because they can accept hydrogen bonds from water at the carbonyl oxygen.
   a. Acetaldehyde and acetone are miscible with water (soluble in all proportions).
   b. The water solubility of aldehydes and ketones along a series diminishes rapidly with increasing molecular mass.
   c. Acetone and 2-butanone are especially valued as solvents because they dissolve not only water but also a wide variety of organic compounds.

## II. Spectroscopy of Aldehydes and Ketones

### A. IR Spectroscopy

1. The principal infrared absorption of aldehydes and ketones is the C=O stretching absorption, a strong absorption that occurs in the vicinity of 1700 $cm^{-1}$.
2. The position of the C=O stretching absorption varies predictably for different types of carbonyl compounds.
   a. 1710–1715 $cm^{-1}$ for simple ketones
   b. 1720–1725 $cm^{-1}$ for simple aldehydes
3. The stretching absorption of the carbonyl-hydrogen bond of aldehydes near 2710 $cm^{-1}$ is another characteristic absorption.
4. Compounds in which the carbonyl group is conjugated with aromatic rings, double bonds, or triple bonds have lower carbonyl stretching frequencies than unconjugated carbonyl compounds.
5. In cyclic ketones with rings containing fewer than six carbons, the carbonyl absorption frequency increases significantly as the ring size decreases.

### B. NMR Spectroscopy

1. The characteristic proton NMR absorption common to both aldehydes and ketones is that of the protons on the carbons adjacent to the carbonyl group ($\alpha$-protons), which occurs in the $\delta$ 2.0–2.5 region of the spectrum.
2. The absorption of the aldehyde proton is quite distinctive ($\delta$ 9–10 region of the NMR spectrum).
   a. The position of this absorption results from deshielding that is similar to the deshielding effects of the carbon-carbon double bond.
   b. A carbonyl group has a greater deshielding effect than a carbon-carbon double bond because of the electronegativity of the carbonyl oxygen.
3. The most characteristic absorption of aldehydes and ketones in CMR spectroscopy is that of the carbonyl carbon, which occurs typically in the $\delta$ 190–220 range.
   a. This large downfield shift is due to the induced electron circulation in the $\pi$ bond and the additional chemical-shift effect of the electronegative carbonyl oxygen.
   b. Because the carbonyl carbon of a ketone bears no hydrogens, its CMR absorption, like that of other quaternary carbons, is characteristically rather weak.
4. The $\alpha$-carbon CMR absorptions of aldehydes and ketones show modest downfield shifts, typically in the $\delta$ 30–50 range, with, as usual, greater shifts for more branched carbons.

### C. UV Spectroscopy

1. The $\pi \rightarrow \pi^*$ absorptions of unconjugated aldehydes and ketones occur at about 150 nm (well below the operating range of common UV spectrometers).
2. A much weaker absorption occurs in the 260–290 nm region in simple aldehydes and ketones.
   a. This absorption is caused by excitation of the unshared electrons (sometimes called the *n* electrons) on oxygen.

b. This high-wavelength absorption is usually referred to as an $n{\to}\pi^*$ absorption.

c. An $n{\to}\pi^*$ absorption arises from promotion of one of the $n$ (unshared) electrons on the carbonyl oxygen to a $\pi^*$ molecular orbital.

   i. $n{\to}\pi^*$ absorptions are weak.

   ii. These absorptions are termed "forbidden" and thus have very low intensities.

3. The $\pi$ electrons of compounds in which carbonyl groups are conjugated with double or triple bonds have strong UV absorptions (large extinction coefficients).

   a. The $\pi{\to}\pi^*$ absorptions of conjugated carbonyl compounds arise from the promotion of a $\pi$ electron from a bonding to an antibonding ($\pi^*$) molecular orbital.

   b. The $\lambda_{max}$ of a conjugated aldehyde or ketone is governed by the same variables that affect the $\lambda_{max}$ of conjugated dienes:

   i. the number of conjugated double bonds.

   ii. substitution on the double bond.

   c. When an aromatic ring is conjugated with a carbonyl group, the typical aromatic absorptions are more intense and shifted to higher wavelengths than those of benzene.

## D. Mass Spectrometry

1. Important fragmentations of aldehydes and ketones arise from

   a. cleavage of the molecular ion at the bond between the carbonyl group and an adjacent carbon by:

   i. inductive cleavage (the alkyl fragment carries the charge and the carbonyl fragment carries the unpaired electron).

   ii. $\alpha$-cleavage (the carbonyl fragment carries the charge and the alkyl fragment carries the unpaired electron).

   b. cleavage at the carbon-hydrogen bond (which accounts for the fact that many aldehydes show a strong M − 1 peak).

   c. McLafferty rearrangement (abstraction of a hydrogen atom from a carbon five atoms away by the oxygen radical in the molecular ion) and subsequent $\alpha$-cleavage. McLafferty rearrangement is a common mechanism for the production of odd-electron fragment ions in the mass spectrometry of carbonyl compounds.

hydrogen abstraction          $\alpha$-cleavage

## III. *Synthesis and Reactivity of Aldehydes and Ketones*

### A. Review of Ketone and Aldehyde Synthetic Methods

1. The three most important preparations of aldehydes and ketones already presented are

   a. Oxidation of alcohols (Sec. 10.6A, text page 467).

   i. Primary alcohols can be oxidized to aldehydes.

   ii. Secondary alcohols can be oxidized to ketones.

   b. Friedel-Crafts acylation (Sec. 16.4E, text page 754).

   c. Hydration (Sec. 14.5A, text page 654) and hydroboration-oxidation of alkynes (Sec. 14.5B, text page 657).

2. Less important synthetic methods that have already been presented are

   a. Ozonolysis of alkenes (Sec. 5.4, text page 188).

   b. Periodate cleavage of glycols (Sec. 10.6C, text page 470).

### B. Introduction to Aldehyde and Ketone Reactions

1. The reactions of aldehydes and ketones can be conveniently grouped into two categories:

   a. Reactions of the carbonyl group.

   b. Reactions involving the $\alpha$-carbon.

2. There are three categories of important carbonyl-group reactions of aldehydes and ketones:

    a. Reactions with acids (the carbonyl oxygen is weakly basic and thus reacts with Lewis and Brønsted acids); carbonyl basicity is important because it plays a role in several other carbonyl-group reactions.

    b. Addition reactions to the C=O double bond (the most important carbonyl-group reaction).

    c. Oxidation of aldehydes (aldehydes can be oxidized to carboxylic acids).

## C. BASICITY OF ALDEHYDES AND KETONES

  1. Aldehydes and ketones are weakly basic and react at the carbonyl oxygen with protons or Lewis acids.

    a. The conjugate acid of an aldehyde or ketone is resonance-stabilized. (In some cases, the conjugate acids of aldehydes and ketones undergo typical carbocation reactions.)

an α-hydroxy carbocation

    b. Closely related to protonated aldehydes and ketones are α-alkoxy carbocations (cations in which the acidic proton of a protonated ketone is replaced by an alkyl group).

an α-alkoxy carbocation

    c. α-Hydroxy carbocations and α-alkoxy carbocations are considerably more stable than ordinary alkyl cations.

      *i.* The greater stability is due to resonance interaction of the electron-deficient carbon with the unshared electrons of the neighboring oxygen.

      *ii.* This resonance effect far outweighs the electron-attracting polar effect of the oxygen which, by itself, would destabilize these carbocations.

  2. Aldehydes and ketones in solution are considerably less basic than alcohols; their conjugate acids are more acidic than those of alcohols.

    a. The relative acidity of protonated alcohols and carbonyl compounds is an example of a solvent effect.

    b. One reason for the greater basicity of alcohols in solution is that protonated alcohols have more O—H hydrogens to participate in hydrogen bonding to solvent than do protonated aldehydes or ketones.

## D. REVERSIBLE ADDITION REACTIONS OF ALDEHYDES AND KETONES

  1. One of the most typical reactions of aldehydes and ketones is addition to the carbon-oxygen double bond; two such reactions are

    a. hydration (addition of water) to give a hydrate or *gem*-diol (occurs more extensively with aldehydes than with ketones).

    b. addition of HCN to give a cyanohydrin, a special type of nitrile (or organic cyanide).

  2. Addition to a carbonyl group is regioselective.

    a. The more electropositive species (the proton of water in the case of hydration) adds to the carbonyl oxygen.

    b. The more electronegative species (—OH in the case of hydration or —CN in the case of HCN addition) adds to the carbonyl carbon.

## E. MECHANISMS OF CARBONYL-ADDITION REACTIONS

  1. Carbonyl-addition reactions occur by two general types of mechanisms.

2. Under basic conditions:
   a. A nucleophile attacks the carbonyl group at the carbonyl carbon, and the carbonyl oxygen becomes negatively charged.

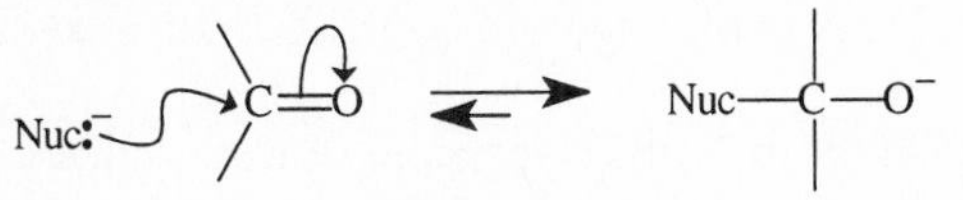

   b. The negatively charged oxygen is a relatively strong base, and is protonated by a weak acid to complete the addition.

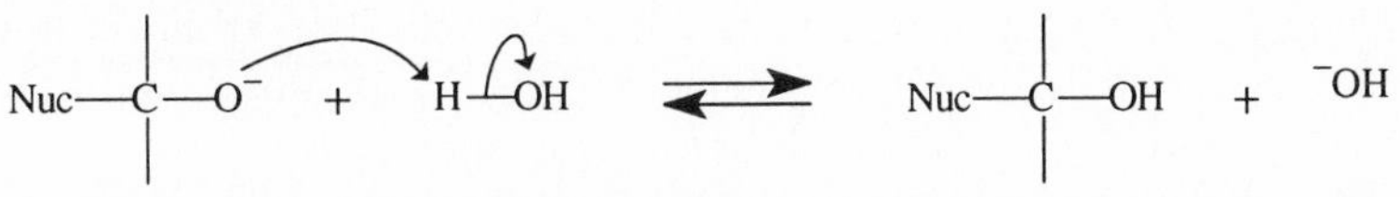

   *i.* This mechanism, nucleophilic addition, has no analogy in the reactions of ordinary alkenes.
   *ii.* This pathway occurs with aldehydes and ketones because, in the transition state, negative charge is placed on oxygen, an electronegative atom.
   *iii.* Attack of the nucleophile occurs on the carbon of the carbonyl group rather than the oxygen for the same reason—the negative charge is "pushed" onto the more electronegative atom—oxygen
   *iv.* The nucleophile is typically the more electronegative partner of the groups that add, and the nucleophile always attacks the carbonyl carbon.

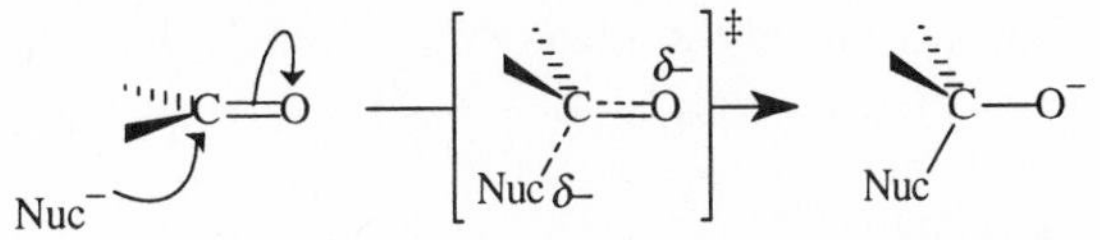

3. Under acidic conditions, the mechanism is closely analogous to the mechanism for the addition of acids to alkenes. Acid-catalyzed hydration of aldehydes and ketones is an example of this mechanism.
   a. The first step in acid-catalyzed addition is usually protonation of the carbonyl oxygen.

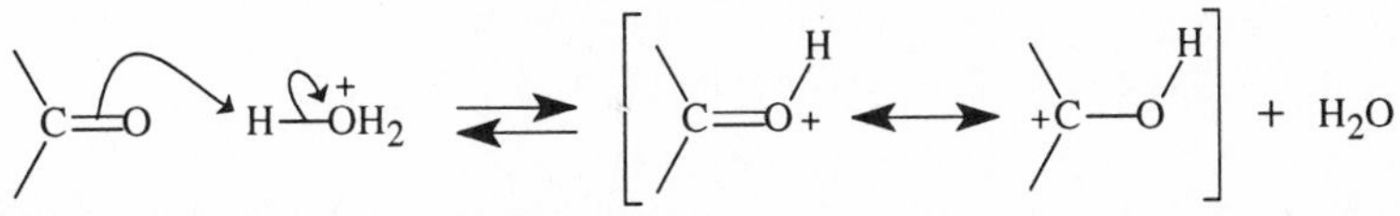

   *i.* The carbon of a protonated carbonyl group is a much stronger Lewis acid than the carbon of an unprotonated carbonyl group.
   *ii.* As a result, relatively weak bases such as $H_2O$ can react at the carbonyl carbon.

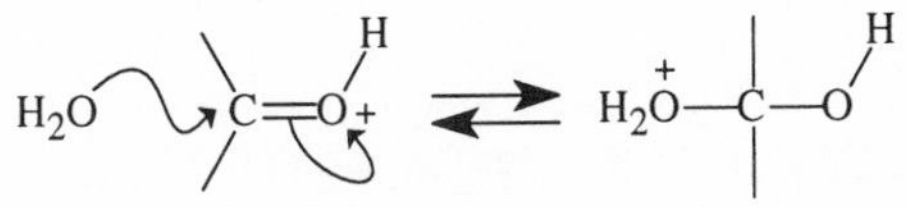

   b. Loss of a proton to solvent completes the reaction.

## F.  EQUILIBRIA AND RATES IN CARBONYL-ADDITION REACTIONS

1. Some carbonyl-addition reactions, such as hydration and cyanohydrin formation, are reversible. (Not all carbonyl additions are reversible.)
2. Whether the equilibrium for a reversible addition favors the addition product or the carbonyl compound depends strongly on the structure of the carbonyl compound.
   a. Addition is more favorable for aldehydes than for ketones.
   b. Electronegative groups near the carbonyl carbon of an aldehyde or ketone make carbonyl addition more favorable.

    c. Addition is less favorable when groups are present that donate electrons by resonance to the carbonyl carbon.

3. The rates of carbonyl-addition reactions (the reactivities of carbonyl compounds) follow similar trends; that is, the more a compound favors addition at equilibrium, the more rapidly it reacts in addition reactions.

    a. The stability of the carbonyl compound relative to that of the addition product governs the $\Delta G°$ for addition.

    b. Added stability in the carbonyl compound increases the energy change ($\Delta G°$), and hence decreases the equilibrium constant, for formation of an addition product.

4. Anything that stabilizes carbocations also tends to stabilize carbonyl compounds.

    a. Electronegative groups such as halogens destabilize carbocations by their polar effect, and for the same reason they destabilize carbonyl compounds.

    b. Groups that are conjugated with the carbonyl group stabilize carbocations by resonance, and hence they stabilize carbonyl compounds.

       *i.* Resonance interaction with the carbonyl group cannot occur in an addition product such as a hydrate because the carbonyl group is no longer present.

       *ii.* Consequently, aryl aldehydes and ketones have relatively unfavorable addition equilibria.

5. Aldehydes are generally more reactive than ketones in addition reactions; formaldehyde is more reactive than many other simple aldehydes.

## G.  PROTECTING GROUPS

1. A common tactic of organic synthesis is the use of protecting groups.

2. Carbonyl groups react with a number of reagents used with other functional groups but can be rendered inert to these reagents by the use of protecting groups.

3. Acetals are commonly used to protect the carbonyl groups of aldehydes and ketones from basic, nucleophilic reagents.

    a. Once the protection is no longer needed, the acetal protecting group is easily removed, and the carbonyl group re-exposed, by treatment with dilute aqueous acid.

    b. Because acetals are unstable in acid, they do not protect carbonyl groups under acidic conditions.

## H.  MANUFACTURE AND USE OF ALDEHYDES AND KETONES

1. Formaldehyde is manufactured by the oxidation of methanol over a silver catalyst.

2. Formaldehyde is used in the synthesis of a class of polymers known as phenol-formaldehyde resins.

    a. A resin is a polymer with a rigid three-dimensional network of repeating units.

    b. Phenol-formaldehyde resins are produced by heating phenol and formaldehyde with acidic or basic catalysts.

    c. One phenol-formaldehyde resin, Bakelite, was the first useful synthetic polymer.

3. The simplest ketone, acetone, is co-produced with phenol by the autoxidation-rearrangement of cumene (Sec. 18.10, text page 851).

## IV. *Introduction to Amines*

### A.  GENERAL

1. A primary amine (general structure $R\ddot{N}H_2$) is an organic derivative of ammonia in which only one ammonia hydrogen is replaced by an alkyl or aryl group.

2. A secondary amine (general structure $R_2\ddot{N}H$) is an organic derivative of ammonia in which two ammonia hydrogens are replaced by alkyl or aryl groups; the nitrogen may be part of a ring.

3. A tertiary amine (general structure $R_3\ddot{N}$) is an organic derivative of ammonia in which all three ammonia hydrogens are replaced by alkyl or aryl groups; the nitrogen may be part of a ring.

H—N⟨H H (ammonia)   H₃C—C(CH₃)(CH₃)—N⟨H H (a primary amine)   Ph—N⟨H CH₂CH₃ (a secondary amine)   N—CH₃ (a tertiary amine)

## B.  AMINE DERIVATIVES

1. An imine is a nitrogen analog of an aldehyde or ketone; the imine functional group is C=N—R, where R = alkyl, aryl, or H.
   a. Imines are sometimes called Schiff bases.
   b. Imines are prepared by the reactions of aldehydes or ketones with primary amines.
   c. Imines revert to the corresponding carbonyl compounds and amines in aqueous acid.

$$\text{C=O} + \text{H}_2\text{NR} \underset{}{\overset{-\text{H}_2\text{O}}{\rightleftharpoons}} \text{C=N—R}$$

an imine

2. A carbinolamine is a compound with an amine group (—NH₂, —NHR, or —NR₂) and a hydroxy group on the same carbon.
   a. Carbinolamines are intermediates in imine formation; most carbinolamines are not isolated.
   b. Carbinolamines undergo acid-catalyzed dehydration to form imines.

$$\underset{}{\overset{\text{O}}{\text{C}}} + \text{H}_2\text{NR} \rightleftharpoons -\underset{\text{HNR}}{\overset{\text{OH}}{\text{C}}} - \underset{}{\overset{\text{H}_3\text{O}^+}{\rightleftharpoons}} \underset{}{\overset{\text{NR}}{\text{C}}} + \text{H}_2\text{O}$$

a carbinolamine          an imine

3. An enamine is the nitrogen analog of an enol in which the hydroxy group is replaced by an amine group bearing two R groups, where R can be alkyl, aryl, or part of a ring.
   a. Formation of an enamine occurs when a secondary amine reacts with an aldehyde or ketone, provided that the carbonyl compound has an α-hydrogen.
   b. Enamines, like imines, revert to the corresponding carbonyl compounds and amines in aqueous acid.

$$\text{H—C—C}(\text{O}) + \text{HNR}_2 \rightleftharpoons \underset{}{\overset{\text{H} \quad \text{NR}_2}{\text{C=C}}}$$

an enamine

2. A hydrazone is a nitrogen analog of an aldehyde or ketone; the hydrazone functional group is C=N—NR₂, where R = alkyl, aryl, or H.
   a. Hydrazones are prepared by the reactions of aldehydes or ketones with hydrazines (H₂N—NR₂).
   b. 2,4-Dinitrophenylhydrazine (2,4-DNP) reacts with aldehydes and ketones to give 2,4-dinitro-phenylhydrazones (2,4-DNP derivatives).  These derivatives are generally solids having characteristic melting points which were used, before the advance of NMR spectroscopy, to aid in the identification of new organic compounds.

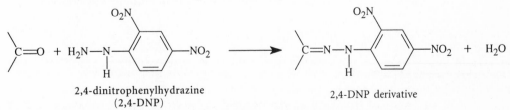

2,4-dinitrophenylhydrazine
(2,4-DNP)                                    2,4-DNP derivative

# Reactions

## I. Carbonyl-Addition Reactions

### A. HYDRATION OF ALDEHYDES AND KETONES

1. Acid-catalyzed hydration of aldehydes and ketones is reminiscent of acid-catalyzed hydration of alkenes.
   a. The first step is protonation of the carbonyl oxygen; the protonated carbonyl compound is a much stronger Lewis acid (electron acceptor) than an unprotonated carbonyl compound.
   b. The electron-deficient carbon is attacked by $H_2O$, a nucleophile.
   c. Loss of a proton to the solvent (water) completes the reaction.

a carbonyl hydrate

2. Hydration of aldehydes and ketones also occurs in neutral and basic solution.

### B. CYANOHYDRIN FORMATIONS

1. Hydrogen cyanide, HCN, reacts with aldehydes and ketones by a nucleophilic addition mechanism to give cyanohydrins.
   a. A cyanide ion, formed by ionization of the weak acid HCN, attacks the carbonyl group at the carbonyl carbon; the carbonyl oxygen becomes negatively charged.
   b. The negatively charged oxygen is a relatively strong base, and is protonated by either water or HCN to complete the addition.

2. Cyanohydrin formation favors the cyanohydrin addition product in the case of aldehydes and methyl ketones, but not in the case of aryl ketones.
3. Cyanohydrin formation is another method of forming carbon-carbon bonds.

### C. REDUCTION OF ALDEHYDES AND KETONES TO ALCOHOLS

1. Aldehydes and ketones are reduced to alcohols with either lithium aluminum hydride ($LiAlH_4$) or sodium borohydride ($NaBH_4$), which serve as sources of nucleophilic hydride ion ($H:^-$).
   a. These reactions, which are not reversible, result in the net addition of the elements of $H_2$ across the C=O bond.
   b. Reduction of an aldehyde gives a primary alcohol.
   c. Reduction of a ketone gives a secondary alcohol.

an aldehyde     $\xrightarrow{\text{hydride reduction}}$     a primary alcohol

a ketone     $\xrightarrow{\text{hydride reduction}}$     a secondary alcohol

2. $LiAlH_4$ and $NaBH_4$ reductions are generally referred to as hydride reductions and are examples of nucleophilic addition.

3. The reaction of LiAlH$_4$ with aldehydes and ketones involves the nucleophilic attack of hydride (delivered from $^-$AlH$_4$) on the carbonyl carbon; all four hydride equivalents of LiAlH$_4$ are active in this reaction.
   a. A lithium ion coordinated to the carbonyl oxygen acts as a Lewis-acid catalyst.
   b. Hydride attacks the carbonyl carbon to give an alkoxide salt.
   c. The alkoxide salt is converted by protonation into the alcohol product; the proton source is water (or an aqueous solution of a weak acid such as NH$_4$Cl), which is added in a separate step.

   d. LiAlH$_4$ reacts violently with water and therefore must be used in dry solvents such as anhydrous ether or THF.
4. The reaction of NaBH$_4$ with aldehydes and ketones involves the nucleophilic attack of hydride on the carbonyl carbon; all four hydride equivalents of NaBH$_4$ are active in this reaction.
   a. The sodium ion of NaBH$_4$ does not form as strong a bond to the carbonyl oxygen as the lithium ion.
   b. NaBH$_4$ reductions are carried out in protic solvents, such as alcohols; hydrogen bonding between the alcohol solvent and the carbonyl group serves as a weak acid catalysis that activates the carbonyl group.
   c. NaBH$_4$ reacts only slowly with alcohols, and it can be used in water if the solution is not acidic.

5. LiAlH$_4$ is a much more reactive reagent than NaBH$_4$.
   a. A number of functional groups react with LiAlH$_4$ but not NaBH$_4$:
      *i.* alkyl halides
      *ii.* alkyl tosylates
      *iii.* esters
      *iv.* nitro groups
   b. The greater selectivity and safety of NaBH$_4$ make it the preferred reagent in many applications.
6. Aldehydes and ketones can also be reduced to alcohols by catalytic hydrogenation.
   a. This reaction is analogous to the catalytic hydrogenation of an alkene.

   b. It is usually possible to use catalytic hydrogenation for the selective reduction of an alkene double bond in the presence of a carbonyl group; palladium catalysts are particularly effective for this purpose.

$$Ph—CH{=}CH—CH{=}O \xrightarrow[\text{5\% Pd/C}]{H_2} Ph—CH_2—CH_2—CH{=}O$$

## D. Reactions of Aldehydes and Ketones with Grignard Reagents

1. The reaction of Grignard reagents with carbonyl groups is the most important application of Grignard reagents in organic chemistry; addition to aldehydes and ketones in an ether solvent, followed by protonolysis, gives alcohols. (See Appendix IV, text page A-7, for a review of other syntheses of alcohols.)
   a. The reaction of Grignard reagents with aldehydes and ketones is another example of carbonyl addition; the addition of Grignard reagents to aldehydes and ketones is not reversible.
   b. The magnesium of the Grignard reagent, a Lewis acid, bonds to the carbonyl oxygen; this bonding makes the carbonyl carbon more electrophilic.

c. The carbon group of the Grignard reagent attacks the carbonyl carbon to form a halomagnesium alkoxide.

d. Addition of dilute acid in a separate step to the reaction mixture gives an alcohol.

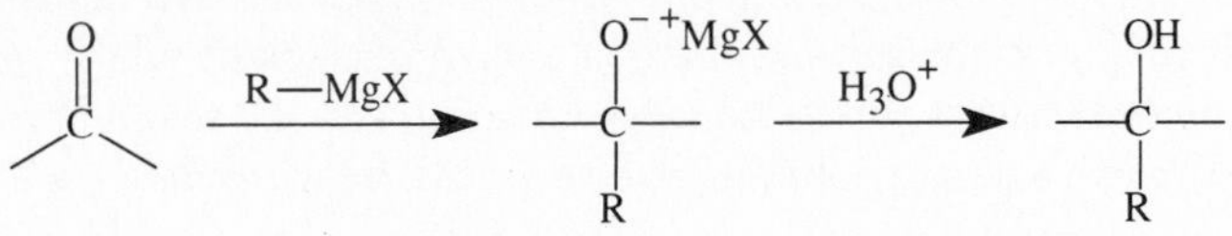

e. The Grignard synthesis of a tertiary alcohol, or in some cases a secondary alcohol, can be extended to an alkene synthesis by dehydration of the alcohol with strong acid during the protonation step.

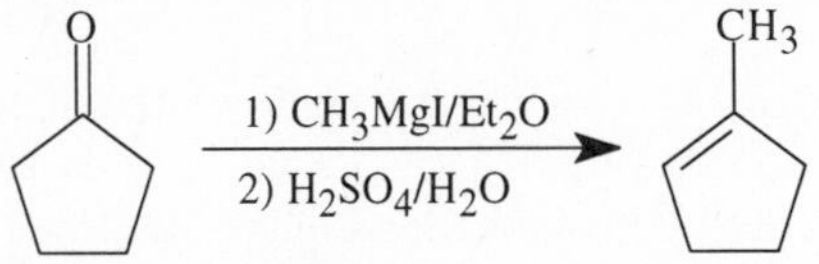

2. The reactions of organolithium and sodium acetylide reagents with aldehydes and ketones are fundamentally similar to the Grignard reaction.
3. The net effect of the Grignard reaction, followed by protonolysis, is addition of R—H (R = an alkyl or aryl group) across the C=O double bond; this addition is not reversible.
   a. Primary alcohols are synthesized by the addition of Grignard reagents to formaldehyde.

formaldehyde

a primary alcohol

   b. Secondary alcohols are synthesized by the addition of Grignard reagents to aldehydes other than formaldehyde.

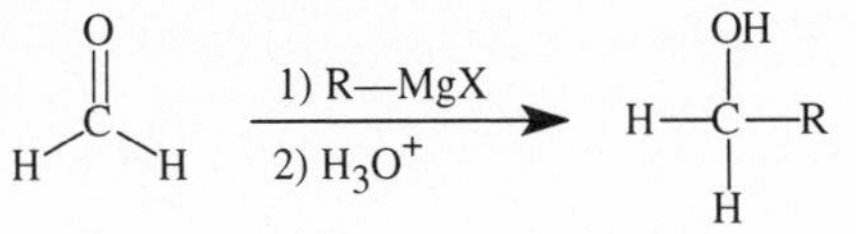

an aldehyde

a secondary alcohol

   c. Tertiary alcohols are synthesized by the addition of Grignard reagents to ketones.

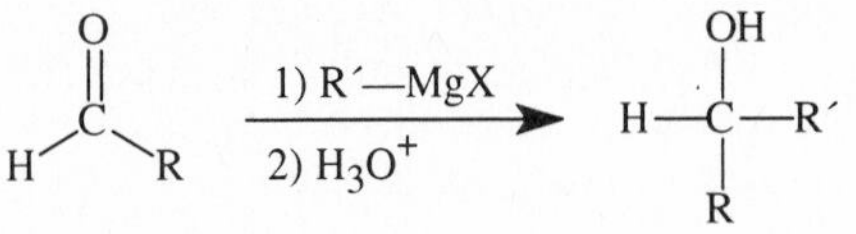

a ketone

a tertiary alcohol

4. The Grignard reaction is an excellent method of carbon-carbon bond formation. (See Appendix V, text page A-11, for a review of other reactions used to form carbon-carbon bonds.)

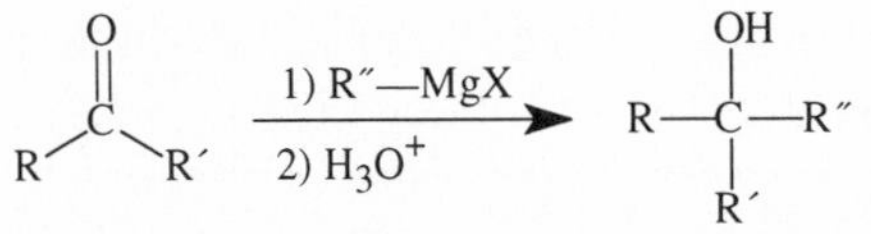

## E. Preparation and Hydrolysis of Acetals

1. When an aldehyde or ketone reacts with a large excess of an alcohol in the presence of a trace of strong acid, an acetal is formed.
   a. An acetal is a di-ether in which both ether oxygens are bound to the same carbon.
   b. Acetals are ethers of hydrates or *gem*-diols.

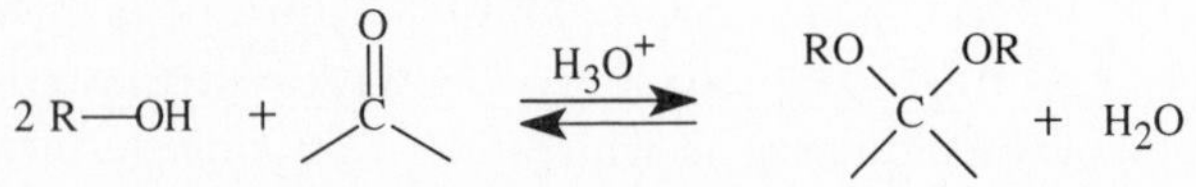

2. Two equivalents of alcohol are consumed in the formation of acetals, but one equivalent of a 1,2- or 1,3-diol can react to form a cyclic acetal, in which the acetal group is part of a five- or six-membered ring, respectively.

3. Acetal formation is reversible and involves an acid-catalyzed carbonyl addition followed by a substitution that occurs by an $S_N1$ mechanism:
   a. The first step in the mechanism of acetal formation is acid-catalyzed addition of the alcohol to the carbonyl group to give a hemiacetal (a compound with an —OR and —OH group on the same carbon).
   b. The hemiacetal is converted into an acetal by substitution of the —OH group by another —OR group.

4. The reaction is driven to the right by applying LeChatelier's principle in one or both of the following ways:
   a. Use of excess alcohol as the solvent.
   b. Removal of the water by-product (usually by azeotropic distillation).
5. Acetals in the presence of acid and excess water are transformed rapidly back into the corresponding carbonyl compounds and alcohols.
   a. This process is called acetal hydrolysis.
   b. By the principle of microscopic reversibility, the mechanism of acetal hydrolysis is the reverse of the mechanism of acetal formation.
   c. Acetal hydrolysis, like hemiacetal formation, is acid-catalyzed.
6. The interconversion of hemiacetals and aldehydes is catalyzed not only by acids, but by bases as well; however, the interconversion of hemiacetals and acetals is catalyzed only by acids. For this reason, the hydrolysis of acetals is catalyzed only by acids; acetals are stable in basic and neutral solutions.
7. Hemiacetals in most cases cannot be isolated because they react further to yield acetals or decompose to aldehydes or ketones plus water.
   a. Simple aldehydes form appreciable amounts of hemiacetals in alcohol solution, just as they form appreciable amounts of hydrates in water.
   b. Five- and six-membered cyclic hemiacetals form spontaneously from the corresponding hydroxy aldehydes, and most are stable, isolable compounds.

## II. Oxidation of Aldehydes and Ketones

### A. OXIDATION OF ALDEHYDES

1. Aldehydes can be oxidized to carboxylic acids.
2. Common oxidants, such as aqueous Cr(VI) reagents, nitric acid, or aqueous $KMnO_4$/NaOH, can be used in the oxidation of aldehydes to carboxylic acids; these oxidizing agents are the same ones used for oxidizing alcohols.
3. Some aldehyde oxidations begin as addition reactions.
   a. In the oxidation of aldehydes by Cr(VI) reagents, the hydrate, not the aldehyde, is actually the species oxidized.
   b. For this reason, some water should be present in solution in order for aldehyde oxidations with Cr(VI) to occur at a reasonable rate.

4. In the laboratory, aldehydes can be conveniently oxidized to carboxylic acids with Ag(I) reagents.
   a. Aldehydes that contain double bonds or alcohol —OH groups (functional groups that react with other oxidizing reagents) are oxidized by $Ag_2O$ without oxidizing these other functional groups.
   b. If the silver ion is solubilized as its ammonia complex, $^+Ag(NH_3)_2$, oxidation of the aldehyde is accompanied by the deposition of a metallic silver mirror on the walls of the reaction vessel. (This observation can be used as a convenient test for aldehydes, known as Tollens' test.)

ammonium salt of
a carboxylic acid

5. Many aldehydes are oxidized by the oxygen in air upon standing for long periods of time; this process is another example of autoxidation.

## B. Oxidation of Ketones

1. Ketones cannot be oxidized without breaking carbon-carbon bonds.
2. Ketones are resistant to mild oxidation with Cr(VI) reagents.

## III. *Reactions of Aldehydes and Ketones with Amines*

### A. Imine Formation with Primary Amines

1. Imines are prepared by the reaction of aldehydes or ketones with primary amines.
   a. Formation of imines is reversible and generally takes place with acid or base catalysis, or with heat.

an imine

   b. Imine formation is typically driven to completion in one or both of the following ways:
      *i.* Precipitation of the imine.
      *ii.* Removal of water.
2. Imine formation consists of a carbonyl-addition reaction followed by $\beta$-elimination.
   a. The first step of the mechanism is a nucleophilic addition to the carbonyl group in which the nucleophile (an amine) reacts with an aldehyde or ketone to give an unstable addition compound called a carbinolamine.
   b. The carbinolamine undergoes acid-catalyzed dehydration to form an imine.
   c. Dehydration of the carbinolamine is typically the rate-limiting step of imine formation.

a carbinolamine                 an imine

3. Certain types of imine adducts sometimes find use as derivatives of aldehydes and ketones because they are in most cases solids with well-defined melting points. (See Table 19.3, text page 906, for the corresponding structures.)
   a. Hydroxylamine $\Rightarrow$ oximes.
   b. Hydrazine $\Rightarrow$ hydrazones.
   c. Phenylhydrazine $\Rightarrow$ phenylhydrazones.
   d. 2,4-Dinitrophenylhydrazine $\Rightarrow$ 2,4-dinitrophenylhydrazones (2,4-DNP derivatives).
   e. Semicarbazide $\Rightarrow$ semicarbazones.

B. ENAMINE FORMATION WITH SECONDARY AMINES

1. Formation of an enamine occurs when a secondary amine reacts with an aldehyde or ketone that has an $\alpha$-hydrogen.
   a. Just as most aldehydes and ketones are more stable than their corresponding enols, most imines are more stable than their corresponding enamines.
   b. Because secondary amines cannot form imines, they form enamines instead.
2. Like imine formation, enamine formation is reversible and must be driven to completion by the removal of one of the reaction products (usually water).
   a. The mechanism of enamine formation begins like the mechanism of imine formation, as a nucleophilic addition to give a carbinolamine intermediate.
   b. Dehydration of a carbinolamine involves proton loss from an adjacent carbon.

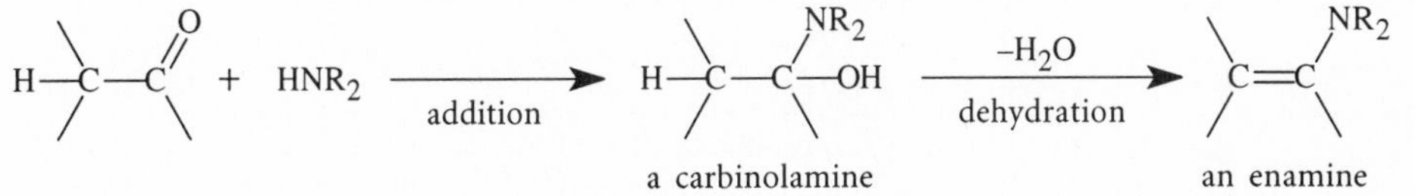

3. Enamines, like imines, revert to the corresponding carbonyl compounds and amines in aqueous acid.

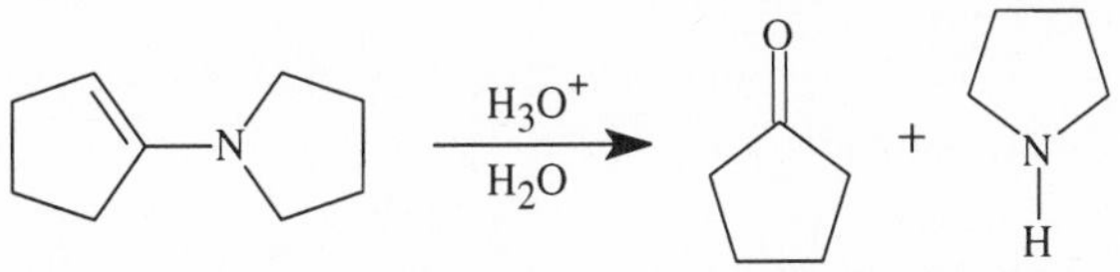

C. REACTIONS OF ALDEHYDES AND KETONES WITH TERTIARY AMINES

1. Tertiary amines do not react with aldehydes and ketones to form stable derivatives.

## IV. Conversion of Ketones and Aldehydes into Alkanes and Alkenes

A. REDUCTION OF CARBONYL GROUPS TO METHYLENE GROUPS

1. A carbonyl group of an aldehyde or ketone can be reduced completely to a methylene ($-CH_2-$) group.
2. One procedure for effecting this transformation involves heating the aldehyde or ketone with hydrazine ($H_2N-NH_2$) and strong base.

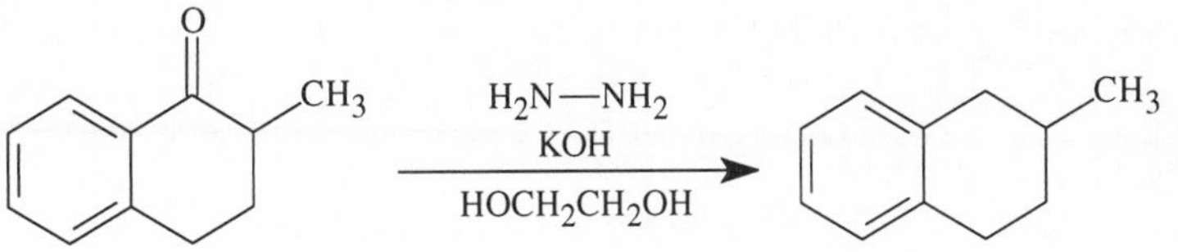

   a. This reaction, called the Wolff-Kishner reduction, typically utilizes ethylene glycol or similar compounds as co-solvents.
   b. The high boiling points of these solvents allow the reaction mixtures to reach the high temperatures required for the reduction to take place at a reasonable rate.
3. The Wolff-Kishner reduction is an extension of imine formation; an intermediate in the reduction is a hydrazone (an imine of hydrazine).
4. The Wolff-Kishner reduction takes place under strongly basic conditions.

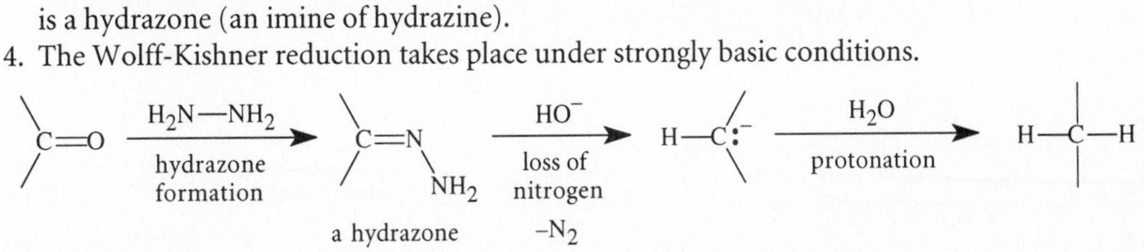

5. The same overall transformation can be achieved under acidic conditions by a reaction called the Clemmensen reduction, which involves reduction of an aldehyde or ketone with zinc amalgam (a solution of zinc metal in mercury) in the presence of HCl.

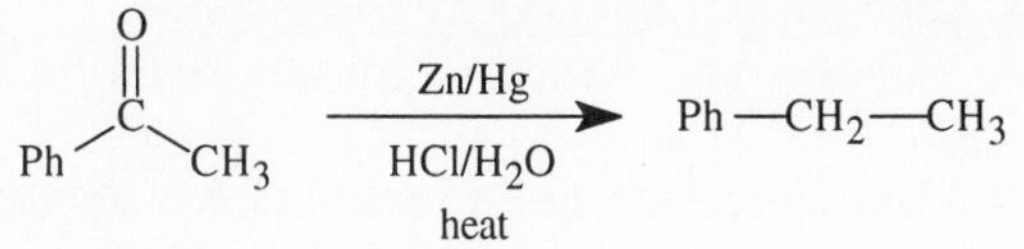

6. The Wolff-Kishner and Clemmensen reactions are particularly useful for the introduction of alkyl substituents into benzene rings by the following sequence:
   a. Friedel-Crafts acylation to give an aryl ketone.
   b. Wolff-Kishner or Clemmensen reduction to yield the corresponding alkyl group.

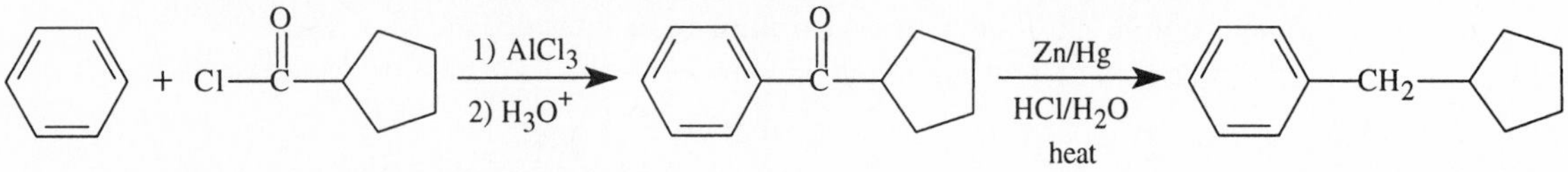

## B. THE WITTIG ALKENE SYNTHESIS

1. Another addition-elimination reaction, called the Wittig reaction, is an important method for preparing alkenes from aldehydes and ketones.
2. The nucleophile in the Wittig reaction is a type of ylid.
   a. An ylid is any compound with opposite charges on adjacent covalently bound atoms, each of which has an electronic octet.
   b. Because phosphorus can accommodate more than eight valence electrons, a phosphorus ylid has an uncharged resonance structure.
   c. Although the structures of phosphorus ylids are sometimes written with phosphorus-carbon double bonds, the charged structures, in which each atom has an octet of electrons, are very important contributors.

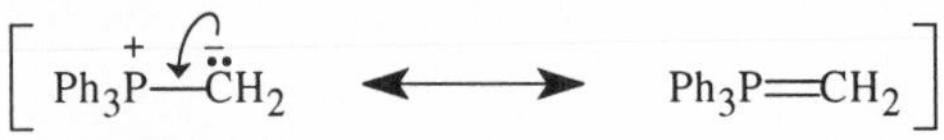

an ylid

3. Preparation of phosphorus ylids:
   a. An alkyl halide reacts with triphenylphosphine ($Ph_3P$) in an $S_N2$ reaction to give a phosphonium salt. (Because the alkylation of triphenylphosphine is a typical $S_N2$ reaction, it is limited for the most part to methyl and primary alkyl halides.)
   b. The phosphonium salt is converted into its conjugate base, the ylid, by reaction with a strong base such as an organolithium reagent.

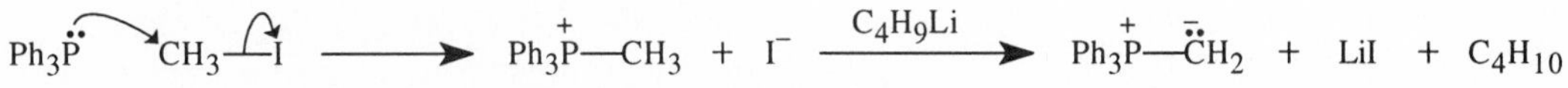

4. The mechanism of the Wittig reaction:
   a. A nucleophile (the anionic carbon of the ylid) attacks the carbonyl carbon.
      i. The anionic oxygen in the resulting species reacts with phosphorous to form an oxaphosphetane intermediate.
      ii. An oxaphosphetane is a saturated four-membered ring containing both oxygen and phosphorus as ring atoms.
   b. Under the usual reaction conditions, the oxaphosphetane spontaneously decomposes to the alkene and the by-product triphenylphosphine oxide.

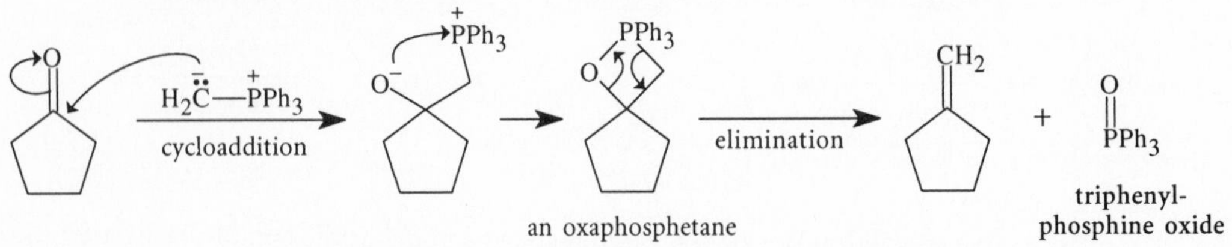

5. The Wittig reaction is especially important because it gives alkenes in which the position of the alkene double bond is unambiguous.

    a. The reaction is thus completely regioselective.

    b. The reaction can be used for the preparation of alkenes that would be difficult to prepare by other reactions.

6. To plan the synthesis of an alkene by the Wittig reaction, consider the origin of each part of the product, and then reason deductively.

    a. One carbon of the alkene double bond originates from the alkyl halide used to prepare the ylid; the other is the carbonyl carbon of the aldehyde or ketone.

    b. In principle, two Wittig syntheses are possible for any given alkene.

    c. Most Wittig syntheses are planned so that the most reactive alkyl halide (a methyl or primary alkyl halide) can be used as one of the starting materials.  [See (3a) above.]

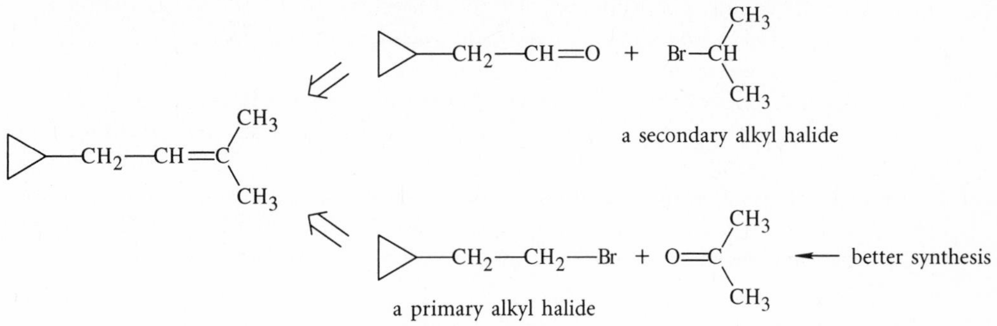

    d. The Wittig reaction in many cases gives mixtures of *E* and *Z* isomers, although certain modifications of the Wittig reaction that avoid this problem have been developed.

## Study Guide Links

### 19.1   IR Absorptions of Cyclic Ketones

The increase of the infrared carbonyl absorption frequency with ring size is a very useful trend that can be used to diagnose the presence of small rings. The main reason for this trend is a coupling phenomenon. In the presentation of IR spectroscopy in Chapter 12, IR absorption is treated as a phenomenon associated with isolated bonds. Even though this is a highly useful approximation, the effects of nearby bonds in some cases cannot be ignored, and the carbonyl stretching frequencies of the cyclic ketones is one of those cases.

Imagine two extreme situations. In the first, the carbonyl group forms an angle of 180° with adjacent C—C bonds (as in ketene, $H_2C{=}C{=}O$). In such a situation, because the vibrations of both bonds occur along the same line, expansion of the carbon-oxygen bond that occurs during a vibration also compresses the carbon-carbon bond; and compression of the carbon-oxygen bond also expands the carbon-carbon bond. The two stretching vibrations are *coupled*:

$$H_2C{=}C{=}O \quad \xleftrightarrow[\text{compression of C=O bond;}]{\begin{array}{c}\text{expansion of C=O bond;}\\\text{compression of C=C bond}\end{array}} \quad H_2C{=}C{=}O$$

$$\text{(compression of C=O bond; expansion of C=C bond)} \updownarrow$$

$$H_2C{=}C{=}O$$

The additional energy required to cause both bonds to vibrate at the same time is reflected in a higher vibrational frequency for both bonds.

Now consider a hypothetical situation in which the $C{=}O$ bond is perpendicular to adjacent bonds. In such a situation, the $C{=}O$ bond can stretch and compress without stretching the adjacent C—C bond:

$$H_2C{-}\underset{\underset{\displaystyle C}{\|}}{C}{-}CH_2 \quad \xleftrightarrow{\text{C=O bond stretch}} \quad H_2C{-}\underset{\underset{\displaystyle C}{\|}}{}{-}CH_2$$

(Remember that bending of the C—C bond requires much less energy than stretching.) At this angle, the stretching vibrations of the two bonds are completely uncoupled. Hence, it requires much less energy to stretch the $C{=}O$ bond in this situation than when the two bonds are collinear.

An intermediate situation occurs at bond angles between 180° and 90°: the stretching vibrations are partially coupled. The more closely the C—C$={}$O angle approaches 180°, the greater the coupling and the higher the vibration frequency. Notice that the bond angle in question is the one between the $C{=}O$ bond and the ring C—C bond, not the internal angle within the ring. This angle varies from 120° in cyclohexanone to greater values for the ketones with smaller rings. As the data in Eq. 19.4 of the text show, the carbonyl stretching frequencies increase toward smaller ring sizes.

Another factor that contributes to carbonyl stretching frequencies is the relative strengths of the $C{=}O$ bonds. The $C{=}O$ bond in ketene involves an *sp*-hybridized carbon atom, whereas the $C{=}O$ bond in cyclohexanone involves an *sp*²-hybridized carbon. The amount of *s* character in the $C{=}O$ $\sigma$ bond is therefore greater in ketene, and the amount of *s* character increases from cyclohexanone through the smaller rings. As with C—H bonds (Eq. 14.25, text p. 664), the strengths of C=O bonds, and hence, their absorption frequencies, increase with increasing *s* character.

The coupling effect discussed above is believed to dominate the trend toward higher frequencies, with the bond-strength effect making a secondary contribution.

A similar trend can be seen with carbon–carbon double bond stretching frequencies:

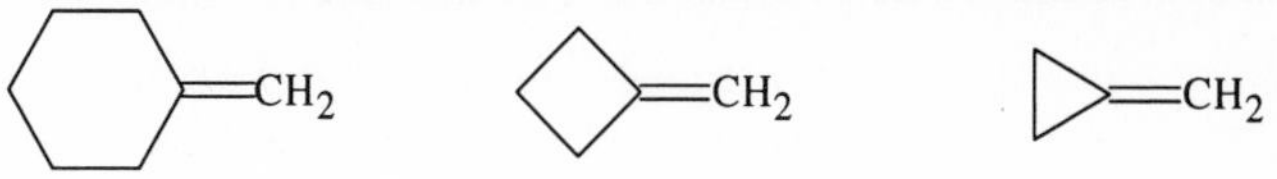

C=C stretching frequencies:　　　　　1650 cm$^{-1}$　　　　　1672 cm$^{-1}$　　　　　1781 cm$^{-1}$

## ✓19.2　Why Nucleophiles React at the Carbonyl Carbon

It is tempting to use resonance structures to rationalize the reactivity of a carbonyl group as follows:

It is sometimes erroneously said that carbonyl compounds react with nucleophiles at the carbonyl carbon because there is positive charge at this carbon, and hence, it is a logical point of attack for electron pairs of nucleophiles.

There are two fallacies in this argument. First, when two species collide, they do so randomly, not from any preferred direction. For example, a nucleophile in solution can collide randomly with any of the atoms in a carbonyl compound. In other words, a nucleophile is not selectively directed to a carbonyl carbon. Rather, reaction occurs the way it does because when the nucleophile, in one of many random collisions, happens to collide with the carbonyl carbon from the proper direction, *electrons (and charge) can be shifted onto the electronegative carbonyl oxygen.*

The second fallacy in the charge-separation argument above is its implication that resonance increases reactivity. In fact, *resonance stabilization has the opposite effect.* Carbonyl compounds are actually *less reactive* than they would be if such resonance stabilization did not exist. Many studies have shown that when molecules are prepared in which resonance stabilization of double bonds cannot occur (for example, molecules in which $\pi$-electron systems are twisted out of coplanarity), these bonds become more reactive than double bonds in similar molecules in which resonance interaction can occur. The carbon-oxygen $\pi$ bond is actually about 105 kJ/mol (25 kcal/mol) *stronger* than a carbon-carbon $\pi$ bond, and this fact, taken alone, means that it should be *less reactive* than a carbon-carbon double bond, exactly as the resonance structures suggest.

If the carbonyl $\pi$ bond is so stable, why does it react? Remember from Sec. 3.6A of the text that when a Brønsted acid-base reaction occurs, it is not only the strength of the bond to the hydrogen, but also *how well the other atom in the bond accepts electrons,* that governs how easily the reaction takes place. Attack on a carbonyl carbon is no different, except that it is a carbon rather than a hydrogen which is attacked by the base (nucleophile), and a $\pi$ bond rather than a $\sigma$ bond which is broken. *Carbonyl compounds react with nucleophiles at the carbonyl carbon because the electronegative oxygen readily accepts negative charge.*

It nevertheless is true that resonance structures *do* suggest ways in which molecules can react. In the case of a carbonyl group, the dipolar resonance structure suggests that the carbon, with its partial positive charge, is the site of electron attack. The reason this works is that the atom which accepts electrons in the dipolar resonance structure (oxygen) is the same one that accepts electrons in the transition state for nucleophilic addition. *Such correspondences between reactivity patterns and resonance structures invariably occur throughout organic chemistry.* For this reason, organic chemists find themselves using resonance structures to predict sites of reactivity in molecules. Resonance structures are undeniably useful for this purpose; but when we use them this way, we must remember that resonance is not the *reason for reactivity.*

## 19.3    Analogy for Nucleophilic Addition

The nucleophilic addition reaction can be thought of as an "$S_N2$ reaction" in which the $\pi$ electrons are displaced from the carbonyl carbon by the nucleophile onto the carbonyl oxygen. A closely related reaction is ring opening of epoxides:

*Nucleophilic ring-opening of epoxides:*

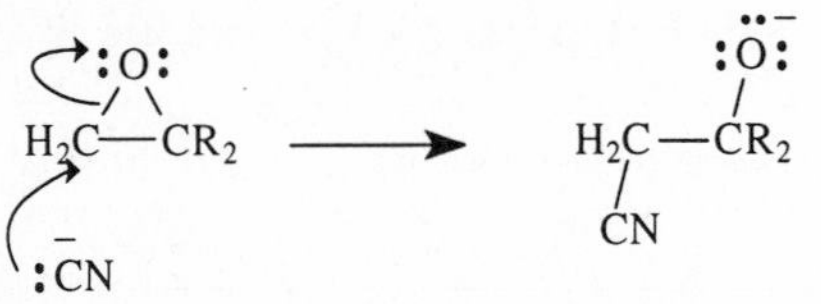

*Nucleophilic attack at a carbonyl group:*

The one apparent difficulty with this analogy is the stereochemistry of the reaction. In epoxide opening, attack of the nucleophile occurs backside to the C—O bond, while backside attack on a $\pi$ bond is hard to envision with Lewis structures. The stereochemical aspect of this reaction becomes clearer when the C=O bond is redrawn as two "bent" bonds (somewhat like those you might find in some model sets):

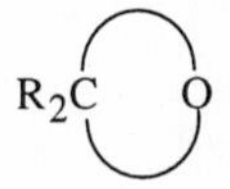

It turns out that such a description of the double bond is completely equivalent physically to the usual $\sigma + \pi$ description! In other words, a double bond can be described *either* as one $\sigma$ bond and one $\pi$ bond, *or* as two identical bonds that are in between these two extremes—bent bonds, or (as they are called) *tau bonds.* Backside attack on one of the tau bonds occurs from the correct direction for carbonyl addition. The other tau bond rehybridizes during the addition to become the C—O $\sigma$ bond.

   The point is that two reactions which seem superficially to be very different—one a substitution at a saturated carbon atom, the other an addition at an unsaturated carbon atom—really are fairly similar, when we get to the heart of the matter.

## √19.4    Acids and Bases in Reaction Mechanisms

Notice that under acidic conditions, as in the hydration mechanism in the text (Eqs. 19.18a–b), the molecule that acts as a base is *water*, not the hydroxide ion. Although hydroxide ion would be a superb base, hydroxide is not present in any significant concentration in acidic solution because of the very small ion-product constant of water.

   Likewise, in basic solution, in which hydroxide is present in significant concentration, significant amounts of hydronium ions cannot exist; hence, when an acid is required for reactions in basic solution, either the conjugate acid of hydroxide—water—or other weak bases must serve the purpose. (In the case of cyanohydrin formation, HCN serves as the acid, because it is weak enough to exist in mildly alkaline solution.)

To summarize: when we write mechanisms for reactions requiring strong acids we have to use weak bases, because such bases are the only ones that can exist in acidic solution. When we write mechanisms requiring strong bases, we must use weak acids, because these are the only acids that can exist in basic solution. Specifically, if a reaction involves $H_3O^+$ as an acidic catalyst, the base involved is water, not $^-OH$; if a reaction involves $^-OH$ as a basic catalyst, the acid involved is water, not $H_3O^+$.

## ✓19.5    Lewis Acid Catalysis

Study Guide Link 16.2 focused on Lewis-acid interactions with leaving groups. What you should notice about both Grignard reagents and lithium aluminum hydride is the important role of Lewis acid catalysis in promoting reactions of these reagents with carbonyl compounds (as well as with epoxides; Sec. 11.4B). Each reagent has its own "built-in" Lewis acid catalyst. As explained in the text, the Lewis acid catalyst in $LiAlH_4$ reductions is the lithium ion, $Li^+$; and the Lewis acid catalyst in Grignard reactions is the magnesium of the Grignard reagent.

Why not use proton (Brønsted) acids as catalysts in these reactions? Don't forget that both $LiAlH_4$ and Grignard reagents react instantaneously with protons of even weak acids such as water and alcohols. Consequently, the use of proton acids is not an option, because such acids destroy the reagents. Only *after* the addition of these reagents takes place can a proton source be added to the reaction mixture to replace the Lewis acid ($Li^+$ or $^+MgBr$) with a hydrogen.

What you should notice here is that the Lewis acid serves the role of a "substitute proton"—or "fat proton," as one of the author's colleagues likes to call it—and fulfills *exactly* the same catalytic role that a proton would fulfill if a proton *could* be used.

## ✓19.6    Reactions that Form Carbon-Carbon Bonds

It's important to pay special attention to the reactions that can be used to form carbon-carbon bonds, for these are the reactions that are used to build up carbon skeletons. A complete list of these reactions, in the order in which they occur in the text, can be found in Appendix V. The three reactions for forming carbon-carbon bonds you either have encountered, or will encounter, in this chapter are cyanohydrin formation (Sec. 19.7), reactions of Grignard and related reagents with aldehydes and ketones (Sec. 19.9), and the Wittig alkene synthesis (Sec. 19.13). How many others can you list? Can you give examples of each? Do you understand the limitations of each, if any, and the reasons for these limitations?

## ✓19.7    Alcohol Syntheses

Study Guide Link 19.6 suggested making special note of reactions that form carbon-carbon bonds. Another approach to reviewing reactions is to classify them by the types of products they give. For example, the reactions of Grignard reagents with aldehydes and ketones, as you've seen, give alcohols as products. The other reaction type you've studied in this chapter that can be used to prepare alcohols is the hydride reduction of aldehydes and ketones (Sec. 19.8). Both reactions are methods for alcohol synthesis of *major* importance. What's more, a great variety of other compounds can be prepared from alcohols. Thus, these methods have importance for the preparation of other types of organic compounds as well.

A complete list of alcohol syntheses, as well as the methods used to prepare every other major functional group, is found in Appendix IV. The reactions listed are given in the order that they are presented in the text. How many of these can you name without looking at the list? What are the limitations of each? Do you understand the reasons for the limitations?

## ✓19.8    Hemiacetal Protonation

Some students ask why, in Eq. 19.47b of the text, the hemiacetal is protonated on the —OH oxygen rather than on the —OCH$_3$ oxygen. This is a reasonable question because ethers and alcohols have similar basicities. The answer is that protonation on the —OCH$_3$ oxygen *does* occur, as does loss of methanol from the resulting protonated species. However, these steps are simply the reverse of hemiacetal formation and lead back to aldehyde or ketone and the starting alcohol. Under conditions of excess alcohol and removal of water, the equilibrium shifts to favor the alcohol.

When writing mechanisms, we usually do not write out reasonable steps that are not important in the formation of the product of interest.

## ✓19.9    Mechanism of Carbinolamine Formation

You are asked to write the mechanism for acid-catalyzed carbinolamine formation because it is another example of *carbonyl addition*. It is important to understand this reaction mechanistically because you'll see it repeatedly in organic chemistry and biochemistry. If you *really* want to make progress, take these suggestions seriously.

Are you having trouble getting started? In any carbonyl addition, *identify the nucleophile.* The nucleophilic atom is the nitrogen of the amine. *The nucleophile attacks at the carbonyl carbon* because electrons can flow onto the electronegative oxygen. (See Study Guide Link 19.2 on page 465 of this manual.)

$$
\begin{array}{ccc}
\ddot{\text{O}}: & & :\ddot{\text{O}}:^{-} \\
\| & & | \\
\text{R}-\text{C}-\text{R} & \rightleftharpoons & \text{R}-\text{C}-\text{R} \\
& & | \\
\text{NH}_2-\text{R} & & {}^{+}\text{NH}_2-\text{R}
\end{array}
\qquad \text{(SG19.1)}
$$

Perhaps you added a proton to the carbonyl oxygen first, and *then* let the amine attack. *This is a reasonable step for a beginning student to take,* and follows the mechanism for acid-catalyzed hydration.

protonated carbonyl group

$$
\begin{array}{ccc}
\text{H}-\overset{+}{\text{O}}: & & \text{H}-\ddot{\text{O}}: \\
\| & & | \\
\text{R}-\text{C}-\text{R} & \rightleftharpoons & \text{R}-\text{C}-\text{R} \\
& & | \\
\ddot{\text{N}}\text{H}_2-\text{R} & & {}^{+}\text{NH}_2-\text{R}
\end{array}
$$

Which is correct? For purposes of *your understanding,* either mechanism represents a reasonable first step. However, let's think about this issue in a little more detail. Remember that carbonyl oxygens are *very weak bases* and amine nitrogens are *fairly strong bases*—about like ammonia. If the acid is strong enough to provide a significant concentration of the protonated carbonyl compound, then surely under the same conditions the amine nitrogen would be completely protonated to an ammonium ion (see Eq. 19.57 on text p. 905) and would no longer be nucleophilic. Hence, the first mechanism—attack of the neutral amine on the neutral carbonyl compound (Eq. SG19.1)—is more likely to be correct.

Once attack of the amine has occured, a proton is transferred *from* $H_3O^+$ to the carbonyl oxygen and *from* the attacking nitrogen to $H_2O$. Use *two separate steps* for these transfers. It doesn't matter which you write first. Notice that the oxygen, because of its negative charge, is an alkoxide, which is a rather strong base, and is efficiently protonated by rather low concentrations of $H_3O^+$.

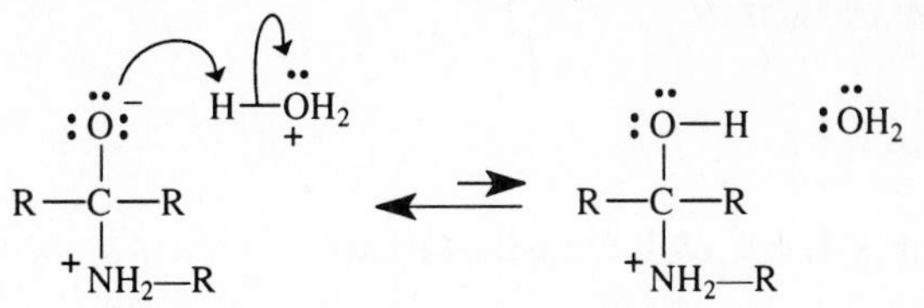

$$\text{(SG19.2)}$$

The final step is deprotonation of the nitrogen. Because the nitrogen of the carbinolamine is a base, this final step is an equilibrium. If the acid concentration is high enough, a significant amount of the protonated carbinolamine could be present.

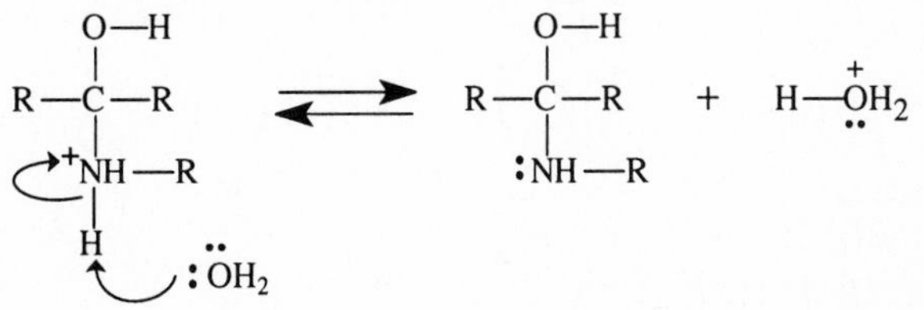

$$\text{(SG19.3)}$$

Notice that when $H_3O^+$ is the acid used in the mechanism, then its conjugate base $H_2O$ must be used as the base.

As illustrated here, the mechanism, including the proton-transfer steps, is written *one step at a time*. Students are often tempted to try to show everything in one step. This is not correct because simultaneous collisions of more than two molecules are highly improbable. More important is that consolidating several steps into one can lead to confusion. Remember to write mechanisms *one step at a time*.

## Solutions

## Solutions to In-Text Problems

**19.1**  (a)

$(CH_3)_2CHCH{=}O$

(c)

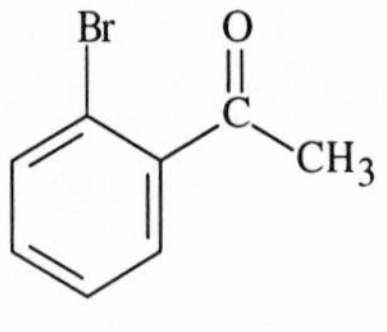

(e)

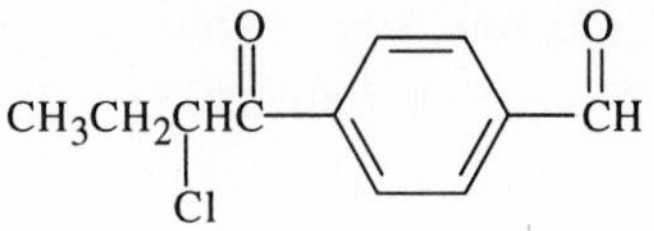

(g)

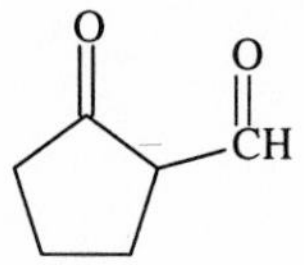

(i)

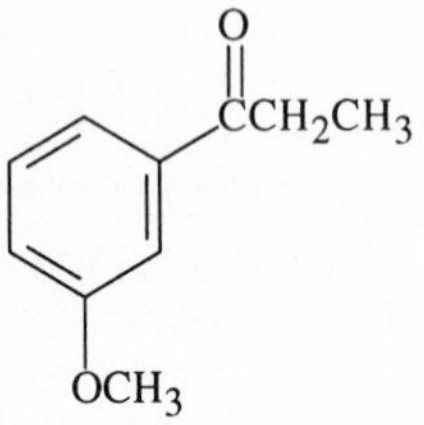

**19.2**  (a) 2,4-dimethyl-3-pentanone
(c) 3-allyl-2,4-pentanedione
(e) 3-methylcyclobutanecarbaldehyde

**19.3**  (a) 2-Cyclohexenone has a lower carbonyl stretching frequency because its two double bonds are conjugated.
(c) 3-Buten-2-ol has both O—H and C=C stretching absorptions and 2-butanone does not; 2-butanone has a carbonyl stretching absorption, and 3-buten-2-ol does not.

**19.4**  (a)

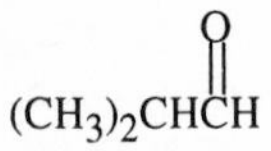

isobutyraldehyde
(2-methylpropanal)

(c)

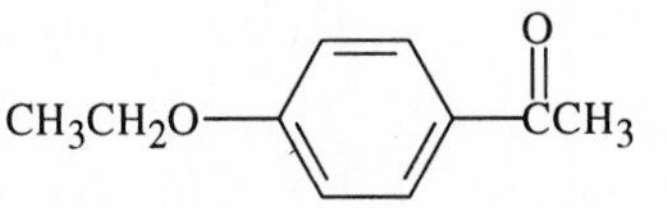

*p*-ethoxyacetophenone

**19.5**  The low intensities of the carbonyl carbon absorption at $\delta$ 212.6 and the $\alpha$-carbon absorption at $\delta$ 44.2 show that these two carbons have no attached hydrogens; therefore this compound must have the following partial structure:

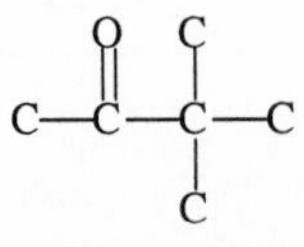

Because all six carbons and their connectivities are accounted for, the only possible choice is to add the appropriate number of hydrogens. The compound is 3,3-dimethyl-2-butanone:

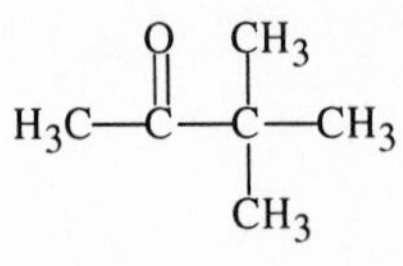

3,3-dimethyl-2-butanone

**19.7**  (a)  The first compound has more conjugated double bonds, and therefore will have the UV spectrum with the greater $\lambda_{max}$.

(c)  In 1-phenyl-2-propanone the aromatic ring is not conjugated with the carbonyl group; in *p*-methylacetophenone the aromatic ring and the carbonyl group are conjugated. Because of its additional conjugation, *p*-methylacetophenone has the UV spectrum with the greater $\lambda_{max}$.

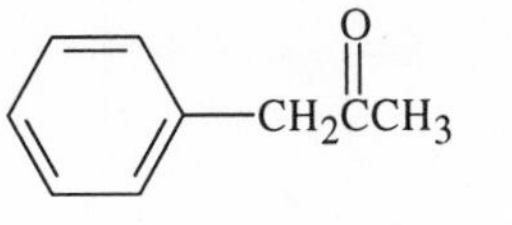

1-phenyl-2-propanone

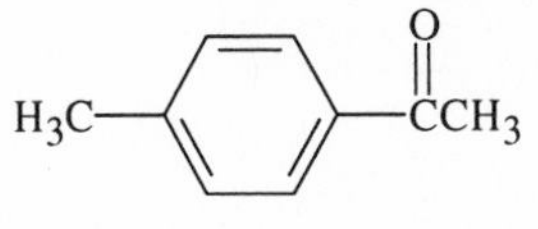

*p*-methylacetophenone
(has the UV spectrum with
the greater $\lambda_{max}$)

**19.8**  In NaOH solution, the phenolic —OH groups of both compounds are ionized. In the conjugate base of vanillin, the electron pair associated with the negative charge can be delocalized into the carbonyl group:

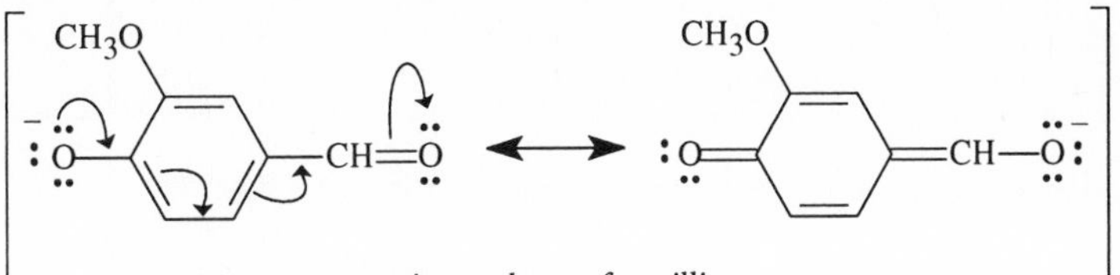

conjugate base of vanillin

In the conjugate base of isovanillin, the negative charge can be delocalized into the ring (as in any phenolate ion) but not into the carbonyl group. The more extensive conjugation (that is, delocalization) in the conjugate base of vanillin leads to a greater $\lambda_{max}$.

**19.10**  (a)  In both ketones, inductive cleavage of the molecular ion to give an acyl radical and a carbocation accounts for the $m/z = 57$ peaks:

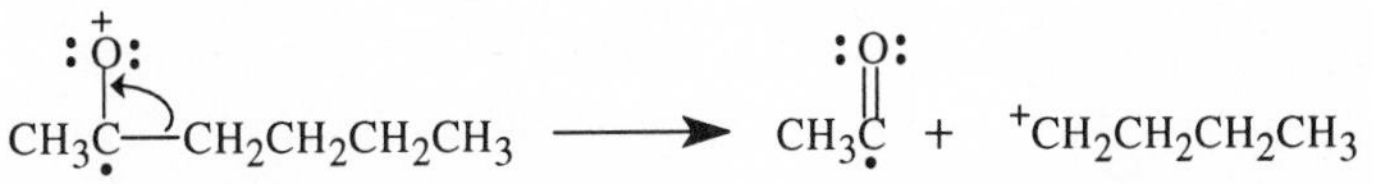

molecular ion of 2-hexanone

butyl cation
($m/z = 57$)

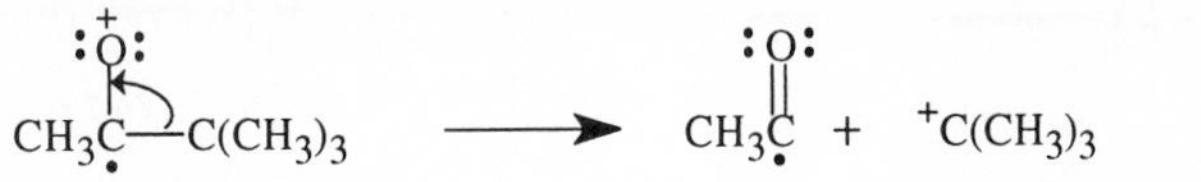

molecular ion of
3,3-dimethyl-2-butanone

*tert*-butyl cation
($m/z = 57$)

Because the carbocation formed from 3,3-dimethyl-2-butanone is more stable (why?), more of this fragmentation mode is observed for 3,3-dimethyl-2-butanone; that is, the $m/z = 57$ peak is more abundant.

(b)  Its even mass suggests that the $m/z = 58$ fragment is an odd-electron ion. Such an ion can be readily produced by a McLafferty rearrangement from the molecular ion of 2-hexanone, but not from the molecular ion of 3,3-dimethyl-2-butanone:

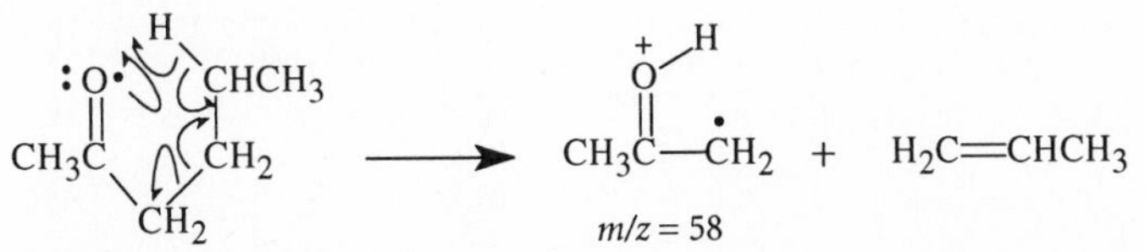

molecular ion of 2-hexanone

$m/z = 58$

**19.12** (a) The solvolysis mechanism of chloro(methoxy)methane in ethanol:

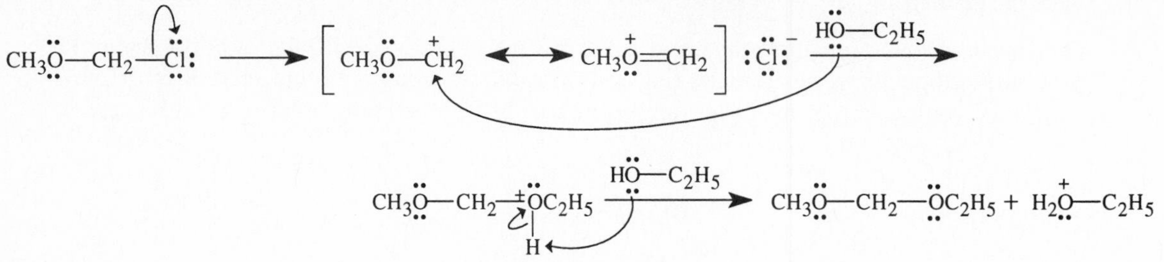

**19.13** The products follow from a consideration of the mechanism in Study Problem 19.2. The first steps of this mechanism consist of protonation of an —OH group and loss of water to form a carbocation intermediate. When the two —OH groups are chemically nonequivalent, the —OH group lost is the one that gives the more stable carbocation intermediate.

(a) The —OH group of the tertiary alcohol is lost because a relatively stable tertiary carbocation is formed.

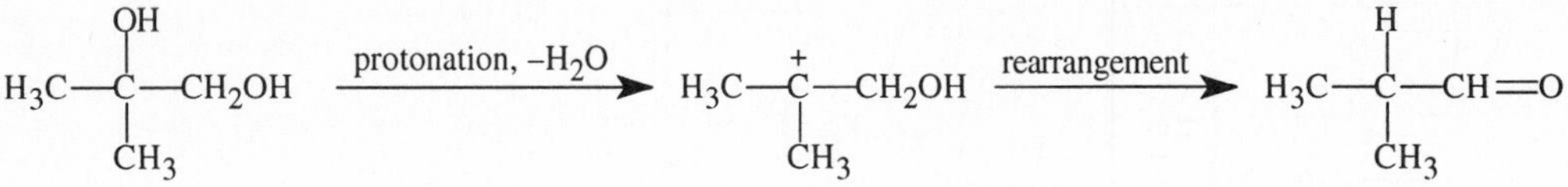

(c) In this case, rearrangement involves a ring expansion. If reasoning through this type of rearrangement gives you difficulty, be sure to consult Study Guide Link 10.2 (p. 226, Vol. 1 of this manual) for assistance. The product is the following ketone:

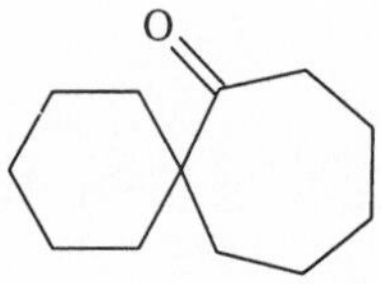

**19.14** (a) Protonated *p*-methoxybenzaldehyde is essentially a benzylic carbocation; as you learned in Chapter 17, benzylic carbocations are stabilized by substituents that can donate electrons by resonance, and this cation is no exception:

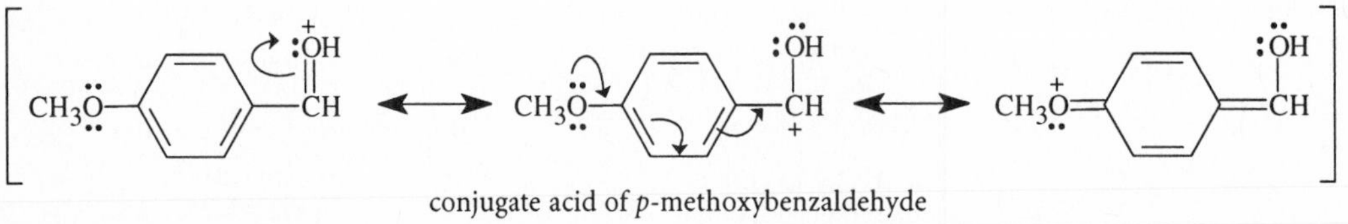

conjugate acid of *p*-methoxybenzaldehyde

Stabilization of the conjugate acid of any compound *increases* the compound's basicity. The *p*-nitro group *destabilizes* the corresponding cation that results from protonation of *p*-nitrobenzaldehyde. (See the solution to Problem 17.24 on pp. 405–406, Vol. 1 of this manual, for a very similar case.) Destabilization of the conjugate acid of a compound *decreases* the compound's basicity.

**19.15** The mechanism of the hydroxide-catalyzed hydration of acetaldehyde:

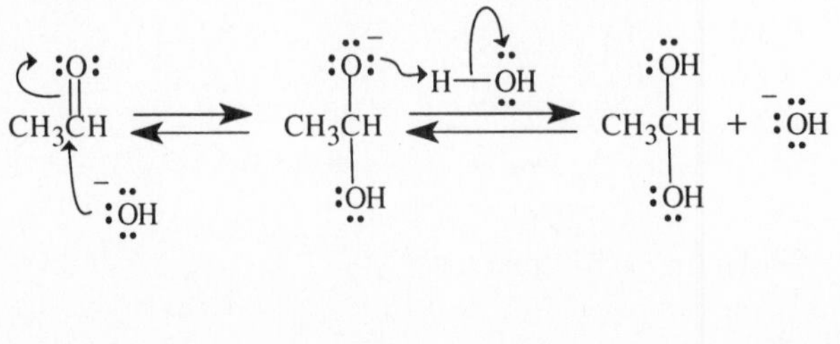

**19.16** (a) The acid-catalyzed addition of methanol involves mechanistic steps like those involved in acid-catalyzed hydration (text Eqs. 19.18a–b).

**19.17** The central carbonyl group is hydrated for the following reason. The carbon of any carbonyl group bears a partial positive charge. The central carbonyl group is therefore bonded to two partially positive carbonyl carbons, and the positive charge on each of these carbons has an unfavorable repulsive interaction with the partial positive charge on the central carbon. In addition, the other two carbonyl groups are conjugated with the benzene ring. As explained in the text, adjacent electronegative groups make carbonyl-addition reactions more favorable, and groups that are conjugated with the carbonyl group make carbonyl-addition reactions less favorable. It follows that hydration of the central carbonyl group is the more favorable process. The structure of the hydrate is as follows:

hydrate of ninhydrin

**19.19** (a) The second compound, bromoacetone, is more reactive because the electronegative atom is closer to the carbonyl group.

(c) The first compound, *p*-nitrobenzaldehyde, is more reactive, because the *p*-nitro group raises the energy of the molecule by an unfavorable interaction of the positive charge on the nitrogen with the positive charge on the carbonyl carbon. In contrast, the *p*-methoxy group in the second compound, *p*-methoxybenzaldehyde, stabilizes the molecule by a resonance interaction with the positive charge on the carbonyl carbon.

some resonance structures of *p*-methoxybenzaldehyde

Because stabilization of a carbonyl compound results in decreased reactivity, *p*-methoxybenzaldehyde is less reactive.

(d) The second compound, cyclopropanone, is more reactive for the following reason. A carbonyl compound is most stable when the bond angles at the carbonyl group can be close to 120°; however, in cyclopropanone they are constrained by the three-membered ring to be 60°. As a result, there is considerable strain in the carbonyl form. In the hydrate, the preferred bond angle is approximately tetrahedral (109.5°). Because this angle is closer to the 60° bond angle enforced by the ring than the bond angle in cyclopropanone is, there is less strain in the hydrate than there is in cyclopropanone. Hence, formation of the hydrate relieves some of the strain in cyclopropanone. In other words, the ketone is destabilized relative to the hydrate by strain; there is a driving force to form the hydrate. Because there is little or no strain in cyclopentanone, such effects do not come into play.

**19.20** (a)

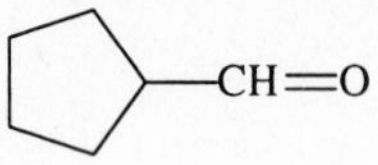


(c)

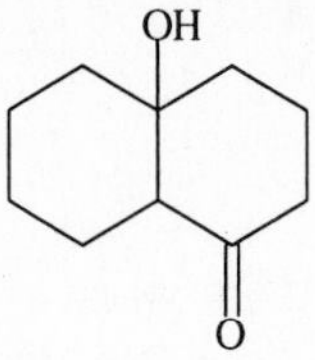


An extra equivalent of $LiAlH_4$ would have to be used with the compound in (c), because the hydride would react vigorously with the tertiary alcohol to form the conjugate-base alkoxide. The tertiary alcohol would be re-formed when a weak acid is added to the reaction mixture.

**19.21** The tertiary alcohol *C* could not be synthesized by a hydride reduction, because only primary and secondary alcohols can be prepared by this method.

**19.22** In each case, ethyl bromide, $CH_3CH_2Br$, reacts with Mg to give ethylmagnesium bromide, $CH_3CH_2MgBr$, which is then allowed to react as shown below.

(a)

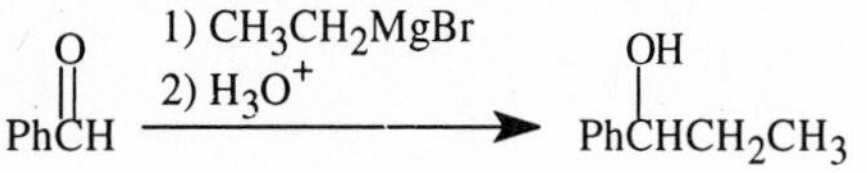


(c)

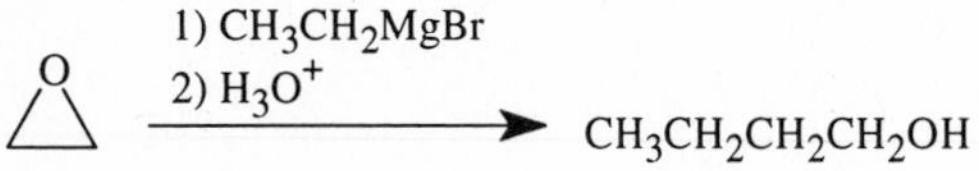


1-butanol

(e)

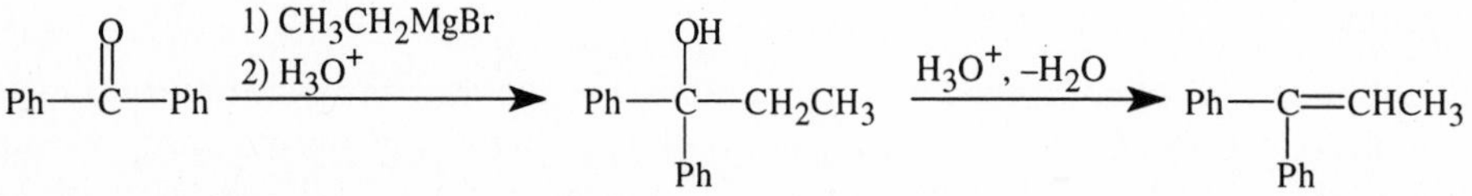


(g)

$$CH_2{=}O \xrightarrow[2)\ H_3O^+]{1)\ CH_3CH_2MgBr} CH_3CH_2CH_2OH \xrightarrow{CrO_3(pyridine)_2} CH_3CH_2CH{=}O$$

**19.23** (a) Either alkyl group bound to the $\alpha$-carbon of the alcohol can in principle originate from the Grignard reagent. Thus, either the reaction of $(CH_3)_2CH—CH{=}O$ with $CH_3MgI$, followed by protonolysis, or the reaction of $(CH_3)_2CHMgBr$ with $CH_3CH{=}O$, followed by protonolysis, will give the desired alcohol.

**19.24** (a)

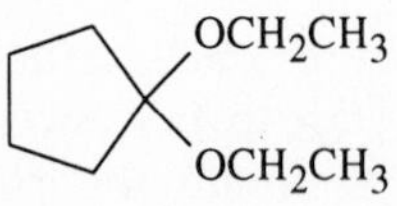


cyclopentanone diethyl acetal
(1,1-diethoxycyclopentane)

**19.25** (a)

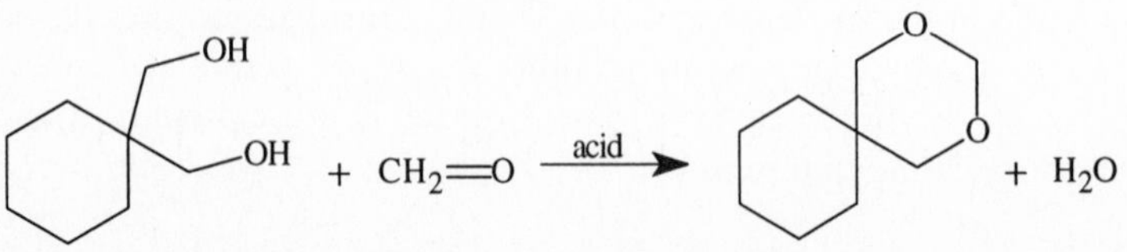

**19.26** (a) The formula indicates addition of two carbon atoms and it indicates one degree of unsaturation. If a diethyl acetal were formed and the ring were opened, the formula of the product would be $C_9H_{20}O_3$. The one degree of unsaturation suggests that the ring is intact. The product is a mixed acetal:

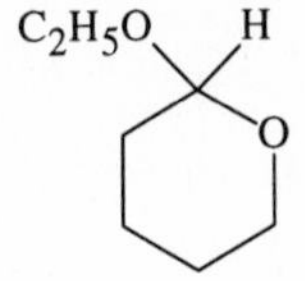

**19.27** (a) In the following synthesis, any common alcohol can be used instead of methanol.

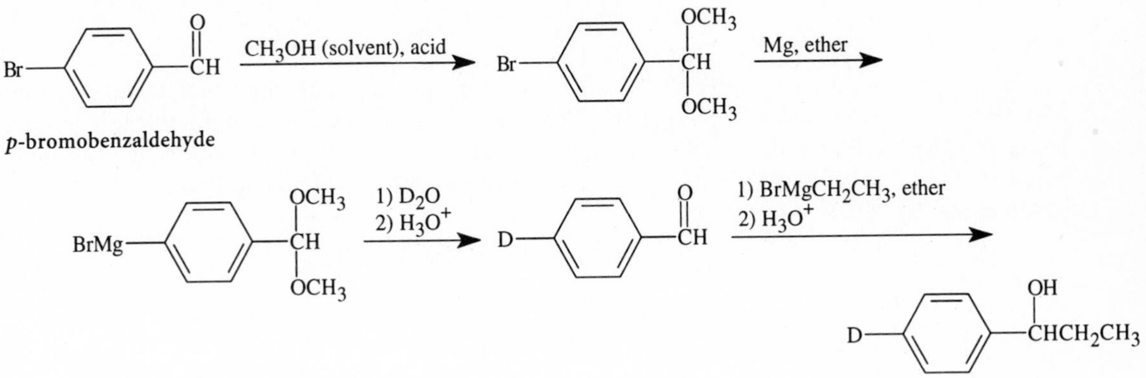

*p*-bromobenzaldehyde

**19.28** (a)                                  (c)

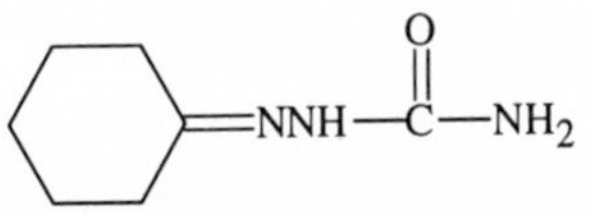

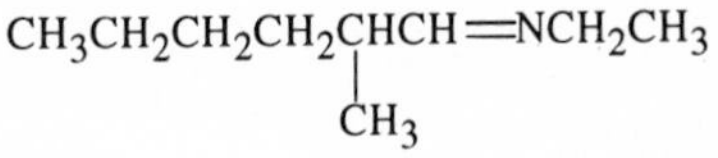

$$CH_3CH_2CH_2CH_2\underset{\underset{CH_3}{|}}{C}HCH=NCH_2CH_3$$

**19.29** First, the carbinolamine intermediate is formed. This intermediate then undergoes acid-catalyzed dehydration to give the hydrazone. (Be sure to consult Study Guide Link 19.9 on p. 468 of this manual.)

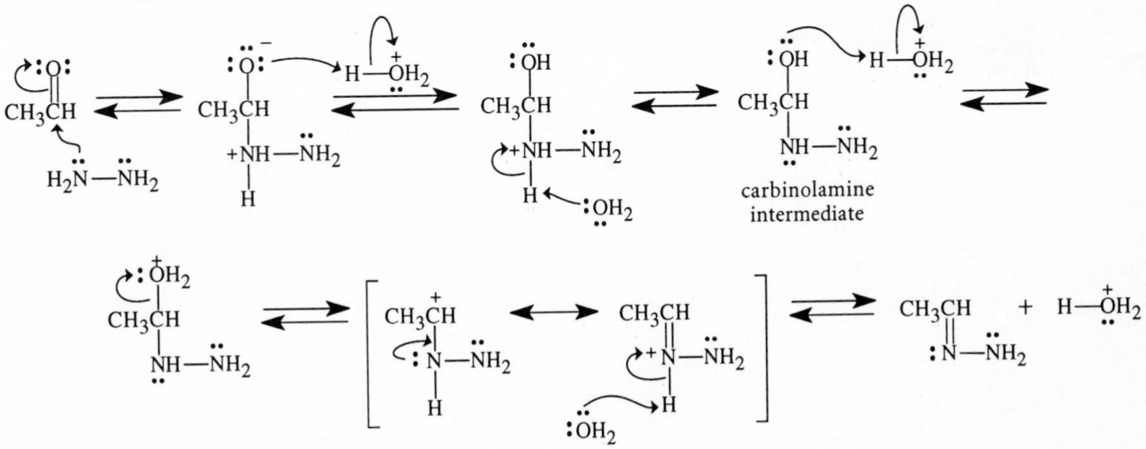

carbinolamine
intermediate

Note that it is equally appropriate to write the loss of water and formation of the carbon-nitrogen double bond as one step, thus avoiding the necessity of drawing resonance structures:

*(solution continues)*

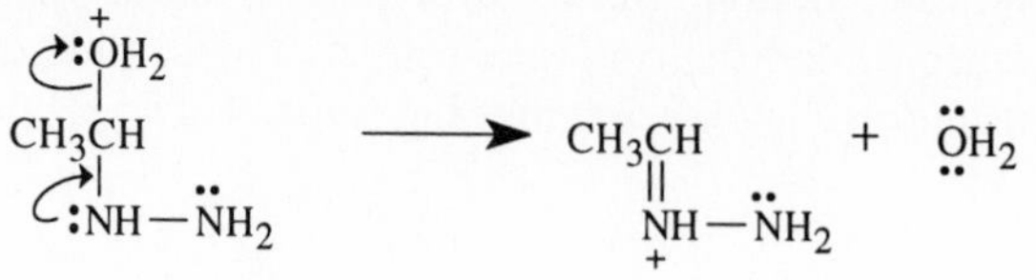

**19.31** (a)

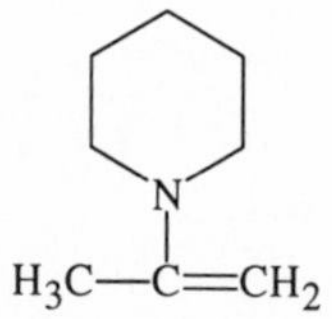

**19.32** Alkylate the phenol, carry out a Friedel-Crafts acylation, and then apply the Wolff-Kishner reaction. Alkylation of the phenol should precede the Friedel-Crafts reaction because of the sluggish reactivity of phenols in the Friedel-Crafts acylation reaction; see text Sec. 18.8. Note that Friedel-Crafts alkylation of the ring with $ClCH_2CH_2CH_2CH_3$ would give some rearrangement product in addition to the desired product; see Eq. 16.19 on text p. 758.

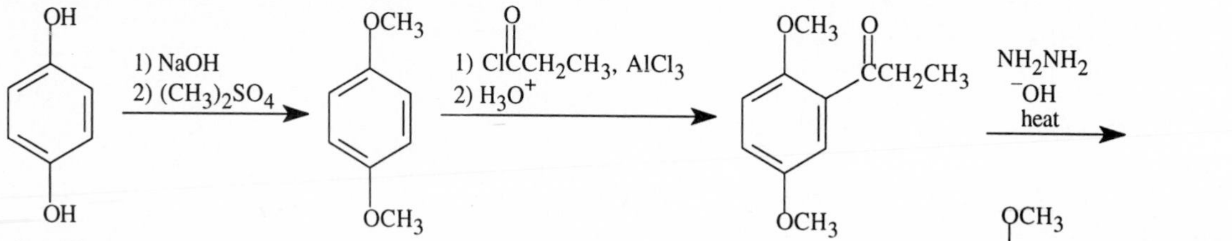

1,4-dimethoxy-2-propylbenzene

**19.34** (a) $(CH_3)_2C=CHCH_3$

**19.35** (a) Either "half" of the alkene can in principle be derived from an aldehyde. The first synthesis:

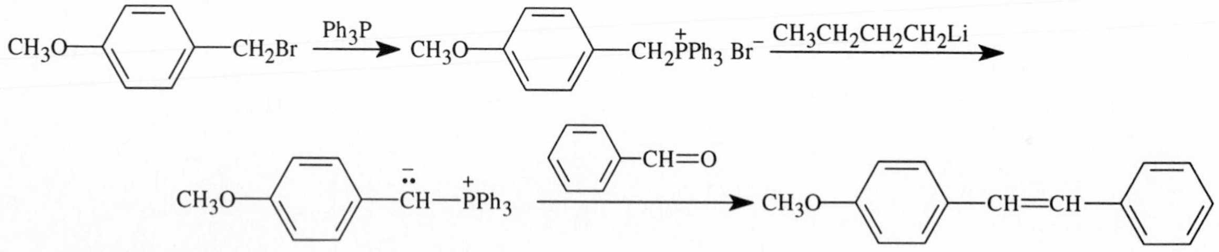

The second synthesis:

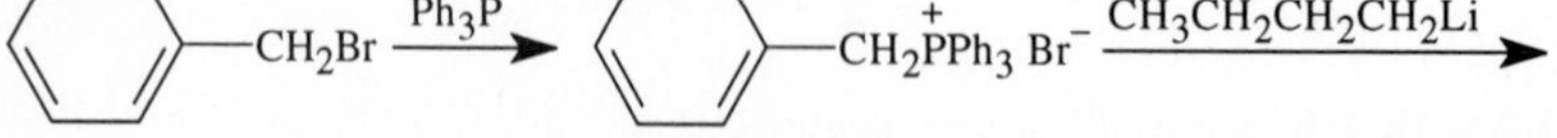

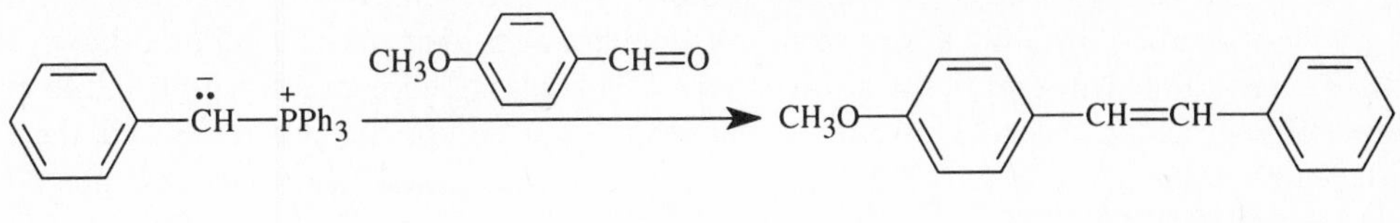

(c)

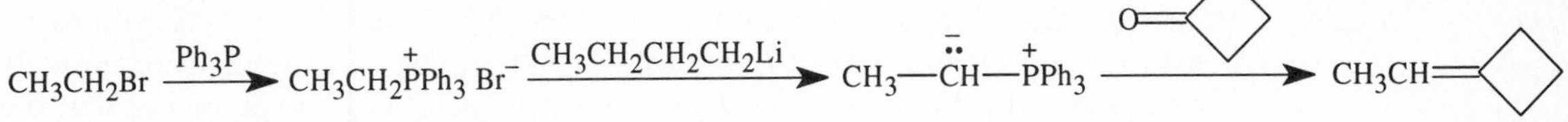

**19.36** The starting material has five degrees of unsaturation, four of which are accounted for by a benzene ring. Consequently, the compound must be both an aldehyde and an alcohol:

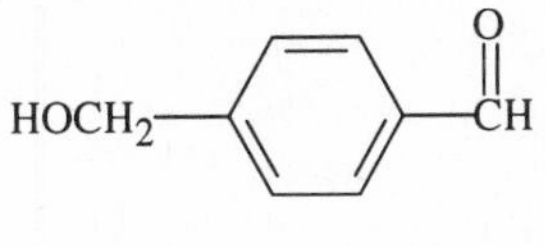

4-(hydroxymethyl)benzaldehyde

## Solutions to Additional Problems

· · · · · · · · · · · · · · · · · · · · · · · · · · · · · · · · · · · · · · · · · · · · · · · · · · · · · · · · · · · · · · · · · · · · · · · · · · ·

**19.38** Only organic products are shown in the answers below.

(a) Acetone is protonated on oxygen to give the conjugate acid shown below, but unless the acid is very strong, this reaction does not occur to a great extent.

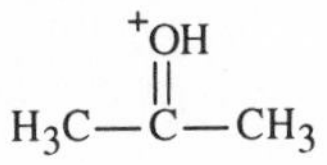

$$\overset{+OH}{\underset{}{H_3C-\overset{\|}{C}-CH_3}}$$

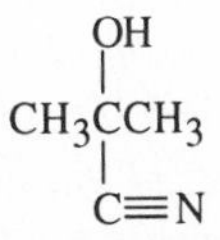

(b) 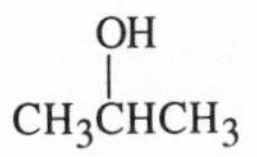              (c) No reaction.  (d)              (e) 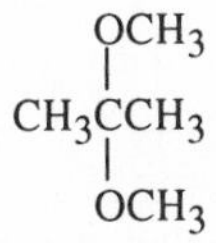              (f) 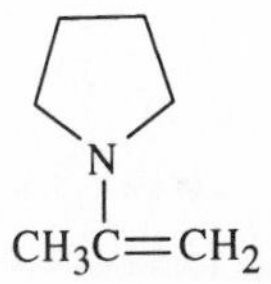              (g)

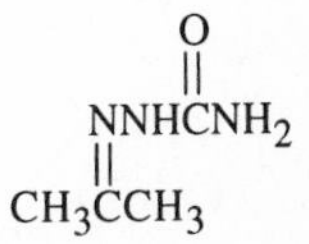

(h) 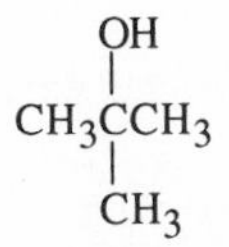              (i) 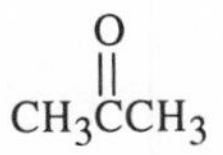              (j) 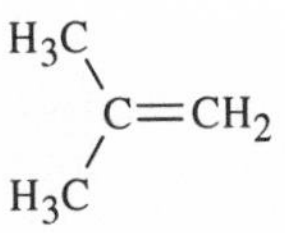              (k) 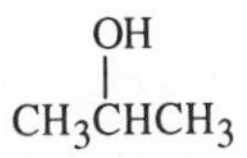              (l)              (m)

**19.40** (a) This is a simple addition reaction in which the sulfur of the bisulfite ion attacks the carbonyl carbon of the aldehyde. In the last step of the mechanism the initially formed addition product *A* ionizes because it is a fairly strong acid.

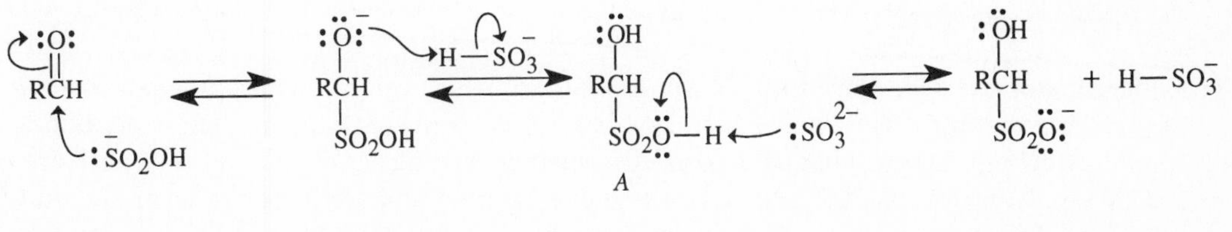

(b) Acid, that is, $H_3O^+$, destroys sodium bisulfite by reacting with it and converting it into $SO_2$, a gas, and $H_2O$. Destruction of bisulfite in this manner eliminates a reactant in the equilibria shown above and pulls the reaction to the left, that is, toward free aldehyde. Hydroxide ion reacts with bisulfite and converts it into sulfite ion, $SO_3^{2-}$; because bisulfite is removed as a reactant by this process as well, the equilibrium is also pulled to the left.

In the mechanisms presented thus far in this Study Guide and for the most part in the text, unshared electron pairs have been drawn in explicitly. Omitting these electron pairs can save a some time and can eliminate a great deal of tedium in writing mechanisms; most chemists do not draw electron pairs explicitly. This shortcut is useful only if you realize that the unshared electron pairs which are not shown are understood to be present. In many of the mechanisms shown subsequently, electron pairs are omitted in the most common situations. (In unusual or uncommon situations, the electron pairs are retained.) Here are some typical situations in which electron pairs should not be necessary:

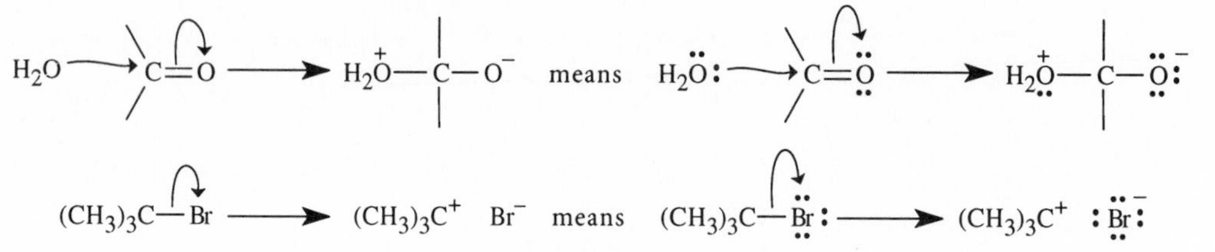

Thus, an oxygen with three bonds and a positive charge has one unshared electron pair; an oxygen with one bond and a negative charge has three unshared pairs; an uncharged nitrogen with three bonds or a negatively charged carbon with three bonds has one unshared pair; and a halide ion has four unshared pairs.

The reason we do not have to show unshared valence electrons is that if the *formal charge* on an atom and the *number of bonds to an atom* are known, *the number of unshared valence electrons are automatically known.* It follows that *the formal charge must be shown;* otherwise, the electronic state of an atom is undetermined.

Most of the mechanisms shown after this point will be written without showing the unshared pairs. If you ever become confused by the absence of electron pairs, *do not hesitate to draw them in explicitly.*

**19.41**   (a)  Two diastereomeric alcohols are formed by attack of hydride at the upper and lower face, respectively, of the carbonyl group followed by protonation of the resulting alkoxides.

from attack of hydride at the
upper face of the carbonyl carbon

from attack of hydride at the
lower face of the carbonyl carbon

**19.42**   This reaction gives two constitutional isomers, each of which can be formed as a pair of diastereomers. Notice that the two diastereomers that contain the six-membered ring are *meso*-compounds and are therefore achiral; the two that contain the five-membered ring are chiral, and thus both are obtained as racemates. Constitutional isomers have different properties and are therefore separable; and diastereomers also have different properties, and are also separable.

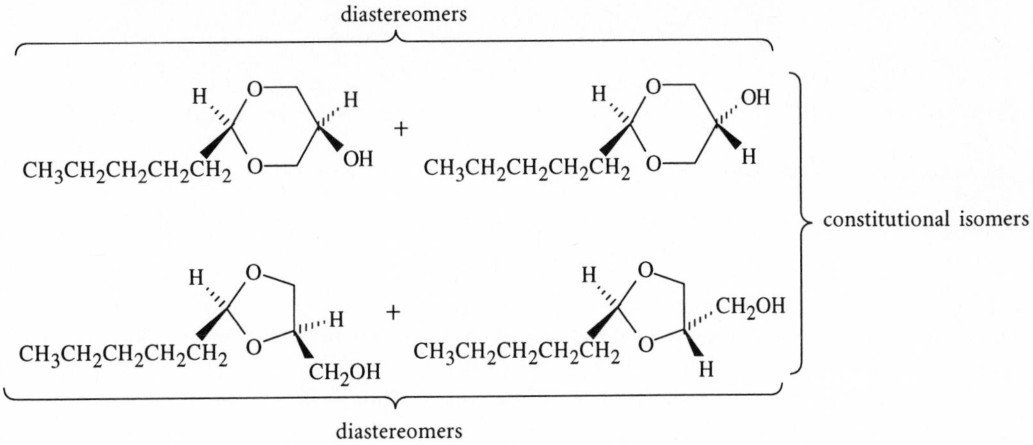

diastereomers

constitutional isomers

diastereomers

**19.44**  The two separable isomers are the diastereomeric acetals:

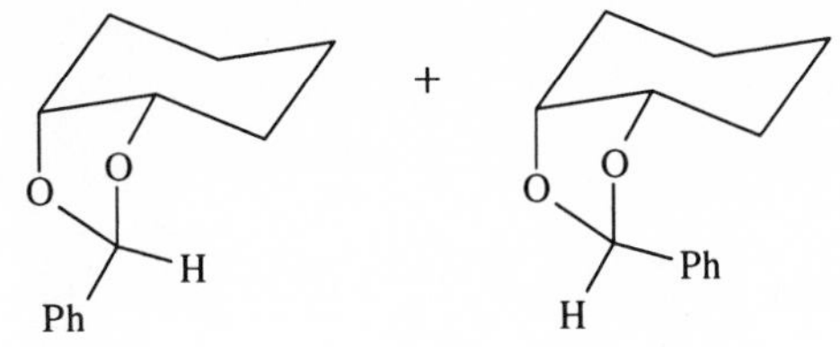

**19.45**  (a)

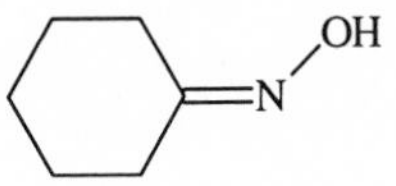

(c)

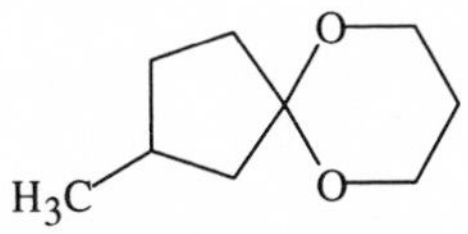

(e)

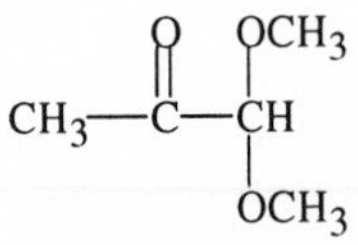

In part (a), the reaction is formation of an oxime, a type of imine; see Table 19.3, text p. 906. Part (c) is formation of a cyclic acetal. In part (e), the formula shows that only one of the carbonyl groups is involved in acetal formation; the aldehyde carbonyl reacts selectively because aldehydes react more rapidly to form addition compounds than do ketones, and because addition reactions of aldehydes are thermodynamically more favorable than those of ketones.

(g)

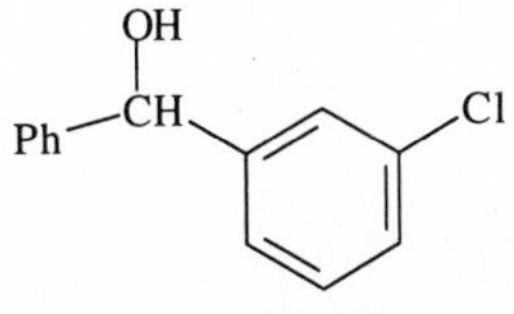

(i)

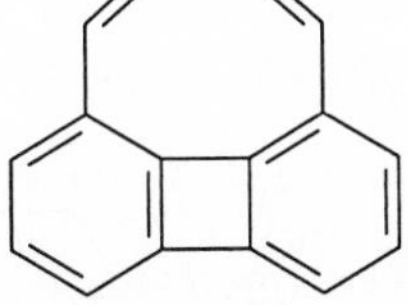

In part (g), a Grignard reagent is formed selectively from the aryl bromide, because only aryl bromides react with magnesium in ether solvent; see Sec. 18.5, text p. 836. In part (i), a "double" Wittig reagent is formed from the two benzylic bromides, and each Wittig reagent reacts at a different carbonyl group of the same aldehyde molecule to form a diene.

**19.46**  (a)  In this reaction, the carbonyl group is reduced to an alcohol, which is subsequently ionized by NaH. The resulting intermediate is essentially a Wittig reaction intermediate, as the hint suggests; this intermediate decomposes to form an alkene.

$$CH_3CH_2CH_2\overset{\overset{\displaystyle Ph_2P}{|}}{CH}-\overset{\overset{\displaystyle OH}{|}}{CH}CH_2CH_3 \xrightarrow{\text{NaH}} CH_3CH_2CH_2\overset{\overset{\displaystyle Ph_2P}{|}}{CH}-\overset{\overset{\displaystyle O^-}{|}}{CH}CH_2CH_3 \longrightarrow CH_3CH_2CH_2\overset{\overset{\displaystyle Ph_2P-O}{|}}{CH}\overset{|}{-}CHCH_2CH_3 \longrightarrow$$

product of NaBH₄ reduction

$$CH_3CH_2CH_2CH=CHCH_2CH_3 \;+\; Ph_2\overset{\overset{\displaystyle O^-}{\|}}{P}=O$$

**19.47** The data indicate that compound *A* is a benzene derivative and that it has two substituents on a benzene ring, one of which is a methyl group and one of which is an aldehyde. Only *p*-methylbenzaldehyde would give, after Clemmensen reduction, a compound (*p*-xylene) that in turn gives, because of its symmetry, one and only one monobromination product, 1-bromo-2,5-dimethylbenzene.

$$H_3C-\underset{\substack{\\ \textbf{\textit{p}-methylbenzaldehyde}\\ \textbf{(compound \textit{A})}}}{\bigcirc}-CH=O \xrightarrow{\text{Zn/Hg, HCl}} H_3C-\underset{\textbf{\textit{p}-xylene}}{\bigcirc}-CH_3 \xrightarrow{\text{Br}_2/\text{Fe}} H_3C-\underset{\underset{\text{Br}}{\bigcirc}}{}-CH_3$$

1-bromo-2,5-dimethylbenzene

**19.49** (a)

$$CH_3CH_2CH_2\overset{\overset{\displaystyle O}{\|}}{CH} \xrightarrow{\text{PhMgBr}} CH_3CH_2CH_2\overset{\overset{\displaystyle OH}{|}}{CH}-\bigcirc \xrightarrow{\text{CrO}_3(\text{pyridine})_2} CH_3CH_2CH_2\overset{\overset{\displaystyle O}{\|}}{C}-\bigcirc$$

butyraldehyde                          butyrophenone

(c)

$$\bigcirc=O \xrightarrow[\text{CH}_3\text{OH}]{\text{NaBH}_4} \bigcirc-OH \xrightarrow[\text{2) CH}_3\text{I}]{\text{1) NaH}} \bigcirc-OCH_3$$

cyclohexanone                              cyclohexyl methyl ether
                                         (methoxycyclohexane)

(e)

$$CH_3\overset{\overset{\displaystyle O}{\|}}{C}\underset{\underset{\displaystyle CH_3}{|}}{CH}CH_2CH_2CH_3 \xrightarrow[\text{2) H}_3\text{O}^+]{\text{1) CH}_3\text{MgI}} (CH_3)_2\overset{\overset{\displaystyle OH}{|}}{C}\underset{\underset{\displaystyle CH_3}{|}}{CH}CH_2CH_2CH_3 \xrightarrow{\text{H}_2\text{SO}_4} (CH_3)_2C=\underset{\underset{\displaystyle CH_3}{|}}{C}CH_2CH_2CH_3$$

3-methyl-2-hexanone                                                      2,3-dimethyl-2-hexene

(g)

$$CH_3O-\bigcirc-Br \xrightarrow{\text{Mg, ether}} CH_3O-\bigcirc-MgBr \xrightarrow[\text{2) H}_3\text{O}^+]{\text{1) H}_2\text{C}=O} CH_3O-\bigcirc-CH_2OH$$

*p*-bromoanisole                                                        *p*-methoxybenzyl alcohol
(1-bromo-4-methoxybenzene)

(i)

$$\bigcirc \xrightarrow{\text{OsO}_4} \overset{\text{OH}}{\underset{\text{OH}}{\bigcirc}} \xrightarrow{\text{periodic acid}} \overset{\text{CH}=O}{\underset{\text{CH}=O}{\diagup}} \xrightarrow{\text{NaBH}_4} \overset{\text{CH}_2-OH}{\underset{\text{CH}_2-OH}{\diagup}}$$

cyclohexene                                                            1,6-hexanediol

Ozonolysis could also be used to prepare the dialdehyde.

(j)  In the following synthesis, the two acyl groups could be introduced in the opposite order.

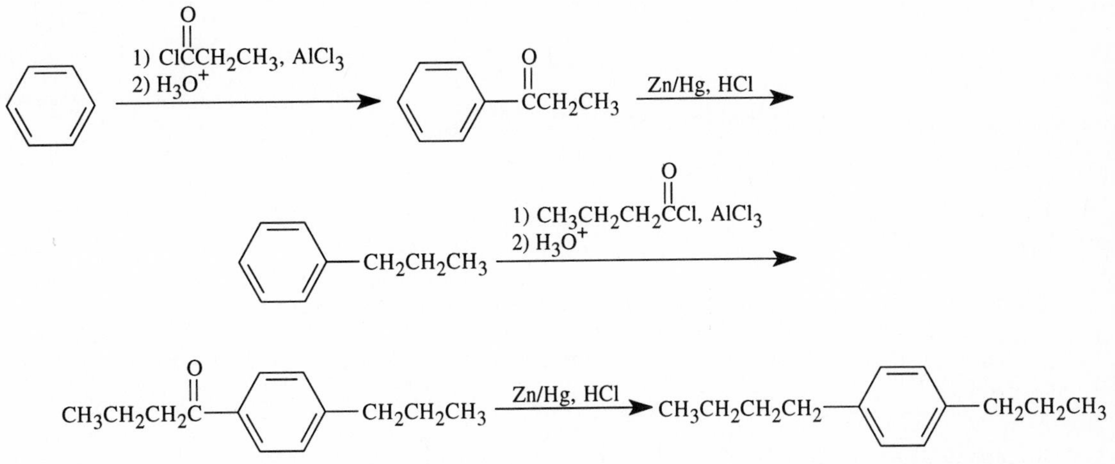

1-butyl-4-propylbenzene

(k)  First prepare benzylmagnesium bromide from benzaldehyde:

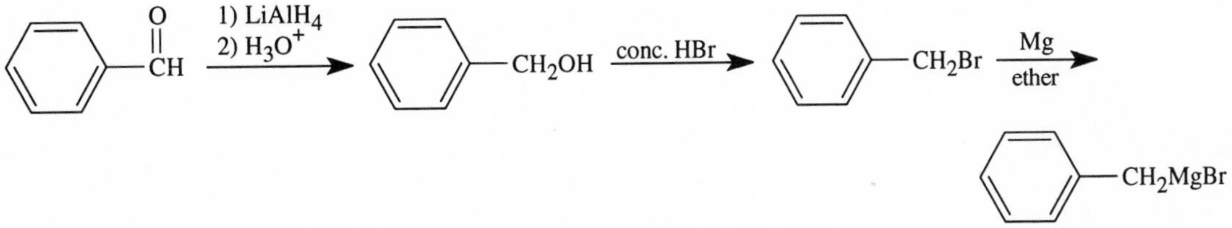

benzylmagnesium bromide

Then let this Grignard reagent react with benzaldehyde and oxidize the resulting alcohol:

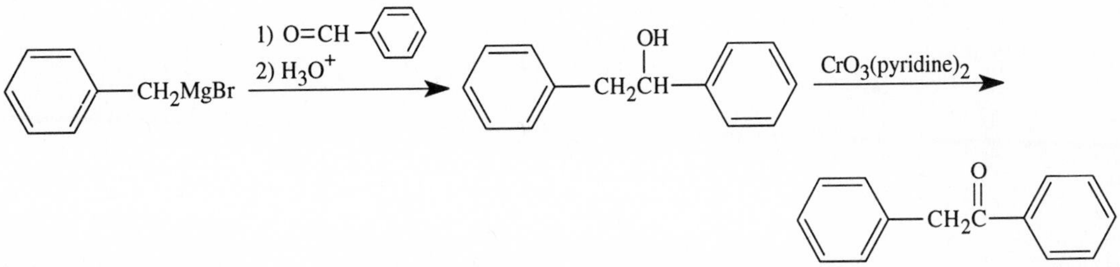

(l)  Following the hint, use a Diels-Alder reaction to form the ring and then reduce the product aldehyde. (It is preferable to carry out the Diels-Alder reaction before the reduction because conjugated carbonyl compounds are particularly reactive as dienophiles in Diels-Alder reactions.)

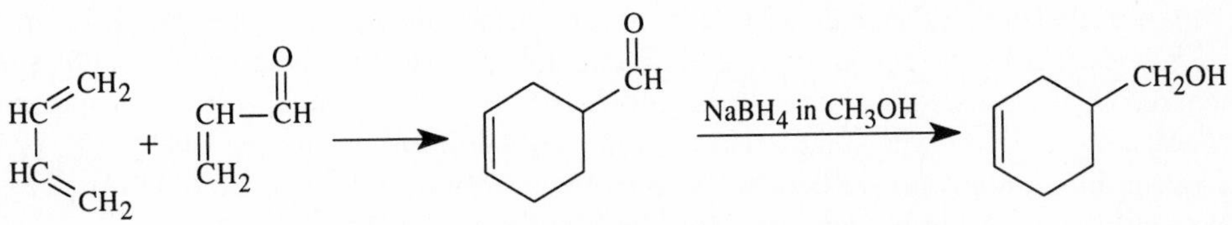

(m)  The first hint indicates that a protecting group must be used. The second hint is meant to call attention to the hemiacetal linkage which, because it is a five-membered *cyclic* hemiacetal, will form spontaneously from the corresponding aldehyde.

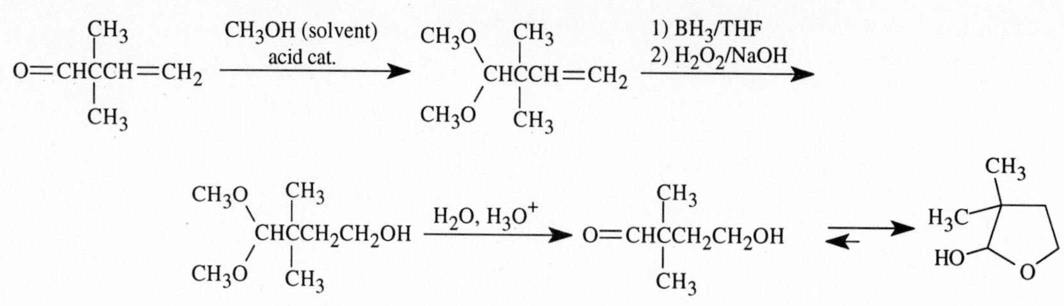

**19.50** (a)

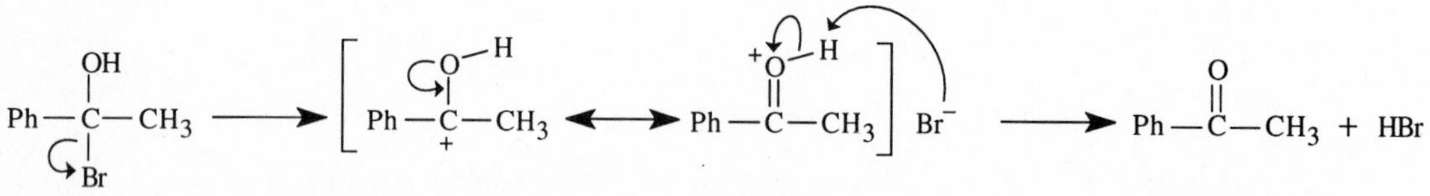

(b) Osmium tetroxide reacts to form the 1,2-glycol *A*, which also has a bromine on carbon 1 and thus decomposes as shown in the mechanism of part (a):

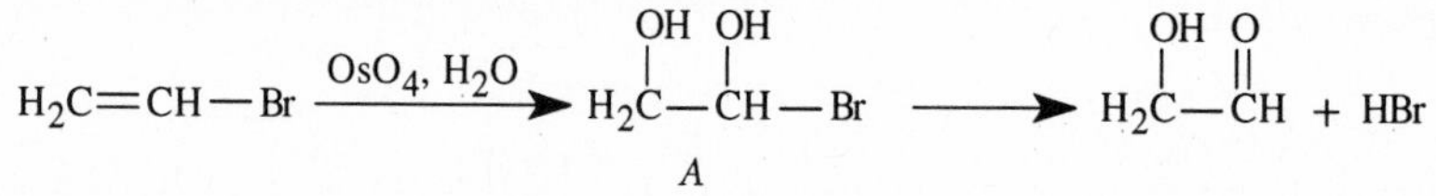

**19.51** The molecular formula of compound *A* shows that two equivalents of methanol are added to the alkyne. The fact that the product hydrolyzes to acetophenone indicates that the two methoxy groups of compound *A* are on the same carbon, that is, that compound *A* is an acetal.

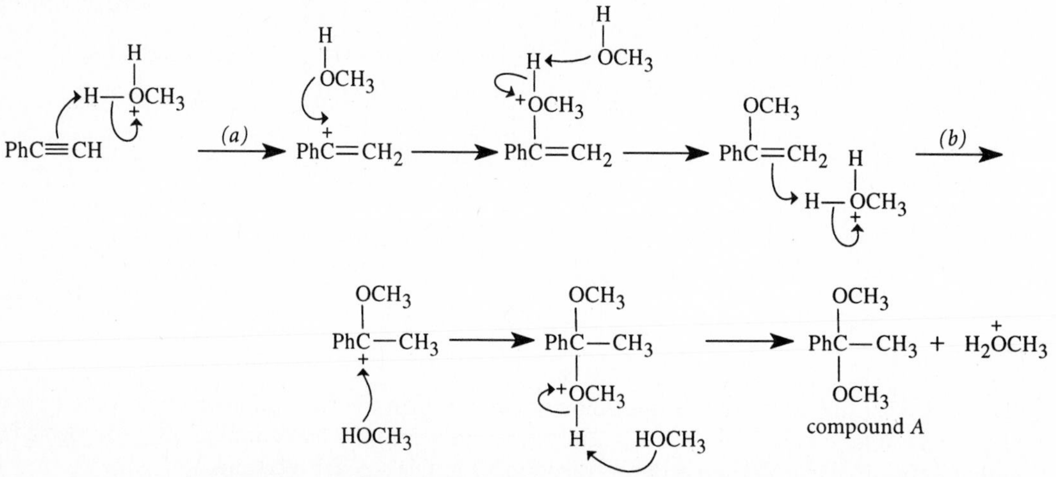

In the step labeled (a), the acid $CH_3\overset{+}{O}H_2$ is used to protonate the alkyne because it is the major acidic species present when $H_2SO_4$ is dissolved in methanol (just as $H_3O^+$ is the major acidic species present when $H_2SO_4$ is dissolved in water). Protonation occurs on the terminal carbon because it gives a carbocation that is benzylic and therefore resonance-stabilized. In the step labeled (b), protonation again occurs on the terminal carbon because the resulting carbocation is resonance-stabilized by electron donation from both the benzene ring and the neighboring oxygen.

**19.52** (a) Because a cyclic imine is formed, the reaction must be intramolecular; that is, the amine and the carbonyl group that react must be in the same molecule.

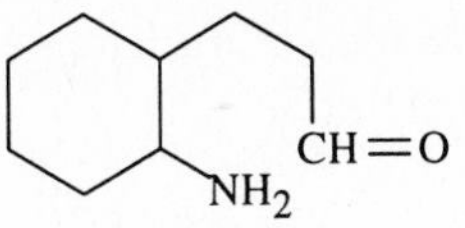

**19.53** Because compound *A* has one carbon more than the product *D*, a carbon is lost at some point; the most likely place is the benzylic oxidation that gives compound *D* itself. The structure of compound *D* shows that all compounds contain benzene rings with four substituent groups. That compound *A* can be regenerated by oxidation of optically active *B* suggests that *B* is an alcohol and *A* is a ketone. Compound *A* cannot be a primary alcohol because such an alcohol could not be optically active. (The chirality has to be associated with the alcohol functionality because optical activity is lost when the alcohol is oxidized.) It cannot be a tertiary alcohol because it could not be formed by LiAlH$_4$ reduction and could not be oxidized. Compound *A* has one degree of unsaturation in addition to the four accounted for by its benzene ring and the one accounted for by its carbonyl group; the only way to accommodate all the data is for compound *A* to contain a ring. The correct structures are as follows:

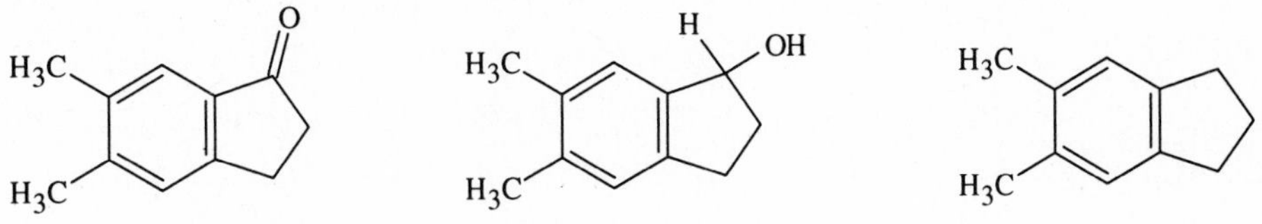

*A*                    *B*                    *C*

**19.55** (a) The mechanism below begins with the protonated ketone, which serves as a carbocation electrophile in the ring alkylation of a phenol. The resulting product, an alcohol, dehydrates under the acidic conditions to give another carbocation that alkylates a second phenol molecule and thus forms the product.

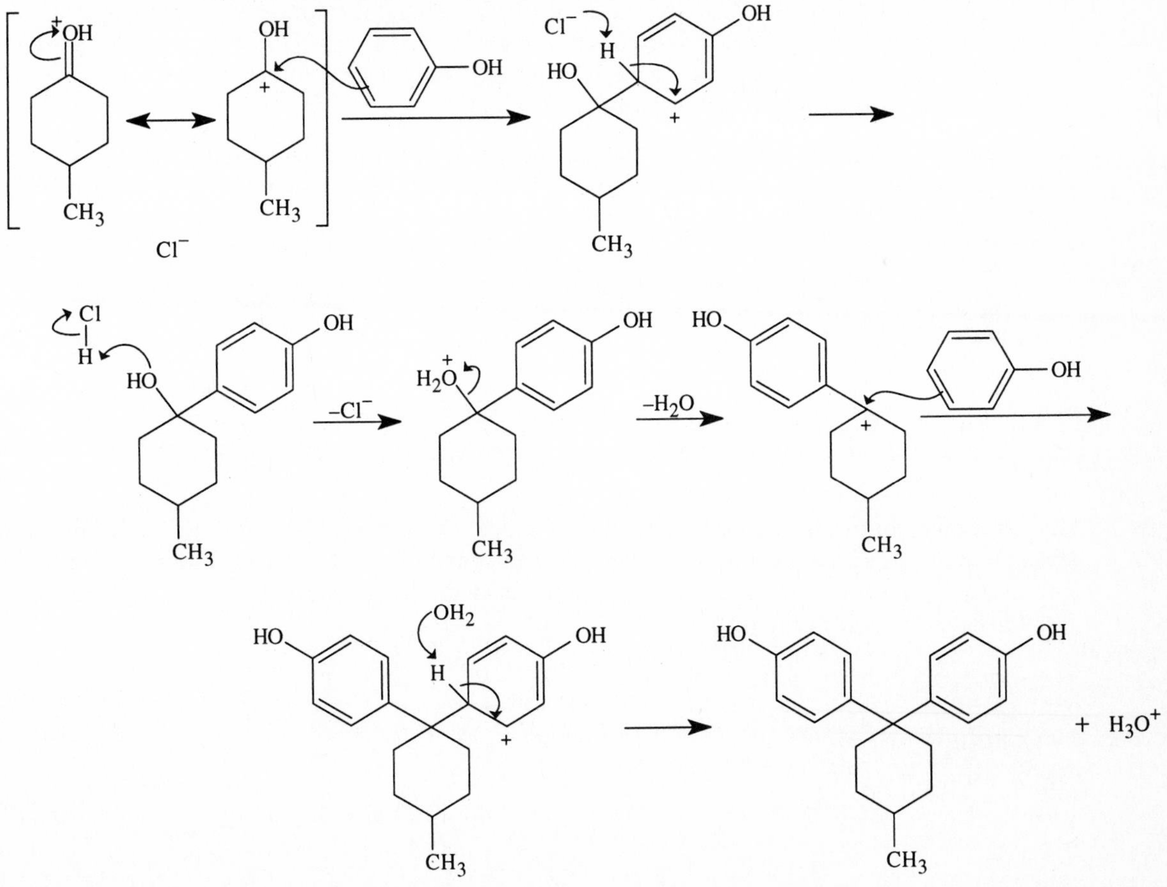

**19.56** (a) In this case, LiAlD$_4$ serves as a source of nucleophilic isotopic hydride (deuteride); deuteride opens the epoxide with inversion of configuration.

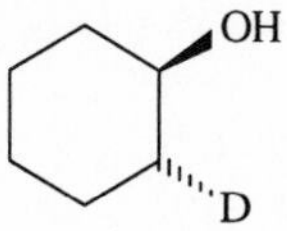

(racemate)

**19.57** (a)

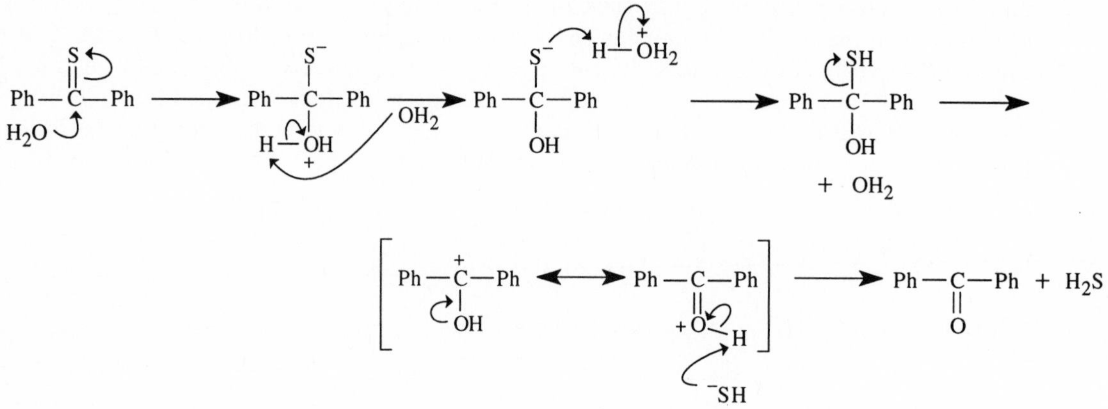

Compare loss of the —SH group, which does not require protonation of the —SH group, to loss of the —OH group in the acid-catalyzed decomposition of a hydrate, which does require protonation of an —OH group. Protonation of the —SH group is not necessary for it to serve as a leaving group for two reasons. First, the C—S bond is substantially weaker than the O—H bond; and second, ⁻SH is a much weaker base than ⁻OH. Furthermore, the —SH group itself is a much weaker base than the —OH group; consequently, much less —SH than —OH is protonated under neutral or dilute-acid conditions.

(c) Opening of the epoxide by the phosphorus compound yields an oxaphosphetane, the same type of compound that serves as an intermediate in the Wittig reaction; this compound collapses to an alkene as it does in the Wittig reaction.

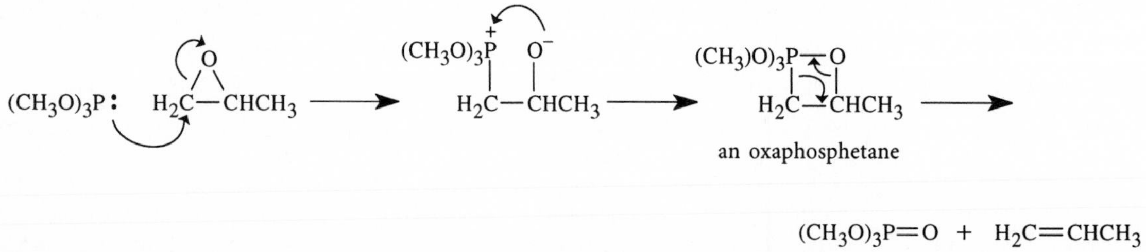

an oxaphosphetane

$$(CH_3O)_3P{=}O \ + \ H_2C{=}CHCH_3$$

(d) The imine, formed by the mechanism shown in Eqs. 19.56a–b on text p. 905, is in equilibrium with a small amount of an enamine (just as an aldehyde is in equilibrium with a small amount of enol). The nitrogen of the enamine serves as a nucleophile to attack the second aldehyde carbonyl group intramolecularly to form the ring.

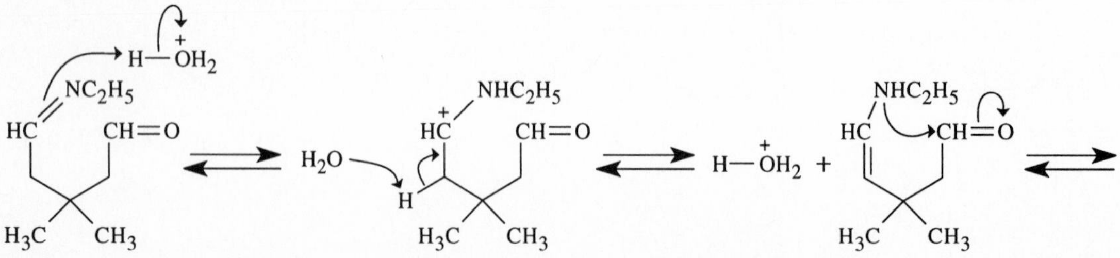

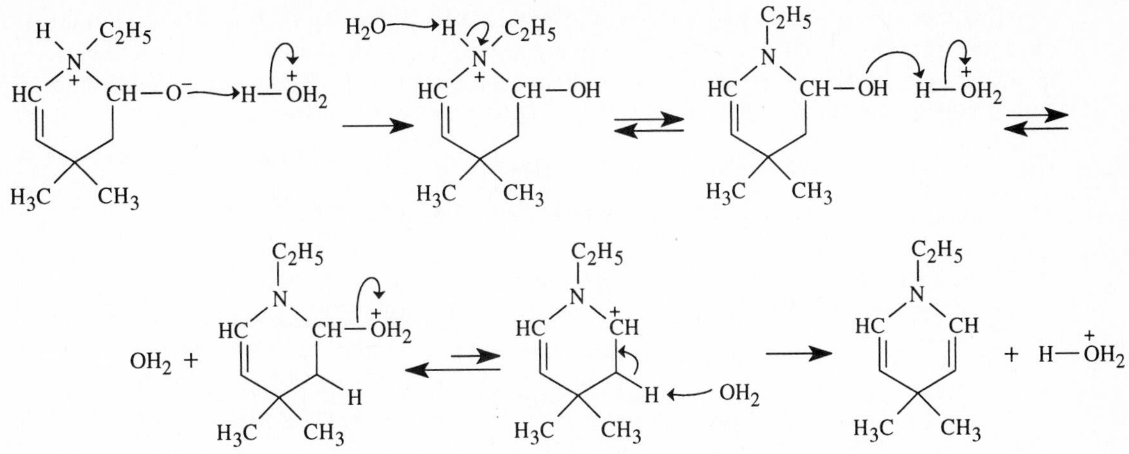

**19.58** (a) Thumbs wants a Grignard reagent to react selectively with a ketone in the presence of an aldehyde. Because aldehydes are more reactive than ketones, the aldehyde, not the ketone, will react most rapidly.

**19.59** (a) The synthesis of 4-methyl-3-heptanol:

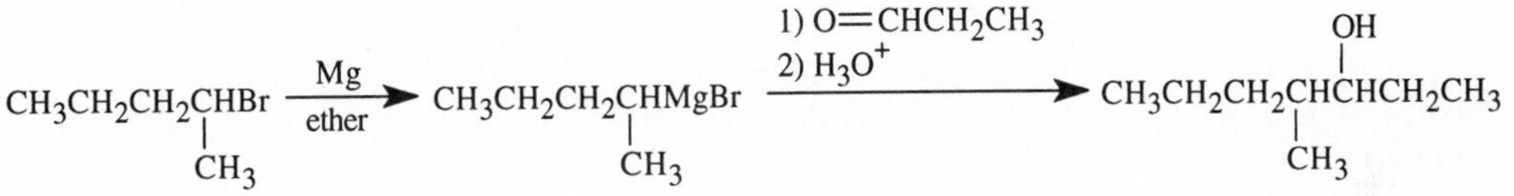

(b) Because 4-methyl-3-heptanol contains two asymmetric carbons, it can exist as diastereomers. Even if the starting alkyl halide were enantiomerically pure, and even if the Grignard reagent could be prevented from racemizing, the product would likely be a mixture of diastereomers.

**19.60** (a) The cyclic trimer can exist as a mixture of diastereomers:

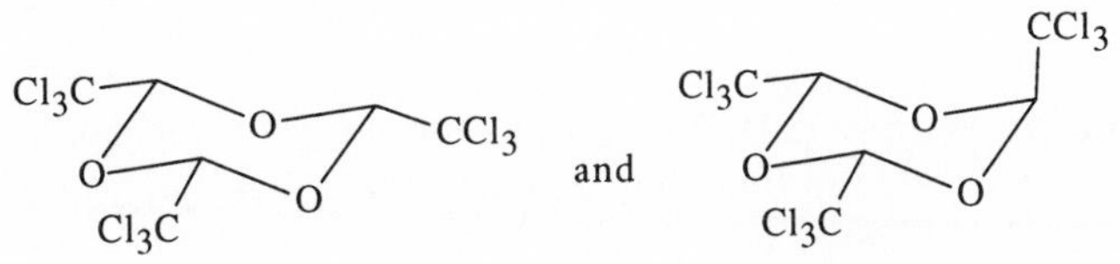

two diastereomers of the cyclic trimer of chloral

(You were not expected to specify which form is which.) Notice that the higher-melting form has the more symmetrical structure.

(b) The α-form has two chemically nonequivalent sets of protons; hence, its proton NMR spectrum has two resonances in a 2:1 ratio. Because all protons in the β-form are chemically equivalent, this form has only one resonance in its NMR spectrum. Related conclusions apply to the CMR spectra: the α-form has four lines in its spectrum, whereas the β-form has only two.

**19.61** (a) Subject the ketone benzyl phenyl ketone to an excess of isotopically enriched water in the presence of an acidic or basic catalyst. Several cycles of formation and decomposition of the hydrate will eventually "wash out" the isotope of lower atomic mass:

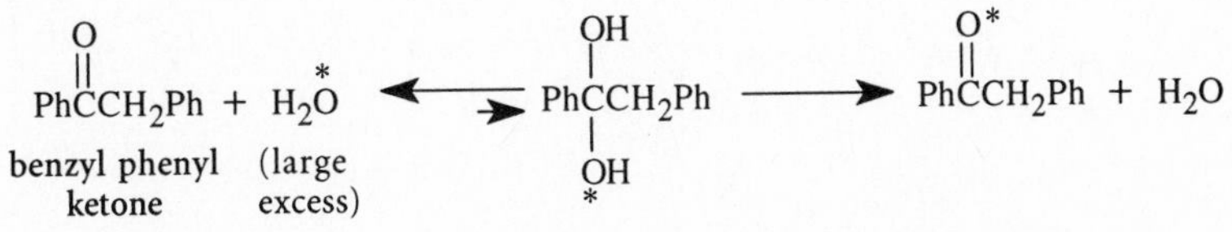

Note that the last step is effectively irreversible only because of the large excess of isotopically enriched water; the unenriched water is present at very low concentration at all times. The ratio of enriched to unenriched ketone will be the same as the ratio of enriched to unenriched water at equilibrium. Once the ketone is formed, reduce it with $LiAlH_4$ or $NaBH_4$ to provide the desired product:

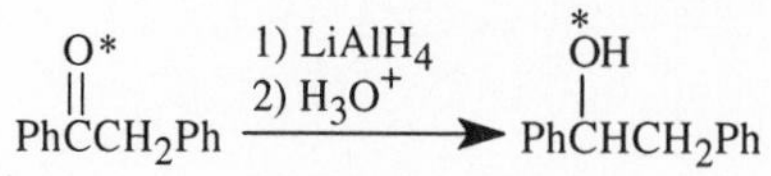

$$\underset{PhCCH_2Ph}{\overset{O*}{\|}} \xrightarrow[\text{2) H}_3O^+]{\text{1) LiAlH}_4} \underset{PhCHCH_2Ph}{\overset{\overset{*}{O}H}{|}}$$

Another reasonable synthesis is oxymercuration-reduction of *cis*- or *trans*-stilbene ("1,2-diphenyl-ethylene") with isotopically enriched water used as the nucleophile in the oxymercuration step.

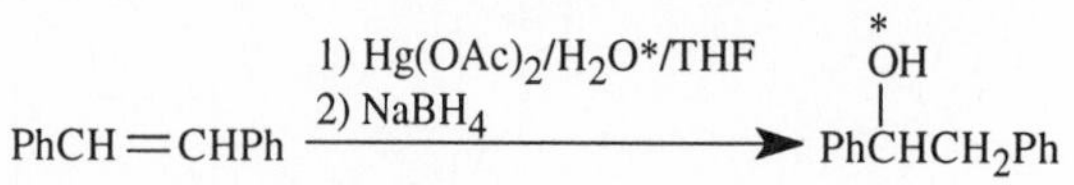

$$\underset{\substack{PhCH=CHPh \\ \textit{cis- or trans-}\text{stilbene}}}{} \xrightarrow[\text{2) NaBH}_4]{\text{1) Hg(OAc)}_2/\text{H}_2\text{O*/THF}} \underset{PhCHCH_2Ph}{\overset{\overset{*}{O}H}{|}}$$

 The $S_N1$ reaction of the corresponding alkyl bromide with isotopic water would not be a satisfactory synthesis because this particular $S_N1$ reaction gives mostly *trans*-stilbene (structure above), the elimination (E1) product. This alkene is the major product because it has an internal conjugated double bond and is therefore very stable.

**19.62**  (a)  In 1,2-cyclopentanedione the C—O bond dipoles are constrained by the geometry of the ring to be aligned at an angle of about 72°. This alignment provides a significant resultant dipole, as shown by diagram *A*:

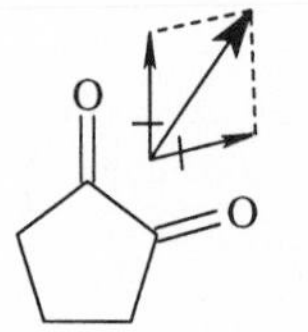

1,2-cyclopentanedione
*A*

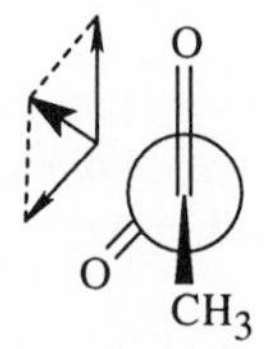

biacetyl
*B*

In biacetyl (2,3-butanedione), rotation about the central carbon-carbon bond is possible. Evidently, the two bond dipoles are oriented so that their resultant is less than the resultant in 1,2-cyclopentanedione. Notice that if their dihedral angle were 180°, their resultant would be zero. Hence, the bond dipoles are aligned at an angle somewhere between 72° and 180°, as shown in the Newman projection of diagram *B* above. (The angle turns out to be about 135°.)

The alignment in biacetyl occurs because it is the alignment of lowest energy; there are two reasons for this reduced energy. First, the alignment reduces the electrostatic repulsion between the negative ends of the bond dipoles. Second, it prevents overlap between the π-electron systems of the two carbonyl groups that would occur at an angle of either 0° or 180°. This overlap is unfavorable because it creates a charge distribution in the molecule that involves an electron-deficient oxygen, as symbolized by the following resonance structures.

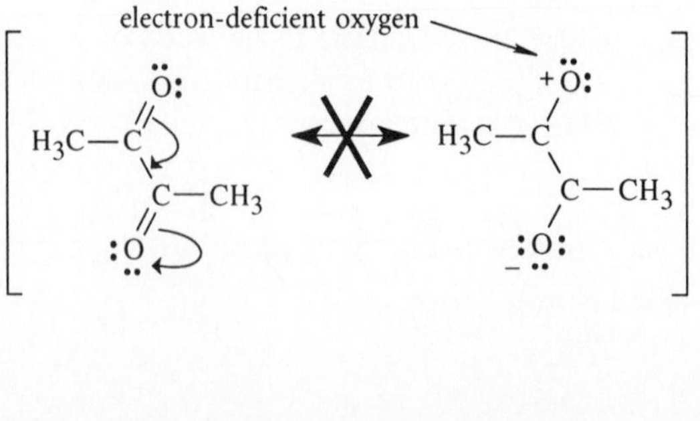

The information given in the problem plus a simplifying assumption that the cyclopentane ring has the shape of a regular pentagon allows you to use the law of cosines to calculate the bond dipole of a C=O bond as well as the angle between the two carbonyl bond dipoles in biacetyl (given above as about 135°). Try it!

(b) The $n \to \pi^*$ absorption is characteristic of the carbonyl group. This absorption disappears because a reaction occurs in which the carbonyl group is converted into another group that does not have this absorption. This reaction is addition of ethanethiol to give the sulfur analog of a hemiacetal:

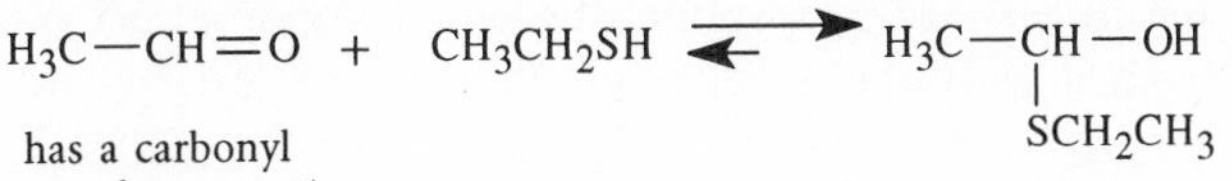

$$H_3C-CH=O \quad + \quad CH_3CH_2SH \; \rightleftharpoons \; H_3C-\underset{\underset{SCH_2CH_3}{|}}{CH}-OH$$

| | | |
|:---:|:---:|:---:|
| has a carbonyl group; has $n \to \pi^*$ absorption | | has no carbonyl group; has no $n \to \pi^*$ absorption |

(c) Tollens' test (oxidation with the ammonia complex of $Ag^+$) is a characteristic reaction of aldehydes. This compound reacts slowly because very little of it is in the aldehyde form; rather, the molecule exists largely as a cyclic hemiacetal.

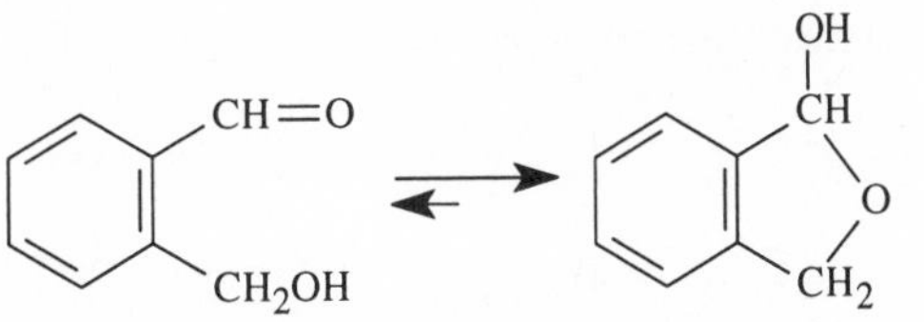

| reacts rapidly with Tollens' reagent | The molecule exists mostly in this form, which is not an aldehyde, and which therefore does not react with Tollens' reagent. |
|:---:|:---:|

**19.63** (a) The NMR spectrum suggests a great degree of symmetry: four aromatic protons, all equivalent, and two methyl groups adjacent to a carbonyl group. The IR spectrum suggests the presence of a carbonyl group conjugated with the aromatic ring.

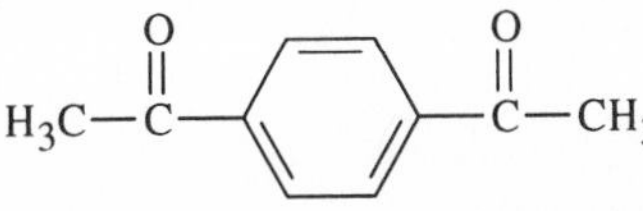

(c) The NMR indicates an aldehyde adjacent to a CH and two vinylic hydrogens. The IR absorption at 970 $cm^{-1}$ shows that the vinylic hydrogens are *trans*. The double bond and the aldehyde account for all of the unsaturation. The UV absorption at 215 nm indicates that the alkene and aldehyde double bonds are conjugated. The allylic $CH_2$ absorption at $\delta$ 2.3 is a quartet, which suggests three hydrogens on adjacent carbons, that is, a vinylic hydrogen (because the absorption is allylic) and two others. The high-field triplet integrates for three protons, and thus represents a methyl group; its splitting suggests an adjacent $CH_2$ group. The following structure is thus defined by the data:

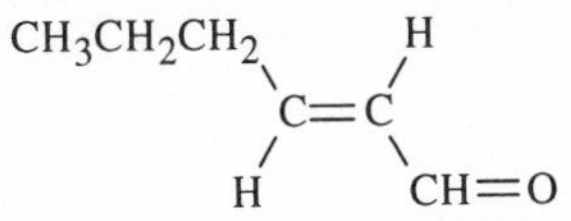

**19.64** The double absorptions of equal intensity in the mass spectrum indicate the presence of one bromine. The IR absorptions indicate a ketone in which the carbonyl group is conjugated to a benzene ring (1678 $cm^{-1}$) as well as aromatic "double bonds" (1600 $cm^{-1}$). The NMR spectrum indicates a *para*-disubstituted benzene ring and a methyl group adjacent to a carbonyl group. The compound has structure *A* below (*p*-bromoacetophenone). This compound, of course, has the molecular mass (198 for $^{79}Br$) indicated by the mass spectrum.

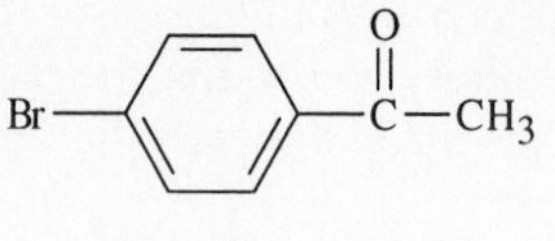

*p*-bromoacetophenone
A

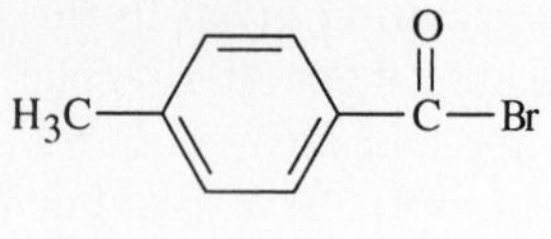

B

If you proposed structure *B*, it is not a bad answer, because you don't know about the spectra of acid halides. However, the IR carbonyl absorption of such a compound occurs at much higher frequency; furthermore, the chemical shift of the methyl group is more consistent with its being adjacent to a carbonyl group.

19.66 The formula indicates one degree of unsaturation which is accounted for by a ketone (IR absorption at 1710 cm$^{-1}$, CMR absorption at $\delta$ 204.9 with zero attached hydrogens). The remaining oxygens cannot be alcohols or other carbonyls. The CMR-DEPT data indicate the presence of at least one methyl ether ($\delta$ 53.5) as well as a carbon bearing one hydrogen ($\delta$ 101.7) that could also be attached to two oxygens; thus, an acetal group is a possibility. The proton NMR data indicate a methyl group attached to a carbonyl. The only carbon unaccounted for is a methylene group with a CMR absorption at $\delta$ 47.2, which places it adjacent to the carbonyl and near, but not adjacent to, the acetal. All the data conspire to define the following structure:

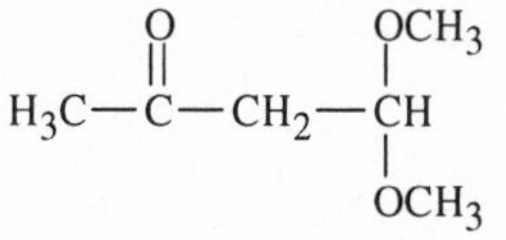

4,4-dimethoxy-2-butanone

# 20

# Chemistry of Carboxylic Acids

## Terms

## Concepts

### I. Introduction to Carboxylic Acids

#### A. COMMON NOMENCLATURE OF CARBOXYLIC ACIDS

1. The characteristic functional group in a carboxylic acid is the carboxy group, —$CO_2H$.
2. Common nomenclature is widely used for the simpler carboxylic acids, some names of which owe their origins to natural sources.
3. In common nomenclature, a carboxylic acid is named by adding the suffix *ic* and the word *acid* to the prefix for the appropriate group.

| | | | |
|---|---|---|---|
| form | H— | isobutyr | $(CH_3)_2CH_2$— |
| acet | $CH_3$— | valer | $CH_3CH_2CH_2CH_2$— |
| propion | $CH_3CH_2$— | isovaler | $(CH_3)_2CHCH_2$— |
| butyr | $CH_3CH_2CH_2$— | benzo | Ph— |

4. Substitution in the common system is denoted with Greek letters rather than numbers; the position adjacent to the carboxy group is designated as $\alpha$.

$$\underset{\delta}{Cl}-CH_2-\underset{\gamma}{CH_2}-\underset{\beta}{CH_2}-\underset{\alpha}{CH_2}-\overset{\displaystyle O}{\overset{\|}{C}}-OH \qquad \delta\text{-chlorovaleric acid}$$

carboxy group

5. Carboxylic acids with two carboxy groups are called dicarboxylic acids.
   a. The unbranched dicarboxylic acids are particularly important, and are invariably known by their common names:

   | | | | |
   |---|---|---|---|
   | oxalic acid | $HO_2CCO_2H$ | glutaric acid | $HO_2C(CH_2)_3CO_2H$ |
   | malonic acid | $HO_2CCH_2CO_2H$ | adipic acid | $HO_2C(CH_2)_4CO_2H$ |
   | succinic acid | $HO_2C(CH_2)_2CO_2H$ | pimelic acid | $HO_2C(CH_2)_5CO_2H$ |

   b. Phthalic acid is an important aromatic dicarboxylic acid.

   phthalic acid

6. Many carboxylic acids were known long before any system of nomenclature existed, and their time-honored traditional names are widely used.

## B.   SYSTEMATIC NOMENCLATURE OF CARBOXYLIC ACIDS

1. A carboxylic acid is named systematically by dropping the final *e* from the name of the hydrocarbon with the same number of carbon atoms and adding the suffix *oic* and the word *acid* (the final *e* is not dropped in the name of dicarboxylic acids).
2. When the carboxylic acid is derived from a cyclic hydrocarbon, the suffix *carboxylic* and the word *acid* is added to the name of the hydrocarbon.  (One exception to this nomenclature is benzoic acid, for which the IUPAC recognizes the common name.)
3. The principal chain in substituted carboxylic acids is numbered by assigning the number 1 to the carbonyl carbon.  (In carboxylic acids derived from cyclic hydrocarbons, numbering begins at the ring carbon bearing the carboxy group).

3,3-dimethylbutanoic acid         3-ethylcyclopentanecarboxylic acid

4. When carboxylic acids contain other functional groups, the carboxy groups receive priority over aldehyde and ketone carbonyl groups, over hydroxy groups, and over mercapto groups for citation as the principal group.  (A complete list of nomenclature priorities for all the functional groups covered in the text is given in Appendix I, text page A-1.)

$$-CO_2H \; > \; -CH{=}O \; > \; \overset{\displaystyle |}{\underset{\displaystyle |}{C}}{=}O \; > \; -OH \; > \; -SH$$

6-hydroxy-2,5-dimethyl-3-oxo-4-hexenoic acid

5. The carboxy group is sometimes named as a substituent.

2-(carboxymethoxy)benzoic acid

## C. STRUCTURE AND PHYSICAL PROPERTIES OF CARBOXYLIC ACIDS

1. Carboxylic acids have trigonal geometry at their carbonyl carbons.
2. The two oxygens of a carboxylic acid are quite different.
   a. The carbonyl oxygen is part of the C=O double bond, which has about the same bond length as the C=O double bond of aldehydes and ketones.
   b. The carboxylate oxygen is part of the C—O single bond and is considerably shorter than the C—O bond in an alcohol or ether.
      i. The C—O bond in a carboxylic acid is an $sp^2$-$sp^3$ single bond, whereas the C—O bond in an alcohol or ether is an $sp^3$-$sp^3$ single bond.
      ii. Carboxylic acids have a resonance structure in which the C—O bond has some double-bond character. (See Eq. 20.2, text p. 934.)
3. The carboxylic acids of lower molecular mass have acrid, piercing odors and considerably higher boiling points than many other organic compounds of about the same molecular weight and shape.
   a. The high boiling points of carboxylic acids can be attributed not only to their polarity, but also to the fact that they form very strong hydrogen bonds.
   b. In the solid state, and under some conditions in both the gas phase and solution, carboxylic acids exist as hydrogen-bonded dimers.
4. Many aromatic and dicarboxylic acids are solids.
5. The simpler carboxylic acids have substantial solubilities in water; the unbranched carboxylic acids below pentanoic acid are miscible with water. (Many dicarboxylic acids also have significant water solubilities.)

## II. Acidity and Basicity of Carboxylic Acids

### A. ACIDITY OF CARBOXYLIC AND SULFONIC ACIDS

1. The acidity of a carboxylic acid is due to ionization of the O—H group to give the conjugate-base carboxylate ion.

a carboxylate ion

   a. Carboxylic acids are among the most acidic organic compounds (more acidic than alcohols or phenols).
   b. The conjugate bases of carboxylic acids are called generally carboxylate ions; carboxylate salts are named by replacing the *ic* in the name of the acid with the suffix *ate*.

acetic acid         acetate

2. The acidity of carboxylic acids is due to two factors.
   a. The first factor is the polar effect of the carbonyl group.
      i. The carbonyl group is much more electron-withdrawing than the phenyl ring of a phenol or the alkyl group of an alcohol.
      ii. The polar effect of the carbonyl group stabilizes charge in the carboxylate ion (stabilization of a conjugate base enhances acidity).
   b. The second factor is the resonance stabilization of their conjugate-base carboxylate ions. (See the resonance structures in item 1 above.)
3. The acidity of carboxylic acids, like that of alcohols, is influenced by the polar effects of substituents.

$$CH_3CO_2H \qquad\qquad FCH_2CO_2H$$

$$pK_a = 4.76 \qquad\qquad pK_a = 2.66$$

4. Sulfonic acids are much stronger than comparably substituted carboxylic acids and are useful as acid catalysts in organic solvents because they are more soluble than most inorganic acids.
   a. The sulfur atom in a sulfonic acid has a high oxidation state.
   b. The octet structure for a sulfonate anion indicates that sulfur has considerable positive charge, which stabilizes the negative charge on the oxygens.

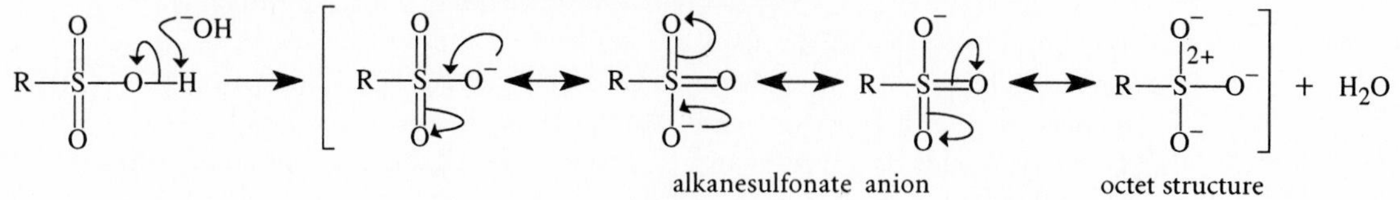

5. Although many carboxylic acids of moderate molecular weight are not soluble in water, their alkali metal salts are ionic compounds that are much more soluble in water.
   a. Many water-insoluble carboxylic acids dissolve in solutions of alkali metal hydroxides (NaOH, KOH) because the insoluble acids are converted completely into their soluble salts.
      i. A typical carboxylic acid can be separated from mixtures with other water-insoluble, non-acidic substances by extraction with NaOH, $Na_2CO_3$, or $NaHCO_3$ solution.
      ii. After isolating the basic aqueous solution containing the carboxylate salt, it can be acidified with a strong acid to yield the carboxylic acid, which may be isolated by filtration or extraction with organic solvents.
   b. A carboxylic acid can also be separated from a phenol by extraction of the acid with 5% $NaHCO_3$, if the phenol is not one that is unusually acidic.

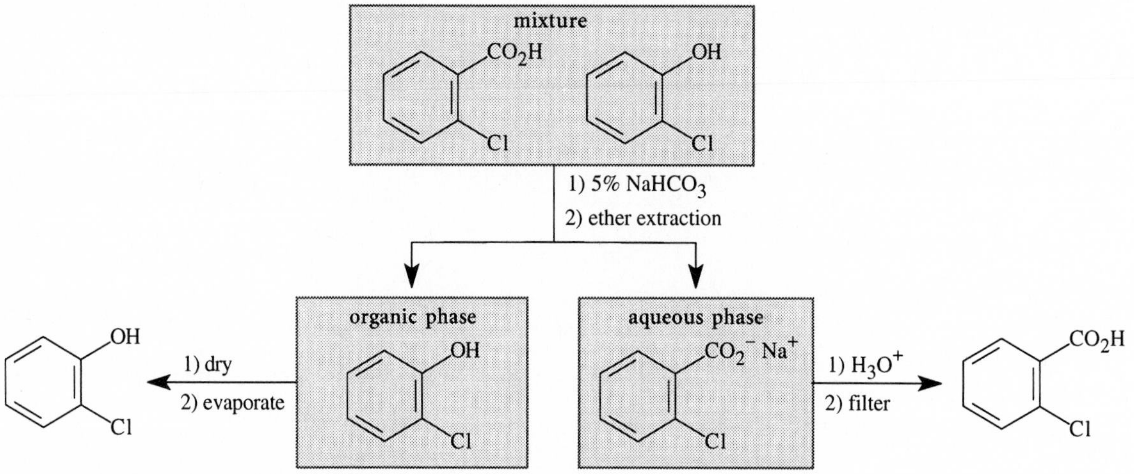

## B. BASICITY OF CARBOXYLIC ACIDS

1. The carbonyl oxygen of a carboxylic acid is weakly basic.
   a. Protonation of a carboxylic acid on the carbonyl oxygen occurs because a resonance-stabilized cation is formed.

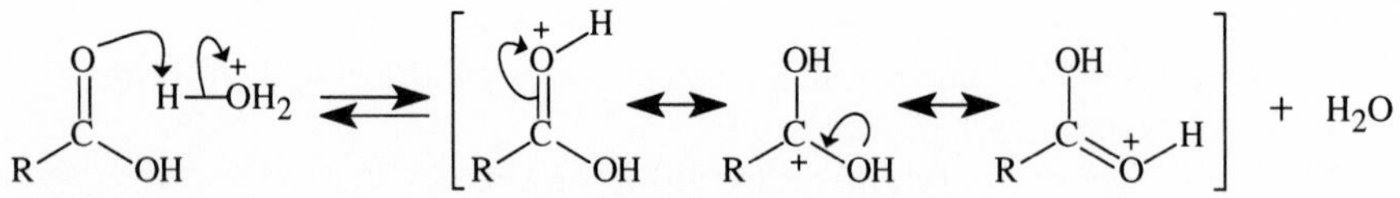

   b. Protonation of the carboxylate oxygen is much less favorable because:
      i. it does not give a resonance-stabilized cation
      ii. the positive charge on oxygen is destabilized by the electron-withdrawing polar effect of the carbonyl group.
2. The basicity of carboxylic acids plays a very important role in many of their reactions.

## C. FATTY ACIDS, SOAPS, AND DETERGENTS

1. Carboxylic acids with long, unbranched carbon chains are called fatty acids.
   a. Fatty acids are liberated from fats and oils by a hydrolytic process called saponification.

c. The sodium and potassium salts of fatty acids, called soaps, are the major ingredients of commercial soaps.

$$CH_3CH_2CH_2CH_2CH_2CH_2CH_2CH_2CH_2CH_2CH_2CH_2CH_2CH_2C\overset{\overset{\displaystyle O}{\|}}{\phantom{C}}-O^- \, Na^+ \quad \text{or} \quad CH_3(CH_2)_{14}CO_2^- \, Na^+$$

sodium hexadecanoate (sodium palmitate), a soap

   *i.* Closely related to soaps are synthetic detergents, some of which are the salts of sulfonic acids.

   *ii.* Many soaps and detergents have not only cleansing properties, but also germicidal characteristics.

d. Hard-water scum is a precipitate of the calcium or magnesium salts of fatty acids.

2. Soaps and detergents are two examples of a larger class of molecules known as surfactants (molecules with two structural parts that interact with water in opposing ways):

   a. They have a polar head group, which is readily solvated by water. In a soap, the polar head group is the carboxylate anion.

   b. They have a hydrocarbon tail, which is not well solvated by water. In a soap, the hydrocarbon tail is the carbon chain.

3. Soaps and detergents are examples of anionic surfactants, that is, surfactants with an anionic polar head group. (Cationic surfactants are also known.)

   a. When the surfactant concentration in water is raised above the critical micelle concentration (CMC), the surfactant molecules spontaneously form micelles, which are approximately spherical aggregates of 50–150 surfactant molecules.

   b. The micellar structure satisfies the solvation requirements of both the polar head groups, which are close to water (on the outside of the micelle), and the nonpolar tails, which associate with each other (on the inside of the micelle).

4. The antiseptic action of some surfactants owes its success to a phenomenon similar to micelle formation.

   a. A cell membrane is made up of molecules, called phospholipids, that are also surfactants.

   b. When the bacterial cell is exposed to a solution containing a surfactant, phospholipids of the cell membrane tend to associate with the surfactant.

   c. In some cases this disrupts the membrane enough that the cell can no longer function, and it dies.

5. Surfactants are also extremely important as components of fuels and lubricating oils.

## III. Spectroscopy of Carboxylic Acids

### A. IR Spectroscopy

1. Two important absorptions in the infrared spectrum are hallmarks of a carboxylic acid:

   a. A strong $C{=}O$ stretching absorption occurs near 1710 cm$^{-1}$ for carboxylic acid hydrogen-bonded dimers. (The $C{=}O$ absorption of carboxylic acid monomers occurs near 1760 cm$^{-1}$ but is rarely observed.)

   b. The O—H stretching absorption of a carboxylic acid is much broader than the O—H stretching absorption of an alcohol or phenol, and covers a very wide region of the spectrum—typically 2400–3600 cm$^{-1}$.

2. A conjugated carbon-carbon double bond affects the position of the carbonyl absorption much less in acids than it does in aldehydes and ketones. (A substantial shift in the carbonyl absorption is observed, however, for acids in which the carboxy group is on an aromatic ring.)

### B. NMR Spectroscopy

1. The $\alpha$-protons of carboxylic acids show NMR absorptions in the $\delta\,2.0$–2.5 chemical shift region.

2. The carboxylic acid O—H proton resonance is typically found far downfield, in the $\delta\,9$–13 region, and in many cases it is broad.

   a. The O—H proton resonances of carboxylic acids occur at positions that depend on the acidity of the acid and on its concentration.

   b. The O—H proton resonances of carboxylic acids are readily distinguished from the resonances of aldehydic protons because only acid protons rapidly exchange with $D_2O$.

3. The carbonyl-carbon CMR absorptions of carboxylic acids are observed at somewhat higher field than those of aldehydes or ketones. (This unusual chemical shift is caused by shielding effects of the uncharged electron pairs on the carboxylate oxygen.)

# Reactions

## I. Synthesis and Reactivity of Carboxylic Acids

### A. SYNTHESIS OF CARBOXYLIC ACIDS—REVIEW

1. Methods for preparing carboxylic acids introduced in previous chapters:
   a. Oxidation of primary alcohols (Secs. 10.6B, text pp. 469–470, and 19.14, text pp. 915–917).
   b. Side-chain oxidation of alkylbenzenes (Sec. 17.5, text pp. 804–805).
   c. Oxidation of aldehydes (Sec. 19.14, text pp. 915–917).
   d. Ozonolysis of alkenes followed by oxidative workup with $H_2O_2$ (Sec. 5.4, text pp. 188–192).

### B. SYNTHESIS OF CARBOXYLIC ACIDS WITH GRIGNARD REAGENTS

1. The reactions of Grignard reagents with carbon dioxide, followed by protonolysis, give carboxylic acids.
   a. Addition of a Grignard reagent to carbon dioxide gives the halomagnesium salt of a carboxylic acid.
   b. When aqueous acid is added to the reaction mixture in a separate reaction step, the free carboxylic acid is formed.

$$R\text{—}MgX \ + \ \underset{O}{\overset{O}{C}} \ \longrightarrow \ R\text{—}\underset{O^- \ ^+MgX}{\overset{O}{C}} \ \xrightarrow{H_3O^+} \ R\text{—}\underset{OH}{\overset{O}{C}} \ + \ HOMgX$$

2. The reaction of Grignard reagents with $CO_2$ is another method for the formation of carbon-carbon bonds. (See Appendix V, text page A-11, for a review of other carbon-carbon bond-forming reactions.)

### C. INTRODUCTION TO CARBOXYLIC ACID REACTIONS

1. Reactions of carboxylic acids can be categorized into four types:
   a. Reactions at the carbonyl group.
      *i.* The most typical reaction at the carbonyl group is substitution at the carbonyl carbon. In such a reaction, the —OH of the carboxy group is typically substituted by a nucleophilic group.
      *ii.* Reaction of the carbonyl oxygen with an electrophile (Lewis acid or Brønsted acid), that is, the reaction of the carbonyl oxygen as a base.
   b. Reactions at the carboxylate oxygen, such as reactions of the carboxylate oxygen as a nucleophile.
   c. Loss of the carboxy group as $CO_2$ (decarboxylation).
   d. Reactions involving the $\alpha$-carbon.
2. Many substitution reactions at the carbonyl carbon are acid-catalyzed; that is, the reactions of nucleophiles at the carbonyl carbon are catalyzed by the reactions of acids at the carbonyl oxygen.

## II. Reaction of Carboxylic Acids at the Carbonyl Group

### A. ACID-CATALYZED ESTERIFICATION OF CARBOXYLIC ACIDS

1. Esters are carboxylic acid derivatives in which the proton on the carboxylate oxygen in effect has been replaced by an alkyl or aryl group.
2. When a carboxylic acid is treated with a large excess of an alcohol in the presence of a strong acid catalyst, an ester is formed; this reaction is called acid-catalyzed esterification (or Fischer esterification).

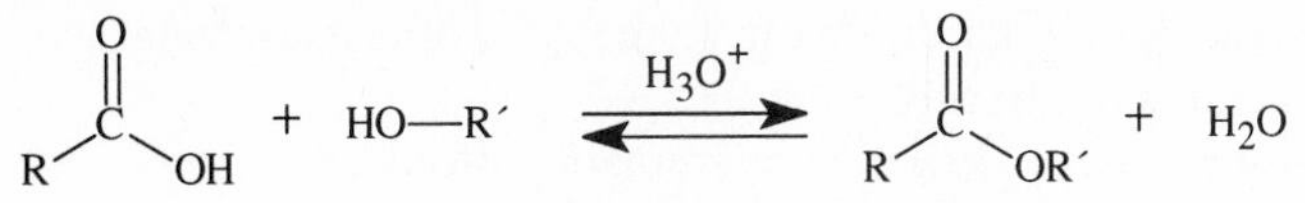

an ester

a. Acid-catalyzed esterification is a substitution of —OH at the carbonyl group of the acid by the OR group of the alcohol; the —OH group leaves as water.

b. The equilibrium constants for esterification with most primary alcohols are near unity.
  i. The reaction is driven toward completion by applying LeChatelier's principle.
  ii. Use of the reactant alcohol as the solvent, which ensures that the alcohol is present in large excess, drives the equilibrium toward the ester product.

c. Acid-catalyzed esterification cannot be applied to the synthesis of esters from phenols or tertiary alcohols.

3. The mechanism of acid-catalyzed esterification serves as a model for the mechanisms of other acid-catalyzed reactions of carboxylic acids and their derivatives.

a. The first step of the mechanism is protonation of the carboxyl oxygen.
  i. The catalyzing acid is the conjugate acid of the solvent.
  ii. Protonation of a carbonyl oxygen makes the carbonyl carbon more electrophilic because the carbonyl oxygen becomes a better electron acceptor.

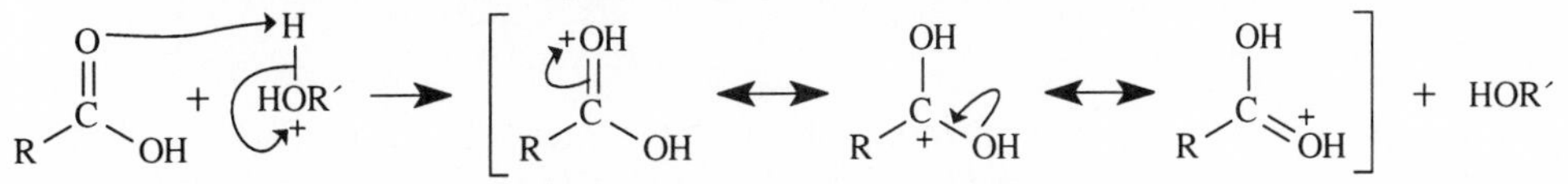

b. Attack of alcohol on the carbonyl carbon of this carbocation, followed by loss of a proton, gives a tetrahedral addition intermediate.

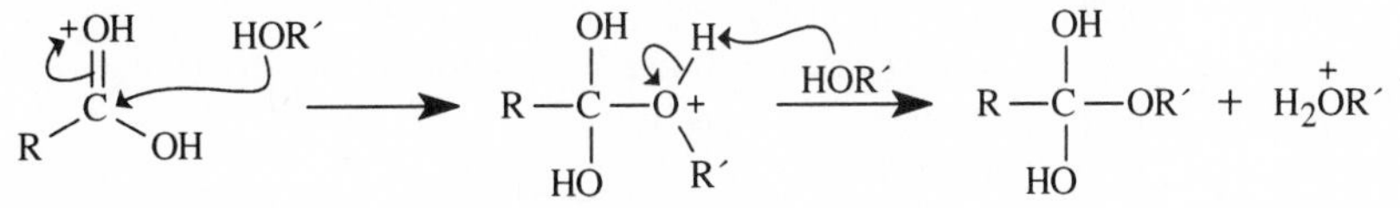

c. The tetrahedral addition intermediate, after protonation, loses water to give the conjugate acid of the ester.

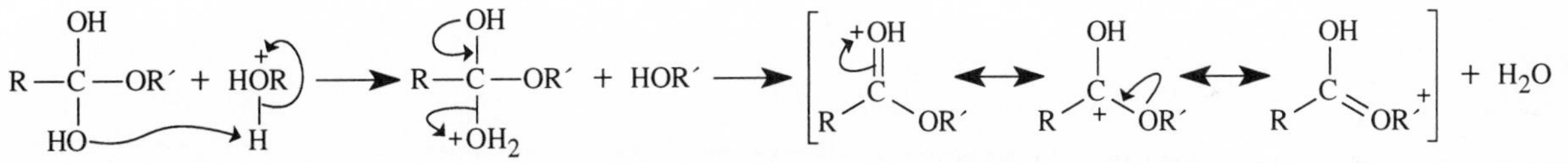

d. Loss of a proton gives the ester product and regenerates the acid catalyst.

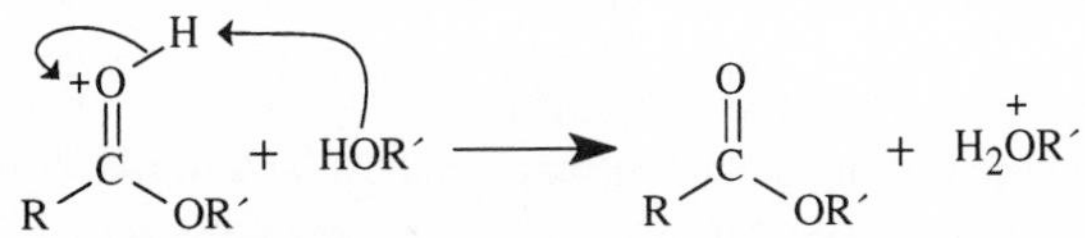

## B. SYNTHESIS OF ACID CHLORIDES

1. Acid chlorides are carboxylic acid derivatives in which the —OH group has been replaced by —Cl.
  a. Acid chlorides are often prepared from the corresponding carboxylic acids.
  b. Two reagents used for this purpose are thionyl chloride, $SOCl_2$, and phosphorus pentachloride, $PCl_5$.

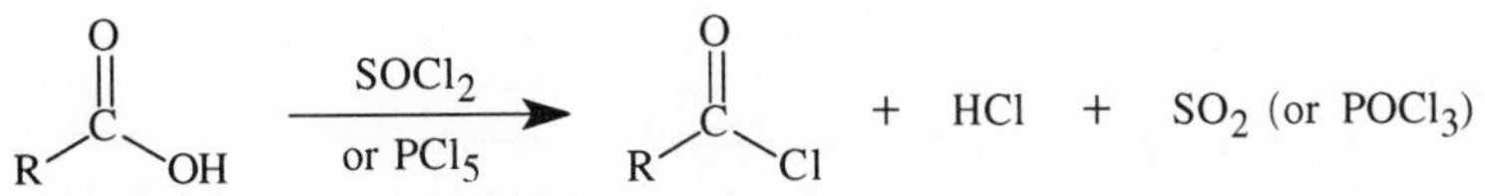

2. Acid chloride synthesis fits the general pattern of substitution reactions at a carbonyl group; in this case, —OH is substituted by —Cl.

3. Sulfonyl chlorides are the acid chlorides of sulfonic acids.
  a. Sulfonyl chlorides are prepared by treatment of sulfonic acids or their sodium salts with $PCl_5$.

b. Aromatic sulfonyl chlorides can be prepared directly by the reaction of aromatic compounds with chlorosulfonic acid, $ClSO_3H$.

    *i.* This reaction is a variation of aromatic sulfonation, an electrophilic aromatic substitution reaction.

    *ii.* Chlorosulfonic acid, the acid chloride of sulfuric acid, acts as an electrophile in this reaction.

    *iii.* The sulfonic acid produced in the reaction is converted into the sulfonyl chloride by reaction with another equivalent of chlorosulfonic acid.

## C. Synthesis of Anhydrides

1. Carboxylic acid anhydrides are carboxylic acid derivatives in which the hydroxy group has been replaced by an acyloxy group.

    a. Anhydrides are prepared by treatment of carboxylic acids with strong dehydrating agents (usually $P_2O_5$).

    b. Most anhydrides may themselves be used as reagents for the preparation of other anhydrides.

2. Some dicarboxylic acids react with acetic anhydride to form cyclic anhydrides—compounds in which the anhydride group is part of a ring.

    a. Cyclic anhydrides containing five- and six-membered rings are readily prepared from the corresponding dicarboxylic acids.

    b. Formation of cyclic anhydrides with five- or six-membered rings is so facile that in some cases it occurs on heating the dicarboxylic acid.

3. Anhydrides, like acid chlorides, are used in the synthesis of other carboxylic acid derivatives.

## D. Reduction of Carboxylic Acids to Alcohols

1. When a carboxylic acid is treated with $LiAlH_4$, then with dilute acid, a primary alcohol is formed.

    a. Before the reduction itself takes place, $LiAlH_4$ reacts with the acidic hydrogen of the carboxylic acid to give the lithium salt of the carboxylic acid and one equivalent of hydrogen gas. (The lithium salt of the carboxylic acid is the species that is actually reduced.)

2. The reduction occurs in two stages:

    a. $AlH_3$ (formed from the reaction of $LiAlH_4$ with the acidic proton of the carboxylic acid) reduces the carboxylate ion to an aldehyde.

    b. The aldehyde is rapidly reduced further to give, after protonolysis, the primary alcohol. (Because the aldehyde is more reactive than the carboxylate salt, it cannot be isolated.)

3. The $LiAlH_4$ reduction of a carboxylic acid incorporates two different types of carbonyl reactions:

    a. A net substitution reaction at the carbonyl group to give an aldehyde intermediate.

    b. An addition to the aldehyde thus formed.

4. Sodium borohydride, $NaBH_4$, does not reduce carboxylic acids, although it does react with the acidic hydrogens of carboxylic acids.

## III. Reaction of Carboxylic Acids at the Oxygen of the —OH Group

### A. ESTERIFICATION BY ALKYLATION

1. When carboxylic acids are treated with diazomethane in ether solution, they are rapidly converted into their methyl esters.
   a. Protonation of diazomethane by the carboxylic acid gives the methyldiazonium ion; this ion has one of the best leaving groups, molecular nitrogen.
   b. An $S_N2$ reaction of the methyldiazonium ion with the carboxylate oxygen results in the displacement of $N_2$ and formation of the ester.

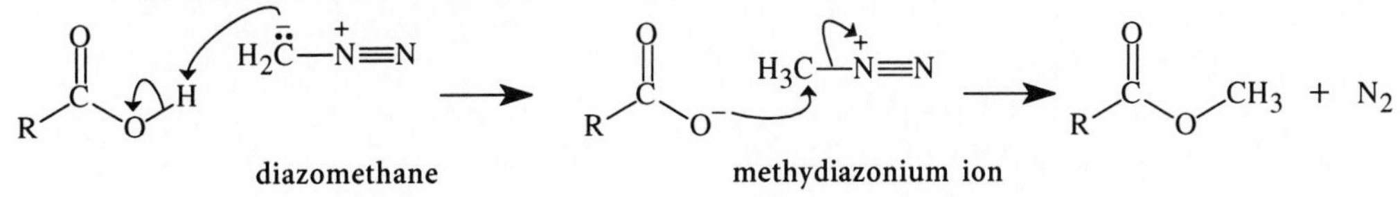

2. The reaction of certain alkyl halides with carboxylate ions give esters.
   a. This is an $S_N2$ reaction in which the carboxylate ion, formed by acid-base reaction of the acid and a base (such as $K_2CO_3$), acts as the nucleophile that attacks the alkyl halide.

   b. This reaction works best on alkyl halides that are especially reactive in the $S_N2$ reaction, such as methyl iodide and benzylic or allylic halides, because carboxylate ions are relatively poor nucleophiles.

## IV. Decarboxylation of Carboxylic Acids

### A. DECARBOXYLATION OF CARBOXYLIC ACIDS

1. The loss of carbon dioxide from a carboxylic acid is called decarboxylation.
2. Certain types of carboxylic acids are readily decarboxylated:
   a. $\beta$-keto acids (carboxylic acids with a keto group in the $\beta$-position).
   b. malonic acid derivatives.
   c. derivatives of carbonic acid.

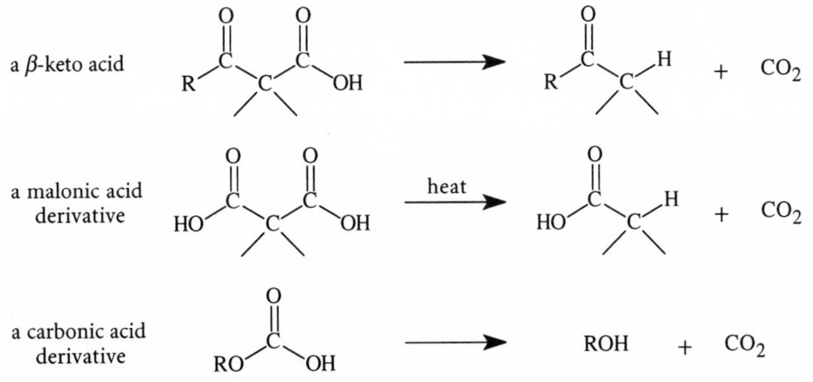

3. $\beta$-Keto acids readily decarboxylate at room temperature in acidic solution.
   a. Decarboxylation of a $\beta$-keto acid involves an enol intermediate that is formed by an internal proton transfer from the carboxylic acid group to the carbonyl oxygen atom of the ketone.
   b. The enol is transformed spontaneously into the corresponding ketone.

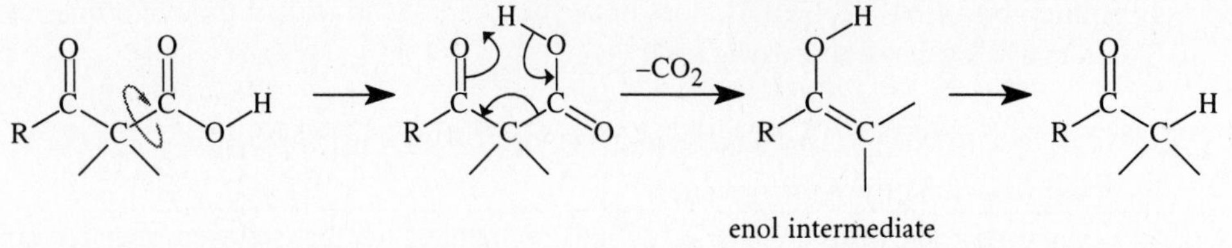

enol intermediate

4. The acid form of the $\beta$-keto acid decarboxylates more readily than the conjugate-base carboxylate form because the latter has no acidic proton that can be donated to the $\beta$-carbonyl oxygen.

5. Malonic acid and its derivatives readily decarboxylate upon heating in acid solution. This reaction, which does not occur in base, bears a close resemblance to the decarboxylation of $\beta$-keto acids, because both types of acids have a carbonyl group $\beta$ to the carboxy group.

6. Carbonic acid is unstable and decarboxylates spontaneously in acid solution to carbon dioxide and water.

   a. Carbonic acid derivatives in which only one carboxylate oxygen is involved in ester or amide formation also decarboxylate under acidic conditions.

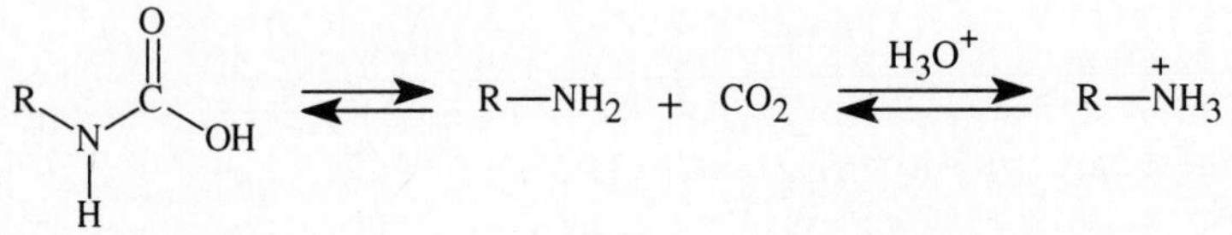

   b. Under basic conditions, carbonic acid is converted into its salts bicarbonate and carbonate, which do not decarboxylate.

   c. Carbonic acid derivatives in which both carboxylic acid groups are involved in ester or amide formation are stable.

      *i.* Carbonate esters are diesters of carbonic acid.

      *ii.* Ureas are diamides of carbonic acid.

## Study Guide Links

### 20.1    Chemical Shifts of Carbonyl Carbons

As pointed out in the text, the electronegativity of oxygen might lead one to expect the chemical shift for the carbonyl group of acetic acid to be *greater* than that of acetone. An additional puzzle is provided by the carbonyl chemical shift of the acetate ion, the conjugate base of acetic acid:

$$CH_3 - \overset{\overset{\textstyle O}{\|}}{C} - OH \qquad \delta\,178 \qquad\qquad CH_3 - \overset{\overset{\textstyle O}{\|}}{C} - O^- \qquad \delta\,182$$

acetic acid            acetate ion

If chemical shifts in CMR spectroscopy correlate with electronegativity, as they do in proton NMR, it seems that the acetate ion, because of its relatively electron-rich —O⁻, should have a *smaller* carbonyl chemical shift than acetic acid. The data above show that this is not the case.

A related phenomenon is the chemical shift of the O—H proton, which was discussed in Sec. 13.6D on text p. 617. In the gas phase (in which hydrogen bonding and other solvent effects are absent), this proton has a chemical shift of $\delta\,0.8$, a much smaller value than would be expected for a proton directly attached to an electronegative atom.

These cases illustrate the point that *chemical shifts are not determined solely by electronegativity of nearby atoms.* You learned, for example, that induced circulation of $\pi$ electrons in the presence of the applied field accounts for the significant downfield shifts of both aromatic and alkene carbons as well as the shifts of aromatic and alkene protons—shifts far greater than would be predicted on the basis of the electronegativity of an $sp^2$ hybridized carbon (Fig. 13.15 on text p. 613 and Fig. 16.2 on text p. 742). In the cases discussed here, induced electron circulation also accounts for the results, except that the circulation involved is that of the oxygen unshared electrons, and it occurs in such a way as to oppose the applied field in the region of the adjacent atoms; as a result, a higher field is required for resonance of these atoms. The result is a smaller chemical shift. The $\alpha$-carbon atoms of acetone have no such contribution, because they have no unshared electrons. Hence, the carbonyl carbon chemical shifts of ketones are greater than those of carboxylic acids. In acetate ion, the contribution of one unshared pair is partially offset by that of another, because the two orbitals are directed differently in space with respect to the applied field; hence, acetate absorbs at a chemical shift that is closer to the "normal" carbonyl value.

### ✓20.2    Bases that React with Carboxylic Acids

In Eq. 20.3 of the text, water is the base that reacts with a carboxylic acid, because water is the most abundant base present in the aqueous solution of an acid. However, you should not forget that carboxylic acids can in principle react with *any* base. (For example, write the reaction of a carboxylic acid with one equivalent of a Grignard reagent.)

Another point worth special note is that the reaction of strong bases not only with carboxylic acids, but also with alcohols and water, is very fast—so fast that a reaction occurs at virtually every encounter between an acid molecule and a base molecule. This becomes very important in situations in which proton transfer competes with other reactions for a limiting amount of a base. Suppose, for example, we want the aldehyde group in the following molecule to react with a Grignard reagent. The first equivalent of Grignard reagent reacts not with the aldehyde, but rather with the carboxylic acid, because proton transfer is such a fast reaction.

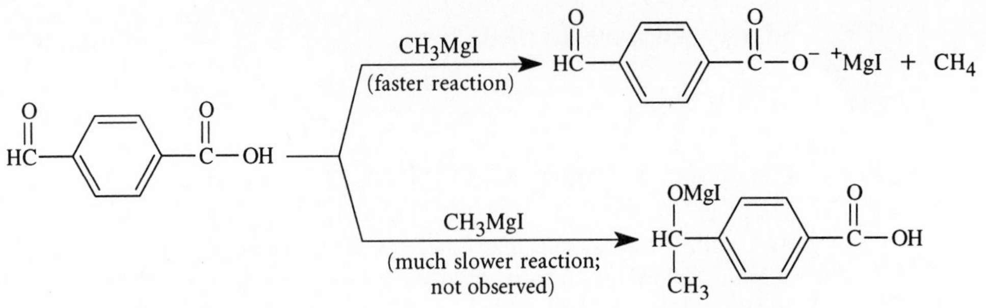

To achieve reaction of the aldehyde, a second equivalent of the Grignard reagent must be added:

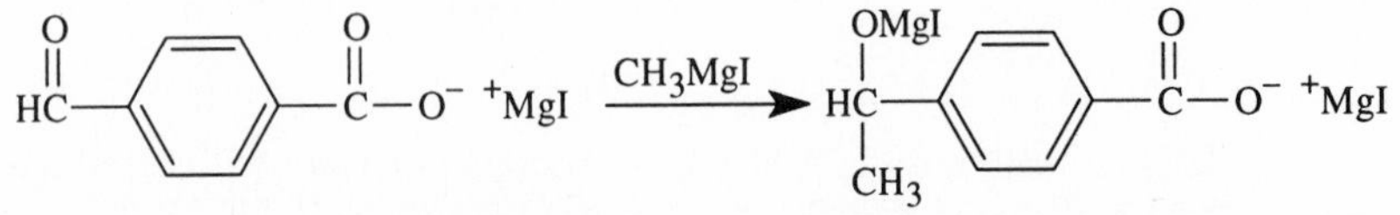

(The aldehyde carbonyl is much more reactive than the carboxylate carbonyl for reasons that are discussed in Chapter 21.)

In summary, proton transfers to strong bases from O—H groups in most cases are much faster than other reactions of strong bases.

## ✓20.3  Resonance Effect on Carboxylic Acid Acidity

The resonance stabilization of carboxylate ions was once considered the *major* effect responsible for the enhanced acidity of carboxylic acids. It is now believed to be less important than originally thought (although still significant) because *carboxylic acids are also resonance-stabilized* (see Eq. 20.2 on text p. 934). Figure 3.2 on text p. 109 shows that stabilizing a conjugate-base carboxylate decreases the p$K_a$ of the corresponding carboxylic acid. It follows that stabilizing a carboxylic acid, that is, lowering its energy relative to that of its conjugate-base carboxylate, *increases* its p$K_a$. Hence, the resonance stabilization of an acid and the resonance stabilization of its carboxylate are opposing effects. However, *resonance is more important in the carboxylate* because the carboxylate resonance structures *disperse* or *delocalize* charge, whereas the carboxylic acid resonance structures *separate* charge. Guideline 3, text p. 710, discusses the fact that resonance structures which separate charge are less important than those which do not. An exactly analogous situation was discussed for phenols in Study Guide Link 18.3 on page 423, Vol. 1, of this manual.

## 20.4  More on Surfactants

The word "surfactant" means "surface-active compound." To say that something is *surface active* means that is lowers the *surface tension* of a liquid. Without getting into a technical discussion of surface tension, we can say that it is the resistance of a surface to penetration by another object. Water has a particularly high surface tension. (If you've ever seen someone float a coin on the surface of a glass of water, you've witnessed the effects of surface tension.) The high surface tension of water is due to the tight network of strong hydrogen bonds at the water surface. Because water cannot form hydrogen bonds to air, water molecules form particularly strong hydrogen bonds to each other at the air-water interface.

When a surfactant is dissolved in water, the surfactant molecules tend to collect at the surface of the water with the hydrocarbon tails pointed toward the water-air interface, as shown in Fig. SG20.1 on the following page.

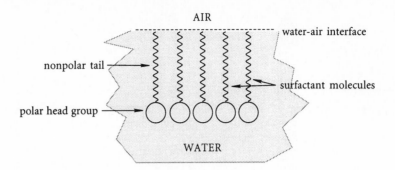

**Figure SG20.1**    *A diagram showing schematically the orientation of surfactant molecules at the interface of an aqueous solution and air. The nonpolar tails are oriented towards the interface.*

This orientation of surfactant molecules at the surface allows the solvent water molecules to interact with the polar head groups and places the nonpolar tails in contact with air, which is composed mostly of other nonpolar molecules. In this situation the network of tight hydrogen bonds at the surface of pure water is no longer present. As a result, the surface tension of the solution is lowered. One perceptible manifestation of this phenomenon is aqueous solutions of surfactants (for example, soap solutions) are slippery to the touch.

## 20.5   Orthoesters

Some students ask whether esters react further under the esterification conditions to form compounds analogous to acetals. (See Sec. 19.10A on text pp. 898–902.)

$$R-\overset{\overset{\displaystyle O}{\|}}{C}-OCH_3 \ + \ 2CH_3OH \ \xrightarrow{\ \ \text{acid}\ \ } \ R-\overset{\overset{\displaystyle OCH_3}{|}}{\underset{\underset{\displaystyle OCH_3}{|}}{C}}-OCH_3 \ + \ H_2O \qquad \text{(not observed)}$$

an orthoester

Such compounds, called orthoesters, are indeed known. However, they *cannot* be formed in this manner because the equilibrium constants for addition to the carbonyl groups of acid derivatives are only about $10^{-8}$ times those for formation of aldehyde addition compounds. Such equilibria are simply too unfavorable to be driven to completion in the usual ways.

The reason that such equilibria are unfavorable follows directly from the discussion of the relationship between the stability of a carbonyl compound and its equilibrium constant for addition. (See Sec. 19.7B on text pp. 887–890.) Stabilization of a carbonyl compound reduces its equilibrium constant for addition. Esters are stabilized by resonance interaction of the carboxylate oxygen with the carbonyl group. Ketones, in contrast, have fewer resonance structures and are less stable relative to their addition products.

$$\left[\ R-\overset{\overset{\displaystyle :\!\ddot{O}:}{\|}}{C}-\ddot{O}CH_3 \ \longleftrightarrow \ R-\overset{\overset{\displaystyle :\ddot{O}:^-}{|}}{\underset{+}{C}}-\ddot{O}CH_3 \ \longleftrightarrow \ R-\overset{\overset{\displaystyle :\ddot{O}:^-}{|}}{C}=\overset{+}{\ddot{O}}CH_3 \ \right]$$

An ester has more resonance structures and is more stable relative to its addition products; addition reactions are less favorable.

$$\left[\ R-\overset{\overset{\displaystyle :\!\ddot{O}:}{\|}}{C}-R \ \longleftrightarrow \ R-\overset{\overset{\displaystyle :\ddot{O}:^-}{|}}{\underset{+}{C}}-R \ \right]$$

An aldehyde or ketone has fewer resonance structures and is less stable relative to its addition products; addition reactions are more favorable.

Hence, the equilibrium constant for an addition to a ketone is more favorable than the equilibrium constant for the same addition to an ester. This point is illustrated in Fig. SG20.2.

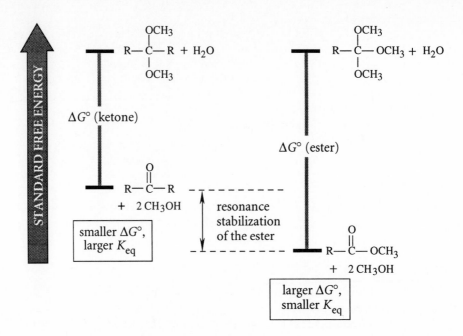

**Figure SG20.2**   *The greater resonance stabilization of esters causes addition reactions of esters to have smaller equilibrium constants than the corresponding addition reactions of ketones.*

## ✓20.6   Mechanism of Acid Chloride Formation

The formation of acid chlorides follows a mechanism that is fundamentally the same as the mechanism for esterification. In esterification, the carbonyl group is activated by protonation. In acid chloride formation, it is activated by reaction with the Lewis acid $SOCl_2$. Notice that chloride ion is displaced from $SOCl_2$ in the process:

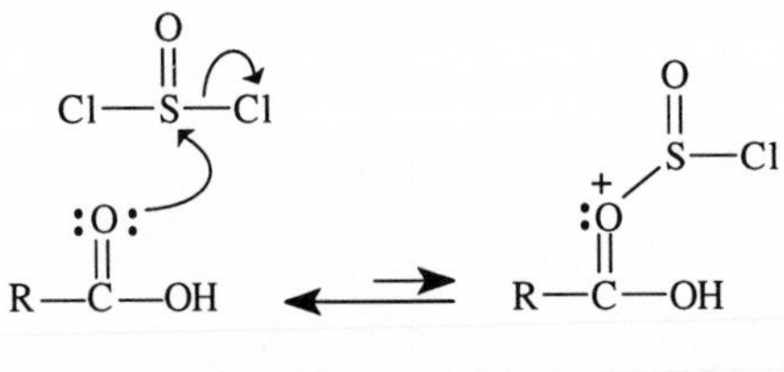

As a result of this reaction the carbonyl oxygen is more electrophilic. In other words, the Lewis acid $SOCl_2$ activates the carbonyl carbon toward nucleophilic attack in much the same sense that a proton does. (You've seen that other Lewis acids, such as the lithium ion of $LiAlH_4$ or the magnesium of a Grignard reagent, can also activate attack at a carbonyl carbon.) The nucleophile chloride ion produced in the step above reacts at the carbonyl carbon:

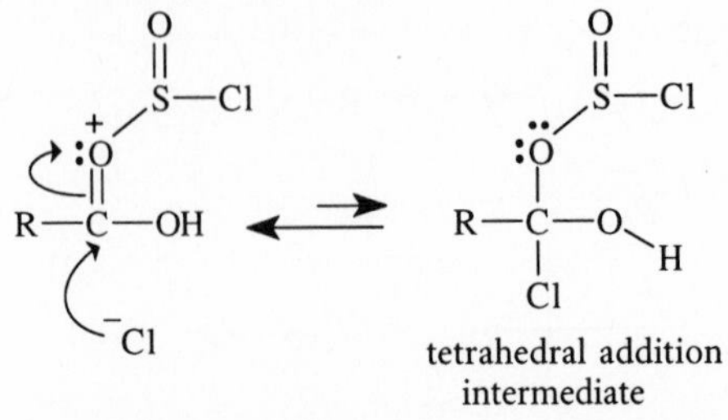

tetrahedral addition
intermediate

Notice the formation of a tetrahedral addition intermediate, as in esterification. *This is a common mechanistic thread that runs throughout carbonyl chemistry.* The —OSOCl group is an excellent leaving group and departs to give a protonated acid chloride. Loss of the —OSOCl group is irreversible, because it decomposes to SO₂, a gas, and chloride ion:

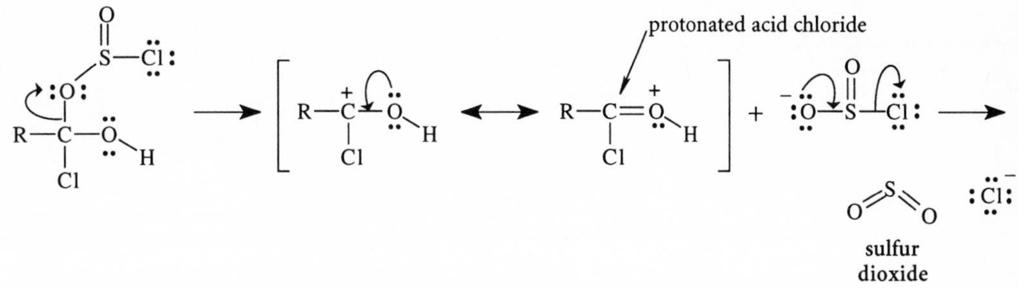

Loss of a proton gives the acid chloride:

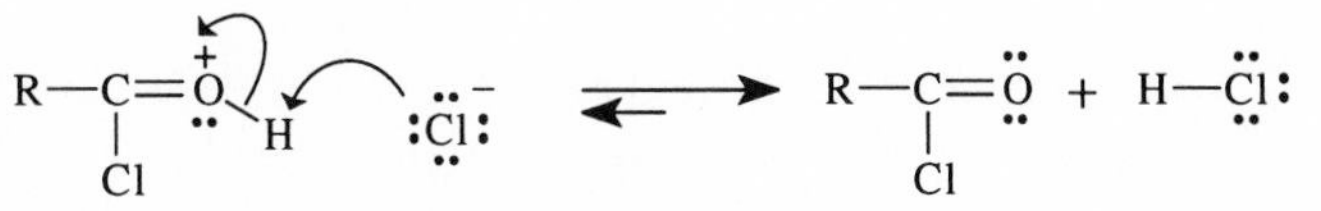

## ✓20.7  More on Synthetic Equivalents

Study Guide Link 17.3 (p. 397, Volume 1 of this manual) discussed the notion of a *synthetic equivalent.* Acid chlorides present another opportunity to think about this concept. Recall that acid chlorides are the acylating agents used in Friedel-Crafts acylation (Sec. 16.4E on text p. 754). Because acid chlorides are often prepared from carboxylic acids, it follows, then, that the acyl group of an aromatic ketone can originate from a carboxylic acid. Thus, a carboxylic acid or acid chloride can be viewed as the synthetic equivalent of the acyl group in an aromatic ketone.

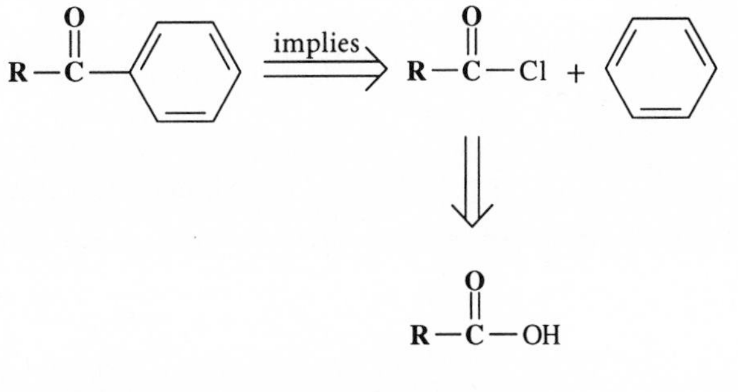

## ✓20.8  Mechanism of Anhydride Formation

Working through the mechanism of anhydride formation will allow you to see again the patterns involved in carbonyl substitution reactions. Consider the mechanism for the formation of the anhydride of a general carboxylic acid, RCO₂H, by acetic anhydride. The acetic anhydride serves as a Lewis acid to activate the carbonyl group of an R—CO₂H molecule:

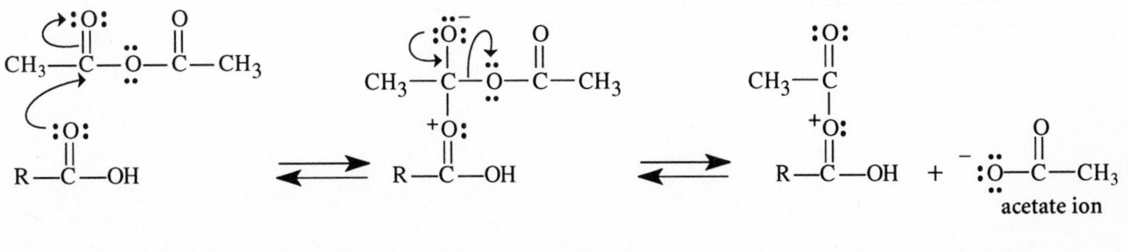

The activated carbonyl group reacts with another molecule of carboxylic acid and loses a proton to the acetate ion formed in the previous step to give a tetrahedral addition intermediate:

tetrahedral
addition
intermediate

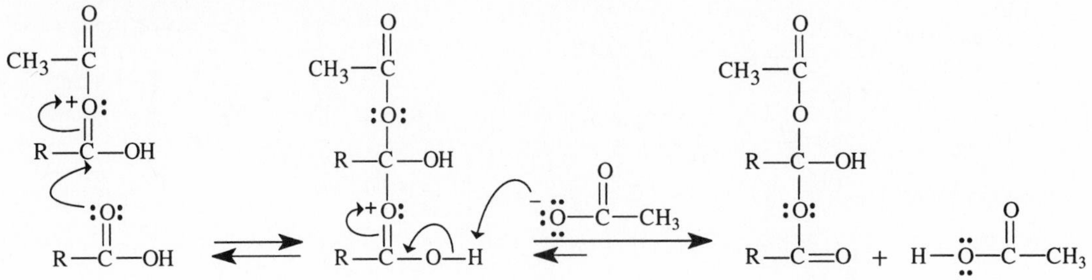

You should now complete the mechanism by showing, first, the loss of another acetate ion as a leaving group, and second, loss of a proton:

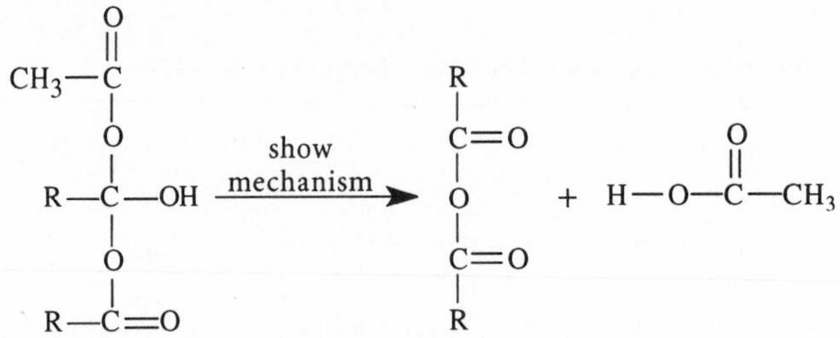

show mechanism

## 20.9    Mechanism of the LiAlH₄ Reduction of Carboxylic Acids

The mechanism of the LiAlH$_4$ reduction of carboxylic acids is presented in the text only in outline. This Study Guide Link fills in the details.

Reduction of the carboxylate ion by AlH$_3$ (step *a* in the following equation) is analogous to the addition of BH$_3$ to the double bond of an alkene. (BH$_3$ also reduces carboxylic acids to primary alcohols.)

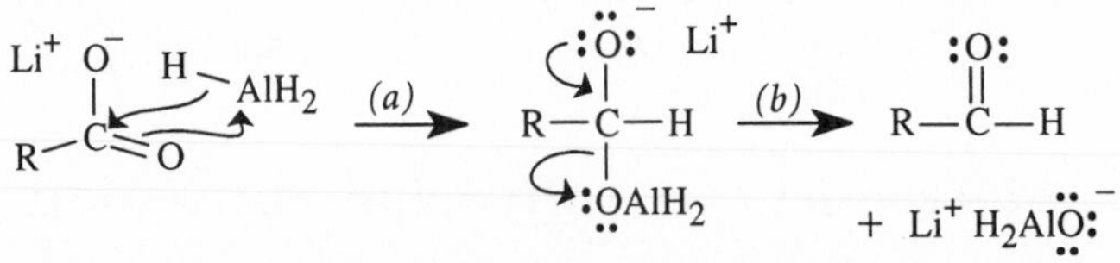

Notice that a tetrahedral addition intermediate is formed in step *a*. Breakdown of this intermediate (step *b*) expels Li$^+$ H$_2$AlO$^-$, which also can act as a reducing agent. Further reduction of the aldehyde to the alcohol (Sec. 19.8, text pp. 890–893) can occur either by this species or by excess LiAlH$_4$ present in the reaction mixture.

P  **Solutions**

## Solutions to In-Text Problems

**20.1** (a)

HOCH$_2$CH$_2$CH$_2$CO$_2$H
γ-hydroxybutyric acid

(c)

CH$_3$CH$_2$    CH$_2$CO$_2$H
        C=C
        H    H
(Z)-3-hexenoic acid

(e)

HO$_2$C—⬡—CO$_2$H
1,4-cyclohexanedicarboxylic acid

(g)

          Cl
          |
HO$_2$CCCH$_2$CH$_2$CH$_2$CO$_2$H
          |
          Cl
α,α-dichloroadipic acid

**20.2** (a) 2,2-dimethylbutanoic acid (common: α,α-dimethylbutyric acid)
(c) 3-hydroxy-4-oxocyclohexanecarboxylic acid
(e) 2-methylpropanedioic acid (common: α-methylmalonic acid)

➠ Note that a numerical prefix for substituents in the names of some compounds, although not "wrong," is unnecessary because the position of the substituent is unambiguous without a number. Thus, 2-methylpropanedioic acid is found in *Chemical Abstracts* as methylpropanedioic acid because a methyl substituent on a propanedioic acid skeleton can be *only* at carbon-2. The common name methylmalonic acid is also acceptable. Similarly, chloroacetic acid (rather than α-chloroacetic acid), methylpropanoic acid (rather than 2-methylpropanoic acid), and methylpropene (rather than 2-methylpropene) are the "official" names of these compounds. However, the numerical prefixes in such cases have been added in the text for the sake of consistency in the use of substitutive nomenclature.

**20.3** The dimer owes its stability to intermolecular hydrogen bonds. The solvent carbon tetrachloride has no donor or acceptor atoms to compete for these hydrogen bonds. However, the solvent water can compete as both a donor and an acceptor for hydrogen bonds. Hence, the dimer has less of an energetic advantage in water and, as a result, there is a lower concentration of dimer in aqueous solution.

**20.4** The IR spectrum indicates the presence of a carboxy group. A carboxy group accounts for 45 mass units; this leaves 43 mass units unaccounted for. The NMR spectrum indicates seven hydrogens in addition to the carboxy hydrogen at $\delta$ 10. A group with seven hydrogens and 43 mass units must be either a propyl or an isopropyl group; the splitting in the NMR spectrum indicates the latter. The compound is 2-methyl-propanoic acid (isobutyric acid), (CH$_3$)$_2$CH—CO$_2$H.

**20.6** First write the structures of the two compounds.

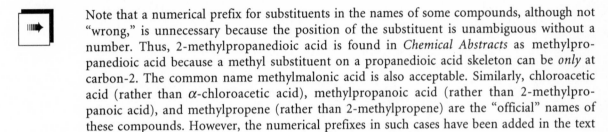

          CH$_3$
          |
HO$_2$C—C—CH$_2$—CO$_2$H        HO$_2$CCH$_2$CH$_2$CH$_2$CH$_2$CO$_2$H
          |                              adipic acid
          CH$_3$
α,α-dimethylsuccinic acid

(a) α,α-Dimethylsuccinic acid should have five resonances in its CMR spectrum because it has five chemically nonequivalent sets of carbons. Adipic acid, in contrast, should have three resonances because it has only three chemically nonequivalent sets of carbons.

(b) Only the proton NMR spectrum of α,α-dimethylsuccinic acid consists of three singlets; the proton NMR spectrum of adipic acid is more complex.

**20.7** (a) The first and second ionizations of succinic acid:

$$HO_2CCH_2CH_2CO_2H \underset{H_3O^+}{\overset{H_2O}{\rightleftharpoons}} {}^-O_2CCH_2CH_2CO_2H \underset{H_3O^+}{\overset{H_2O}{\rightleftharpoons}} {}^-O_2CCH_2CH_2CO_2^-$$

$$pK_a = 4.21 \qquad\qquad pK_a = 5.64$$

(b) The first $pK_a$ is lower because the electron-withdrawing polar effect of an un-ionized carboxy group stabilizes the anion formed by ionization of the other carboxy group. The second $pK_a$ is higher because the negative charge of an ionized carboxy group destabilizes the anion formed by the second ionization. This destabilization is the result of the electrostatic repulsion of like charges.

**20.9** Extract an ether solution of the two compounds with an aqueous solution of $NaHCO_3$, $Na_2CO_3$, or $NaOH$. The acid will ionize and its conjugate-base anion will dissolve in the aqueous layer as the sodium salt; *p*-bromotoluene will remain in the ether layer. After isolating the aqueous layer, acidify it with concentrated HCl; neutral *p*-bromobenzoic acid will precipitate.

**20.10** (a)

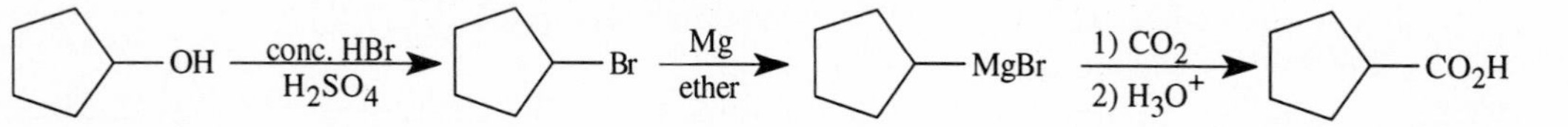

cyclopentanol                                                                                cyclopentanecarboxylic acid

**20.11** (a) Both carboxy groups are esterified; the product is dipropyl adipate:

$$CH_3CH_2CH_2OC(CH_2)_4COCH_2CH_2CH_3 \quad \text{dipropyl adipate}$$

**20.12** (a) Follow the reverse of the steps shown in Eqs. 20.18a–c of the text with R— = Ph—.

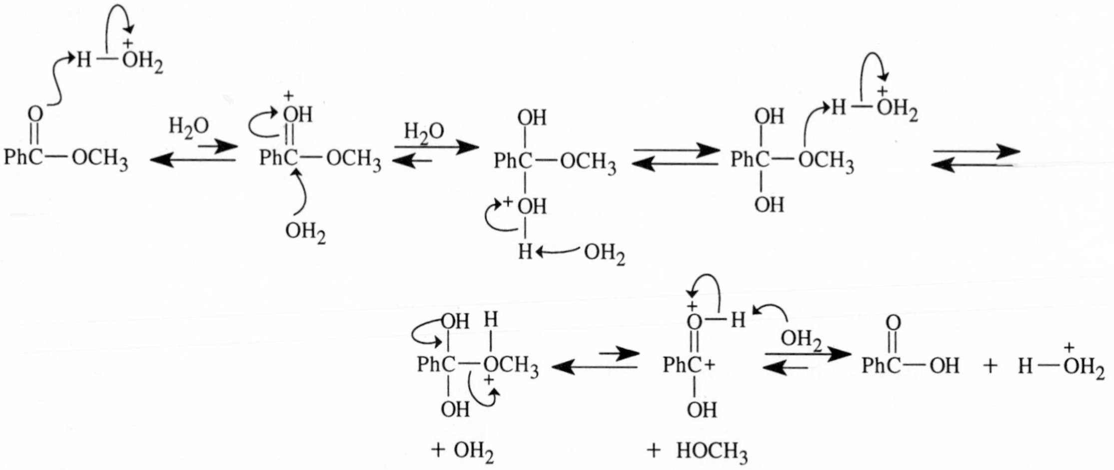

(b) To favor ester hydrolysis rather than ester formation, use a large excess of water as solvent rather than an alcohol. By LeChatelier's principle, this drives the carboxylic acid-ester equilibrium toward the carboxylic acid.

(c) The mechanism of hydroxide-promoted ester hydrolysis:

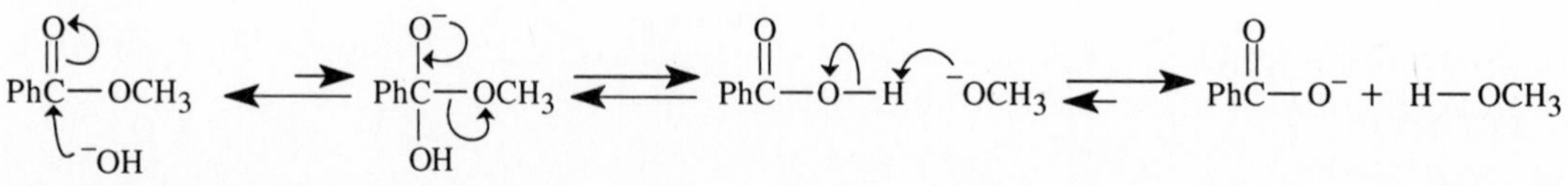

Notice that under the basic conditions the carboxylic acid product is ionized in the last step. This drives base-promoted ester hydrolysis to the right.

(d) As shown in the last step of the mechanism in the solution to part (c), methoxide reacts with the *acidic proton* of benzoic acid to form the benzoate ion much more rapidly than it reacts at the carbonyl carbon. (See Study Guide Link 20.2 on p. 499 of this manual for a discussion of this point.)

**20.13** (a)                                                    (c)

$(CH_3)_2CHC{-}OCH_3$

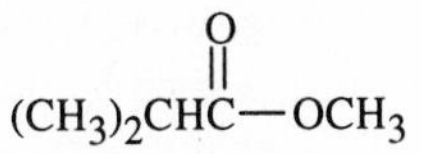

methyl isobutyrate

$(CH_3)_2CHC{-}OCH_2Ph$

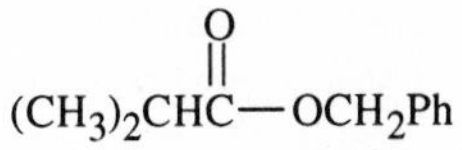

benzyl isobutyrate

**20.14** (a) In this reaction the carboxylic acid is alkylated by the *tert*-butyl cation, which is formed by protonation of 2-methylpropene.

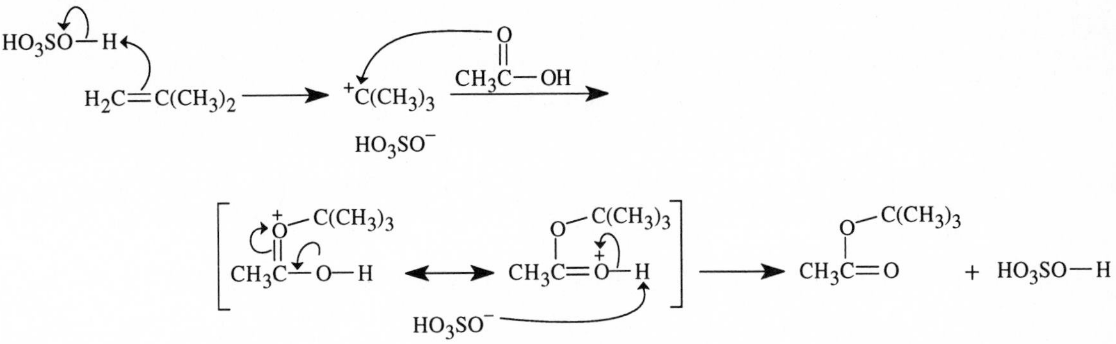

A variation on this mechanism proposed by many students is alkylation of the carboxylate oxygen rather than the carbonyl oxygen by the *tert*-butyl cation:

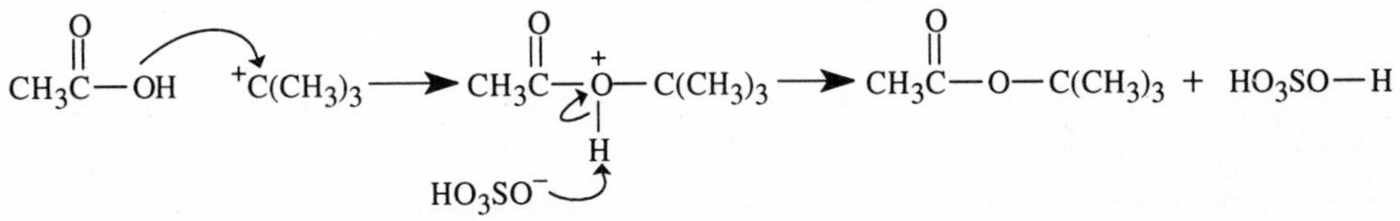

Although this appears to be a more direct route to the ester, the mechanism given above is more likely to be correct for the following reason. Just as a carboxylic acid is *protonated* on the carbonyl oxygen rather than the carboxylate oxygen (why? see text p. 941), it should be *alkylated* on the carbonyl oxygen for the same reason.

(b) This reaction is mechanistically more like the alkylation reactions in Sec. 20.8B because a bond is formed between oxygen and the alkyl group of the ester. In acid-catalyzed esterification, the bond is formed between an oxygen and the *carbonyl carbon*. Furthermore, there is no tetrahedral addition intermediate in this reaction as there is in acid-catalyzed esterification.

**20.15** (a)                                                    (c)

$CH_3CH_2CH{-}C{-}Cl$
           $|$
          $CH_3$

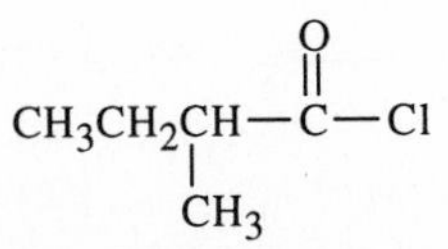

2-methylbutanoyl chloride

$CH_3CH_2CH_2{-}S{-}Cl$

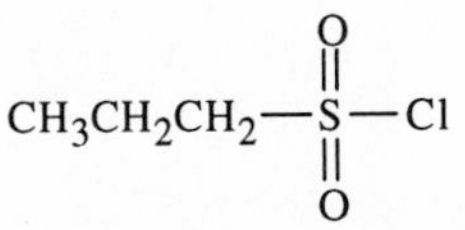

1-propanesulfonyl chloride

**20.16** (a)

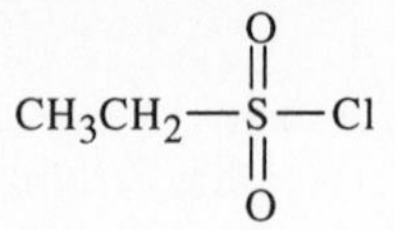

ethanesulfonyl chloride

(c)

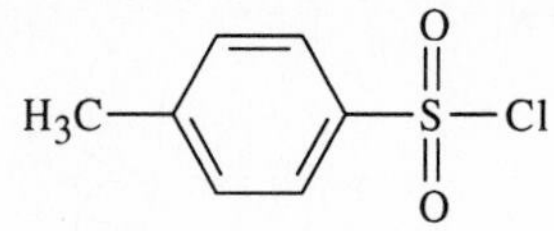

*p*-toluenesulfonyl chloride
(tosyl chloride)

**20.17** Study Guide Link 20.7 points out that a carboxylic acid is the synthetic equivalent of an acyl group in an aromatic ketone. This concept is realized by applying the Friedel-Crafts acylation reaction to the corresponding acid chloride. This synthesis takes advantage of the fact that the methoxy group is an *ortho, para*-directing substituent in electrophilic aromatic substitution.

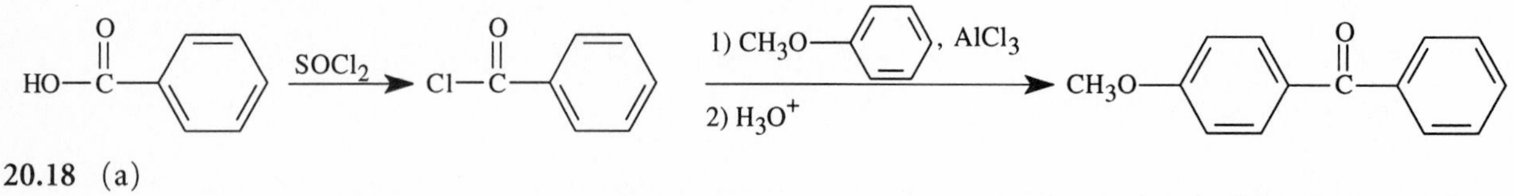

**20.18** (a)

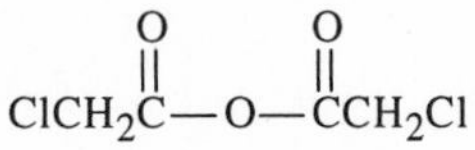

chloroacetic anhydride

**20.19** (a) Maleic anhydride forms a cyclic anhydride because its carboxy groups are *cis*. A cyclic anhydride of fumaric acid would require that a five-membered ring contain a *trans* double bond.

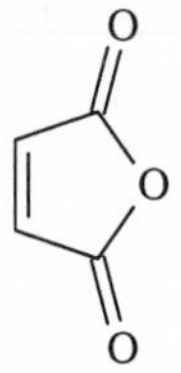

maleic anhydride

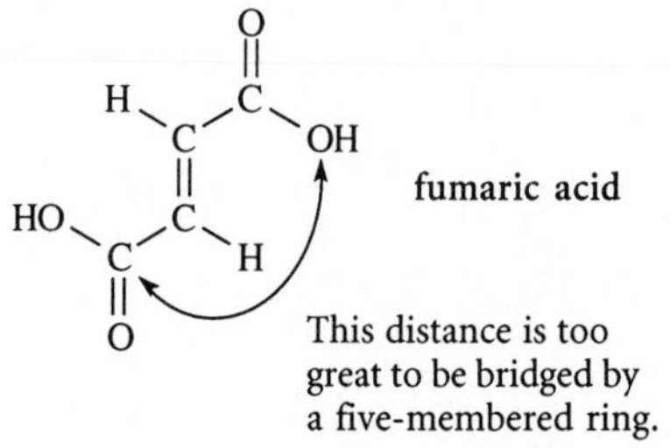

fumaric acid

This distance is too great to be bridged by a five-membered ring.

**20.20** (a) The compound must contain three oxygens and two degrees of unsaturation in addition to the benzene ring. *p*-Formylbenzoic acid meets the indicated criteria.

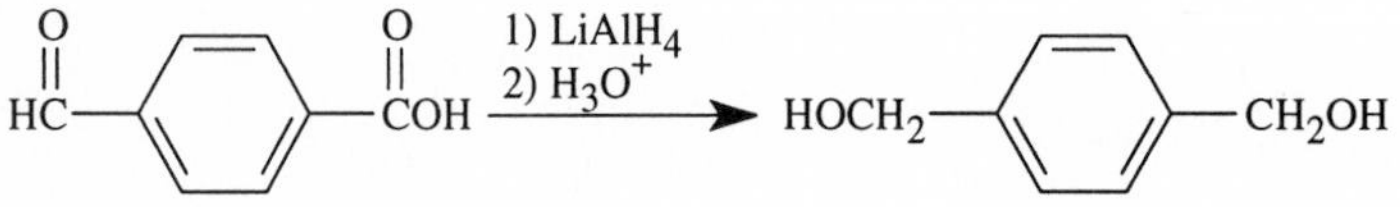

*p*-formylbenzoic acid

**20.21** (a) This synthesis requires the addition of two carbons, which can be accomplished by the reaction of a Grignard reagent with oxirane (text Sec. 11.4C).

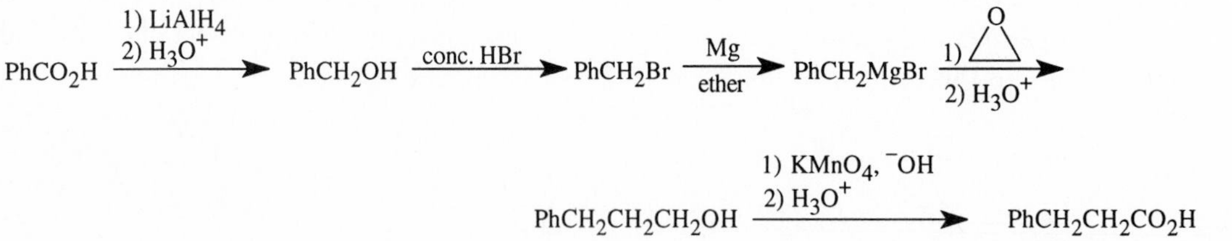

3-phenylpropanoic acid

**20.22** Each compound decarboxylates; the net effect is the replacement of the carboxy group by a hydrogen. (In part (c), the amine product is protonated under the acidic conditions.)

(a)

$$Ph-\overset{\overset{\displaystyle O}{\|}}{C}-\underset{\underset{\displaystyle CH_3}{|}}{\overset{\overset{\displaystyle CH_3}{|}}{CH}}$$

(b)

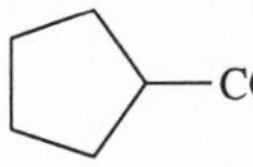

cyclopentane—$CO_2H$

(c)

$C_2H_5NH_2 \underset{\longleftarrow}{\overset{H_3O^+}{\longrightarrow}} C_2H_5\overset{+}{N}H_3$

**20.24** Were this carboxylic acid to decarboxylate by the mechanism shown in Eq. 20.39 on p. 958 of the text, the following enol would be formed as an intermediate:

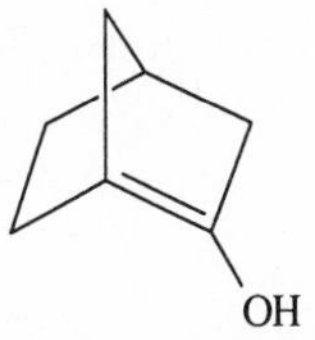

This enol is very unstable because it violates Bredt's rule; that is, the double bond is twisted to such an extent that the *p* orbitals on each carbon cannot overlap.

## Solutions to Additional Problems

· · · · · · · · · · · · · · · · · · · · · · · · · · · · · · · · · · · · · · · · · · · · · · · · · · · · · · · · · · · · · · · · · · · ·

**20.25** (a)

$$CH_3CH_2CH_2\overset{\overset{\displaystyle O}{\|}}{C}OCH_2CH_3$$

(b)

$$CH_3CH_2CH_2\overset{\overset{\displaystyle O}{\|}}{C}O^- \, Na^+$$

(c) $CH_3CH_2CH_2CH_2OH$

(d) No reaction

(e)

$$CH_3CH_2CH_2\overset{\overset{\displaystyle O}{\|}}{C}Cl$$

(f)

$$CH_3CH_2CH_2\overset{\overset{\displaystyle O}{\|}}{C}OCH_3$$

(g)

$(CH_3CH_2CH_2CH_2O)_2CHCH_3$

(h)

$$CH_3CH_2CH_2\overset{\overset{\displaystyle O}{\|}}{C}Ph$$

(i)

$CH_3CH_2CH_2CH_2Ph$

**20.27** The dicarboxylic acids with the formula $C_6H_{10}O_4$:

$HO_2C(CH_2)_4CO_2H$
hexanedioic acid
(adipic acid)
(1)

$HO_2CCH_2CH_2\underset{\underset{\displaystyle CO_2H}{|}}{CHCH_3}$
2-methylpentanedioic acid
($\alpha$-methylglutaric acid)
(2)

$HO_2CCH_2\underset{\underset{\displaystyle CO_2H}{|}}{CH}CH_2CH_3$
2-ethylbutanedioic acid
($\alpha$-ethylsuccinic acid)
(3)

$HO_2C\underset{\underset{\displaystyle CO_2H}{|}}{CH}CH_2CH_2CH_3$
2-propylpropanedioic acid
(propylpropanoic acid,
$\alpha$-propylmalonic acid,
propylmalonic acid)
(4)

$HO_2C\underset{\underset{\displaystyle CH_2CH_3}{|}}{\overset{\overset{\displaystyle CH_3}{|}}{C}}CO_2H$
2-ethyl-2-methylpropanedioic acid
(ethylmethylpropanedioic acid,
$\alpha$-ethyl-$\alpha$-methylmalonic acid,
ethylmethylmalonic acid)
(5)

$HO_2C\underset{\underset{\displaystyle CO_2H}{|}}{CH}CH(CH_3)_2$
2-isopropylpropanedioic acid
(isopropylpropanedioic acid,
$\alpha$-isopropylmalonic acid,
isopropylmalonic acid)
(6)

*(solution continues)*

HO$_2$CCH$_2$CHCH$_2$CO$_2$H
|
CH$_3$

**3-methylpentanedioic acid**
(β-methylglutaric acid)
(7)

HO$_2$CCHCHCO$_2$H
|      |
H$_3$C  CH$_3$

**2,3-dimethylbutanedioic acid**
(α,β-dimethylsuccinic acid)
(±) and *meso*
(8)      (9)

CH$_3$
|
HO$_2$CCCH$_2$CO$_2$H
|
CH$_3$

**2,2-dimethylbutanedioic acid**
(α,α-dimethylsuccinic acid)
(10)

Compounds (2), (3), and (8) are chiral. Compounds (2), (3), (7), (8), (9), and (10) would readily form cyclic anhydrides on heating, because those anhydrides contain five- or six-membered rings. Compounds (4), (5), and (6) would readily decarboxylate on heating because they are malonic acid derivatives.

**20.28**  (a)  First calculate the number of moles of 0.1 *M* NaOH required for the neutralization:

$$\text{mol NaOH} = (0.100 \text{ mol NaOH/L})(8.61 \times 10^{-3} \text{ L}) = 8.61 \times 10^{-4}$$

The same molar amount of the carboxylic acid must be present because one mole of NaOH is required to neutralize one mole of a monocarboxylic acid. The molecular mass is

$$\text{molecular mass of the acid} = (100 \times 10^{-3} \text{ g}) \div (8.61 \times 10^{-4} \text{ mol}) = 116 \text{ g/mol}$$

**20.29**  In each part, acetic acid is converted by the base into its conjugate base, acetate ion.

(a)

$$CH_3C\text{—}O\text{—}H \quad ^-OCH_3 \text{ Na}^+ \longrightarrow CH_3C\text{—}O^- \text{ Na}^+ + H\text{—}OCH_3$$

(c)

$$CH_3C\text{—}O\text{—}H \quad BrMg\text{—}CH_2CH_3 \longrightarrow CH_3C\text{—}O^- \text{ }^+MgBr + CH_3CH_3$$

(e)

$$CH_3C\text{—}O\text{—}H \quad \ddot{N}H_3 \longrightarrow CH_3C\text{—}O^- \text{ }^+NH_4$$

(g)

$$CH_3C\text{—}O\text{—}H \quad ^-O\text{—}C\text{—}OH \rightleftharpoons CH_3C\text{—}O^- + H\text{—}O\text{—}C\text{—}OH \longrightarrow H_2O + CO_2$$

bicarbonate ion                                          carbonic acid

**20.30**  Only enough NaOH is present to ionize one of the carboxy groups. The question, then, is which if any of the carboxy groups is more acidic. Because polar effects of substituents decrease with increasing distance of the substituents from the reaction center, the acid-strengthening polar effect of the chlorines has a greater effect on the carboxy group to which they are closer. Therefore, in the presence of one molar equivalent of NaOH, it is this carboxy group that is principally ionized.

$$^-O\text{—}C\text{—}C\text{—}CH_2\text{—}C\text{—}OH \xleftarrow{\hspace{1cm}} HO\text{—}C\text{—}C\text{—}CH_2\text{—}C\text{—}O^-$$

major species present                    little of this anion is present

**20.32**  Carry out a calculation of pH using the dissociation-constant expression for each acid. (See Eq. 20.7a, text p. 940.) Let the concentration of the undissociated acid be $0.1 - x$; the concentrations of both H$_3$O$^+$ and

$CH_3CO_2^-$ are then $x$.

$$K_a = 10^{-4.76} = 1.74 \times 10^{-5} = \frac{[H_3O^+][CH_3CO_2^-]}{[CH_3CO_2H]} = \frac{x^2}{0.1 - x}$$

Because the dissociation constant of acetic acid is small, use the approximation that $x$ is small compared to the amount of acetic acid present.

$$\frac{x^2}{0.1} = 1.74 \times 10^{-5}$$
$$x = [H_3O^+] = 1.32 \times 10^{-3}$$
$$-\log x = pH = 2.88.$$

Thus, the pH is less than 3, and acetic acid therefore turns litmus paper red.

An analogous calculation reveals that the pH of a 0.1 $M$ solution of phenol ($pK_a = 9.95$) is 5.48. Since this pH is greater than 3, a 0.1 $M$ solution of phenol ("carbolic acid") does not turn litmus paper red.

**20.34** (a)  Consider the following equilibrium, in which AH is an acid, and S is the solvent:

$$AH + S \rightleftharpoons A^- + \overset{+}{S}H$$

Suppose that AH is more acidic than $^+SH$. Then this equilibrium lies to the right, and $^+SH$, the conjugate acid of the solvent, is the major species present. That is, because AH cannot survive, the strongest acid that can exist is the conjugate acid of the solvent.

**20.35**  The proton and the $\pi$ bond move back and forth between the two oxygens so rapidly that the two oxygens retain their identities for only a tiny fraction of a second. The mechanism varies with the situation. In a solvent (for example, water) in which the acid readily dissociates, even to a small extent, the two oxygens in the conjugate-base carboxylate ion are equivalent and are protonated with essentially equal frequency; these reactions are very fast. (In the equations that follow, one of the oxygens is marked with an asterisk (*) so that its fate can be traced throughout the reaction.)

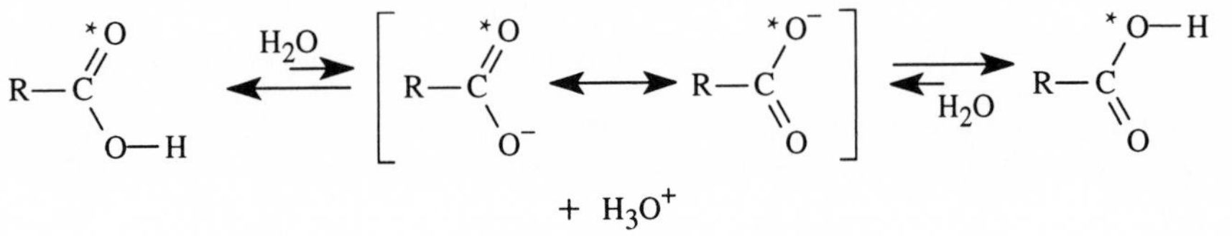

In solvents in which the carboxylic acid molecule cannot readily dissociate, protons can jump back and forth within a carboxylic acid hydrogen-bonded dimer (text p. 935):

**20.36** (a)

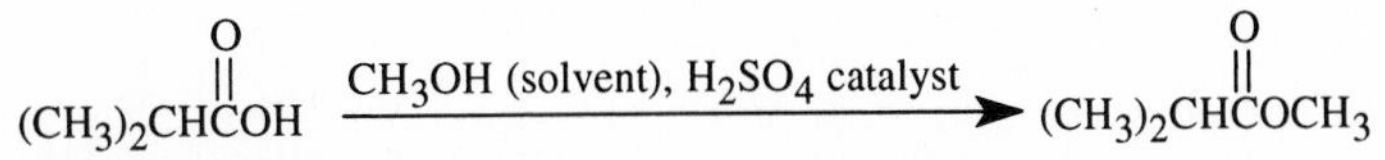

isobutyric acid
(methylpropanoic acid,
2-methylpropanoic acid)

(c)  Notice in this synthesis than an additional carbon must be added. Follow the strategy in Study Problem 20.2 on text p. 957.

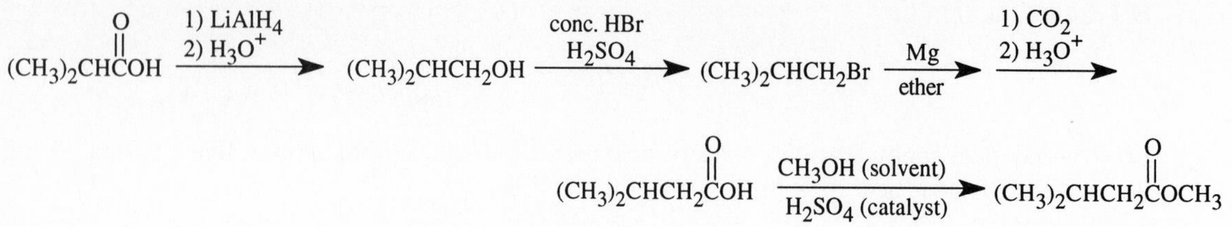

$(CH_3)_2CHCOH \xrightarrow[\text{2) } H_3O^+]{\text{1) LiAlH}_4} (CH_3)_2CHCH_2OH \xrightarrow[H_2SO_4]{\text{conc. HBr}} (CH_3)_2CHCH_2Br \xrightarrow[\text{ether}]{\text{Mg}} \xrightarrow[\text{2) } H_3O^+]{\text{1) CO}_2}$

$(CH_3)_2CHCH_2COH \xrightarrow[H_2SO_4 \text{ (catalyst)}]{CH_3OH \text{ (solvent)}} (CH_3)_2CHCH_2COCH_3$

(e)

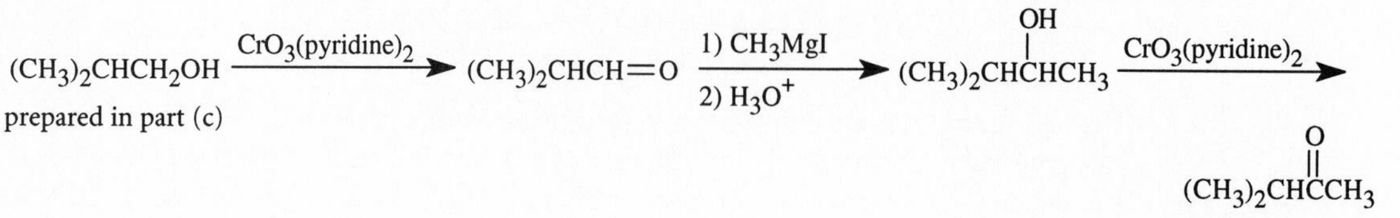

$(CH_3)_2CHCH_2OH \xrightarrow{CrO_3(\text{pyridine})_2} (CH_3)_2CHCH=O \xrightarrow[\text{2) } H_3O^+]{\text{1) CH}_3MgI} (CH_3)_2CHCHCH_3 \xrightarrow{CrO_3(\text{pyridine})_2}$

prepared in part (c)

above the final alcohol: OH

$(CH_3)_2CHCCH_3$

**3-methyl-2-butanone**

(g)

$(CH_3)_2CHCH=O \xrightarrow[\text{(Wittig reaction)}]{\overset{-}{C}H_2-\overset{+}{P}Ph_3} (CH_3)_2CHCH=CH_2$

prepared in part (e)

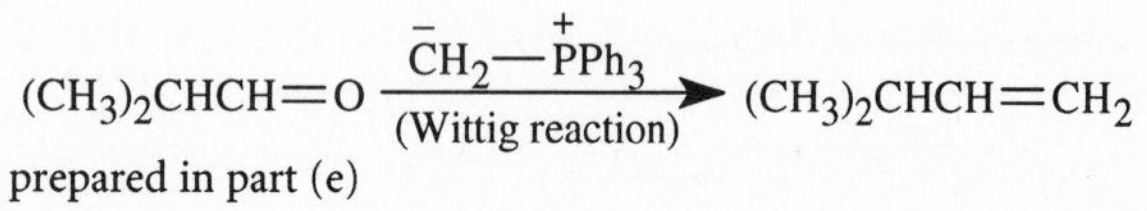

**20.37** Convert each enantiomer of 4-methylhexanoic acid into 3-methylhexane by any process that does not involve breaking a bond to the asymmetric carbon. Because each enantiomer of 4-methylhexanoic acid has known absolute configuration, the configuration of the corresponding enantiomer of 3-methylhexane will be determined. Suppose, for example, that (S)-4-methylhexanoic acid gives (−)-3-methylhexane. Then the two molecules should have corresponding configurations, and the configuration of (−)-3-methylhexane would then be S:

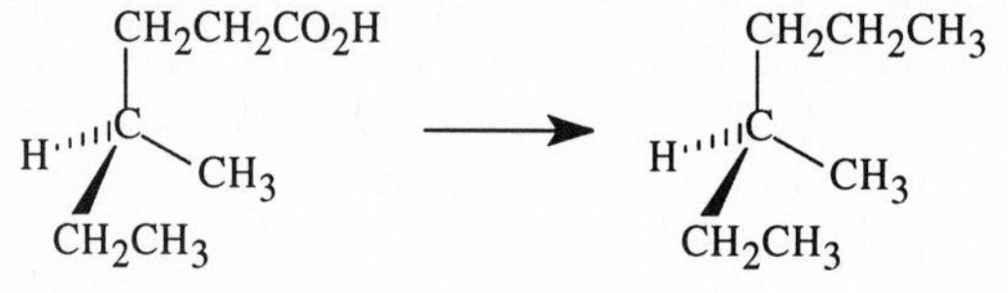

(S)-4-methylhexanoic acid          (S)-3-methylhexane

The following sequence of reactions would work for such a conversion:

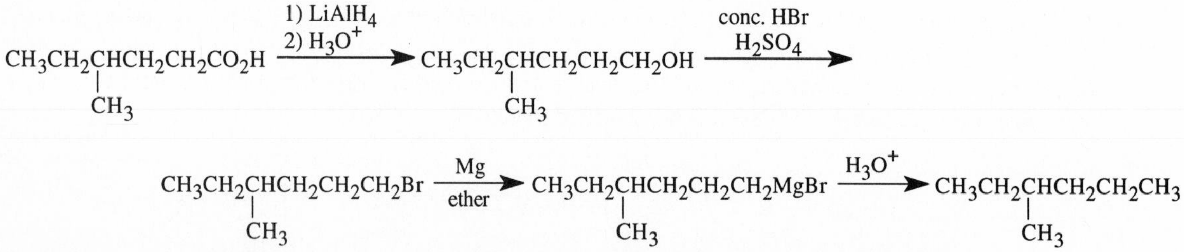

$CH_3CH_2\underset{\underset{CH_3}{|}}{C}HCH_2CH_2CO_2H \xrightarrow[\text{2) } H_3O^+]{\text{1) LiAlH}_4} CH_3CH_2\underset{\underset{CH_3}{|}}{C}HCH_2CH_2CH_2OH \xrightarrow[H_2SO_4]{\text{conc. HBr}}$

$CH_3CH_2\underset{\underset{CH_3}{|}}{C}HCH_2CH_2CH_2Br \xrightarrow[\text{ether}]{\text{Mg}} CH_3CH_2\underset{\underset{CH_3}{|}}{C}HCH_2CH_2CH_2MgBr \xrightarrow{H_3O^+} CH_3CH_2\underset{\underset{CH_3}{|}}{C}HCH_2CH_2CH_3$

**20.39** To confirm the labeling pattern, oxidize phenylacetic acid vigorously with $KMnO_4$ or other oxidizing agent that will convert it into benzoic acid.

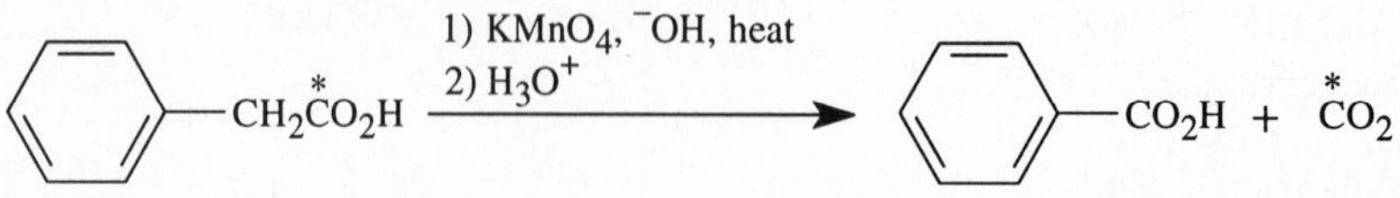

If all the radiolabel of phenylacetic acid is at the carbonyl carbon, the product benzoic acid should be completely devoid of radioactivity. However, if there is labeling elsewhere, the fraction of radioactivity located in carbons other than the carbonyl carbon is equal to the amount of radioactivity in the product benzoic acid divided by the total radioactivity in the starting phenylacetic acid.

**20.41** Dissolve the mixture in a suitable solvent; since the company specializes in chlorinated organic compounds, methylene chloride might be readily available for the purpose. Extract with 5% $NaHCO_3$ solution. *p*-Chlorobenzoic acid is extracted as its conjugate-base anion into the aqueous layer. Acidification of the aqueous layer will give *p*-chlorobenzoic acid itself. Then extract the methylene chloride solution with 5% aqueous NaOH solution. *p*-Chlorophenol will be extracted into the aqueous layer as its conjugate-base , *p*-chlorophenolate anion; acidification will afford *p*-chlorophenol. The basis for this separation is Eq. 20.7c on text p. 940. When the p$K_a$ of an acid is much lower than the pH of the solution, the acid is converted into its conjugate base; that is, the ratio of $[RCO_2^-]$ to $[RCO_2H]$ is large. The p$K_a$ of *p*-chlorobenzoic acid, which is about 4 (text Table 20.2), is much lower than the pH of aqueous sodium bicarbonate, which is about 8.5. Hence, *p*-chlorobenzoic acid is more than 99.99% ionized by the sodium bicarbonate solution. The p$K_a$ of *p*-chlorophenol, 9.4, is higher than the pH of aqueous sodium bicarbonate, and thus this phenol is not appreciably ionized. However, the p$K_a$ of the phenol is much lower than the pH of 0.1 *M* aqueous NaOH solution, which is about 13; hence, the phenol is ionized by, and thus extracted into, this solution.

Once the acid and the phenol are removed, evaporation of the methylene chloride gives a mixture of 4-chlorocyclohexanol and chlorocyclohexane. The boiling point of the alcohol is *much* higher because of hydrogen bonding; consequently, these two remaining substances can be separated by fractional distillation.

 Many students confuse p$K_a$, which is a property of a *compound*, with pH, which is a property of the *solution* in which the compound is dissolved. As Eq. 20.7c of the text shows, the ratio of concentrations of an acid and its conjugate base in solution depends on the relationship of the pH of the *solution* to the p$K_a$ of the *acid*. This relationship is exploited in the extractions described in the foregoing discussion.

**20.43** For this problem, let

$$\text{Ar} - = (CH_3)_2N - \!\!\!\!\!\bigcirc\!\!\!\!\! -$$

(a) The color of crystal violet is due to the overlap of the empty *p* orbital of the positively charged carbon with the $\pi$ orbitals of the benzene ring. The bleaching is due to the reaction of the carbocation with hydroxide to form the alcohol. In the alcohol, there is no longer an empty *p* orbital; hence, the conjugation and the color associated with the carbocation are absent.

$$\underset{\substack{\text{crystal violet}\\\text{(colored)}}}{Ar_3C^+} \quad {}^-OH \longrightarrow \underset{\text{(colorless)}}{Ar_3C-OH}$$

(b) SDS, a typical detergent, forms micelles. Despite its charge, crystal violet is incorporated into the interior of the SDS micelles because the "greasy" benzene rings interact favorably with the "greasy" hydrocarbon tails of SDS. (Think of this as a solubility phenomenon; benzene is more soluble in hydrocarbons than it is in water.) In the interior of a micelle, crystal violet is no longer accessible to hydroxide ion. Because hydroxide is strongly solvated by water, hydroxide ion cannot easily enter the interior of the micelles, and therefore it cannot react with the entrapped crystal violet. The very slow reaction that does occur is probably due to hydroxide reacting with the very small amount of crystal violet in solution.

*(solution continues)*

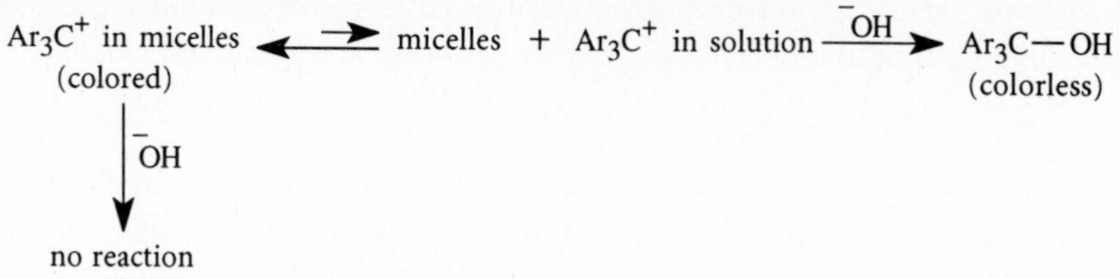
Ar3C+ in micelles
(colored)
micelles + Ar3C+ in solution
OH
Ar3C—OH
(colorless)
OH
no reaction

**20.44** In parts (c)–(d), the cyclohexane ring must undergo a chair flip to bring the two carboxy groups close enough together to form the anhydride ring.

(a)

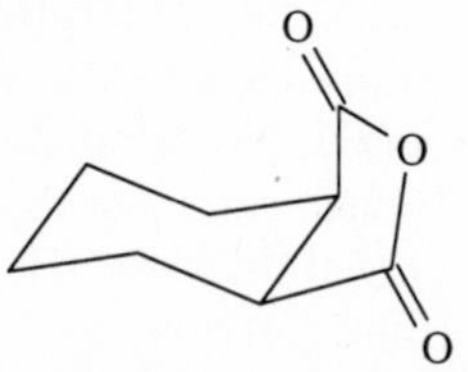

(c)

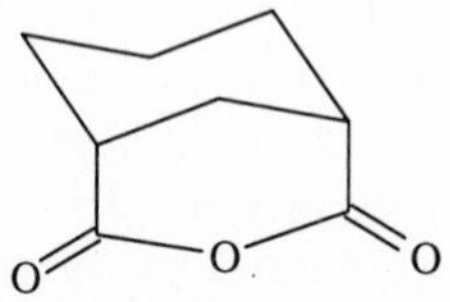

(d) Three identical anhydride molecules are in equilibrium:

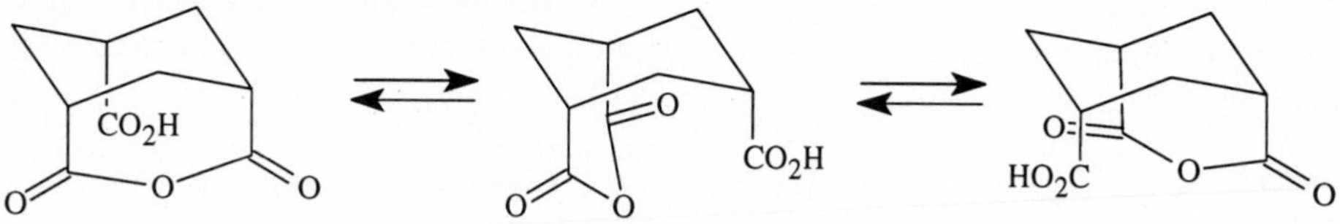

**20.45** Always remember that correct syntheses other than the one given may be possible. Be sure to ask your professor or teaching assistant if your synthesis differs from the one given and you are not sure whether it is correct.

(a)

$$CH_3CH_2CO_2H \xrightarrow[\text{2) } H_3O^+]{\text{1) } LiAlH_4} CH_3CH_2CH_2OH \xrightarrow[H_2SO_4]{\text{conc. HBr}} CH_3CH_2CH_2Br \xrightarrow[\text{ether}]{Mg}$$

propionic acid

$$CH_3CH_2CH_2MgBr \xrightarrow[\text{2) } H_3O^+]{\text{1) } CH_3CH{=}O} CH_3\overset{OH}{\overset{|}{C}}HCH_2CH_2CH_3$$

2-pentanol

(c)

$$(CH_3)_2CHCH_2CH_2CO_2H \xrightarrow[\text{2) } H_3O^+]{\text{1) } LiAlH_4} (CH_3)_2CHCH_2CH_2CH_2OH \xrightarrow[H_2SO_4]{\text{conc. HBr}}$$

4-methylpentanoic acid

$$(CH_3)_2CHCH_2CH_2CH_2Br \xrightarrow[\text{ether}]{Mg} (CH_3)_2CHCH_2CH_2CH_2MgBr \xrightarrow{H_3O^+} (CH_3)_2CHCH_2CH_2CH_3$$

2-methylpentane

(e)

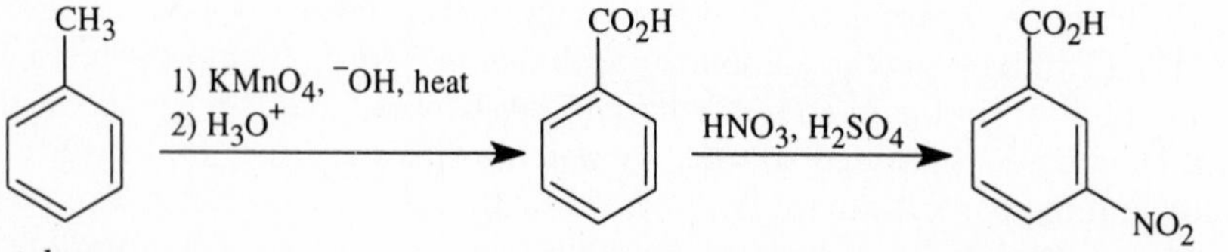
CH3
1) KMnO4, OH, heat
2) H3O+
CO2H
HNO3, H2SO4
CO2H
NO2

toluene

*m*-nitrobenzoic acid

(g)

PhCO$_2$H $\xrightarrow[\text{2) H}_3\text{O}^+]{\text{1) LiAlH}_4}$ PhCH$_2$OH $\xrightarrow{\text{conc. HBr}}$ PhCH$_2$Br $\xrightarrow[\text{ether}]{\text{Mg}}$ PhCH$_2$MgBr $\xrightarrow[\text{2) H}_2\text{SO}_4]{\text{1) O=CHCH}_2\text{Ph}}$

benzoic acid

PhCH=CHCHPh $\xrightarrow{\text{H}_2, \text{cat.}}$ PhCH$_2$CH$_2$CH$_2$Ph

1,3-diphenylpropane

⯮  Although most chemists would call the product in part (g) 1,3-diphenylpropane, the rigorously correct substitutive name for this compound is (3-phenylpropyl)benzene.

(h)

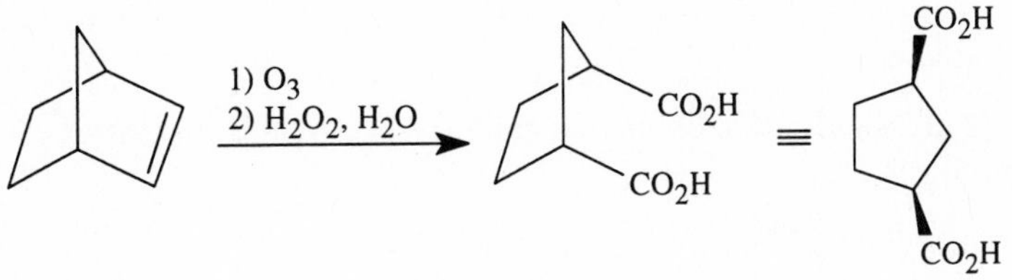

(i)

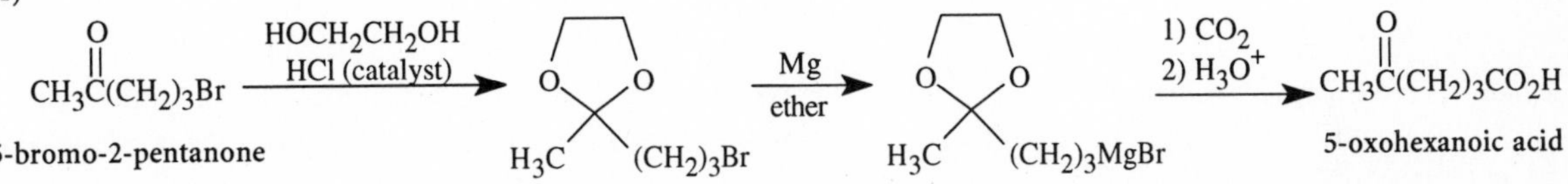

CH$_3$C(CH$_2$)$_3$Br $\xrightarrow[\text{HCl (catalyst)}]{\text{HOCH}_2\text{CH}_2\text{OH}}$   ...   $\xrightarrow[\text{ether}]{\text{Mg}}$   ...   $\xrightarrow[\text{2) H}_3\text{O}^+]{\text{1) CO}_2}$ CH$_3$C(CH$_2$)$_3$CO$_2$H

5-bromo-2-pentanone          H$_3$C   (CH$_2$)$_3$Br          H$_3$C   (CH$_2$)$_3$MgBr          5-oxohexanoic acid

**20.46** (a) The reagent involved is essentially a diazomethane molecule in which the hydrogens of diazomethane have been replaced by phenyl groups. Just as diazomethane forms methyl esters, this reagent forms a diphenylmethyl ester.

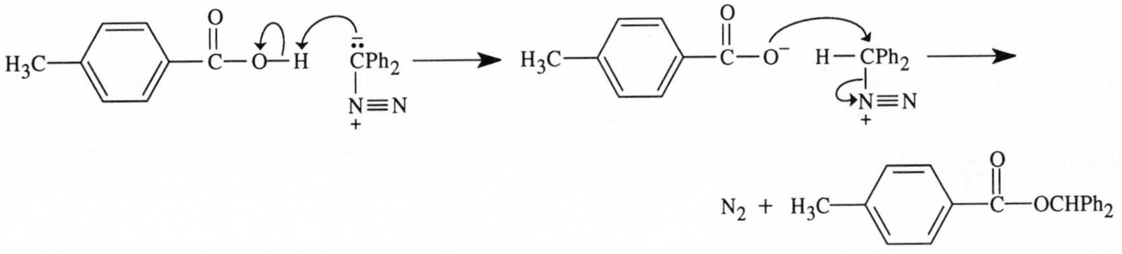

diphenylmethyl 4-methylbenzoate

(c) Ethylene glycol is a diol and the acid, terephthalic acid (1,4-benzenedicarboxylic acid), is a dicarboxylic acid. Each end of both the diacid and the diol can be esterified; the result is a polymeric ester, polyethylene terephthalate. You may be wearing some of this compound, which is known in the industrial world as *polyester*.

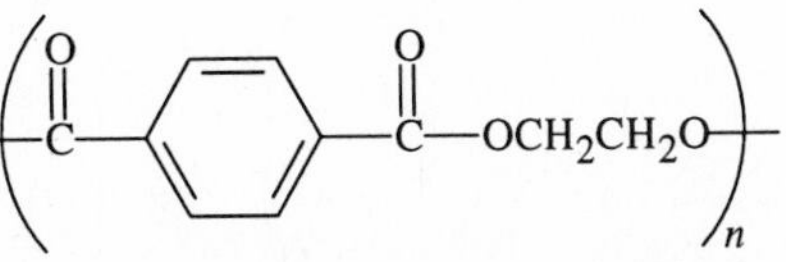

polyethylene terephthalate
(polyester)

(e) (In early printings of the text, this problem was not asterisked but should have been.) The conjugate-base carboxylate of propionic acid acts as the nucleophile in a ring-opening reaction of the epoxide. Because the conditions are basic, attack occurs at the less branched carbon of the epoxide. Note that KOH reacts *much* more rapidly with the acidic hydrogen of the acid than it does with the epoxide; because only one equivalent of KOH is present, reaction of hydroxide ion with the epoxide is not a competing reaction.

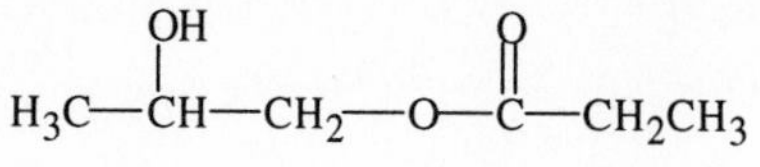

$$H_3C-\overset{\overset{\displaystyle OH}{|}}{CH}-CH_2-O-\overset{\overset{\displaystyle O}{\|}}{C}-CH_2CH_3$$

(f)  This is an intramolecular variation of the esterification shown in Eq. 20.22, text p. 951. In this case, potassium carbonate converts the carboxylic acid into its conjugate-base potassium carboxylate, which is then intramolecularly alkylated to form the cyclic ester (lactone).

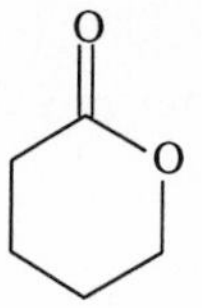

**20.47**  (a)  The two different compounds are the diastereomers that result respectively from loss of each of the two carboxy groups from compound *A*.

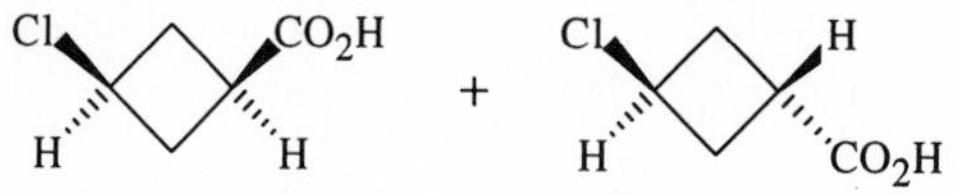

**20.48**  (a)  The first ionization of squaric acid has the lower p$K_a$, just as the first ionization of a dicarboxylic acid has a lower p$K_a$ than the second. (For the reason, see the solution to Problem 20.7(b) on p. 506 of this manual.)

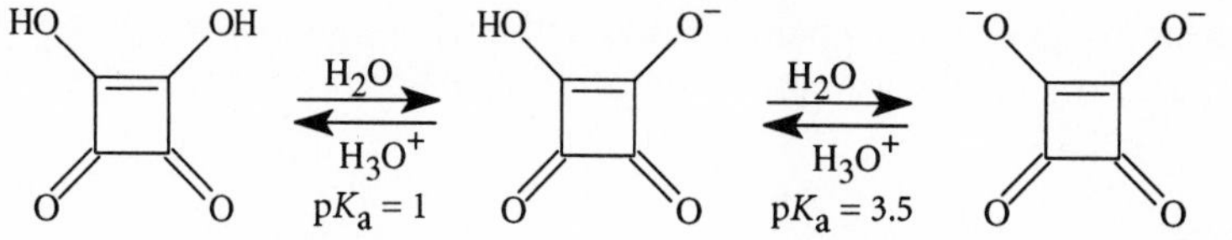

(b)  One reason for the greater acidity of squaric acid is the electron-withdrawing polar effect of the two carbonyl groups, which stabilizes the anions. Another reason for the greater acidity of squaric acid is the resonance stabilization of the conjugate-base anions that results from interaction of the unshared pairs on each of the oxygens with a carbonyl group.

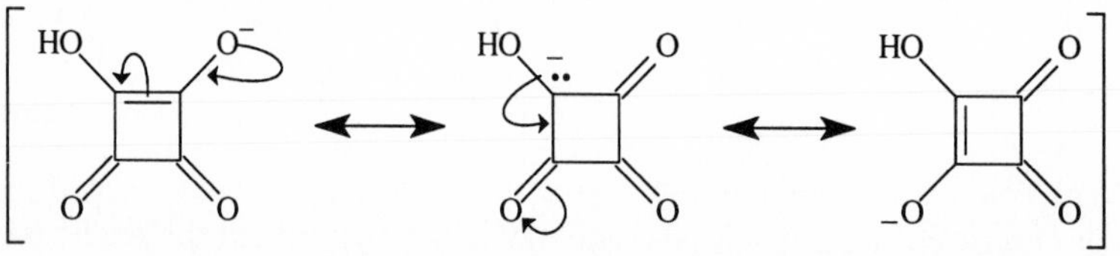

In the conjugate base of an enol, in contrast, the stabilizing polar effect of a carbonyl oxygen is not present, and the only resonance stabilization results from interaction of the electron pair with the double bond.

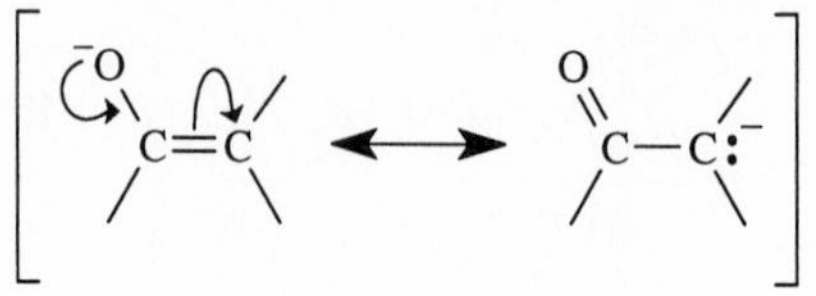

conjugate base of an enol

**20.49**  The hint refers to the fact that organolithium reagents are strong bases and react instantaneously with carboxylic acids to give their conjugate-base lithium carboxylate salts:

$$R-\overset{\overset{\displaystyle O}{\|}}{C}-O-H \quad CH_3-Li \longrightarrow R-\overset{\overset{\displaystyle O}{\|}}{C}-O^-\,Li^+ \;+\; CH_4$$

A second equivalent of organolithium reagent undergoes a carbonyl addition with the carboxylate salt to give compound *A*, which is the conjugate-base di-anion of a ketone hydrate. Since compound *A* does not contain a good leaving group, it is stable until acid is added. Protonation forms the hydrate, which then decomposes to the ketone by the reverse of the mechanism shown for hydrate formation in Eqs. 19.18a–b on text pp. 885–886.

$$R-\overset{\overset{\displaystyle O}{\|}}{\underset{\underset{\displaystyle CH_3-Li}{|}}{C}}-O^-\,Li^+ \longrightarrow R-\overset{\overset{\displaystyle O^-\,Li^+}{|}}{\underset{\underset{\displaystyle CH_3}{|}}{C}}-O^-\,Li^+ \xrightarrow{H-\overset{+}{O}H_2} R-\overset{\overset{\displaystyle O^-\,Li^+}{|}}{\underset{\underset{\displaystyle CH_3}{|}}{C}}-OH \xrightarrow{H-\overset{+}{O}H_2}$$

*A*

$$+\;OH_2$$

$$OH_2 \;+\; R-\overset{\overset{\displaystyle OH}{|}}{\underset{\underset{\displaystyle CH_3}{|}}{C}}-OH \xrightarrow{\text{(see Eqs. 19.18a–b)}} R-\overset{\overset{\displaystyle O}{\|}}{C}-CH_3 \;+\; H_2O$$

**20.51** (a) Orthoesters have the same relationship to esters that acetals have to ketones, and the hydrolysis mechanisms of both orthoesters and acetals are virtually identical.

$$CH_3\overset{\overset{\displaystyle OC_2H_5}{|}}{\underset{\underset{\displaystyle OC_2H_5}{|}}{C}}-OC_2H_5 \xrightarrow{H-\overset{+}{O}H_2} CH_3\overset{\overset{\displaystyle +OC_2H_5}{|}}{\underset{\underset{\displaystyle OC_2H_5}{|}}{C}}-OC_2H_5 \longrightarrow CH_3\overset{\overset{\displaystyle OH_2}{}}{\underset{\underset{\displaystyle OC_2H_5}{|}}{\overset{+}{C}}}-OC_2H_5 \longrightarrow CH_3\overset{\overset{\displaystyle +OH}{|}}{\underset{\underset{\displaystyle OC_2H_5}{|}}{C}}-OC_2H_5 \xrightarrow{OH_2}$$

$$+\;OH_2 \qquad\qquad +\;HOC_2H_5$$

$$CH_3\overset{\overset{\displaystyle OH}{|}}{\underset{\underset{\displaystyle OC_2H_5}{|}}{C}}-OC_2H_5 \xrightarrow{H-\overset{+}{O}H_2} CH_3\overset{\overset{\displaystyle OH\;H}{|\;|}}{\underset{\underset{\displaystyle OC_2H_5}{|}}{C}}-\overset{+}{O}C_2H_5 \longrightarrow CH_3\overset{\overset{\displaystyle +O-H}{|}}{\underset{\underset{\displaystyle OC_2H_5}{|}}{C}} \quad OH_2 \longrightarrow CH_3\overset{\overset{\displaystyle O}{\|}}{C}-OC_2H_5$$

$$+\;OH_2 \qquad\qquad +\;HOC_2H_5 \qquad\qquad\qquad +\;H-\overset{+}{O}H_2$$

(c) Sulfuric acid protonates the alkene to give a carbocation, which is attacked by carbon monoxide to give an acylium ion (Eq. 16.13, text p. 754). This ion, in turn, is attacked by water to give, after appropriate proton transfers, the carboxylic acid.

$$(CH_3)_2C=CH_2 \xrightarrow{H-OSO_3H} (CH_3)_3\overset{+}{C} \quad :\!\overset{-}{C}\!\equiv\!\overset{+}{O}\!: \longrightarrow (CH_3)_3C-\overset{+}{C}\!\equiv\!O\!: \xrightarrow{:\overset{..}{O}H_2}$$

$$^-OSO_3H \qquad\qquad \text{an acylium ion}$$

$$(CH_3)_3C-\overset{\overset{\displaystyle H\overset{..}{O}-H}{|}}{\underset{}{\overset{+}{C}}}=\overset{..}{O}\!: \xrightarrow{:\overset{..}{O}H_2} (CH_3)_3C-\overset{\overset{\displaystyle \overset{..}{H}O\!:}{|}}{C}=\overset{..}{O}\!: \;+\; H-\overset{+}{O}H_2$$

(e) Acid-promoted opening of the epoxide gives a tertiary and benzylic carbocation, which decarboxylates to the enol. The enol then forms the aldehyde by a mechanism like that shown in Eqs. 14.6d–e on text p. 655.

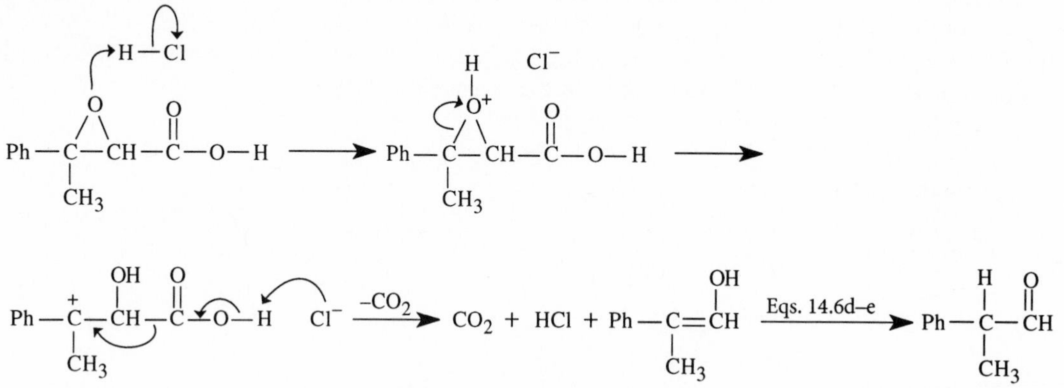

The other C—O bond of the epoxide does not rupture because (1) the carbocation that would result is secondary, and (2) carbocations $\alpha$ to carbonyl groups are very unstable. (The latter point is considered in Chapter 22 on text p. 1054.)

20.52 The analytical data dictate an empirical formula of $C_3H_5$, or a minimum molecular formula of $C_6H_{10}$. A hydrocarbon cannot have an odd mass; hence, the peak in the mass spectrum at $m/z = 67$ cannot be the molecular ion. Given that the unsaturation number of compound *A* is 2, and that compound *B* is a dicarboxylic acid, compound *A* must contain a double bond within a ring. Calculate the number of moles of carboxy groups titrated as follows:

$$\text{moles of carboxy groups} = (13.7\ \text{mL})(0.100\ \text{mol/L})(1.00 \times 10^{-3}\ \text{L/mL}) = 1.37 \times 10^{-3}$$

Noting that there are two carboxy groups per mole of *B*, the molecular mass of compound *B* is calculated as follows:

$$\text{molecular mass of } B = \frac{(100 \times 10^{-3}\ \text{g})(2.00 \text{ moles of carboxy groups/mole of } B)}{(1.37 \times 10^{-3} \text{ moles of carboxy groups})} = 146\ \text{g/mole}$$

From the following transformations, the atomic masses of the four oxygens introduced, and the molecular mass of compound *B*, the molecular mass of compound *A* can be deduced as follows:

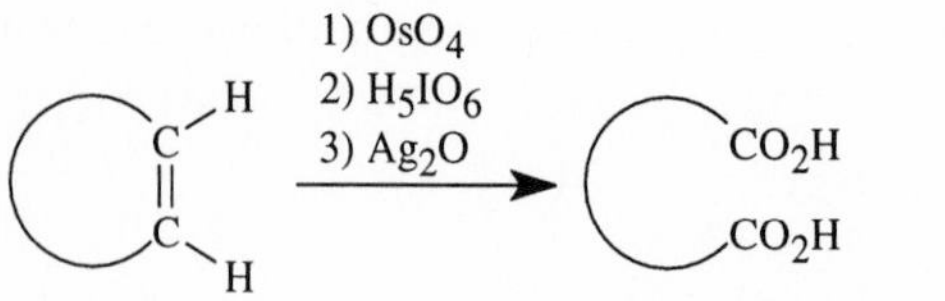

*A*
molecular mass = $x$

*B*
molecular mass = $x + 64 = 146$

$x = 146 - 64 = 82$

If the molecular mass of compound *A* is 82, then the $m/z = 67$ peak in its mass spectrum corresponds to loss of 15 mass units, which corresponds to the loss of a methyl group. Evidently, compounds *A* and *B* contain a ring with a methyl branch. This branch is positioned such that there are three allylic hydrogens (from the NMR integration data) and it is also positioned such that dicarboxylic acid *B* is chiral. Compound *A* is 3-methylcyclopentene.

*A*                                      *B*

3-methylcyclopentene          2-methylpentanedioic acid

**20.53**  (a)  The presence of a carboxy group indicated by the IR data is confirmed by the absorption for the —OH proton at $\delta$ 11.34. The two triplets indicate —$CH_2CH_2$—; the chemical shift of the more downfield triplet suggests that one carbon is bound to an oxygen, and the chemical shift of the other suggests that it is bound either to the carboxylic acid group or to the benzene ring and that it is probably $\beta$ to the oxygen. The resonances near $\delta$ 7 integrate for five hydrogens; hence, the compound contains a monosubstituted benzene ring. The unknown is 3-phenoxypropanoic acid.

3-phenoxypropanoic acid

(c)  The IR and NMR spectra indicate that this compound is a carboxylic acid, and the UV spectrum indicates that the carboxy group is conjugated with a double or triple bond. If the $\delta$ 11.79 resonance corresponds to one proton, then the resonances in the $\delta$ 5.5–7.5 region integrate for 2 protons; hence, there are two vinylic protons. The mass accounted for by a carboxy group (45 mass units) and a carbon-carbon double bond with two vinylic protons (26 mass units) is 71 units. Given that the $m/z = 86$ peak in the mass spectrum is the molecular ion, only 15 mass units remain to be accounted for. This can be accounted for by a methyl group and, indeed, the doublet of doublets near $\delta$ 2.0 integrates for three protons. The unknown is (*E*)-2-butenoic acid; notice that the *trans*, or *E*, stereochemistry at the double bond is indicated both by the 974 cm$^{-1}$ absorption in the IR spectrum and by the 15 Hz separation between the two packets of lines in the NMR spectrum centered at $\delta$ 5.9.

(*E*)-2-butenoic acid

**20.54**  (a)  The even mass of the $m/z = 74$ ion shows that the ion is an odd-electron ion. McLafferty rearrangement is a common mechanism for formation of such ions in carbonyl compounds. (See Sec. 19.3E, text pp. 878–879.)

molecular ion

$m/z = 74$

**20.55**  (a)  The presence of two exchangeable hydrogens suggests that compound *A* is a dicarboxylic acid. The CMR chemical shifts suggest the presence of a methyl group, a carbonyl group, and a carbon bound to a carbonyl group. Because compound *A* is a dicarboxylic acid, each CMR resonance must correspond to two carbons. Compound *A* is 2,3-dimethylbutanedioic acid; the fact that it can be resolved into enantiomers shows that it is the racemate rather than the *meso* diastereoisomer.

*(solution continues)*

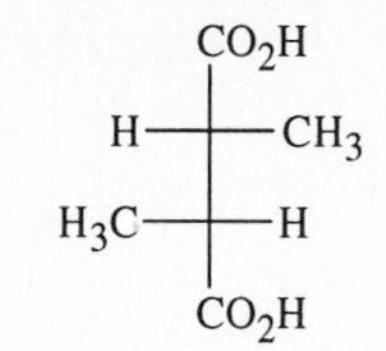

$$CO_2H$$
$$H\text{——}CH_3$$
$$H_3C\text{——}H$$
$$CO_2H$$

compound *A*
(±)-2,3-dimethylbutanedioic acid
(one of two enantiomers)

20.56    The titration data suggest that compound *A* is both a carboxylic acid and a phenol, and the IR spectrum shows both —OH and carbonyl absorptions. The change in the UV spectrum at high pH is also consistent with the presence of a group that ionizes at high pH and is conjugated with a benzene ring, namely, the phenolic —OH group. The resonances near $\delta$ 7 indicate a *para*-disubstituted benzene ring. The packet of resonances near $\delta$ 2.7 integrates for four hydrogens, and they must have very similar chemical shifts. Add the masses of the groups known to be present: the *para*-disubstituted benzene ring (76 units), the phenol —OH group (17 units), and the carboxy group (45 units). On the assumption that the $m/z = 166$ peak in the mass spectrum is the molecular ion, 28 mass units remain to be accounted for; either a $H_3C$—CH group or a —$CH_2CH_2$— group is consistent with this remaining mass. Since the NMR spectrum gives no indication of a methyl group (that is, a three-proton resonance), the latter alternative is more reasonable. Compound *A* is 3-(4-hydroxyphenyl)propanoic acid.

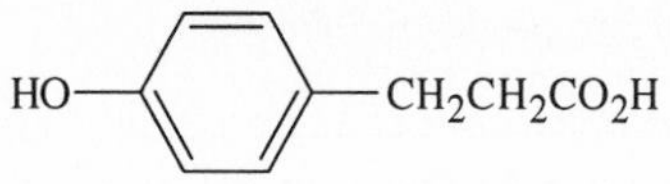

$$HO\text{——}\langle\bigcirc\rangle\text{——}CH_2CH_2CO_2H$$    3-(4-hydroxyphenyl)propanoic acid
(compound *A*)

We might have expected two triplets for the —$CH_2CH_2$— group; evidently, the chemical shifts of the two sets of protons are sufficiently similar that extensive non-first order splitting is observed. (See the discussion of non-first order splitting in Sec. 13.4B on text pp. 607–609.)

   *Rationalization of the mass spectrum:* The peak at $m/z = 166$ is the molecular ion, as hypothesized above. The peak at $m/z = 107$ corresponds to loss of 59 mass units, which corresponds to loss of a —$CH_2CO_2H$ group as a neutral fragment to leave a *p*-hydroxybenzyl cation:

$$HO\text{——}\langle\bigcirc\rangle\text{——}CH_2\text{—}CH_2\text{—}\overset{+}{\underset{\parallel}{C}}\text{—}OH \longrightarrow HO\text{——}\langle\bigcirc\rangle\text{——}\overset{+}{C}H_2 \ + \ CH_2{=}\overset{\cdot\cdot\,}{\underset{}{C}}\text{—}OH$$

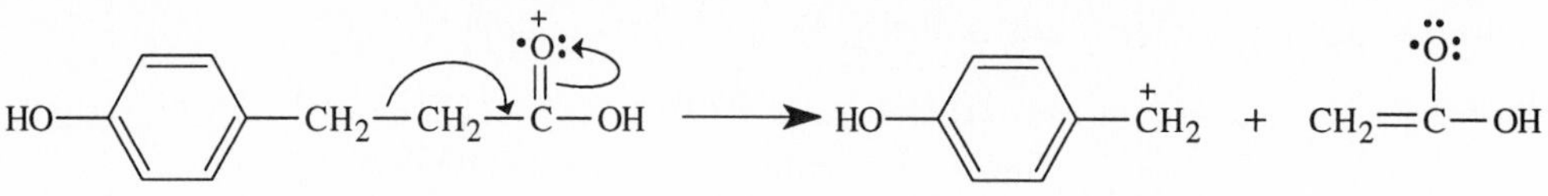

molecular ion

*p*-hydroxybenzyl cation
$m/z = 107$

# Chemistry of Carboxylic Acid Derivatives

## Terms

## Concepts

### I. Nomenclature and Classification of Carboxylic Acid Derivatives

#### A. INTRODUCTION TO CARBOXYLIC ACID DERIVATIVES

1. Carboxylic acid derivatives are compounds that can be hydrolyzed under acidic or basic conditions to give the related carboxylic acids.
2. Carboxylic acids and their derivatives have both structural and chemical similarities.

    a. With the exception of nitriles, all carboxylic acid derivatives contain a carbonyl group.

    b. Many important reactions of these compounds occur at the carbonyl group.

    c. The cyano group (—C≡N) of nitriles has reactivity that resembles that of a carbonyl group.

## B. ESTERS AND LACTONES

1. In both common and systematic nomenclature, esters are named as derivatives of their parent carboxylic acids by applying a variation of the system used in naming carboxylate salts.

    a. The group attached to the carboxylate oxygen is named first as a simple alkyl or aryl group.

    b. This name is followed by the name of the parent carboxylate constructed by dropping the final *ic* from the name of the acid and adding the suffix *ate*.

2. In common nomenclature:

    a. Substitution is indicated by numbering the acid portion of the ester as in carboxylic acid nomenclature, beginning with the adjacent carbon as the α-position.

    b. The alkyl or aryl group is numbered (using Greek letters) from the point of attachment to the carboxylate oxygen.

3. In substitutive nomenclature:

    a. Substitution is indicated by numbering the acid portion of the ester as in carboxylic acid nomenclature, beginning with the carbonyl as carbon-1.

    b. The alkyl or aryl group is numbered (using numbers) from the point of attachment to the carboxylate oxygen.

4. Cyclic esters are called lactones. In common nomenclature:

    a. The name of a lactone is derived from the acid with the same number of carbons in its principal chain.

    b. The ring size is denoted by a Greek letter corresponding to the point of attachment of the lactone ring oxygen to the carbon chain.

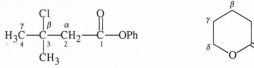

*common name:* phenyl β-chloroisovalerate      δ-valerolactone

*substitutive name:* phenyl 3-chloro-3-methylbutanoate

## C. ACID HALIDES

1. Acid halides are named in any system of nomenclature by replacing the *ic* ending of the acid with the suffix *yl*, followed by the name of the halide.

2. When the acid halide group is attached to a ring, the compound is named as an alkanecarbonyl halide.

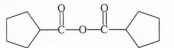

*common name:* δ-chloro-β,β-dimethylvaleryl chloride

*substitutive name:* 5-chloro-3,3-dimethylpentanoyl chloride     3-methoxy-1-cyclopentanecarbonyl bromide

## D. ANHYDRIDES

1. To name an anhydride, the name of the parent acid is replaced by the word *anhydride*.

2. Mixed anhydrides, derived from two different acids, are named by citing the two parent acids in alphabetical order.

cyclopentanecarboxylic anhydride      acetic benzoic anhydride (a mixed anhydride)

## E. NITRILES

1. In the common system, nitriles are named by dropping the *ic* or *oic* from the name of the acid with the same number of carbon atoms (counting the nitrile carbon) and adding the suffix *onitrile* (the name of the three-carbon nitrile is shortened in common nomenclature to propionitrile).

2. In substitutive nomenclature, the suffix *nitrile* is added to the name of the hydrocarbon with the same number of carbon atoms.

3. When the nitrile group is attached to a ring, a special *carbonitrile* nomenclature is used.

I—CH₂—CH₂—C≡N

*common name:* β-iodopropionitrile

*substitutive name:* 3-iodopropanenitrile

1-methylcyclopropanecarbonitrile

## F. AMIDES, LACTAMS, AND IMIDES

1. Simple amides are named in any system by replacing the *ic* or *oic* suffix of the acid name with the suffix *amide*.

2. When the amide functional group is attached to a ring, the suffix *carboxamide* is used.

3. Amides are classified by the number of hydrogens attached to the amide nitrogen:
   a. Primary—two hydrogens attached.
   b. Secondary—one hydrogen attached.
   c. Tertiary—no hydrogens attached.

4. Substitution on nitrogen in secondary and tertiary amides is designated with the letter *N* (italicized or underlined).

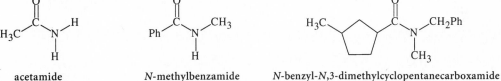

**acetamide**
(a primary amide)

*N*-methylbenzamide
(a secondary amide)

*N*-benzyl-*N*,3-dimethylcyclopentanecarboxamide
(a tertiary amide)

5. Cyclic amides are called lactams. In common nomenclature:
   a. The name of a lactam is derived from the acid with the same number of carbons in its principal chain.
   b. The ring size is denoted by a Greek letter corresponding to the point of attachment of the lactam ring nitrogen to the carbon chain.

penicillin G

a β-lactam

6. Imides are formally the nitrogen analogs of anhydrides (cyclic imides are of greater importance than open-chain imides).

succinimide
(an imide)

## G. NOMENCLATURE OF SUBSTITUENT GROUPS

1. The priorities for citing principal groups in a carboxylic acid derivative are as follows (a complete list of functional group priorities is given in Appendix I, text page A-1):

acid > anhydride > ester > acid halide > amide > nitrile

2. The names used for citing these groups as substituents are:

| | | |
|---|---|---|
| —C(=O)—OH  carboxy | —C(=O)—OCH₃  methoxycarbonyl | —C(=O)—OCH₂CH₃  ethoxycarbonyl |
| —C(=O)—Cl  chloroformyl | —C(=O)—NH₂  carbamoyl | —C≡N  cyano |
| —CH₂—C(=O)—OH  carboxymethyl | —O—C(=O)—CH₃  acetoxy or acetyloxy[a] | —NH—C(=O)—CH₃  acetamido or acetylamino[a] |

[a]Used by Chemical Abstracts.

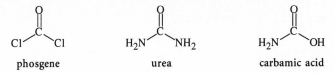

N≡C—CH₂—CH₂—C(=O)—N(CH₃)—CH₂—CH₂—CH₃    3-cyano-*N*-methyl-*N*-propylpropanamide

## H. Carbonic Acid Derivatives

1. Esters of carbonic acid are named like other esters.
2. Other important carbonic acid derivatives have special names that must be learned:

| | | |
|---|---|---|
| Cl—C(=O)—Cl | H₂N—C(=O)—NH₂ | H₂N—C(=O)—OH |
| phosgene | urea | carbamic acid |

# II. Structure, Basicity, and Physical Properties of Carboxylic Acid Derivatives

## A. Basicity of Carboxylic Acid Derivatives

1. Carboxylic acid derivatives are weakly basic at the carbonyl oxygen (conjugate-acid $pK_a \approx -4$), and nitriles are weakly basic at nitrogen (conjugate-acid $pK_a \approx -10$).
   a. The basicity of an ester is about the same as the basicity of the corresponding carboxylic acid.
   b. Amides are considerably more basic than other carboxylic acid derivatives (a reflection of the reduced electronegativity of nitrogen relative to oxygen).
   c. Both esters and amides protonate on the carbonyl oxygen.
2. Nitriles are very weak bases.

## B. Physical Properties of Carboxylic Acid Derivatives

1. Esters are polar molecules, but the ester functional group lacks the capability of acids to donate hydrogen bonds.
   a. Most esters are not soluble in water.
   b. Lower esters are typically volatile, fragrant liquids that have lower densities than water.
2. Most of the lower anhydrides and acid chlorides are dense, water-insoluble liquids with acrid, piercing odors.
   a. The boiling points are not very different from those of other polar molecules of about the same molecular mass and shape.
   b. The simplest anhydride, formic anhydride, and the simplest acid chloride, formyl chloride, are unstable and cannot be isolated under ordinary conditions.
3. Nitriles are among the most polar organic compounds.
   a. Their polarities are reflected in their boiling points, which are rather high, despite the absence of hydrogen bonding.
   b. Acetonitrile is miscible with water and propionitrile has a moderate solubility in water, whereas higher nitriles are insoluble in water.
4. Primary and secondary amides, like carboxylic acids, tend to associate into hydrogen-bonded dimers or higher aggregates in the solid state, in the pure liquid state, or in solvents that do not form hydrogen bonds.
   a. This property is of substantial biological importance in the structure of proteins.

b. With increased substitution at nitrogen:
   i. The capacity for hydrogen bonding is reduced.
   ii. Boiling points decrease in spite of the increase in molecular mass.
c. The lower amides are water-soluble, polar molecules with high boiling points; a number of amides have high dielectric constants.

## C. Structure of Amides

1. In an amide, the bonds to nitrogen have essentially trigonal geometry, which can be understood on the basis of resonance structures in which the bond between the nitrogen and the carbonyl carbon has significant double-bond character.

$$\left[ \underset{NR_2}{\overset{O}{\underset{\shortmid}{C}}} \longleftrightarrow \underset{\overset{+}{N}R_2}{\overset{O^-}{\underset{\shortmid}{C}}} \right]$$

2. Secondary and tertiary amides can exist in both *E* and *Z* conformations about the carbonyl-nitrogen bond.
   a. The *Z* conformation predominates in most secondary amides because, in this form, van der Waals repulsions between the largest groups are avoided.
   b. The interconversion of *E* and *Z* forms of amides is too rapid at room temperature to permit their separate isolation, but is very slow compared to the rotation about ordinary carbon-carbon single bonds.

*N*-ethylpropionamide
*Z* conformation

*N*-ethylpropionamide
*E* conformation

   c. The relatively low rate of internal rotation about the carbon-nitrogen bond is caused by the significant double-bond character in this bond.

## III. Spectroscopy of Carboxylic Acid Derivatives

## A. IR Spectroscopy

1. The most important feature in the IR spectra of most carboxylic acid derivatives is the C=O stretching absorption or, for nitriles, the C≡N stretching absorption.
2. Some noteworthy IR absorption trends of carboxylic acid derivatives are as follows (see Table 21.3, text page 983, for a complete listing):
   a. Esters are readily differentiated from carboxylic acids, aldehydes, or ketones by the unique ester carbonyl absorption at 1735 cm$^{-1}$.
   b. Lactones, lactams, and cyclic anhydrides, like cyclic ketones, have carbonyl absorption frequencies that increase dramatically as the ring size decreases.
   c. Anhydrides and some acid chlorides have two carbonyl absorptions; the two carbonyl absorptions of anhydrides are due to the symmetrical and unsymmetrical stretching vibrations of the carbonyl group.
   d. The carbonyl absorption of amides occurs at much lower frequency than that of other carbonyl compounds.
   e. The C≡N stretching absorption of nitriles generally occurs in the triple-bond region of the spectrum; this absorption is stronger, and occurs at higher frequency, than the C≡C absorption of an alkyne.
   f. Primary and secondary amides show an N—H stretching absorption in the 3200–3400 cm$^{-1}$ and 1640 cm$^{-1}$ regions of the spectrum.
      i. Many primary amides show two N—H absorptions.
      ii. Secondary amides show a single strong N—H absorption.
      iii. A strong N—H bending absorption occurs in the vicinity of 1640 cm$^{-1}$.
      iv. Tertiary amides lack both of these N—H absorptions.

### B. NMR Spectroscopy

1. The $\alpha$-proton resonances of all carboxylic acid derivatives are observed in the $\delta\,1.9$–3 region of the NMR spectrum.
2. In esters, the chemical shifts of protons on the carbon adjacent to the carboxylate oxygen occur at about 0.6 ppm lower field than the analogous protons in alcohols and ethers; this shift is attributed to the electron-attracting character of the carbonyl group.
3. NMR characteristics of amides:
   a. The *N*-alkyl protons of amides have chemical shifts in the $\delta\,2.6$–3 region; the N—H proton resonances of primary and secondary amides (sometimes broad) are observed in the $\delta\,7.5$–8.5 region.
   b. The broadening of these protons signals is caused by a slow chemical exchange with the protons of other protic substances and by unresolved splitting with $^{14}N$, which has a nuclear spin.
   c. Amide N—H resonances can be washed out by exchange with $D_2O$.
4. In carbon NMR (CMR) spectra, the carbonyl chemical shifts of carboxylic acid derivatives are in the range $\delta\,165$–180, very much like those of carboxylic acids.
   a. The chemical shifts of nitrile carbons are considerably smaller, occurring in the $\delta\,115$–120 range.
   b. These shifts are much greater, however, than those of acetylenic carbons.

## IV. *Use and Occurrence of Carboxylic Acids and Their Derivatives*

### A. Nylon and Polyesters

1. Two of the most important polymers produced on an industrial scale are nylon and polyesters.
2. Nylon is the general name given to a group of polymeric amides.
   a. Nylon is an example of a condensation polymer, a polymer formed in a reaction that liberates a small molecule, in this case $H_2O$.
   b. In contrast, an addition polymer such as polystyrene is formed by the simple addition of one molecule to another.
3. Polyesters are condensation polymers derived from the reactions of diols and dicarboxylic acids.

### B. Waxes, Fats, and Phospholipids

1. Waxes, fats, and phospholipids are all important naturally occurring ester derivatives of fatty acids.
2. A wax is an ester of a fatty acid and a fatty alcohol (an alcohol with a long unbranched carbon chain).

$$CH_3CH_2CH_2CH_2CH_2CH_2CH_2CH_2CH_2CH_2CH_2CH_2CH_2CH_2CH_2\underset{\underset{CH_3CH_2CH_2CH_2CH_2CH_2CH_2CH_2CH_2CH_2CH_2CH_2CH_2CH_2}{\overset{O}{|}}}{\overset{\displaystyle O}{\underset{\displaystyle}{C}}}$$

a wax

3. A fat is an ester derived from a molecule of glycerol and three molecules of fatty acid.
   a. The three acyl groups in a fat may be the same or different, and they may contain unsaturation, which is typically in the form of one or more *cis* double bonds.

a fat

   b. Fats with no double bonds, termed saturated fats, are typically solids.
   c. Fats containing double bonds, termed unsaturated fats, are in many cases oily liquids.
4. Phospholipids are closely related to fats, because they too are esters of glycerol.
   a. Phospholipids differ from fats in that one of the terminal oxygens of glycerol in a phospholipid is esterified to a special type of organic phosphate derivative that constitutes the polar head group of the molecule.

b. Phospholipids closely resemble soaps because both types of molecules are amphipathic; that is, they both have polar and nonpolar ends.

c. Phospholipids, along with cholesterol (a steroid) and imbedded proteins, make up the phospholipid bilayers of cell membranes, the envelopes that surround living cells.

$$CH_3CH_2CH_2CH_2CH_2CH_2CH_2CH_2CH_2CH_2CH_2CH_2CH_2{-}\overset{\displaystyle O}{\overset{\displaystyle \|}{C}}{-}O$$

$$CH_3CH_2CH_2CH_2CH_2CH_2CH_2CH_2CH_2CH_2CH_2CH_2CH_2{-}\overset{\displaystyle O}{\overset{\displaystyle \|}{C}}{-}O{-}CH$$

$$(CH_3)_3\overset{+}{N}CH_2CH_2{-}O{-}\overset{\displaystyle O}{\underset{\displaystyle O^-}{\overset{\displaystyle \|}{P}}}{-}O$$

a phospholipid

## V. *Reactivity of Carboxylic Acid Derivatives*

### A. GENERAL

1. The reactions of carboxylic acid derivatives can be categorized as follows:
   a. Reactions of the carbonyl group (or cyano group of a nitrile).
      i. Reactions at the carbonyl oxygen or cyano nitrogen.
      ii. Reactions at the carbonyl carbon or cyano carbon, the most common being reactions with nucleophiles, which include:

      > water (hydrolysis) or hydroxide ion (saponification)
      >
      > alcohols or alkoxides (alcoholysis, or transesterification [in the case of esters])
      >
      > amines (aminolysis)
      >
      > hydrides (reduction)
      >
      > organometallic reagents

   b. Reactions involving the $\alpha$-carbon.
   c. Reactions at the nitrogen of amides.

2. One carbonyl-group reaction of acid derivatives is the reaction of the carbonyl oxygen (or the nitrile nitrogen) as a base, which often serves as the first step in acid-catalyzed reactions.

3. The major carbonyl-group reaction of carboxylic acid derivatives is substitution at the carbonyl carbon, also called acyl substitution.

4. The $C{\equiv}N$ bond behaves chemically much like a carbonyl group.

5. All carboxylic acid derivatives have in common the fact that they undergo hydrolysis to yield carboxylic acids.

### B. MECHANISMS OF NUCLEOPHILIC ACYL SUBSTITUTION REACTIONS

1. Nucleophilic acyl substitution reactions occur by two types of mechanisms:
   a. Under basic conditions:
      i. A nucleophile attacks the carbonyl carbon to give a tetrahedral addition intermediate in which the carbonyl oxygen assumes a negative charge.
      ii. The leaving group —X is then expelled from the tetrahedral intermediate.

$$\underset{\displaystyle Nuc:}{R{-}\overset{\displaystyle O}{\overset{\displaystyle \|}{C}}{-}X} \longrightarrow \underset{\displaystyle Nuc}{R{-}\overset{\displaystyle O^-}{\overset{\displaystyle |}{C}}{-}X} \longrightarrow R{-}\overset{\displaystyle O}{\overset{\displaystyle \|}{C}}{-}Nuc \ + \ X^-$$

   b. Under acidic conditions:
      i. The carbonyl is first protonated.
      ii. This carbon is then attacked by a nucleophile to form a tetrahedral addition intermediate.
      iii. After proton transfer to the leaving group —X, it is expelled as H—X.

$$R-\overset{O}{\overset{\|}{C}}-X \rightleftarrows R-\overset{\overset{+}{O}H \quad A^-}{\underset{:Nuc-H}{\overset{\|}{C}}}-X \rightleftarrows R-\overset{OH}{\underset{\overset{+}{Nuc}-H \quad A^-}{\overset{|}{C}}}-X \rightleftarrows R-\overset{OH}{\underset{Nuc}{\overset{|}{C}}}-X + H-A \rightleftarrows$$

$$A^- + R-\overset{OH}{\underset{Nuc}{\overset{|}{C}}}-\overset{+}{X}-H \rightleftarrows R-\overset{\overset{+}{O}-H \quad A^-}{\overset{\|}{C}}-Nuc \rightleftarrows R-\overset{O}{\overset{\|}{C}}-Nuc + HA$$
$$+ HX$$

## C. Relative Reactivities in Nucleophilic Acyl Substitution Reactions

1. The conditions under which the different carboxylic acid derivatives are hydrolyzed differ considerably.
   a. Hydrolysis reactions of amides and nitriles require heat as well as acid or base.
   b. Hydrolysis reactions of esters require acid or base but require shorter periods of heating.
   c. Hydrolysis reactions of acid chlorides and anhydrides occur rapidly at room temperature even in the absence of acid or base.
2. The trend in carbonyl reactivity of carboxylic acid derivatives parallels the trend in leaving group effectiveness; the practical significance of this reactivity order is that selective reactions are possible:

   carboxylic acid derivative reactivity:  acid chlorides  >  anhydrides  >>  esters, acids  >  amides  >  nitriles

   leaving group effectiveness (basic conditions):  $Cl^-$  >  $RCO_2^-$  >>  $RO^-, HO^-$  >  $R_2N^-$

   leaving group effectiveness (acidic conditions):  $Cl^-$  >  $RCO_2H$  >>  $ROH, HOH$  >  $R_2NH$

3. Relative reactivity is determined by the stability of each type of compound relative to its transition state for addition or substitution:
   a. The more a compound is stabilized, the less reactive it is.
      i. The major factor affecting the stability of the carbonyl compound is the resonance interaction of the unshared electron pairs of the group X with the carbonyl $\pi$ electrons.
      ii. Because oxygen is more electronegative than nitrogen, this resonance interaction is less important in an ester than in an amide; hence, amides are stabilized by resonance more than esters are, and are less reactive than esters.
   b. Reactivity is governed by the standard free energy of activation $\Delta G^{\circ\ddagger}$, the difference in the standard free energies of the transition state and the reactants.
      i. The major factor accounting for the differences in transition-state stability is the relative base strength of the different leaving groups —X; the best leaving groups are the weakest bases.
      ii. The more a transition state for nucleophilic addition or substitution is stabilized, the more reactive is the compound.
4. Reactions of nitriles in base are slower than those of other acid derivatives; the nitrogen is less electronegative than oxygen and accepts negative charge less readily.
5. Reactions of nitriles in acid are slower because of their extreme low basicities; the protonated form of a nitrile reacts with nucleophiles in acid solution, but there is little of this form present.

# Reactions

## I. *Hydrolysis of Carboxylic Acid Derivatives*

### A. Hydrolysis of Acid Chlorides and Anhydrides

1. Acid chlorides and anhydrides react rapidly with water, even in the absence of acids or bases.
2. The hydrolysis reactions of acid chlorides and anhydrides are almost never used for the preparation of carboxylic acids because these derivatives are themselves usually prepared from acids.

## B. BASIC HYDROLYSIS (SAPONIFICATION) OF ESTERS AND LACTONES

1. Ester hydrolysis in aqueous hydroxide is called saponification; the term saponification is sometimes used to refer to hydrolysis in base of any acid derivative.
2. The mechanism of ester saponification involves:
   a. Attack by the nucleophilic hydroxide anion to give a tetrahedral addition intermediate from which an alkoxide ion is expelled.
   b. The alkoxide ion thus formed reacts with the carboxylic acid to give the carboxylate salt and the alcohol.

3. The equilibrium in this reaction lies far to the right because the carboxylic acid is a much stronger acid than the liberated alcohol; saponification is effectively irreversible.
4. Many esters can be saponified with just one equivalent of ⁻OH, although an excess of ⁻OH is often used as a matter of convenience.
5. Saponification converts a lactone completely into the salt of the corresponding hydroxy acid; upon acidification, the hydroxy acid forms.

   a. If a hydroxy acid is allowed to stand in acidic solution, it comes to equilibrium with the corresponding lactone. This reaction is an acid-catalyzed, intramolecular esterification.
   b. Lactones containing five- and six-membered rings are favored at equilibrium over their corresponding hydroxy acids; those with ring sizes smaller than five or larger than six are less stable than their corresponding hydroxy acids.
6. Ester and lactone saponification are examples of acyl substitution.
   a. The mechanisms of these reactions are classified as nucleophilic acyl substitution mechanisms.
   b. In a nucleophilic acyl substitution reaction, the substituting group attacks the carbonyl carbon as a nucleophile.

## C. ACID-CATALYZED HYDROLYSIS OF ESTERS

1. Esters can be hydrolyzed to carboxylic acids in aqueous solutions of strong acids; in most cases this reaction is slower than base-promoted hydrolysis and must be carried out with an excess of water (in which esters are insoluble).
2. By the principle of microscopic reversibility, the mechanism of acid-catalyzed hydrolysis is the exact reverse of acid-catalyzed esterification.
   a. The ester is first protonated by the acid catalyst; protonation makes the carbonyl carbon more electrophilic by making the carbonyl oxygen a better acceptor of electrons.
   b. Water, as the nucleophile, attacks the carbonyl carbon and then loses a proton to give the tetrahedral intermediate.
   c. Protonation of the leaving oxygen converts it into a better leaving group.
   d. Loss of the protonated leaving group gives a protonated carboxylic acid, from which a proton is removed to give the carboxylic acid itself.

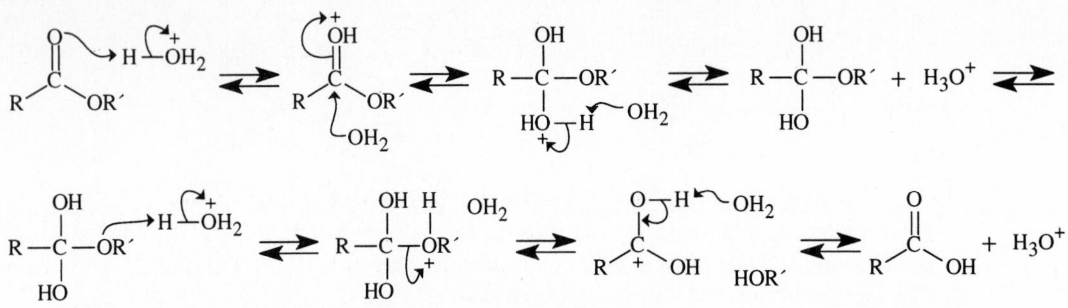

[Reaction mechanism scheme for acid-catalyzed ester hydrolysis]

3. Saponification, followed by acidification, is a much more convenient method for hydrolysis of most esters than acid-catalyzed hydrolysis because:
   a. It is faster.
   b. It is irreversible.
   c. It can be carried out not only in water but also in a variety of solvents, even alcohols.
4. Ester hydrolysis is an example of nucleophilic acyl substitution.

## D. HYDROLYSIS OF AMIDES

1. Amides can be hydrolyzed to carboxylic acids and ammonia or amines by heating them in acidic or basic solution.
2. In acid, protonation of the ammonia or amine by-product drives the hydrolysis equilibrium to completion; the amine can be isolated, if desired, by addition of base to the reaction mixture following hydrolysis.

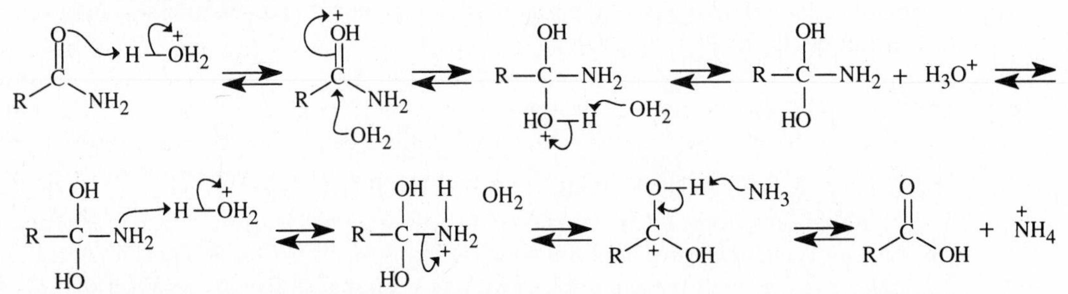

[Reaction mechanism scheme for acid-catalyzed amide hydrolysis]

3. Hydrolysis of amides in base is analogous to saponification of esters; the reaction is driven to completion by formation of the carboxylic acid salt.
4. The conditions of both acid- and base-promoted amide hydrolysis are considerably more severe than the corresponding reactions of esters.
5. The mechanisms of amide hydrolysis are typical nucleophilic acyl substitution mechanisms.

## E. HYDROLYSIS OF NITRILES

1. Nitriles are hydrolyzed to carboxylic acids and ammonia by heating them in acidic or basic solution.
2. The conditions of nitrile hydrolysis are considerably more severe than the corresponding reactions of esters and amides.
3. The mechanism of nitrile hydrolysis in acidic solution involves:
   a. Protonation of the nitrogen making the nitrile carbon more electrophilic.
   b. Attack of the nucleophile water on the nitrile carbon and loss of a proton give an intermediate called an imidic acid.
   c. Conversion of the unstable imidic acid under the reaction conditions into an amide.
   d. Hydrolysis of the amide to a carboxylic acid and ammonium ion. (See Eqs. 21.20a–c on text p. 994 for the detailed mechanism. Note that in early printings of the text, the carbonyl oxygen in the second structure of Eq. 21.20c is misplaced.)

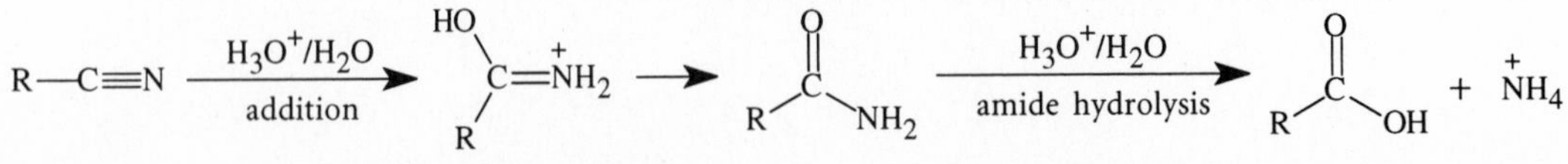

[Reaction scheme: R—C≡N with H₃O⁺/H₂O (addition) → protonated imidic acid → amide → R—C(=O)—NH₂ with H₃O⁺/H₂O (amide hydrolysis) → R—C(=O)—OH + NH₄⁺]

a protonated imidic acid

4. In base, the nitrile group is attacked by basic nucleophiles and, as a result, the electronegative nitrogen assumes a negative charge.
   a. Proton transfer gives an imidic acid which ionizes in base.
   b. The imidic acid reacts further to give the corresponding amide, which in turn, hydrolyzes under the reaction conditions to the carboxylate salt of the corresponding carboxylic acid. (See Eqs. 21.21a–b on text p. 995 for the detailed mechanism.)

$$R\text{—}C\equiv N \xrightarrow[\text{addition}]{^-OH/H_2O} \underset{\substack{\text{conjugate acid of} \\ \text{an imidic acid}}}{\overset{^-O}{\underset{R}{C}}=NH_2} \longrightarrow \underset{R}{\overset{O}{C}}{}_{NH_2} \xrightarrow[\text{amide hydrolysis}]{^-OH/H_2O} \underset{R}{\overset{O}{C}}{}_{O^-} + NH_3$$

5. The hydrolysis of nitriles is a useful way to prepare carboxylic acids because nitriles, unlike many other carboxylic acid derivatives, are generally synthesized from compounds other than the acids themselves.

$$R\text{—}Br \xrightarrow{^-C\equiv N} R\text{—}C\equiv N \xrightarrow{\text{hydrolysis}} R\text{—}CO_2H$$

## II. Reactions of Carboxylic Acid Derivatives with Alcohols

### A. Reaction of Acid Chlorides and Anhydrides with Alcohols and Phenols

1. Esters are formed rapidly when acid chlorides react with alcohols or phenols, usually in the presence of a tertiary amine such as pyridine, or a related base.

$$\underset{R}{\overset{O}{C}}{}_{Cl} \xrightarrow[\substack{\text{pyridine} \\ \text{addition}}]{HOR'} \underset{R}{\overset{HO\quad OR'}{C}}{}_{Cl} \xrightarrow[\text{elimination}]{\text{proton transfer}} \underset{R}{\overset{O}{C}}{}_{OR'} + \underset{\text{pyridinium chloride}}{\overset{+\ Cl^-}{N\text{—}H}}$$

2. Esters of tertiary alcohols and phenols, which cannot be prepared by acid-catalyzed esterification, can be prepared by this method.
3. Anhydrides react with alcohols and phenols in much the same way as acid chlorides. Cyclic anhydrides react with alcohols and phenols to give half-esters.

$$R\text{—}\overset{O}{\overset{\|}{C}}\text{—}O\text{—}\overset{O}{\overset{\|}{C}}\text{—}R + HOR' \longrightarrow R\text{—}\overset{O}{\overset{\|}{C}}\text{—}OR' + HO\text{—}\overset{O}{\overset{\|}{C}}\text{—}R$$

4. Sulfonate esters are prepared by the analogous reactions of sulfonyl chlorides with alcohols.

### B. Reaction of Esters with Alcohols

1. When an ester reacts with an alcohol under acidic conditions, or with an alkoxide under basic conditions, a new ester is formed.

$$\underset{R}{\overset{O}{C}}{}_{OCH_3} \xrightleftharpoons[]{\substack{HOCH_2CH_3 \\ H_3O^+ \text{ (cat.)}}} \underset{R}{\overset{O}{C}}{}_{OCH_2CH_3} + HOCH_3$$

2. This type of reaction, called transesterification, typically has an equilibrium constant near unity.
3. The reaction is driven to completion by the use of an excess of the displacing alcohol or by removal of a relatively volatile alcohol by-product as it is formed.

### III. *Reactions of Carboxylic Acid Derivatives with Amines*

### A. REACTION OF ACID CHLORIDES AND ANHYDRIDES WITH AMMONIA AND AMINES

1. Acid chlorides react rapidly and irreversibly with ammonia or amines by a nucleophilic acyl substitution reaction mechanism to give amides.
   a. Reaction with ammonia yields a primary amide.
   b. Reaction with a primary amine yields a secondary amide.
   c. Reaction with a secondary amine yields a tertiary amide.
2. The amine attacks the carbonyl group to form a tetrahedral intermediate, which expels chloride ion; a proton-transfer step yields the amide.
   a. An important aspect of amide formation is the proton transfer in the last step of the mechanism.
   b. Unless another base is added to the reaction mixture, the starting amine acts as the base in this step; if the only base present is the amine nucleophile, then at least two equivalents must be used:
      i. one equivalent as the nucleophile.
      ii. one equivalent as the base in the final proton-transfer step.

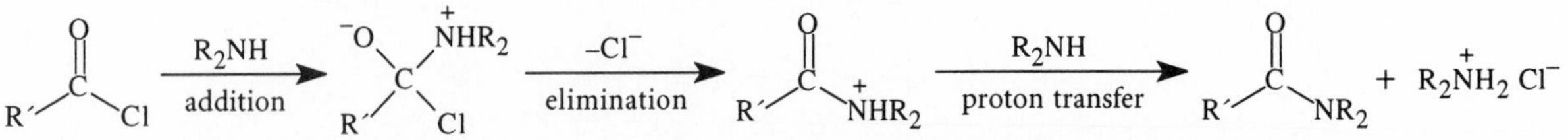

3. In the Schotten-Baumann technique for amide formation, the reaction is run with a water-insoluble acid chloride and an amine in a separate layer over an aqueous solution of NaOH.
4. The presence of a tertiary amine, such as triethylamine or pyridine, does not interfere with amide formation by another amine because a tertiary amine itself cannot form an amide.
5. Anhydrides react with amines in much the same way as acid chlorides.

   a. Half-amides of dicarboxylic acids are produced in analogous reactions of amines and cyclic anhydrides.
   b. These compounds can be cyclized to imides by treatment with dehydrating agents, or in some cases just by heating.

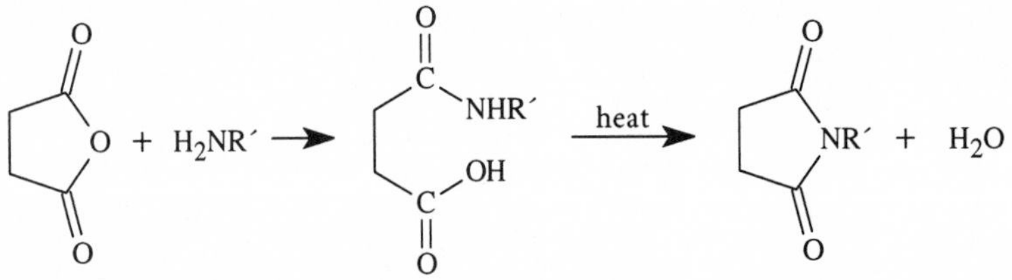

### B. REACTION OF ESTERS WITH AMINES

1. The reaction of esters with ammonia or amines yields amides.

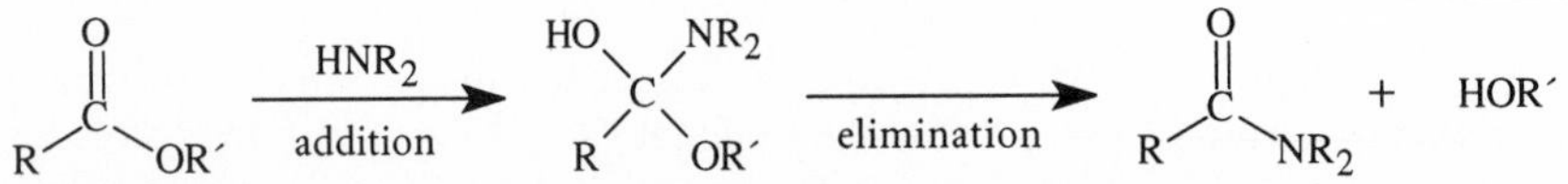

2. The reaction of esters with hydroxylamine ($NH_2OH$) gives *N*-hydroxyamides; these compounds are known as hydroxamic acids.
   a. This chemistry forms the basis for the hydroxamate test, used mostly for esters.
   b. The hydroxamic acid products are easily recognized because they form highly colored complexes with ferric ion.

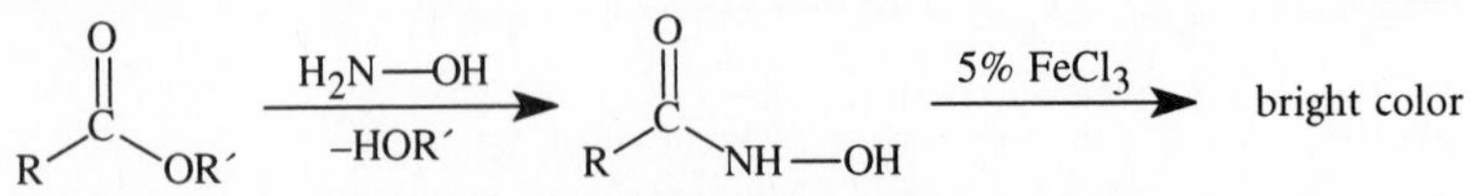

a hydroxamic acid

## IV. *Reactions of Carboxylic Acid Derivatives with Hydrides*

### A. REDUCTION OF ACID CHLORIDES TO ALDEHYDES AND ALCOHOLS

1. Acid chlorides can be reduced to aldehydes by either of two procedures:
   a. Hydrogenation over a catalyst that has been deactivated, or poisoned, with an amine, such as quinoline, that has been heated with sulfur.

   i. This reaction is called the Rosenmund reduction.
   ii. The poisoning of the catalyst prevents further reduction of the aldehyde product.
   b. Reaction at low temperature with lithium tri(*tert*-butoxy)aluminum hydride.

   i. The hydride reagent used in this reduction is derived by the replacement of three hydrogens of $LiAlH_4$ by *tert*-butoxy groups.
   ii. Because acid chlorides are more reactive than aldehydes toward nucleophiles, the reagent reacts preferentially with the acid chloride reactant rather than with the product aldehyde.
2. Acid chlorides (and anhydrides) react with $LiAlH_4$ to give primary alcohols.

### B. REDUCTION OF ESTERS TO PRIMARY ALCOHOLS

1. The reduction of esters to alcohols involves a nucleophilic acyl substitution reaction followed by a carbonyl addition reaction.
   a. The active nucleophile in $LiAlH_4$ reductions is the hydride ion, $H:^-$, which replaces alkoxide at the carbonyl group of the ester to give an aldehyde.
   b. The aldehyde reacts rapidly with $LiAlH_4$ to give, after protonolysis, the alcohol.
   c. Two alcohols are formed:
      i. one derived from the acyl group of the ester
      ii. one derived from the alkoxy group (usually methanol or ethanol which is discarded).

2. Sodium borohydride reacts very sluggishly or not at all with most esters.

### C. REDUCTION OF AMIDES TO AMINES

1. Amides are reduced to amines with $LiAlH_4$ and can be used to prepare:
   a. Primary amines from primary amides.
   b. Secondary amines from secondary amides.
   c. Tertiary amines from tertiary amides.
2. The mechanism of the reaction of a secondary amide with lithium aluminum hydride involves:
   a. Formation of the lithium salt of the amide by reaction of the acidic amide proton with an equivalent of hydride, a strong base.
   b. Reaction of the lithium salt of the amide, a Lewis base, with $AlH_3$, a Lewis acid.
   c. Delivery of hydride to the C=N double bond.
   d. Loss of $^-OAlH_2$, a fairly good leaving group (better than the nitrogen group), to form an imine.

$$
\underset{\text{an imine}}{\overset{\displaystyle H}{\underset{R}{\overset{|}{C}}}=NR'} \quad \xrightarrow[\text{2) } H_3O^+/H_2O]{\text{1) } LiAlH_4} \quad RCH_2NH_2R'
$$

3. The reductions of primary and tertiary amides involve somewhat different mechanisms, but they too involve loss of oxygen rather than nitrogen as a leaving group.

## D.  REDUCTION OF NITRILES TO AMINES

1. Nitriles are reduced to primary amines by reaction with $LiAlH_4$, followed by protonolysis in a separate step.

$$
R-C\equiv N \quad \xrightarrow[\text{2) } H_2O]{\text{1) } LiAlH_4} \quad R-CH_2-NH_2
$$

2. Nitriles are also reduced to primary amines by catalytic hydrogenation. An intermediate in the reaction is the imine, which is not isolated but is hydrogenated to the amine product.

$$
R-C\equiv N \quad \xrightarrow{H_2, \text{ (cat.)}} \quad R-CH=NH \quad \xrightarrow{H_2, \text{ (cat.)}} \quad R-CH_2-NH_2
$$

## V.  *Reactions of Carboxylic Acid Derivatives with Organometallic Reagents*

### A.  REACTION OF ACID CHLORIDES WITH LITHIUM DIALKYLCUPRATES

1. The reaction of lithium dialkylcuprates with acid chlorides gives ketones in excellent yield; lithium dialkylcuprates typically react with acid chlorides and aldehydes, very slowly with ketones, and not at all with esters.

$$
\underset{R}{\overset{O}{\overset{\|}{C}}}\diagdown{Cl} \quad + \quad Li^+\ ^-CuR'_2 \quad \longrightarrow \quad \underset{R}{\overset{O}{\overset{\|}{C}}}\diagdown{R'} \quad + \quad LiCl \quad + \quad CuR'
$$

2. Lithium dialkylcuprates react much like Grignard or lithium reagents, but are less reactive; a lithium dialkylcuprate can be considered conceptually as an alkyl anion complexed with copper.
   a. Lithium dialkylcuprate reagents are prepared by the reaction of two equivalents of an organolithium reagent with one equivalent of a cuprous halide such as CuBr.
   b. The first equivalent forms an alkylcopper compound; the driving force for this reaction is the preference of lithium, the more electronegative metal, to exist as an ion ($Li^+$).
   c. The copper of an alkylcopper reagent is a Lewis acid and reacts accordingly with "alkyl anion" from a second equivalent of the organolithium reagent; the product of this reaction is a lithium dialkylcuprate.

$$
RLi \quad + \quad CuI \quad \longrightarrow \quad LiI \quad + \quad CuR \quad \xrightarrow{RLi} \quad LiCuR_2
$$

### B.  REACTION OF ESTERS WITH GRIGNARD REAGENTS

1. In the reaction of esters with Grignard reagents, tertiary alcohols are formed after protonolysis. (Secondary alcohols are formed from esters of formic acid after protonolysis.)
   a. Two equivalents of organometallic reagent react per mole of ester.
   b. A second alcohol (derived from the alkoxy group, usually methanol or ethanol) is produced in the reaction and discarded.

$$
\underset{R'}{\overset{O}{\overset{\|}{C}}}\diagdown{OR''} \quad \xrightarrow[\text{addition}]{RMgX} \quad \underset{R'}{\overset{R}{\overset{|}{C}}}\diagup{\overset{O^-\ ^+MgX}{}}\diagdown{OR''} \quad \xrightarrow[\text{elimination}]{-R''OMgX} \quad \underset{R'}{\overset{O}{\overset{\|}{C}}}\diagdown{R} \quad \xrightarrow[\text{2) } H_3O^+]{\text{1) } RMgX} \quad R'-\underset{\underset{R}{|}}{\overset{\overset{OH}{|}}{C}}-R
$$

2. This reaction is a nucleophilic acyl substitution followed by an addition.
   a. A ketone is formed in the substitution step.
   b. The ketone intermediate is not isolated because ketones are more reactive than esters toward nucleophilic reagents.

c. The ketone reacts with a second equivalent of the Grignard reagent to form a magnesium alkoxide, which, after protonolysis, gives the alcohol.

3. This reaction is a very important method for the synthesis of alcohols in which at least two of the groups on the $\alpha$-carbon of the alcohol product are identical.

## VI.  Reactions of Carboxylic Acid Derivatives with Other Nucleophiles

### A.  REACTION OF ACID CHLORIDES WITH CARBOXYLATE SALTS

1. Even though carboxylate salts are weak nucleophiles, acid chlorides are reactive enough to be attacked by carboxylate salts to give anhydrides.

2. The reaction of acid chlorides with carboxylate salts can be used to prepare mixed anhydrides.

$$\underset{R}{\overset{O}{\underset{\|}{C}}}\!-\!Cl \;+\; Na^+ \;{}^-O\!-\!\underset{R'}{\overset{O}{\underset{\|}{C}}} \;\xrightarrow{\text{heat}}\; \underset{R}{\overset{O}{\underset{\|}{C}}}\!-\!O\!-\!\underset{R'}{\overset{O}{\underset{\|}{C}}} \;+\; NaCl$$

## VII.  Synthesis of Carboxylic Acid Derivatives—Review

### A.  SYNTHESIS OF ACID CHLORIDES

1. Reaction of carboxylic acids with $SOCl_2$ or $PCl_5$ (Sec. 20.9A, text page 952).

$$R\!-\!\overset{O}{\overset{\|}{C}}\!-\!OH \;\xrightarrow[\text{or } PCl_5]{SOCl_2}\; R\!-\!\overset{O}{\overset{\|}{C}}\!-\!Cl$$

### B.  SYNTHESIS OF ANHYDRIDES

1. Reaction of carboxylic acids with dehydrating agents (Sec. 20.9B, text page 954).

$$2\,R\!-\!\overset{O}{\overset{\|}{C}}\!-\!OH \;\xrightarrow[-H_2O]{P_2O_5}\; R\!-\!\overset{O}{\overset{\|}{C}}\!-\!O\!-\!\overset{O}{\overset{\|}{C}}\!-\!R$$

2. Reaction of acid chlorides with carboxylate salts (Sec. 21.8A, text page 1003).

$$R\!-\!\overset{O}{\overset{\|}{C}}\!-\!O^- \;+\; Cl\!-\!\overset{O}{\overset{\|}{C}}\!-\!R' \;\longrightarrow\; R\!-\!\overset{O}{\overset{\|}{C}}\!-\!O\!-\!\overset{O}{\overset{\|}{C}}\!-\!R'$$

### C.  SYNTHESIS OF ESTERS

1. Acid-catalyzed esterification of carboxylic acids (Sec. 20.8A, text page 946).

$$R\!-\!CO_2H \;+\; HOR' \;\underset{\xleftarrow{\hspace{1.2cm}}}{\overset{H_3O^+}{\rightleftharpoons}}\; R\!-\!CO_2R' \;+\; H_2O$$

2. Alkylation of carboxylic acids or carboxylates (Sec. 20.8B, text page 950).

$$R\!-\!CO_2H \;+\; CH_2N_2 \;\longrightarrow\; R\!-\!CO_2CH_3 \;\xleftarrow{K_2CO_3}\; R\!-\!CO_2H \;+\; CH_3\!-\!I$$

3. Reaction of acid chlorides and anhydrides with alcohols or phenols (Sec. 21.8A, text page 1002).

$$R\!-\!\overset{O}{\overset{\|}{C}}\!-\!Cl \;+\; HOR' \;\xrightarrow{\text{base}}\; R\!-\!\overset{O}{\overset{\|}{C}}\!-\!OR'$$

4. Transesterification of other esters (Sec. 21.8C, text page 1005).

$$R\!-\!CO_2R' \;+\; HOR'' \;\rightleftharpoons\; R\!-\!CO_2R'' \;+\; HOR'$$

### D.  SYNTHESIS OF AMIDES

1. Reaction of acid chlorides, anhydrides, or esters with amines (Sec. 21.8, text pp. 1001–1005).

$$R\!-\!\overset{O}{\overset{\|}{C}}\!-\!X \;+\; HNR'R'' \;\longrightarrow\; R\!-\!\overset{O}{\overset{\|}{C}}\!-\!NR'R'' \qquad\qquad -X = -Cl,\; -O\!-\!\overset{O}{\overset{\|}{C}}\!-\!R,\; \text{or } -OR$$

## E.  Synthesis of Nitriles

1. $S_N2$ reaction of cyanide ion with alkyl halides or sulfonate esters.  In this reaction, primary or unbranched secondary alkyl halides or sulfonate esters are required (Sec. 9.2, text page 389).

$$R-CH_2-X \quad + \quad ^-C\equiv N \quad \longrightarrow \quad R-CH_2-C\equiv N \; + \; X^- \qquad \qquad X = \text{halide ion, } ^-OTs$$

2. Cyanohydrin formation (Sec. 19.7A, text page 884).

$$
\underset{\substack{\| \\ R-C-R'}}{\overset{O}{}} \; + \; H-C\equiv N \quad \longrightarrow \quad \underset{\substack{| \\ R-C-R' \\ | \\ C\equiv N}}{\overset{OH}{}}
$$

## Study Guide Links

## 21.1   NMR Evidence for Internal Rotation in Amides

Internal rotation about the carbonyl-nitrogen bond of amides can be conveniently studied by NMR using the principles discussed in Sec. 13.7 on p. 619 of the text. Consider, for example, internal rotation in *N,N*-dimethylacetamide:

$$H_3C-\overset{\overset{\textstyle O}{\|}}{C}-\underset{\underset{\textstyle CH_3}{|}}{N}-CH_3 \left.\right\} \text{these methyl groups are diastereotopic and hence chemically nonequivalent}$$

The *N*-methyl groups in this compound are diastereotopic; one is *cis* to the carbonyl oxygen, and the other is *trans*. At low temperature, rotation about the C—N bond is slow enough that these methyl groups are observed as separate singlets in the NMR spectrum. However, when the temperature is raised, these signals broaden and, at about 60 °C, they coalesce into one singlet at a chemical shift that is the average of the two individual methyl chemical shifts. This is illustrated in Fig. SG21.1 on the following page.

## ✓21.2   Solving Structure Problems Involving Nitrogen-Containing Compounds

Amides and nitriles are the first nitrogen-containing functional groups that are considered in the text. When you solve structure problems you should remember certain things about compounds that contain nitrogen. First, compounds containing an odd number of nitrogen atoms have *odd molecular mass*. (If you studied mass spectrometry (Chapter 12), you will appreciate that this means that the parent ion in the mass spectrum of such a compound will occur at an odd *m/z* value.) Second, the special formula given below (see also Eq. 4.8 on text p. 137) is used to calculate the unsaturation number *U* of a compound containing nitrogen. If *C* is the number of carbons, *H* is the number of hydrogens, *X* is the number of halogens, and *N* is the number of nitrogens, then *U* is given by

$$U = \frac{(2C + 2) + N - (H + X)}{2}$$

This formula means that every nitrogen increases the number of hydrogens in a fully saturated molecule by one relative to the number of hydrogens in the corresponding hydrocarbon. Finally, when a compound contains nitrogen, you can't tell whether an ion in its mass spectrum is an odd-electron ion or an even-electron ion simply by its mass, because the number of nitrogens in the ion determines whether its mass is odd or even. When a compound contains zero or an even number of nitrogens, odd-electron ions have odd masses; when a compound contains an odd number of nitrogens, odd-electron ions have even masses.

*(text continues on p. 539)*

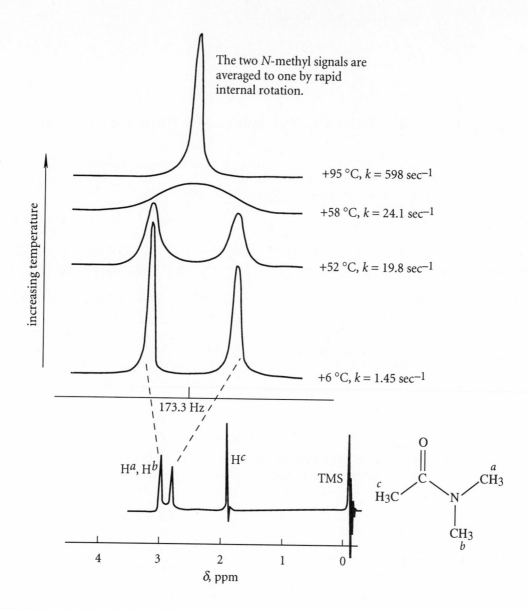

**Figure SG21.1**    *The 60 MHz NMR spectrum of N,N-dimethylacetamide changes with increasing temperature as the rate of rotation about the carbonyl-nitrogen bond increases. The full spectrum of this compound is shown in the lower part of the figure, and the pair of singlets near δ 2.9 (173.3 Hz) is expanded in the upper part of the figure and shown as a function of temperature. The two N-methyl groups are individually observable at temperatures at or below 52° (lower two spectra) because internal rotation about the carbonyl-nitrogen bond is relatively slow at these temperatures. As the temperature is raised, the two resonances broaden and coalesce into one because the internal rotation becomes too rapid for the NMR experiment to resolve the individual lines. (See Fig. 13.19 on text p. 619 for a related phenomenon.) The k values are the first-order rate constants for internal rotation; a k of 1.45 sec[-1] corresponds to an approximate lifetime of 0.5 sec for the individual conformations.*

## √21.3 Basicity of Nitriles

Let's place the basicity of nitriles within the context of other basicities that you've studied. Consider the following series:

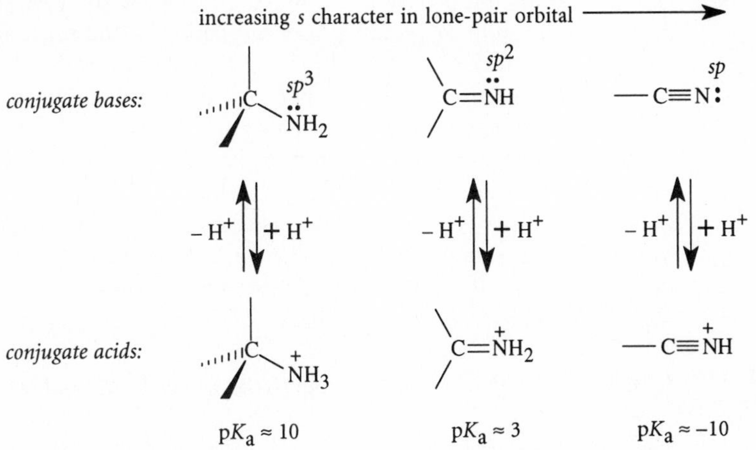

increasing *s* character in lone-pair orbital ⟶

Notice that as the amount of *s* character in the lone-pair orbital of the conjugate base increases, the $pK_a$ of the conjugate acid decreases. Although the $pK_a$ values are much different, this is exactly the same trend observed in the acidities of hydrocarbons (Sec. 14.7A, text pp. 662–664):

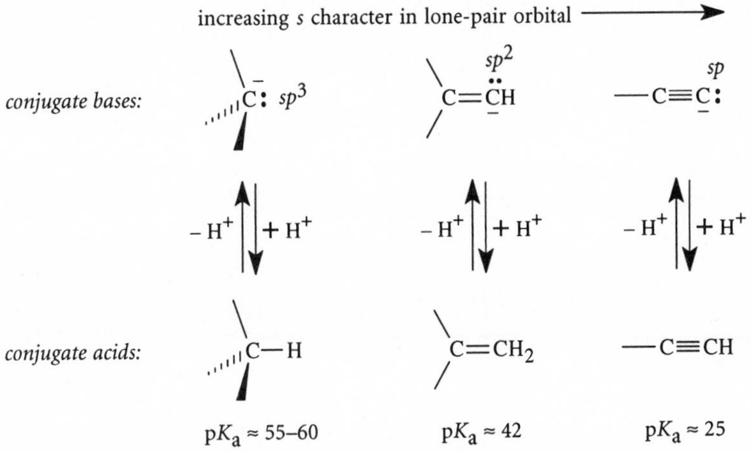

increasing *s* character in lone-pair orbital ⟶

In other words, just as an acetylenic anion, with its lone pair in an *sp* orbital, is the least basic hydrocarbon anion, a nitrile, also with a lone pair in an *sp* orbital, is the least basic among the analogous series of nitrogen compounds.

## √21.4 Mechanism of Ester Hydrolysis

Some students ask: Why not combine the features of the acid-catalyzed hydrolysis and saponification by using strong acid to protonate the carbonyl oxygen and hydroxide to act as the nucleophile?

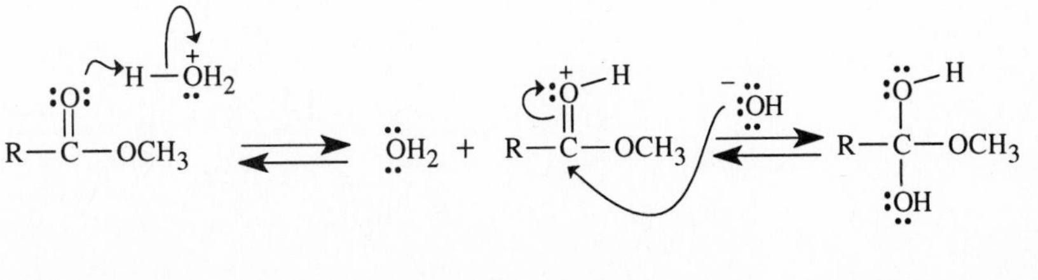

Implicit in such a question is the incorrect notion that when the base is added to the acidic solution, the carbonyl oxygen of the ester would remain protonated and that all hydronium ions would not dissociate. However, both protonated esters and hydronium ions are strong acids. Remember: *the deprotonation of strong acids by strong bases is instantaneous.* (See Study Guide Link 20.2 on p. 499 of this manual.) The very first events that would occur on addition of base to the solution would be deprotonation of the protonated carbonyl oxygen and rapid neutralization of the acid ($H_3O^+$) in solution. Attack of $^-OH$ on the ester carbonyl carbon is a much slower reaction. All of the hydronium ions in solution would react with the hydroxide ion before hydroxide could react to a significant extent at the carbonyl carbon.

In summary: *a solution cannot at the same time be acidic enough to protonate a carbonyl oxygen and basic enough to contain substantial hydroxide ion.* Acid-catalyzed reactions generally involve relatively weak bases such as water as nucleophiles; reactions involving hydroxide as a base generally involve weak acids such as water. The bottom line: *Don't try to use $H_3O^+$ as an acid and $^-OH$ as a base simultaneously in the same solution.* This same point was addressed in Study Guide Link 19.4 (p. 466 of this manual).

## 21.5    Cleavage of Tertiary Esters and Carbonless Carbon Paper

The hydrolysis of tertiary esters has some noteworthy differences from the hydrolysis of other esters discussed in Sec. 21.7A. First, the saponification of tertiary esters is considerably slower than that of primary and secondary esters. For example, the saponification of *tert*-butyl acetate occurs at about 0.01 times the rate of the saponification of methyl acetate. The reason is that in the transition state for saponification, the methyl branches in the *tert*-butyl group are involved in van der Waals repulsions with the attacking base. A practical consequence of this rate difference is that a methyl or ethyl ester can generally be saponified without affecting a *tert*-butyl ester in the same molecule.

A second difference is that the acid-catalyzed hydrolysis of tertiary esters occurs by a mechanism that is completely different from that of primary and secondary esters. The first step in the hydrolysis mechanism is the same—protonation of the carbonyl oxygen. But there the similarity ends. The protonated ester dissociates by an $S_N1$ mechanism into a tertiary carbocation and the carboxylic acid.

Attack of water (or any other nucleophile that might be present) occurs on the tertiary carbocation.

Notice that hydrolysis of tertiary esters by this mechanism involves breaking the *alkyl-oxygen bond.* In contrast, the mechanism of hydrolysis of primary and secondary esters involves breaking the *carbonyl-oxygen bond.* Consequently, the hydrolysis of tertiary esters is *not* a nucleophilic acyl substitution reaction. Rather, it is more like the $S_N1$ mechanism for substitution of a tertiary alcohol (see Eqs. 10.9b–c on text p. 447). Esters of primary and secondary alcohols in most cases do not cleave by an $S_N1$ mechanism because the carbocation intermediates that would be involved are much less stable.

"Carbonless carbon paper" (sometimes called NCR paper, for "No Carbon Required") is a very ingenious commercial application of tertiary ester cleavage as applied to lactones. The active component in "carbonless carbon paper" is the colorless lactone *A*, which is encapsulated within microscopic particles of a phenol–formaldehyde resin (Sec. 19.15 on text p. 917) and deposited on a paper base that also contains an acidic layer. When the particles are broken by the pressure of a pen or typewriter, the lactone is brought into contact with the acidic layer. This causes an $S_N1$-like opening of the lactone ring to give a relatively stable carbocation *B.* This carbocation has a blue-violet color.

*A* (colorless)

*B* (blue-violet)

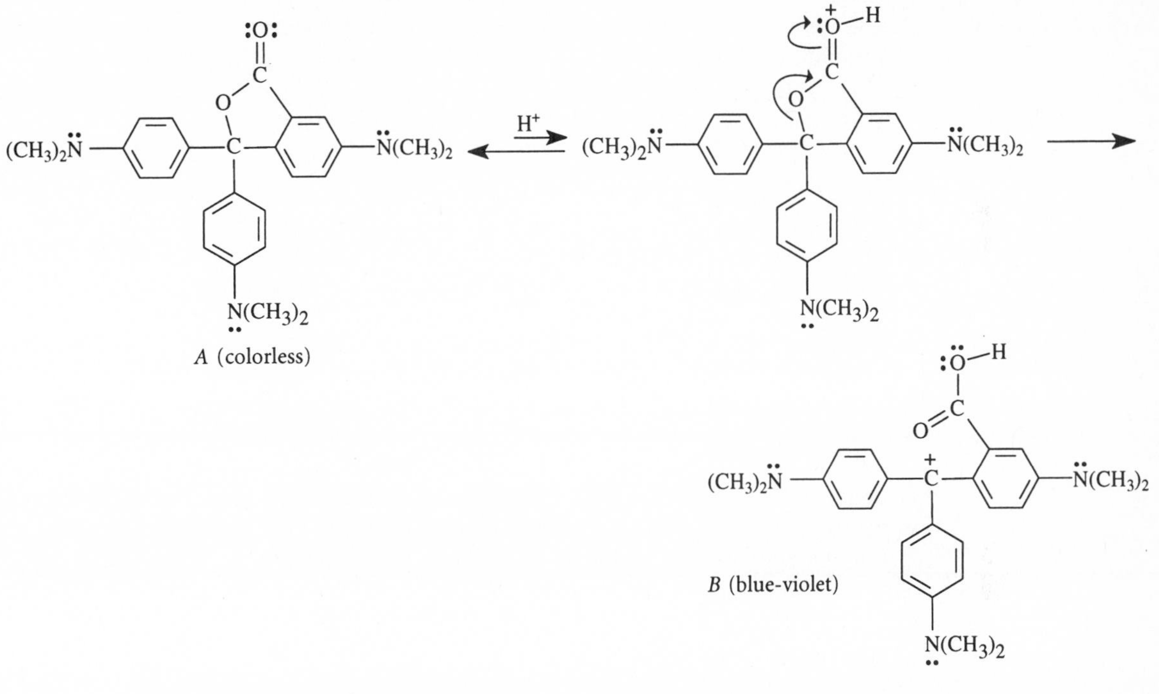

It takes some time for color to develop because the reaction is not instantaneous. As you may know from experience, the color in carbonless carbon paper intensifies with time.

## 21.6 Reaction of Tertiary Amines with Acid Chlorides

Tertiary amines react with acid chlorides to form *acylammonium salts.*

pyridine
(a tertiary amine)

an acylammonium salt

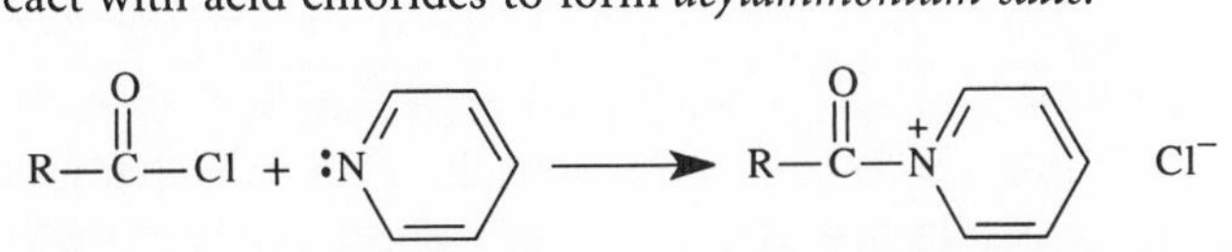

Acylammonium salts, because of their positive charge adjacent to the carbonyl carbon, are very reactive and are converted into amides by reaction with primary or secondary amines.

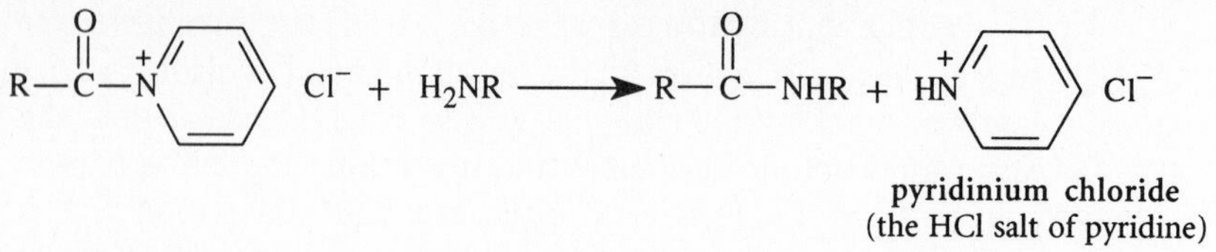

pyridinium chloride<br>(the HCl salt of pyridine)

If the acid chloride has acidic $\alpha$-hydrogens, another possible reaction is the formation of a *ketene* by a $\beta$-elimination:

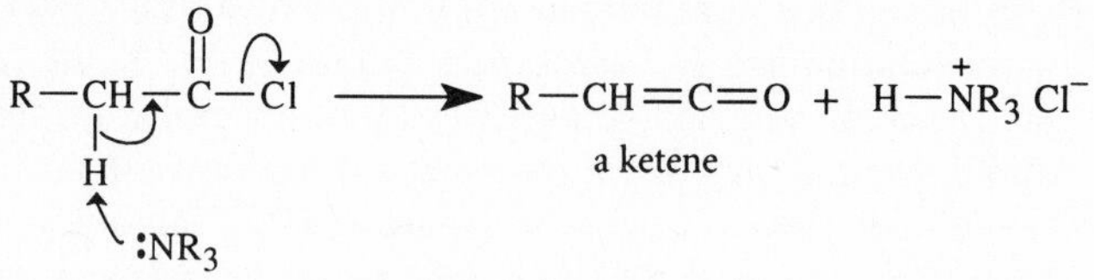

Ketenes also react rapidly with primary and secondary amines to yield amides.

Despite these side reactions, then, tertiary amines can be used as catalysts for amide formation with primary and secondary amines because the products of these side reactions—acylammonium salts and ketenes—themselves react with amines to form amides.

## ✓21.7 Another Look at Friedel-Crafts Acylation

The Friedel-Crafts acylation reaction is first classified in the text as an *electrophilic aromatic substitution* reaction (Sec. 16.4E on text p. 754), because the reaction is first considered as a reaction of benzene derivatives. However, in terms of what happens to the acid chloride, the reaction can also be viewed as a nucleophilic acyl substitution reaction.

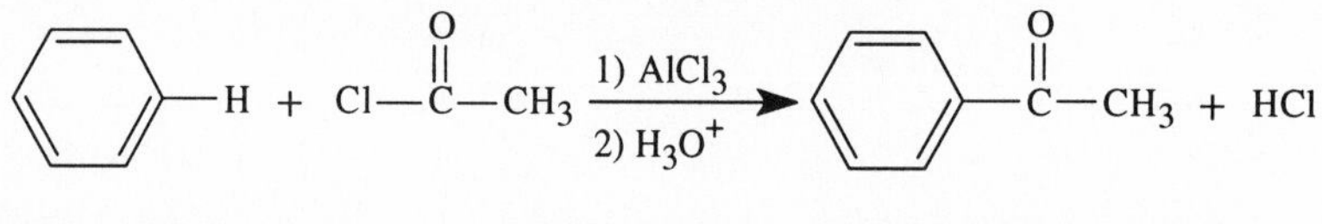

If the Friedel-Crafts acylation is considered to be a nucleophilic acyl substitution reaction, the nucleophilic electrons are the $\pi$ electrons of the aromatic ring. Because benzene and its derivatives are very weak nucleophiles, a strong Lewis acid such as $AlCl_3$ is required to activate the acid chloride. Review again the mechanism of Friedel-Crafts acylation, paying particular attention to what happens to the acid chloride.

## ✓21.8 Esters and Nucleophiles

Notice that most esters of *carboxylic acids* react with nucleophiles at the carbonyl carbon—that is, they undergo nucleophilic acyl substitution reactions. The result is that the bond between the carbonyl group and the ester oxygen—the *acyl-oxygen bond*—is cleaved. (Study Guide Link 21.5 discusses one exception to this generalization.)

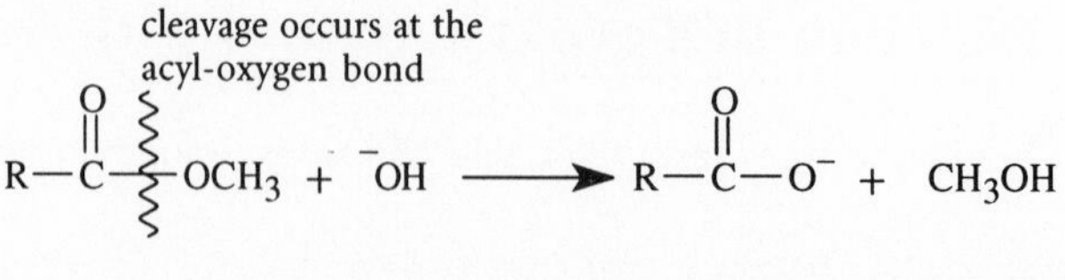

Esters of *sulfonic acids* react in a fundamentally different manner. These reactions are presented in Sec. 10.3A (text p. 452–453), where the focus is on the sulfonate ester groups essentially as equivalents of halide leaving groups. As the discussion in the text shows, most sulfonate esters react with nucleophiles not by attack of the nucleophile on the sulfur, but by $S_N2$ attack of the nucleophile on the *alkyl group* (if the alkyl group is primary or secondary); in such a reaction, the bond between the alkyl group and the ester oxygen is cleaved. (See Eq.

10.15 on text p. 453.)

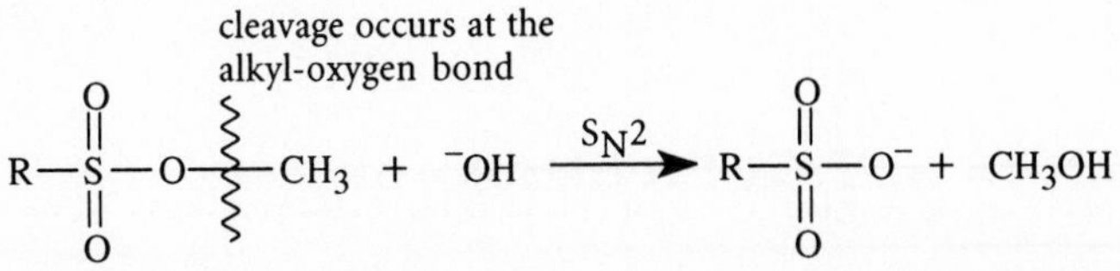

cleavage occurs at the
alkyl-oxygen bond

$$R-\overset{\overset{\displaystyle O}{\|}}{\underset{\underset{\displaystyle O}{\|}}{S}}-O\overset{\xi}{\underset{\xi}{|}}CH_3 \; + \; \;^-OH \; \xrightarrow{\;S_N2\;} \; R-\overset{\overset{\displaystyle O}{\|}}{\underset{\underset{\displaystyle O}{\|}}{S}}-O^- \; + \; CH_3OH$$

The reason sulfonate esters react differently is that they are better leaving groups than carboxylates; hence, they are more easily displaced.

You may have noticed that we are returning with increasing frequency to earlier parts of the text and looking at reactions you've studied earlier in somewhat different contexts. If you willingly participate in this type of exercise, you'll gain a lot of insight about organic chemistry. In particular, you'll see how the parts begin to "fit together," that is, you'll begin to see the "forest" and not just the individual "trees."

## Solutions

### Solutions to In-Text Problems

**21.1**  (a)

$N{\equiv}CCH_2CH_2CH_2CH_2CO_2H$

5-cyanopentanoic acid

(c)

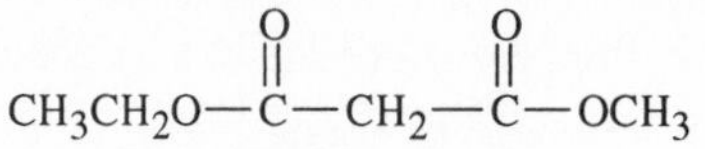

$CH_3CH_2O-\overset{O}{\overset{\|}{C}}-CH_2-\overset{O}{\overset{\|}{C}}-OCH_3$

ethyl methyl malonate

(e)

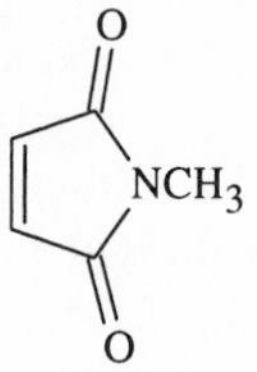

*N*-methylmaleimide

(g)

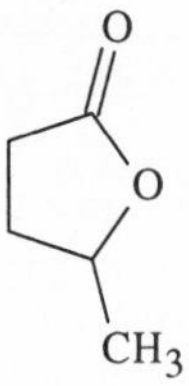

$\gamma$-valerolactone

(i)

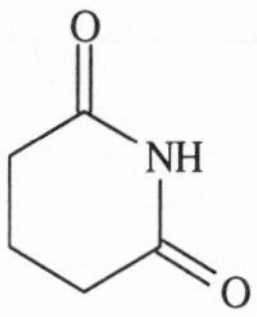

glutarimide

(j)

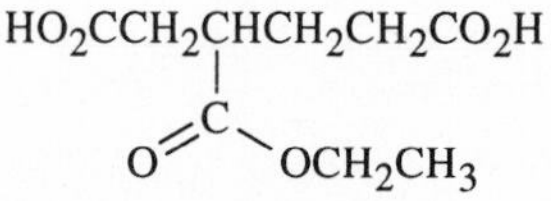

$HO_2CCH_2CHCH_2CH_2CO_2H$

3-ethoxycarbonylhexanedioic acid

**21.2**  (a)  Cyclopropanecarbonyl chloride
(c)  (*Z*)-2-Butenedioic anhydride (*common name more often used:* maleic anhydride)
(e)  Ethyl 3-oxopentanoate
(g)  1-Methyl-3-butenyl propanoate

**21.3**  Any lactam containing a relatively small ring (fewer than eight members) must exist in an *E* conformation. An example is $\delta$-valerolactam. A small-ring lactam cannot exist in a *Z* conformation, because such a conformation would result in a transoid bond within the ring, a very strained and therefore unstable arrangement. (This is the same reason that small-ring (*E*)-cycloalkenes are unstable.) If you try to construct a model of a small-ring lactam with its amide bond in a *Z* conformation, the strain of such a conformation should be obvious.

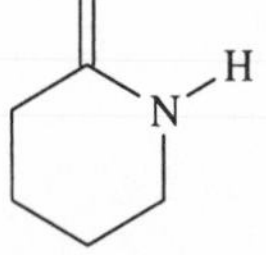

$\delta$-valerolactam

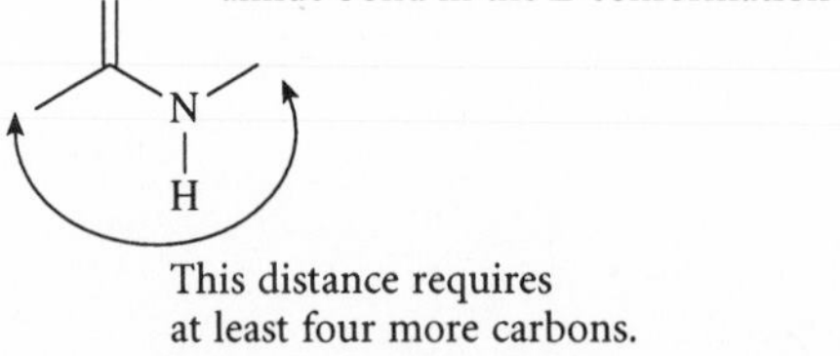

amide bond in the *Z* conformation

This distance requires
at least four more carbons.

**21.5**  (a)  2-Butanone should have the higher dipole moment, because, other things being equal, molecules with higher dipole moments have higher boiling points.

(b)  The dipole-moment data require that esters, like amides, exist predominantly in the *Z* conformation. In this conformation of an ester, the resultant of the bond dipoles of the two carbon-oxygen single bonds opposes the bond dipole of the C=O bond; the overall dipole moment of the ester is thus reduced. In the ketone, the C=O bond dipole is not opposed by any other significant bond dipole.

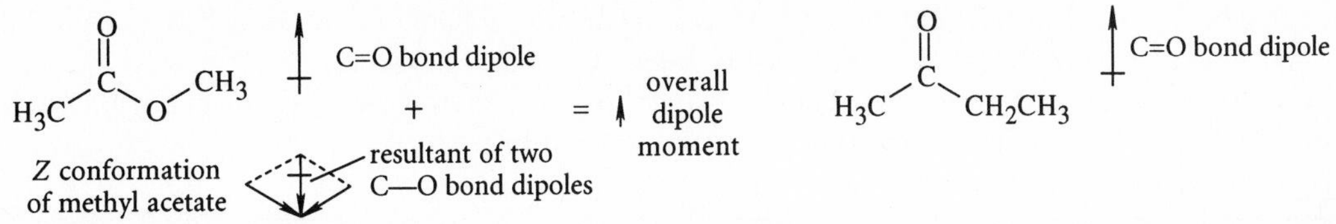

If the ester were in an *E* conformation, the resultant bond dipoles of the two carbon-oxygen single bonds would *augment* the bond dipole of the C=O bond. In this case, the dipole moment of the ester would be larger than that of the ketone.

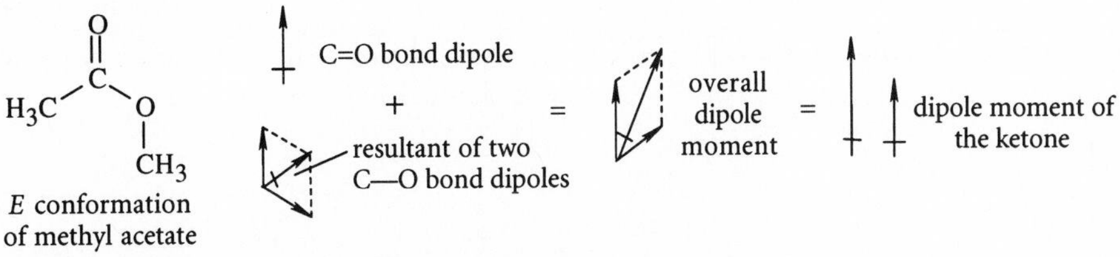

The experimental facts bear out these predictions! $\gamma$-Butyrolactone, an "ester" that is locked into an *E* conformation, has a greater dipole moment than 2-butanone, which has a greater dipole moment than methyl acetate, an ester which exists mostly in the *Z* conformation.

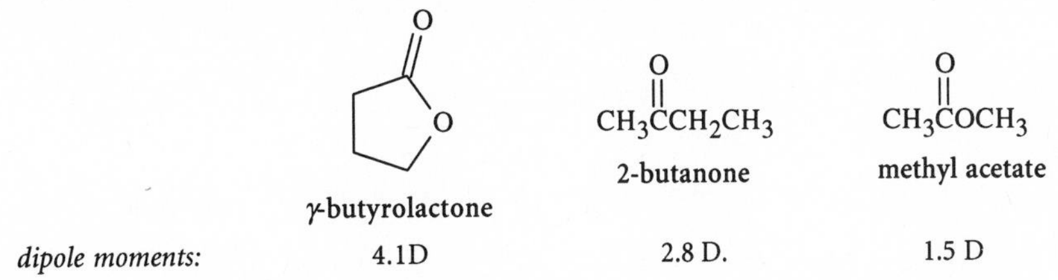

| | $\gamma$-butyrolactone | 2-butanone | methyl acetate |
|---|---|---|---|
| *dipole moments:* | 4.1D | 2.8 D. | 1.5 D |

**21.7** (a) Ethyl butyrate ($CH_3CH_2O_2CCH_2CH_2CH_3$), the ester, has a carbonyl absorption near 1735 cm$^{-1}$, whereas the ether has no carbonyl absorption.

(c) In the proton NMR spectra, there is a single methyl resonance that is a quartet in the amide, whereas in the nitrile, there are two methyl resonances that are singlets. In addition, the integral of the aromatic proton absorptions in the nitrile is smaller in proportion to the integral of the other absorptions because there are fewer aromatic protons.

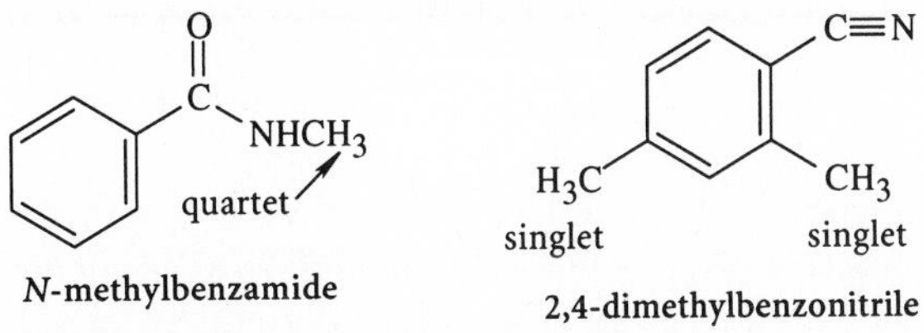

(e) Ethyl isobutyrate, $(CH_3)_2CHCO_2CH_2CH_3$, has five CMR resonances, whereas ethyl butyrate, $CH_3CH_2CH_2CO_2CH_2CH_3$, has six.

**21.8** The odd molecular mass indicates that the compound contains an odd number of nitrogens; that the compound is an amide is confirmed by the N—H absorption at 3300 cm$^{-1}$ and by the carbonyl absorption at 1650 cm$^{-1}$. The compound is *N*-ethylacetamide; the —CH$_2$— resonance of the ethyl group is split by both the adjacent methyl protons and the N—H proton.

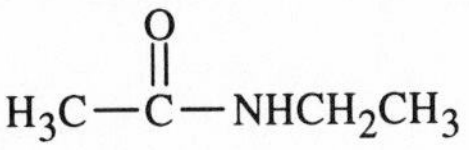

**21.9**   (a)  The first ester is more basic because its conjugate acid is stabilized not only by resonance interaction with the ester oxygen, but also by resonance interaction with the double bond; that is, the conjugate acid of the first ester has one more important resonance structure than the conjugate acid of the second. (See Problem 19.14(b) on text p. 884 and its solution for a similar situation.)

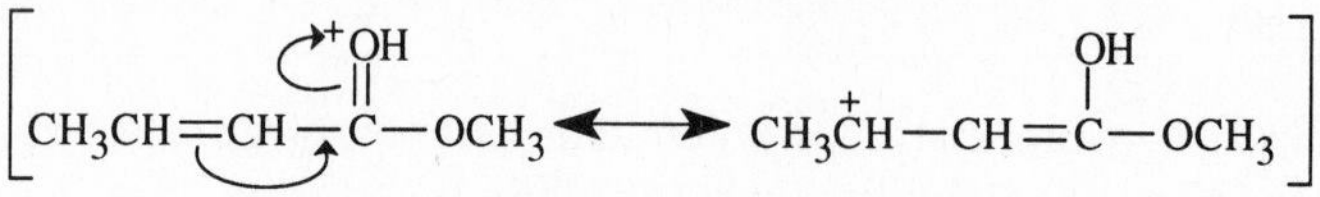

$$\left[ CH_3CH{=}CH{-}\overset{\overset{+OH}{\|}}{C}{-}OCH_3 \longleftrightarrow CH_3\overset{+}{C}H{-}CH{=}\overset{\overset{OH}{|}}{C}{-}OCH_3 \right]$$

resonance interaction of the protonated carbonyl
group with the double bond

**21.10**   (a)  The mechanism of hydrolysis in acidic solution is much like that for the hydrolysis of an ester. (See the solution to Problem 20.12(a) on p. 506 of this manual.)

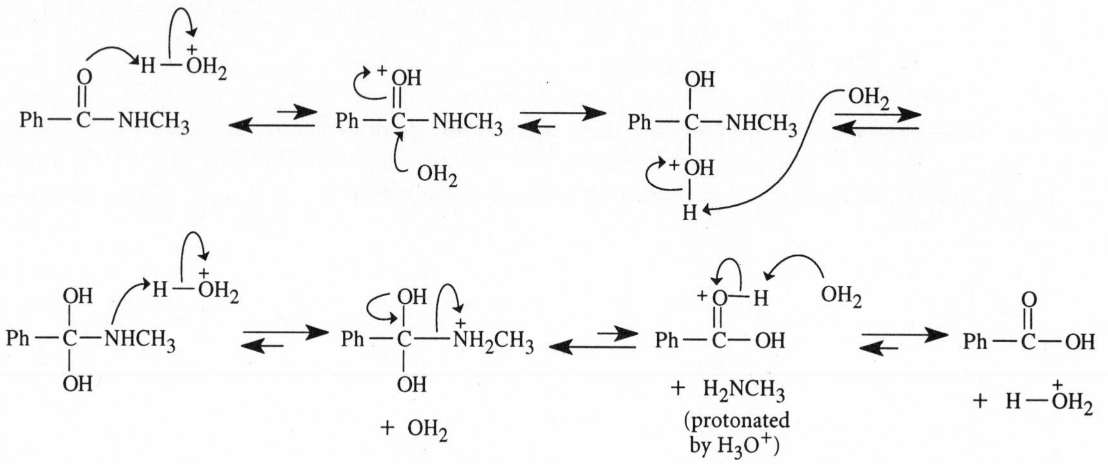

**21.11**   (a)

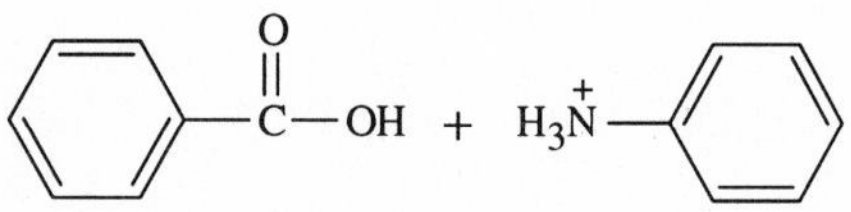

**21.12**   *Resonance effects:* The application of the resonance effect for predicting the relative rates of acid-catalyzed ester and amide hydrolyses is exactly the same as it is for the corresponding base-promoted hydrolyses. (See Eqs. 21.26 and 21.27 and the discussion on text pp. 997–998.)

   *Leaving-group basicities:* In the acid-catalyzed hydrolysis of an ester, the leaving group is an alcohol; in the acid-catalyzed hydrolysis of an amide, the leaving group is ammonia or an amine. Alcohols are much weaker bases (conjugate-acid $pK_a = -2$ to $-3$) than amines or ammonia (conjugate-acid $pK_a = 9–10$). Because weaker bases are better leaving groups, esters should hydrolyze more rapidly.

**21.14**   (a)  Because esters hydrolyze much more rapidly than nitriles, one equivalent of base brings about hydrolysis of only the ester.

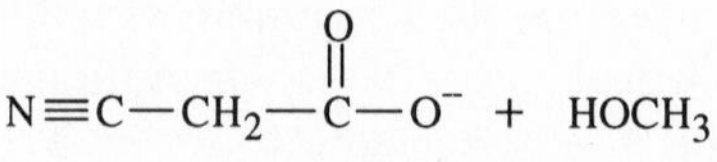

$$N{\equiv}C{-}CH_2{-}\overset{\overset{O}{\|}}{C}{-}O^- + HOCH_3$$

   (c)  Hydrolysis of one amide gives carbamic acid, which spontaneously decarboxylates; see Eq. 20.43 on text p. 959.

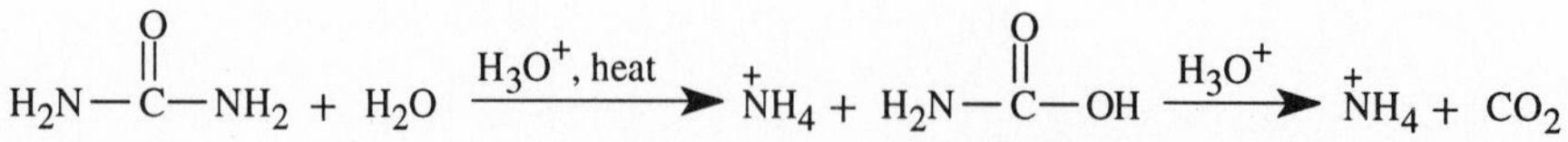

$$H_2N{-}\overset{\overset{O}{\|}}{C}{-}NH_2 + H_2O \xrightarrow{H_3O^+,\ heat} \overset{+}{N}H_4 + H_2N{-}\overset{\overset{O}{\|}}{C}{-}OH \xrightarrow{H_3O^+} \overset{+}{N}H_4 + CO_2$$

carbamic acid

**21.15** (a)

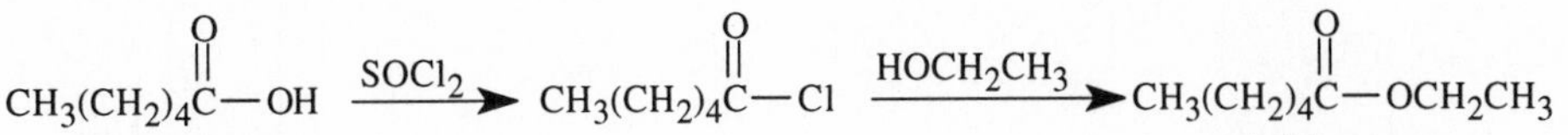

CH₃(CH₂)₄C—OH → [SOCl₂] → CH₃(CH₂)₄C—Cl → [HOCH₂CH₃] → CH₃(CH₂)₄C—OCH₂CH₃

hexanoic acid                                           ethyl hexanoate

**21.16** (a)        (c) 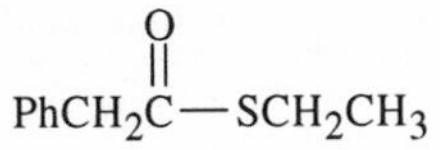        (e) 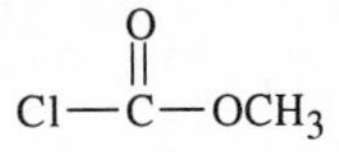        (g)

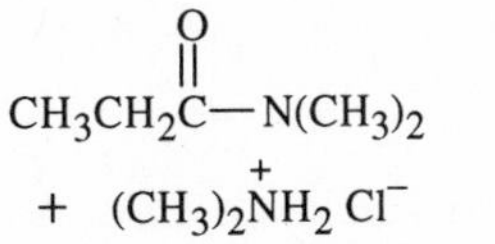

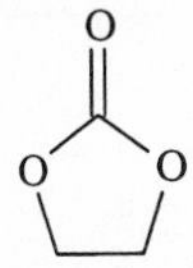

CH₃CH₂C—N(CH₃)₂          PhCH₂C—SCH₂CH₃        Cl—C—OCH₃

 + (CH₃)₂ṄH₂ Cl⁻

In part (e), there is insufficient methanol to react with both acid chlorides. The product is an ester-acid chloride, which is less reactive than the starting material, a double acid chloride, for the same reasons that esters are less reactive than acid chlorides. Consequently, methanol consumes one equivalent of phosgene (the acid chloride) before it can react with the acid chloride group in the product.

   The reaction in part (g) is a transesterification by one —OH group of the diol followed by an intramolecular transesterification by the second —OH group to form a cyclic di-ester.

**21.17** Because sulfonate esters behave like alkyl halides, cleavage in (a) occurs at the alkyl-oxygen bond. (See Study Guide Link 21.8 on p. 542 of this manual for a discussion.) Consequently, the oxygen isotope ends up in the product alcohol. Cleavage in (b) occurs at the acyl-oxygen bond; therefore, the isotope ends up in the product acetate ion. The products are therefore as follows: (*O = ¹⁸O)

(a)                                     (b)

PhCH₂*OH  +  ⁻OSO₂CH₃          PhCH₂OH  +  ⁻*O₂CCH₃

**21.18** (a)

Ph
CH₃CHOH  +  Cl—S⟨  ⟩—CH₃ → [pyridine] → CH₃CHO—S⟨  ⟩—CH₃

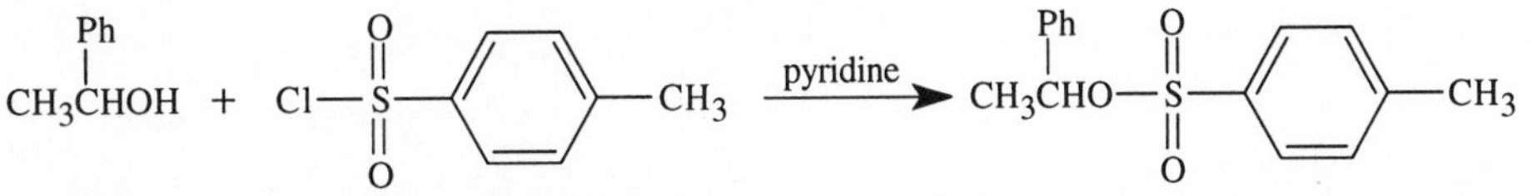

*p*-toluenesulfonyl chloride

(c) (See Problem 21.16(g) for a similar case.)

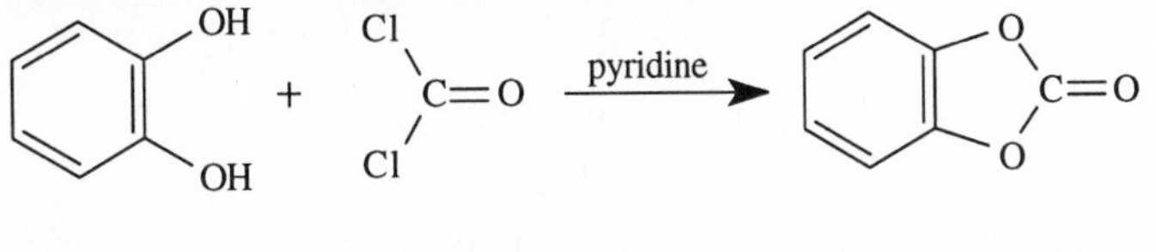

⟨benzenediol⟩ OH, OH  +  Cl, Cl C=O → [pyridine] → ⟨benzodioxole⟩ C=O

**21.19** (a)

PhC—Cl → [1) LiAlH₄  2) H₃O⁺] → PhCH₂OH

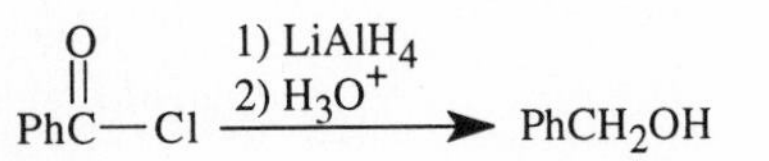

benzoyl chloride          benzyl alcohol

(c)

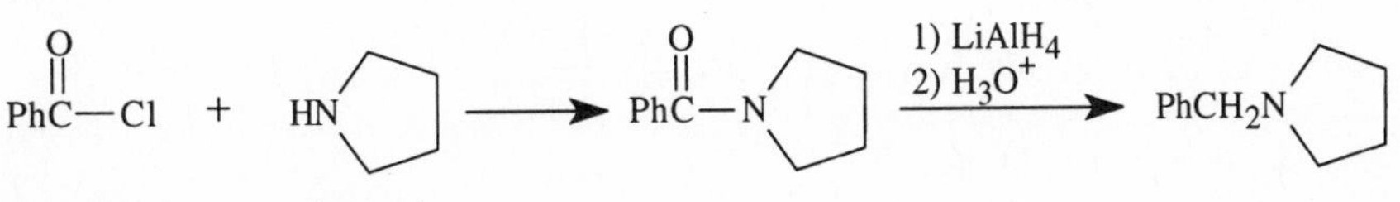

PhC—Cl  +  HN⟨ ⟩ → PhC—N⟨ ⟩ → [1) LiAlH₄  2) H₃O⁺] → PhCH₂N⟨ ⟩

(excess)

**21.20** (a) Although the ester reacts most rapidly, the nitrile also reacts because an excess of LiAlH₄ is present. The product is the amino alcohol HOCH₂CH₂CH₂NH₂ plus the by-product ethanol, C₂H₅OH.

(b) PhCH$_2$CH$_2$NH$_2$

(c) Both esters are reduced.

$$2 \text{ CH}_3\text{CH}_2\text{OH} + \text{PhCHCH}_2\text{OH} \overset{\text{OH}}{|}$$

(from reduction
of both the acetoxy
group and the ethyl ester)

**21.22** (a) As shown in Eq. 21.55, text p. 1011, the reduction involves an imine intermediate. This intermediate *A* can be attacked by a molecule of amine product to form the nitrogen analog of an acetal, which then can lose ammonia to form a different imine *B*, which in turn is reduced to the by-product. (You should be able to provide the curved-arrow formalism for the reactions comprising the *A* ⇄ *B* equilibrium in the overall process outlined below.)

$$\text{RC}{\equiv}\text{N} \xrightarrow{\text{H}_2,\text{ cat.}} \underset{A}{\text{RCH}{=}\text{NH}} \underset{\xleftarrow{\hspace{1cm}}}{\xrightarrow{\text{H}_2\text{NCH}_2\text{R}}} \underset{\underset{\text{NHCH}_2\text{R}}{|}}{\text{RCH}-\text{NH}_2} \rightleftharpoons \text{NH}_3 + \underset{\underset{\text{NCH}_2\text{R}}{\|}}{\text{RCH}} \xrightarrow{\text{H}_2,\text{ cat.}} \underset{\underset{\text{NHCH}_2\text{R}}{|}}{\text{RCH}_2}$$

$$B$$

$$\downarrow \text{H}_2, \text{ cat.}$$

$$\text{RCH}_2\text{NH}_2$$

(b) By LeChatelier's principle, ammonia drives the equilibria in the above equation away from imine *B* and back towards imine *A*, which is eventually reduced.

**21.23** (a)

$$\text{PhCO}_2\text{H} \xrightarrow{\text{C}_2\text{H}_5\text{OH, H}_2\text{SO}_4} \text{PhCO}_2\text{CH}_2\text{CH}_3 \xrightarrow[\text{2) H}_3\text{O}^+]{\text{1) PhMgBr (excess)}} \underset{\text{triphenylmethanol}}{\text{Ph}_3\text{C}-\text{OH}}$$

(c) First, the nitrile is hydrolyzed to isobutyric acid, which is converted into its acid chloride:

$$\underset{\text{isobutyronitrile}}{(\text{CH}_3)_2\text{CHC}{\equiv}\text{N}} \xrightarrow{\text{H}_2\text{O, H}_3\text{O}^+,\text{ heat}} \underset{\text{isobutyric acid}}{(\text{CH}_3)_2\text{CHC}\overset{\text{O}}{\overset{\|}{{}}}-\text{OH}} \xrightarrow{\text{SOCl}_2} \underset{\text{isobutyryl chloride}}{(\text{CH}_3)_2\text{CHC}\overset{\text{O}}{\overset{\|}{{}}}-\text{Cl}}$$

One synthesis involves reaction of the acid chloride with lithium dimethylcuprate :

$$(\text{CH}_3)_2\text{CHC}\overset{\text{O}}{\overset{\|}{{}}}-\text{Cl} \xrightarrow[\text{2) H}_3\text{O}^+]{\text{1) Li}^+\,{}^-\text{Cu(CH}_3)_2} (\text{CH}_3)_2\text{CHC}\overset{\text{O}}{\overset{\|}{{}}}-\text{CH}_3 \xrightarrow[\text{2) H}_3\text{O}^+]{\text{1) CH}_3\text{MgI}} \underset{\underset{\text{CH}_3}{|}}{(\text{CH}_3)_2\text{CHC}\overset{\text{OH}}{\overset{|}{{}}}-\text{CH}_3}$$

**2,3-dimethyl-2-butanol**

The second synthesis involves the Grignard reaction of an ester:

$$(\text{CH}_3)_2\text{CHC}\overset{\text{O}}{\overset{\|}{{}}}-\text{OH} \xrightarrow[\text{H}_2\text{SO}_4]{\text{C}_2\text{H}_5\text{OH (solvent)}} (\text{CH}_3)_2\text{CHC}\overset{\text{O}}{\overset{\|}{{}}}-\text{OC}_2\text{H}_5 \xrightarrow[\text{2) H}_3\text{O}^+]{\text{1) CH}_3\text{MgI (excess)}} \underset{\underset{\text{CH}_3}{|}}{(\text{CH}_3)_2\text{CHC}\overset{\text{OH}}{\overset{|}{{}}}-\text{CH}_3}$$

**2,3-dimethyl-2-butanol**

Although you have not studied this reaction explicitly, a more direct route to the alcohol is the reaction of the acid chloride with excess $CH_3MgI$ followed by protonolysis.

$$(CH_3)_2CHC\overset{O}{\overset{\|}{-}}-Cl \quad \xrightarrow[\text{2) } H_3O^+]{\text{1) } CH_3MgI \text{ (excess)}} \quad (CH_3)_2CH\overset{OH}{\underset{CH_3}{\overset{|}{C}}}-CH_3$$

2,3-dimethyl-2-butanol

**21.24**  (a)  The reaction of a Grignard reagent with ethyl formate gives a secondary alcohol in which the two alkyl groups at the $\alpha$-carbon are identical.

$$CH_3CH_2O\overset{O}{\overset{\|}{C}}H \;+\; 2\,RMgBr \;\longrightarrow\; \xrightarrow{H_3O^+} \; R\overset{OH}{\overset{|}{-}}CH-R \;+\; CH_3CH_2OH$$

ethyl formate

**21.25**  (a)  Because diethyl carbonate has two ester groups, it reacts first to form a mono-ester, then a ketone, and finally, a tertiary alcohol of the form $R_3C$—OH. (You should fill in the mechanistic details.)

$$CH_3CH_2O\overset{O}{\overset{\|}{-}}\overset{}{C}-OCH_2CH_3 \xrightarrow[-CH_3CH_2O^-]{RMgBr} CH_3CH_2O\overset{O}{\overset{\|}{-}}C-R \xrightarrow[-CH_3CH_2O^-]{RMgBr} R\overset{O}{\overset{\|}{-}}C-R \xrightarrow{RMgBr}$$

diethyl carbonate

$$R\overset{O^-\,{}^+MgBr}{\underset{R}{\overset{|}{\overset{|}{C}}}}R \xrightarrow{H_3O^+} R\overset{OH}{\underset{R}{\overset{|}{\overset{|}{C}}}}R$$

**21.26**  (a)  Because only the acid chloride reacts, the product is the following keto nitrile.

$$N\equiv C(CH_2)_{10}\overset{O}{\overset{\|}{C}}CH_3 \quad \text{12-oxotridecanenitrile}$$

**21.27**  The first method involves carbonation of the corresponding Grignard reagent; the second involves hydrolysis of the nitrile.

$$(CH_3)_2CH(CH_2)_3Br$$

1-bromo-4-methylpentane

$$\xrightarrow[\text{ether}]{Mg} \xrightarrow[\text{2) } H_3O^+]{\text{1) } CO_2} (CH_3)_2CH(CH_2)_3CO_2H$$

5-methylhexanoic acid

$$\xrightarrow[\text{DMSO}]{Na^+\,{}^-CN} (CH_3)_2CH(CH_2)_3C\equiv N \xrightarrow{H_2O,\, H_3O^+,\, \text{heat}}$$

**21.29**  (a)–(b)  The required nitrile is 2-hydroxypropanenitrile. This is a cyanohydrin, and this cyanohydrin can be prepared by addition of cyanide to acetaldehyde.

$$CH_3\overset{O}{\overset{\|}{C}}H \xrightarrow{NaCN,\, H_2O} CH_3\overset{OH}{\underset{C\equiv N}{\overset{|}{C}}}H \xrightarrow{H_2O,\, H_3O^+,\, \text{heat}} CH_3\overset{OH}{\underset{CO_2H}{\overset{|}{C}}}H \quad \text{2-hydroxypropanoic acid}$$

*(solution continues)*

Notice that the alcohol does *not* dehydrate under the rather severe conditions of nitrile hydrolysis. The reason is that the carbocation intermediate that would be involved is destabilized by both the cyano group and the carboxy group of the product. Can you think of a reason why?

**21.30**  (a)  The polyester would be reduced to two diols by LiAlH₄:

$$\text{polyester} \xrightarrow[\text{2) H}_3\text{O}^+]{\text{1) LiAlH}_4} \quad \text{HOCH}_2\text{CH}_2\text{OH} \; + \; \text{HOCH}_2\text{—} \langle \text{aryl ring} \rangle \text{—CH}_2\text{OH}$$

**21.31**  Process (*a*) is catalytic hydrogenation. Process (*b*) is ether cleavage with HCl, ZnCl₂ (or other acidic catalyst) and heat. Process (*c*) consists of the S$_N$2 reactions of cyanide ion with the dichloride. Process (*d*) is catalytic hydrogenation. Process (*e*) is nitrile hydrolysis with aqueous acid and heat. Finally, process (*f*) is to mix the amine and the carboxylic acid and heat, or to form the di-acid chloride and allow it to react with the amine.

**21.32**  First, water opens the lactam to an amino acid *A*. The amino group of this compound serves as a nucleophile to open another lactam molecule; and the resulting amino group thus liberated repeats the process, thus growing the polymer chain:

(reaction mechanism scheme showing ε-caprolactam opening)

$$\cdots \xrightarrow{\text{proton transfers}} \cdots \longrightarrow \text{HOC(CH}_2)_5\text{NH}_2$$

*A*

$$\cdots \xrightarrow{\text{proton transfers}} \cdots \longrightarrow$$

$$\text{HOC(CH}_2)_5\text{NH} - \text{C(CH}_2)_5\text{NH}_2 \xrightarrow{\text{similar reactions}} \text{nylon-6}$$

↑
this amino group attacks another
ε-caprolactam molecule

## Solutions to Additional Problems

**21.33** (a)

PhCO$_2$H

+ CH$_3$CH$_2$OH

(b)

PhCO$_2^-$ Na$^+$

+ CH$_3$CH$_2$OH

(c)

$$\underset{\parallel}{PhC}-NH_2$$
(O on top)

+ CH$_3$CH$_2$OH

(d)

PhCH$_2$OH

+ CH$_3$CH$_2$OH

(e)

OH
|
PhCCH$_2$CH$_2$CH$_3$
|
CH$_2$CH$_2$CH$_3$

+ CH$_3$CH$_2$OH

(f)

O
‖
OCCH$_3$
|
PhCCH$_2$CH$_2$CH$_3$   +   (pyridinium) $\overset{+}{N}$H  Cl$^-$
|
CH$_2$CH$_2$CH$_3$

(g)

O   O
‖   ‖
O—S—Ph
|
PhCCH$_2$CH$_2$CH$_3$
|
CH$_2$CH$_2$CH$_3$

(h) In this reaction, ethoxide displaces ethoxide, but the reaction is "invisible" because the product is the same as the starting material. Thus, there is no *net* reaction.

**21.35** (a)

CH$_3$CH$_2$C≡N

propionitrile

(c)

O
‖
CH$_3$COCH$_2$CH$_2$CH$_2$CH$_3$

butyl acetate

(e)

N≡CCH$_2$C≡N

malononitrile

Malononitrile (propanedinitrile) in part (e) hydrolyzes in aqueous acid to malonic acid and ammonium ion, and the malonic acid decarboxylates to acetic acid and carbon dioxide.

**21.36** The easiest way to work this problem is to start with the structure and work backwards. Because carbonation of a Grignard reagent gives a carboxylic acid, compound *D* must be the corresponding alkyl bromide. Compound *E* must be the hydrocarbon that gives compound *D* on benzylic bromination. The Wolff-Kishner reaction produces compound *E* from compound *B*, which must be a ketone. Reduction of ketone *B* with sodium borohydride gives alcohol *C*, which is converted into alkyl bromide *D*. Finally, the structure of compound *B* shows that reaction *A* must be a Friedel-Crafts acylation. In summary:

O
‖
ClCCH$_2$CH$_3$, AlCl$_3$,
then H$_3$O$^+$

*A*

O
‖
PhCCH$_2$CH$_3$

*B*

OH
|
PhCHCH$_2$CH$_3$

*C*

Br
|
PhCHCH$_2$CH$_3$

*D*

PhCH$_2$CH$_2$CH$_3$

*E*

**21.37** The compound is (*R*)-(–)-1-phenyl-1,2-ethanediol. Notice that the asymmetric carbon is unaffected by these transformations.

OH
|
H⋯C—CO$_2$H
|
Ph

(*R*)-(–)-mandelic acid

→ (CH$_3$OH, H$_2$SO$_4$) →

OH
|
H⋯C—CO$_2$CH$_3$
|
Ph

→ (1) LiAlH$_4$  2) H$_3$O$^+$) →

OH
|
H⋯C—CH$_2$OH
|
Ph

(*R*)-(–)-1-phenyl-1,2-ethanediol
(a 1,2-glycol; reacts with periodic acid)

**21.39** (a) Compound *A* has the formula C$_{11}$H$_{13}$NO. The formula of compound *B* indicates that it differs from *A* by the addition of the elements of water. The fact that compound *C* is a lactone with an oxygen at carbon-4 and an additional hydrogen at carbon-3 suggests that addition of water to the double bond of compound *A*, that is, alkene hydration, occurs metabolically to give an alcohol *B*. The lactone *C* is formed by cyclization of compound *B*, that is, an intramolecular esterification. The structure of compound *B* is given in the mechanism of part (b).

(b) A curved-arrow mechanism for the formation of compound *C* from compound *B*:

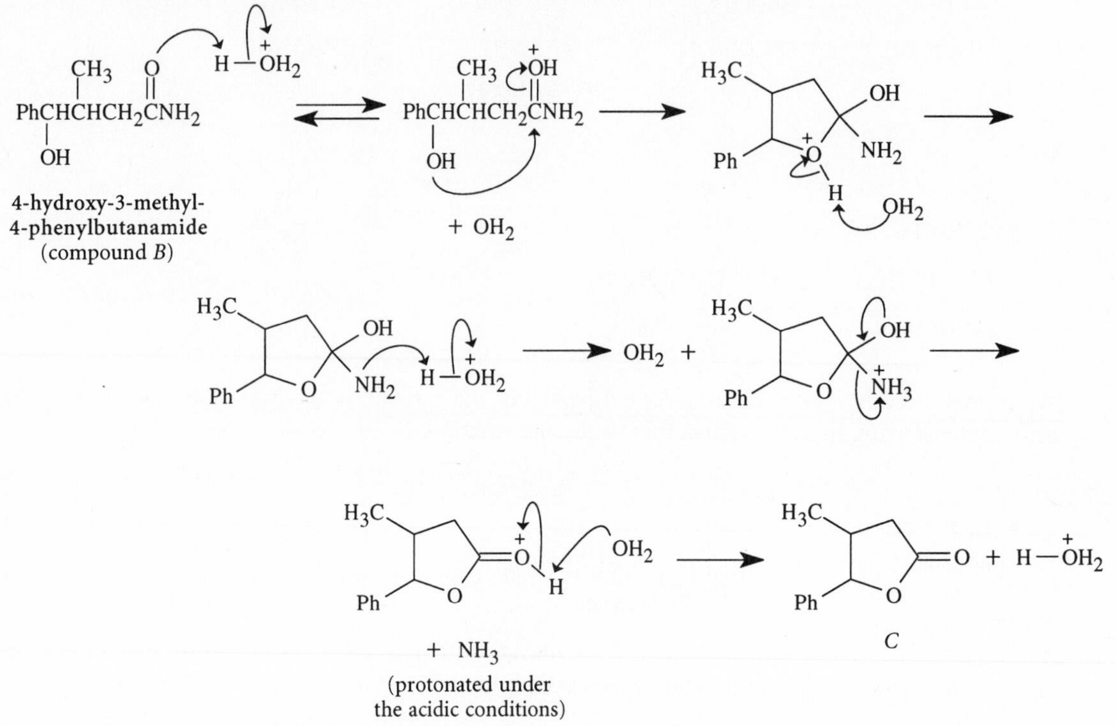
4-hydroxy-3-methyl-4-phenylbutanamide
(compound B)
+ OH2
+ NH3
(protonated under the acidic conditions)
C

21.41 (a)

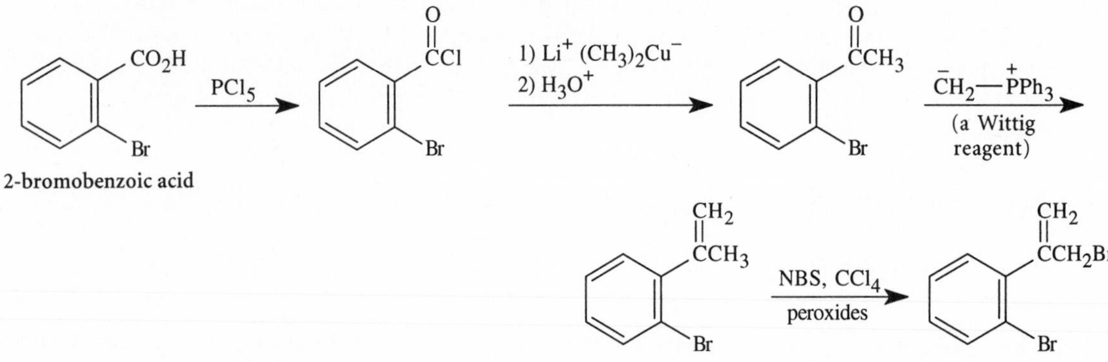
2-bromobenzoic acid
PCl5
1) Li+ (CH3)2Cu−
2) H3O+
(a Wittig reagent)
NBS, CCl4
peroxides

21.42 (a)

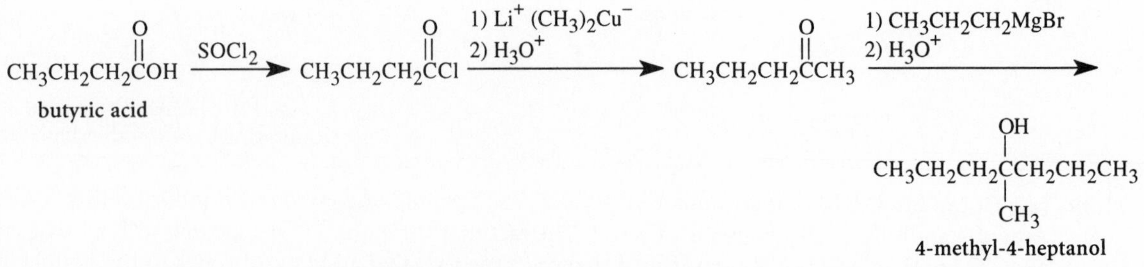
butyric acid
SOCl2
1) Li+ (CH3)2Cu−
2) H3O+
1) CH3CH2CH2MgBr
2) H3O+
4-methyl-4-heptanol

(c)

$$CH_3CH_2CH_2\overset{O}{\overset{\|}{C}}OH \xrightarrow{SOCl_2} CH_3CH_2CH_2\overset{O}{\overset{\|}{C}}Cl \xrightarrow[\text{Pd/C}]{H_2,\ \text{quinoline, S}} CH_3CH_2CH_2\overset{O}{\overset{\|}{C}}H \xrightarrow{\text{1) } CH_3CH_2CH_2MgBr \quad \text{2) } H_3O^+} CH_3CH_2CH_2\overset{OH}{\overset{|}{C}}HCH_2CH_2CH_3$$

**4-heptanol**

(d)

$$CH_3CH_2CH_2\overset{O}{\overset{\|}{C}}OH \xrightarrow{\text{1) } LiAlH_4 \quad \text{2) } H_3O^+} CH_3CH_2CH_2CH_2OH \xrightarrow{\text{conc. HBr},\ H_2SO_4} CH_3CH_2CH_2CH_2Br \xrightarrow[\text{ether}]{Mg}$$

$$CH_3CH_2CH_2CH_2MgBr \xrightarrow[\text{2) } H_3O^+]{\text{1) ethylene oxide}} CH_3CH_2CH_2CH_2CH_2CH_2OH \xrightarrow{\text{1) } KMnO_4,\ ^-OH \quad \text{2) } H_3O^+}$$

$$CH_3CH_2CH_2CH_2CH_2CO_2H \xrightarrow{SOCl_2} CH_3CH_2CH_2CH_2CH_2\overset{O}{\overset{\|}{C}}Cl \xrightarrow{NH_3\ \text{(excess)}}$$

$$CH_3CH_2CH_2CH_2CH_2\overset{O}{\overset{\|}{C}}NH_2 \xrightarrow{\text{1) } LiAlH_4 \quad \text{2) } H_2O} CH_3CH_2CH_2CH_2CH_2CH_2NH_2$$

(e)

$$CH_3CH_2CH_2\overset{O}{\overset{\|}{C}}OH \xrightarrow{\text{1) } LiAlH_4 \quad \text{2) } H_3O^+} CH_3CH_2CH_2CH_2OH \xrightarrow{\text{conc. HBr},\ H_2SO_4} CH_3CH_2CH_2CH_2Br \xrightarrow[\text{ether}]{Mg}$$

$$CH_3CH_2CH_2CH_2MgBr \xrightarrow{\text{1) } CO_2 \quad \text{2) } H_3O^+} CH_3CH_2CH_2CH_2CO_2H \xrightarrow{SOCl_2} CH_3CH_2CH_2CH_2\overset{O}{\overset{\|}{C}}Cl \xrightarrow{NH_3\ \text{(excess)}}$$

$$CH_3CH_2CH_2CH_2\overset{O}{\overset{\|}{C}}NH_2 \xrightarrow{\text{1) } LiAlH_4 \quad \text{2) } H_2O} CH_3CH_2CH_2CH_2CH_2NH_2$$

Notice that parts (d) and (e) involve strategies for lengthening carbon chains.

**21.43** (a) First of all, two sets of constitutional isomers can be obtained as a result of the attack of 1-phenylethanol at either of the nonequivalent carbonyl groups of the anhydride:

α-phenylglutaric anhydride    *A*    *B*

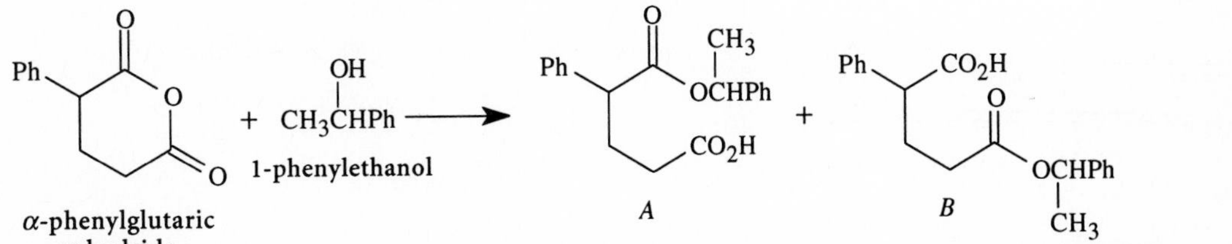

Each of these products has two asymmetric carbon stereocenters and can thus exist as four stereoisomers. All stereoisomers of *A* can in principle be separated from all stereoisomers of *B* because constitutional isomers have different properties. The four stereoisomers of *A* can be separated into two pairs of enantiomers; the same is true of *B*. Consequently, four compounds can in principle be obtained from this reaction mixture without enantiomeric resolution.

(b) The same two constitutional isomers would be obtained, and each would be obtained as a pair of diastereomers. However, only one enantiomer of each diastereomer (the one with the *S* configuration at the carbon derived from the alcohol) would be formed.

 If the diastereomeric esters from part (b) were separated from each other and saponified, the α-phenylglutaric acid formed would be enantiomerically pure. The process described here would therefore be useful as an enantiomeric resolution of α-phenylglutaric acid (or any other chiral dicarboxylic acid from which a cyclic anhydride can be readily formed).

**21.45** Thiol esters are less stabilized by resonance than esters because the resonance interaction of sulfur unshared electron pairs with a carbonyl group is less effective than the resonance interaction of oxygen unshared electron pairs. The reason for this difference is that the orbitals on sulfur are derived from quantum level 3, and such orbitals overlap poorly with the $\pi$ orbitals of a carbonyl group, which are derived from quantum level 2. (See Fig. 16.9 on text p. 770 and the accompanying discussion of this point.) Thus, resonance arguments lead to the conclusion that a thiol ester should hydrolyze more rapidly than an ester. In addition, a thiolate ion is a better leaving group than an alkoxide because thiolates are less basic than alkoxides. Whether resonance stabilization or leaving-group effectiveness is considered, thiol esters are predicted to be more reactive than esters.

   Because both sulfur and chlorine are in the third period of the periodic table, the resonance stabilizations of acid chlorides and thiol esters are comparable. However, chlorine is much more electronegative than sulfur; consequently, attack of base at the carbonyl carbon of an acid chloride is faster. In addition, chloride ion is a much better leaving group than a thiolate ion, because chloride ion is a weaker base than a thiolate ion. These points argue for increased reactivity of acid chlorides relative to thiol esters.

   In summary, the correct answer is (2): thiol esters are less reactive than acid chlorides but more reactive than esters.

**21.47** (a) Synthesis of nylon-4,6 requires mixing 1,4-butanediamine and adipoyl dichloride.

$$H_2N(CH_2)_4NH_2 \;+\; \underset{\text{adipoyl dichloride}}{Cl\overset{\overset{\displaystyle O}{\|}}{C}(CH_2)_4\overset{\overset{\displaystyle O}{\|}}{C}Cl} \longrightarrow \text{nylon-4,6}$$

$$\underset{\text{1,4-butanediamine}}{H_2N(CH_2)_4NH_2}$$

(Alternatively, 1,4-butanediamine can be heated with adipic acid; see Eq. 21.69 on text p. 1021.) The preparations of these materials are as follows:

$$\underset{\text{succinic acid}}{HO\overset{\overset{\displaystyle O}{\|}}{C}CH_2CH_2\overset{\overset{\displaystyle O}{\|}}{C}OH} \xrightarrow{SOCl_2} \underset{}{Cl\overset{\overset{\displaystyle O}{\|}}{C}CH_2CH_2\overset{\overset{\displaystyle O}{\|}}{C}Cl} \xrightarrow{NH_3\text{ (excess)}} \underset{}{H_2N\overset{\overset{\displaystyle O}{\|}}{C}CH_2CH_2\overset{\overset{\displaystyle O}{\|}}{C}NH_2} \xrightarrow[\text{2) H}_2\text{O}]{\text{1) LiAlH}_4} \underset{\text{1,4-butanediamine}}{H_2N(CH_2)_4NH_2}$$

$$\underset{\text{adipic acid}}{HO\overset{\overset{\displaystyle O}{\|}}{C}(CH_2)_4\overset{\overset{\displaystyle O}{\|}}{C}OH} \xrightarrow{SOCl_2} \underset{\text{adipoyl dichloride}}{Cl\overset{\overset{\displaystyle O}{\|}}{C}(CH_2)_4\overset{\overset{\displaystyle O}{\|}}{C}Cl}$$

 Why do you think heating the dicarboxylic acid with the diamine is preferred industrially to reaction of the diamine with the dicarboxylic acid dichloride? (There are at least two reasons.)

**21.48** The IR spectrum of compound *A* indicates that it is an anhydride, and probably a cyclic anhydride containing a five-membered ring. (Compare the carbonyl absorptions of compound *A* with those in Table 21.3 on text p. 983; the 1050 cm⁻¹ absorption is a C—O stretching absorption). Addition of methanol to a cyclic anhydride should give a methyl half-ester, compound *B*. Indeed, the IR spectrum of compound *B* indicates the presence of both a carboxylic acid and an ester group. Subtracting the elements of methanol (CH₄O) from the formula of compound *B* gives the formula of compound *A*, C₄H₄O₃. Compound *A* is succinic anhydride. There are two points about the NMR spectra that are worth noting. First, the methylene protons of compound *B* are detected as a *singlet* at $\delta$ 2.7; thus, they are accidentally equivalent, although they are in principle chemically nonequivalent. However, this is a reasonable observation, because both sets of protons are in very similar electronic environments; both are on carbons that are $\alpha$ to carbonyl groups. The second point is that the proton of the carboxy group of compound *B* is not observed because the spectrum is taken in $D_2O$. Recall (Sec. 13.6D, text pp. 616–618) that O—H protons are rapidly exchanged for deuterium in when $D_2O$ is present, and that deuterium nuclei are "silent" in proton NMR.

succinic anhydride
(compound *A*)

methyl hydrogen succinate
(compound *B*)

**21.50** (a) This is a reprise of Problem 10.56 on text p. 496, which is answered on p. 243, Vol. 1, of this manual. Obviously, Klutz McFingers has been talking to Buster Bluelip. The very weak acid HCN is not significantly dissociated. Consequently, the solution is insufficiently acidic to effect protonation of the —OH group of the alcohol. This protonation is necessary to convert this group into a good leaving group. Furthermore, there is vitrually no cyanide ion ($^-$C≡N) present, and hence virtually no nucleophile to displace the —OH group.

(b) First of all, an excess of an alcohol is generally required to drive acid-catalyzed esterification to completion. However, even if some of the adipic acid is converted into its ester, there is no reason why this mono-ester would not be essentially as reactive as adipic acid itself, and the reaction mixture would ultimately contain a mixture of adipic acid, its monomethyl ester, and its dimethyl ester. The yield of the desired mono-ester would be poor, and it would have to be separated from both the di-ester and the unreacted adipic acid.

(d) The hydroxide ion reacts *much* more rapidly with the O—H proton of the phenol than it does with the ester. (See Study Guide Link 20.2 on p. 499 of this manual for a discussion of this point.) Consequently, the one equivalent of hydroxide is consumed by this reaction, and no base is left to saponify the ester.

(e) There are several functional groups in the $\beta$-lactam molecule that are more reactive toward acid hydrolysis than the amide indicated. Because esters are more reactive than amides, the acetate ester will undoubtedly also hydrolyze under the reaction conditions. The $\beta$-lactam ring itself is an amide, and because hydrolysis of this amide relieves substantial strain, this amide should also hydrolyze more rapidly. Once the $\beta$-lactam hydrolyzes, the resulting enamine also should hydrolyze readily in acid, and the N—C—S linkage is also unstable toward hydrolysis in the same sense that an acetal is unstable. With so many faster competing processes, Klutz has no hope for a selective reaction.

**21.51** (a) Because phenols are very reactive in electrophilic substitution reactions, bromine in $CCl_4$ will effect bromination of the phenol ring as well as addition to the alkene double bond. (See Eq. 18.40 on text p. 846.)

(b) 5% aqueous NaOH will bring about ionization of the phenol to its conjugate-base phenolate ion. The amide is unaffected, because amide hydrolysis requires heat and strong base.

*(solution continues)*

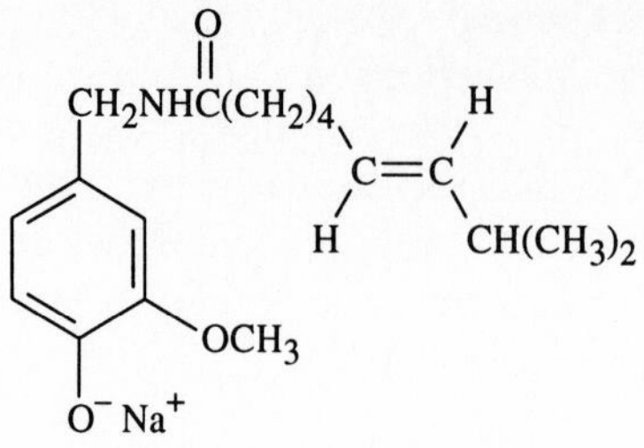

(c) There is no significant reaction with 5% aqueous HCl. (The carbonyl oxygen of the amide is protonated to a small extent.)

(d) Catalytic hydrogenation brings about addition of hydrogen to the alkene double bond.

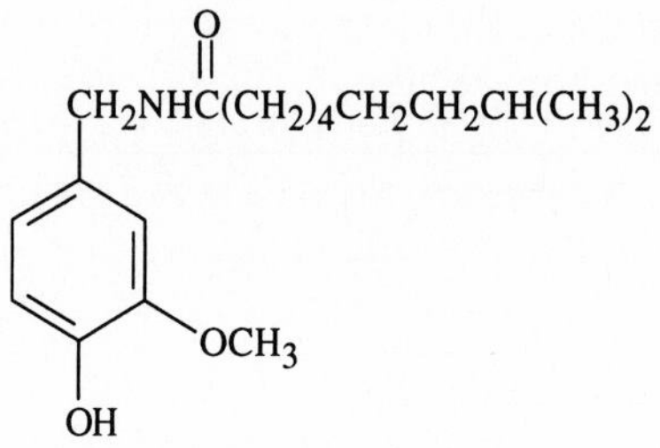

⇒ Depending on the conditions, hydrogenation of the *single bond* between the benzylic $CH_2$ and the nitrogen can also occur; such a reaction is called *hydrogenolysis*. Because you have not studied this reaction, it is ignored.

(e) Heating the product of (d) in 6 *M* HCl brings about amide hydrolysis.

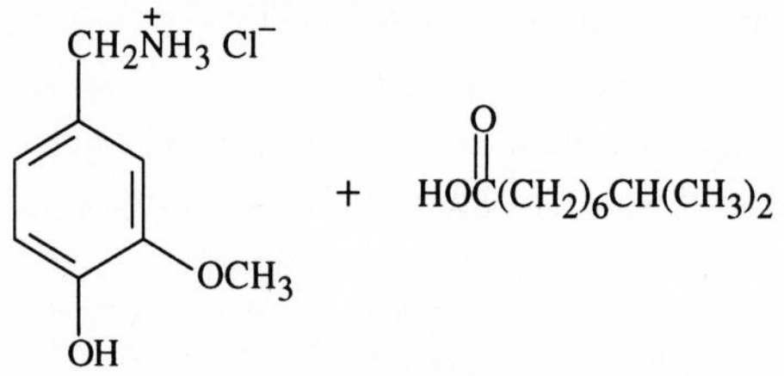

(It is also conceivable that some cleavage of the methyl ether will occur to give methyl chloride and the diphenol.)

(f) The phenolate ion formed in part (b) is alkylated to give a second methyl ether group.

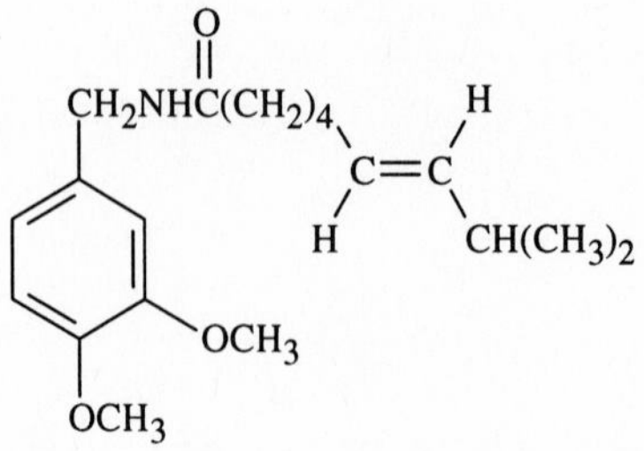

(g) Except for protonation of the amide, no reaction will occur unless heat is applied. However, on heating, the same amide hydrolysis observed in part (e) will take place, and the methyl ether will cleave.

$$\text{CH}_3\text{Br} \; + \; \text{[ring with CH}_2\overset{+}{\text{N}}\text{H}_3\text{ Br}^-\text{, OH, OH]} \; + \; \text{HO}\overset{\text{O}}{\overset{\|}{\text{C}}}(\text{CH}_2)_6\text{CH(CH}_3)_2$$

**21.52** Use the titration data to calculate the molecular mass of the ester. The amount of 1 $M$ NaOH solution consumed in the saponification reaction is (15.00 mL − 5.30 mL) = 9.70 mL. Make the provisional assumption that there is one ester group per molecule of compound $A$. Then one mole of NaOH is consumed per mole of ester saponified, and the molecular mass of compound $A$ is then

$$\text{molecular mass of } A = \frac{2.00 \text{ g}}{(9.70 \times 10^{-3} \text{ L})(1.00 \text{ mol/L})} = 206 \text{ g/mol}$$

Next, try to deduce a structure from the data in the problem. The structure of compound $C$, the alcohol that results from the saponification of compound $A$, can be determined; it is the alcohol that can be oxidized to acetophenone. Therefore, compound $C$ is 1-phenylethanol.

$$\underset{\substack{\text{1-phenylethanol} \\ \text{(compound } C)}}{\text{Ph}-\overset{\overset{\text{OH}}{|}}{\text{CH}}-\text{CH}_3} \quad \xrightarrow{\text{K}_2\text{Cr}_2\text{O}_7} \quad \underset{\text{acetophenone}}{\text{Ph}-\overset{\overset{\text{O}}{\|}}{\text{C}}-\text{CH}_3}$$

The molecular mass of compound $C$ is 122 g/mol. Now, from the overall transformation of ester hydrolysis, the molecular mass of ester $A$ must equal the sum of the molecular masses of compounds $B$ and $C$ minus the molecular mass of $H_2O$, that is,

molecular mass of $A$ = 206 g/mol = molecular mass of $B$ + 122 g/mol − 18 g/mol

or

molecular mass of $B$ = 206 g/mol + 18 g/mol − 122 g/mol = 102 g/mol

Compound $B$ is a carboxylic acid; its carboxy group accounts for 45 molecular mass units. Therefore, the remainder of the carboxylic acid has a mass of 57 units, which is the mass of a butyl group. Because acid $B$ is optically active, it is chiral; therefore the butyl group must contain an asymmetric carbon stereocenter. Only a *sec*-butyl group meets this criterion. The structures of compounds $B$ and $A$ are therefore

$$\underset{\substack{\text{2-methylbutanoic acid} \\ \text{(compound } B)}}{\text{CH}_3\text{CH}_2\overset{\overset{\overset{\text{CH}_3}{|}}{}}{\text{CH}}\text{CO}_2\text{H}} \qquad \underset{\substack{\text{1-phenylethyl 2-methylbutanoate} \\ \text{(compound } A)}}{\text{CH}_3\text{CH}_2\overset{\overset{\text{CH}_3}{|}}{\text{CH}}-\overset{\overset{\text{O}}{\|}}{\text{C}}-\text{O}\underset{\underset{\text{CH}_3}{|}}{\text{CHPh}}}$$

As noted in the problem, the absolute configurations of the asymmetric carbons cannot be determined from the data. Hence, compounds $B$ and $C$ are each one of two possible enantiomers, and compound $A$ is the particular stereoisomer with the corresponding absolute configurations of its asymmetric carbons. Compound $D$ is the racemate of compound $C$, and compound $E$ is the diastereomer of compound $A$ with the following configurations:

$$\text{CH}_3\text{CH}_2\overset{\overset{\text{CH}_3}{|}}{\text{CH}}-\overset{\overset{\text{O}}{\|}}{\text{C}}-\text{O}\underset{\underset{\text{CH}_3}{|}}{\text{CHPh}}$$

same configuration as compound $A$

opposite configuration to compound $A$

compound $E$

**21.54** (a) The "alcohol" formed as one of the saponification products is an enol, which spontaneously reverts to the corresponding ketone. (See Eq. 14.5c on text p. 654.)

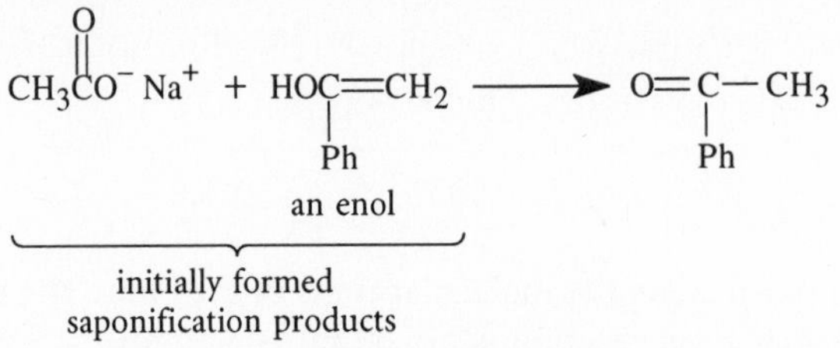

$$CH_3\overset{O}{\underset{}{C}}O^-\ Na^+\ +\ HOC{=}CH_2 \longrightarrow O{=}C{-}CH_3$$

$$\underset{Ph}{\quad\quad\quad\quad\quad} \underset{Ph}{}$$

an enol

initially formed<br>saponification products

(c) The formula shows that the product includes the carbonyl group of phosgene (the acid chloride) minus two HCl molecules. Because the product has an unsaturation number of 7, a ring must be formed. Because both amines and carboxylic acids can react with acid chlorides, the product, called *isatoic anhydride,* is simultaneously a cyclic anhydride and a cyclic amide (lactam).

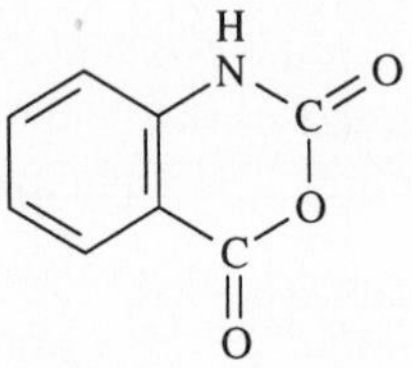

isatoic anhydride

(e) Isotopically labeled methyl benzenesulfonate is formed in the first reaction, and unlabeled methanol is formed in the second by an $S_N2$ reaction of hydroxide at the methyl carbon. (Notice that, because of resonance in the by-product benzenesulfonate anion, the three oxygens are equivalent and indistinguishable.)

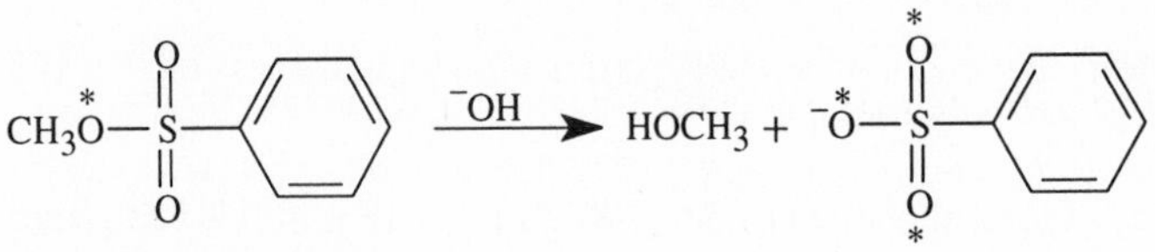

$$CH_3\overset{*}{O}{-}\overset{O}{\underset{O}{S}}{-}\ \ \ \ \overset{^-OH}{\longrightarrow}\ HOCH_3\ +\ {}^-\overset{*}{O}{-}\overset{\overset{*}{O}}{\underset{\underset{*}{O}}{S}}{-}$$

methyl benzenesulfonate            benzenesulfonate ion

(g) Both of the amide bonds hydrolyze; the carboxylic acid product, carbonic acid, is unstable and decomposes spontaneously to $CO_2$ and $H_2O$, and the amine and ammonia are protonated under the acidic reaction conditions.

$$(CH_3)_3C\overset{+}{N}H_3\ +\ \overset{+}{N}H_4\ +\ CO_2\ +\ H_2O$$

(i) The cyclic amide is reduced to a cyclic amine.

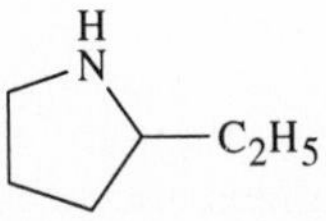

$${-}C_2H_5$$

(k) See the solution to Problem 21.25(a) on p. 549 of this manual for a rationale for this answer. (Et = ethyl = $CH_3CH_2{-}$)

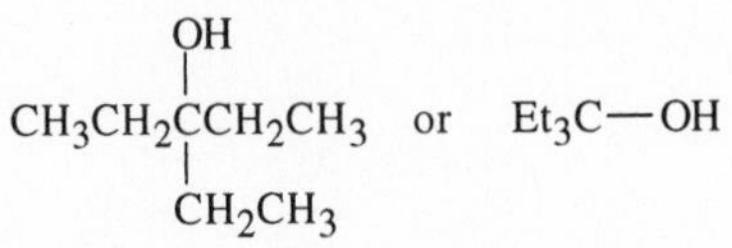

$$\underset{\underset{CH_2CH_3}{|}}{CH_3CH_2\overset{\overset{OH}{|}}{C}CH_2CH_3}\quad or\quad Et_3C{-}OH$$

(m) Because aldehydes are much more reactive than esters, only the aldehyde reacts when only one equivalent of the Grignard reagent is used.

$$O \qquad OH$$
$$CH_3OC(CH_2)_3CHPh$$

(o) The ester group undergoes intramolecular aminolysis to form a γ-lactam. Such intramolecular nucleophilic substitution reactions are very common when five- and six-membered rings can be formed.

(image of γ-lactam structure)   +   $CH_3OH$

21.55 (a) The carbonyl absorption at 1721 cm$^{-1}$ and the slow dissolution (due to saponification) in NaOH identify compound *A* as an ester. The absorption at 3278 cm$^{-1}$ indicates a 1-alkyne.

$$O$$
$$HC\equiv C-C-OCH_2CH_3 \qquad \text{ethyl propynoate (compound } A\text{)}$$

(c) The odd molecular mass indicates the presence of an odd number of nitrogens, and the hydroxamate test as well as the IR carbonyl absorption at 1733 cm$^{-1}$ indicates the presence of an ester. The IR absorption at 2237 cm$^{-1}$ indicates a nitrile. The triplet-quartet pattern in the NMR spectrum clearly indicates that compound *B* is an ethyl ester. Subtracting the masses of all the atoms accounted for leaves 14 mass units, the mass of a CH$_2$ group. Compound *C* is ethyl cyanoacetate.

$$O$$
$$N\equiv C-CH_2-C-OCH_2CH_3 \qquad \text{ethyl cyanoacetate (compound } C\text{)}$$

(e) The IR absorption indicates the presence of an ester. The doubled peaks in the mass spectrum indicate the presence of a single bromine. Consequently, compound *E* contains the elements CO$_2$ (from the ester) and Br. The remaining mass could correspond to C$_4$H$_9$. Therefore, adopt C$_5$H$_9$O$_2$Br as a provisional formula. In the NMR spectrum, the triplet and the doublet at high field cannot be splitting each other, because their integrations are not consistent with such a situation. That is, a mutually split triplet (which indicates two adjacent hydrogens) and doublet (which indicates one adjacent hydrogen) must have an integral ratio of 1:2, not 1:1 as observed. Evidently, the complex absorption near $\delta$ 4.3 contains both of the resonances that are split by the resonances at higher field. From their chemical shifts, some of these protons are probably $\alpha$ to the Br and some are probably $\alpha$ to the oxygen. If the ester is an ethyl ester, then the high-field triplet is the resonance of the methyl group, and, from its integration, the doublet is also due to a methyl group on a carbon adjacent to a CH. These deductions conspire to identify compound *E* as ethyl 2-bromopropanoate.

$$O$$
$$CH_3CH-C-OCH_2CH_3 \qquad \text{ethyl 2-bromopropanoate (compound } E\text{)}$$
$$|$$
$$Br$$

(g) The presence of six carbons in the formula and three resonances in the CMR spectrum indicate a symmetrical structure; the IR absorption indicates an ester. Two structures are worth considering:

$$O \quad O \qquad\qquad\qquad O \qquad\qquad O$$
$$CH_3CH_2O-C-C-OCH_2CH_3 \qquad CH_3O-C-CH_2CH_2-C-OCH_3$$
$$\text{diethyl oxalate} \qquad\qquad\qquad \text{dimethyl succinate}$$
$$\qquad\qquad\qquad\qquad\qquad\qquad \text{(compound } G\text{)}$$

According to the CMR of ethyl acetate on text p. 985, the resonance at $\delta$ 51.8 is at too high a field to be the CH$_2$ group of an ethyl ester, but it is consistent with the methyl group of a methyl ester. Similarly, the resonance at $\delta$ 28.9 is at too high a field to be the methyl group of an ethyl ester, but it could be a $CH_2$ group $\alpha$ to a carbonyl. Thus, the spectral data are more consistent with dimethyl succinate which is, in fact, compound *G*.

21.56 (a)

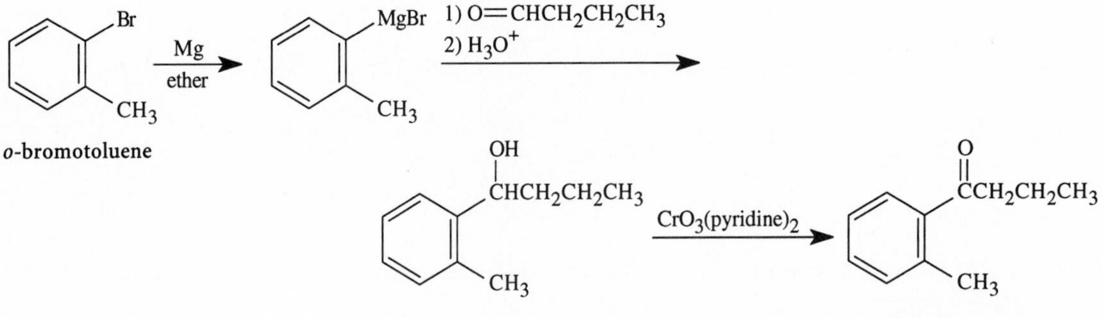


*o*-bromotoluene

*o*-methylbutyrophenone

(c)

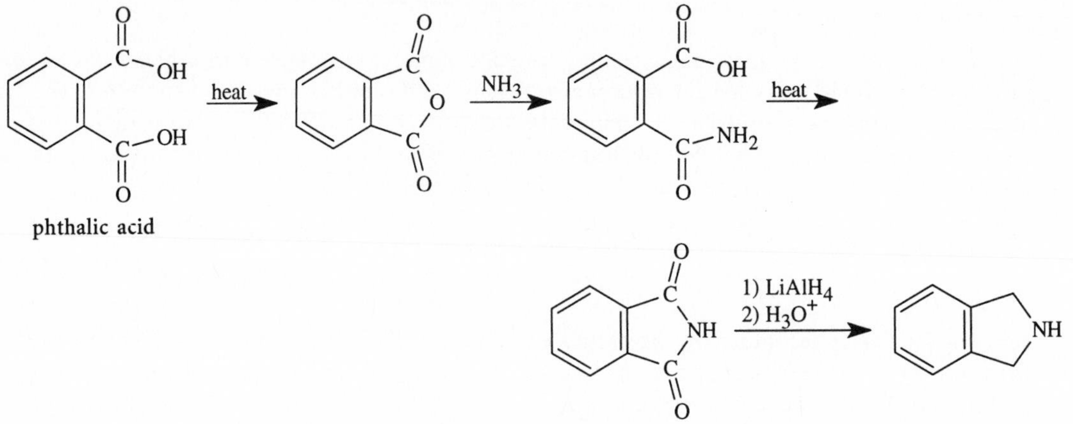


phthalic acid

(e)

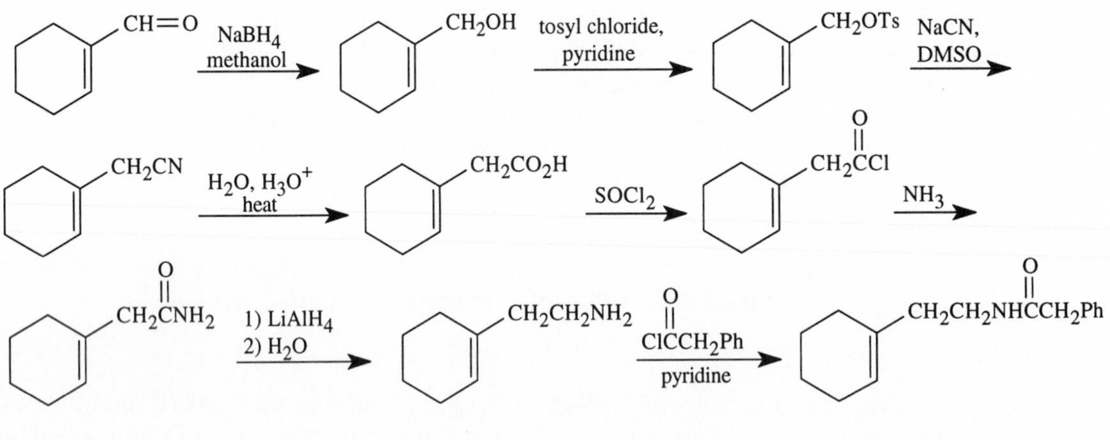


(g)

$$\text{HO—}\overset{\text{O}}{\overset{\|}{\text{C}}}\text{CH}_2\text{CH}_2\overset{\text{O}}{\overset{\|}{\text{C}}}\text{—OH} \xrightarrow{\text{SOCl}_2} \text{Cl—}\overset{\text{O}}{\overset{\|}{\text{C}}}\text{CH}_2\text{CH}_2\overset{\text{O}}{\overset{\|}{\text{C}}}\text{—Cl} \xrightarrow{\text{H}_2\overset{*}{\text{O}}} \overset{*}{\text{HO}}\text{—}\overset{\overset{*}{\text{O}}}{\overset{\|}{\text{C}}}\text{CH}_2\text{CH}_2\overset{\overset{*}{\text{O}}}{\overset{\|}{\text{C}}}\text{—}\overset{*}{\text{OH}}$$

Another synthesis would be to treat succinic anhydride with isotopically labeled water. However, this would not give as great an amount of isotope incorporated (why?). Notice that the isotope at any one oxygen of a carboxy group is rapidly scrambled to both positions; see Problem 20.35 on text p. 963.

(h)

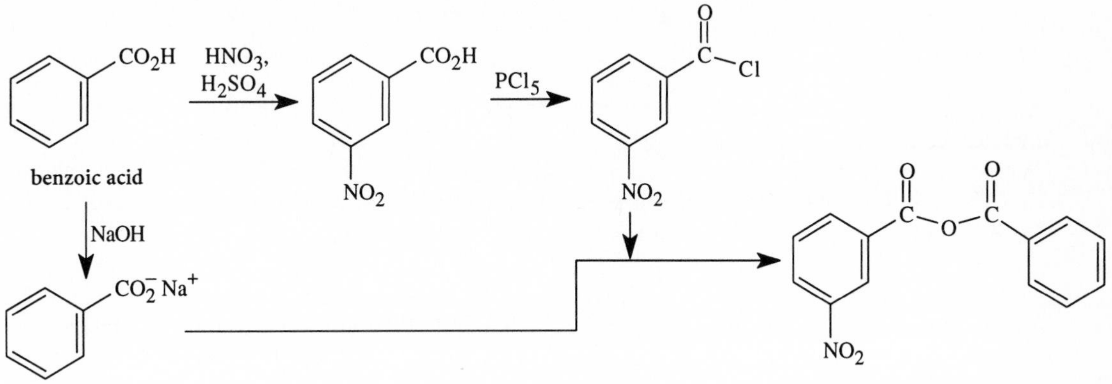

(j)

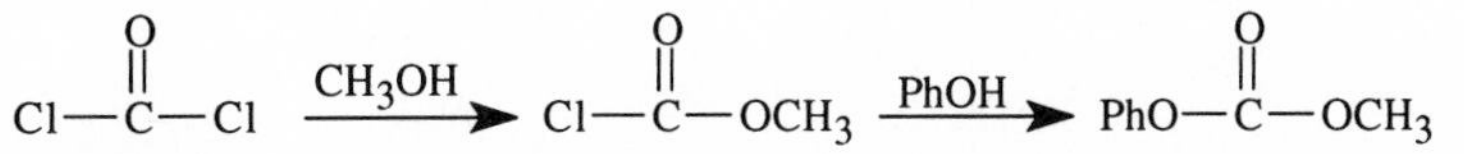

An excess of phosgene is used in the first step to ensure that only one chlorine is displaced. It is probably best to use methanol as the first nucleophile because phenyl esters are fairly reactive and could be trans-esterified by methanol if the order were reversed.

**21.57** (a) Compound *A* is an *O*-substituted oxime, which is formed by the mechanism summarized in Eqs. 19.56a–b on text p. 905. Compound *A* undergoes elimination in base to give the nitrile. (In the following mechanism, the 2,4-dinitrophenyl group is abbreviated Ar.)

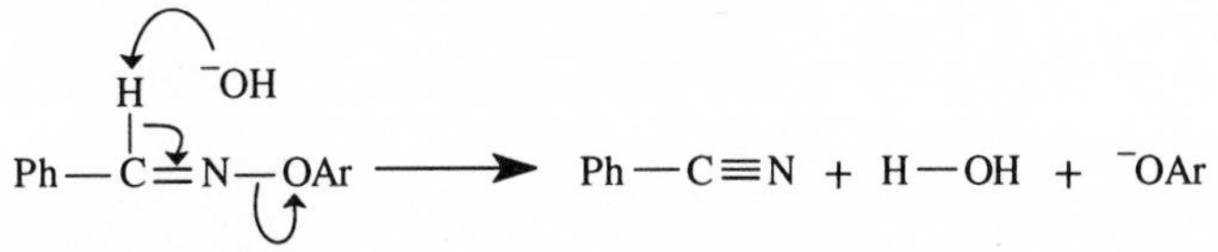

(b) The first step in this transformation is formation of a Grignard reagent. (See Eqs. 8.20a–b on text p. 369.) Because of its anionic character, the Grignard reagent undergoes an elimination to give the conjugate-base carboxylate ion of the product, which is protonated when acid is added. The Grignard reagent is written as a carbanion in the following mechanism to emphasize its anionic character.

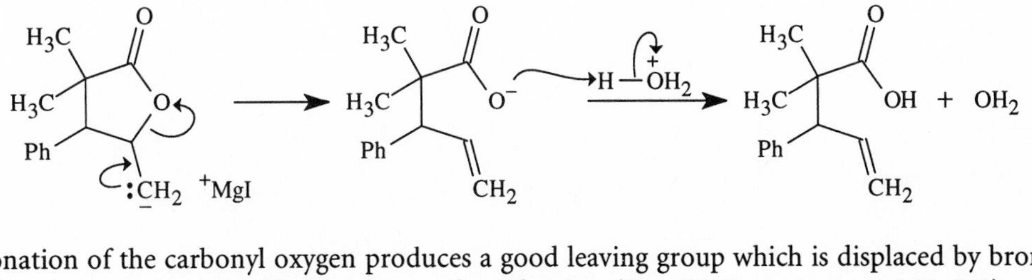

(c) Protonation of the carbonyl oxygen produces a good leaving group which is displaced by bromide ion. The carboxy group is then esterified by the usual mechanism (Eqs. 20.18a–c, text pp. 947–948).

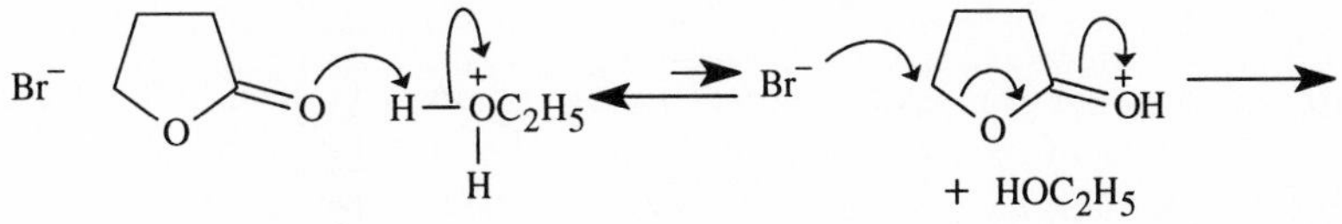

*(solution continues)*

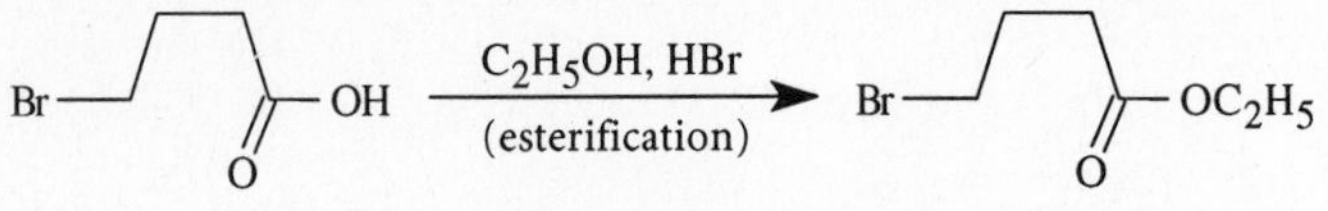

An alternative reasonable mechanism involves transesterification of the lactone by ethanol to give a hydroxy ester, followed by conversion of the alcohol group to a bromide by an acid-catalyzed $S_N2$ reaction. (See Eq. 10.10b, text p. 447.) However, it is likely that the mechanism shown above is the correct one, because the equilibrium between lactone and hydroxy ester is likely to strongly favor the lactone, and because a protonated carboxy group is an excellent leaving group.

(e) This reaction is a variation of oxymercuration. The mercurinium ion formed at one double bond is attacked by the $\pi$ electrons of the second double bond. The resulting carbocation is attacked by the carboxy group. The mechanism below begins with the mercurinium ion. (For the mechanism of its formation, see Eq. 5.12a on text p. 181.)

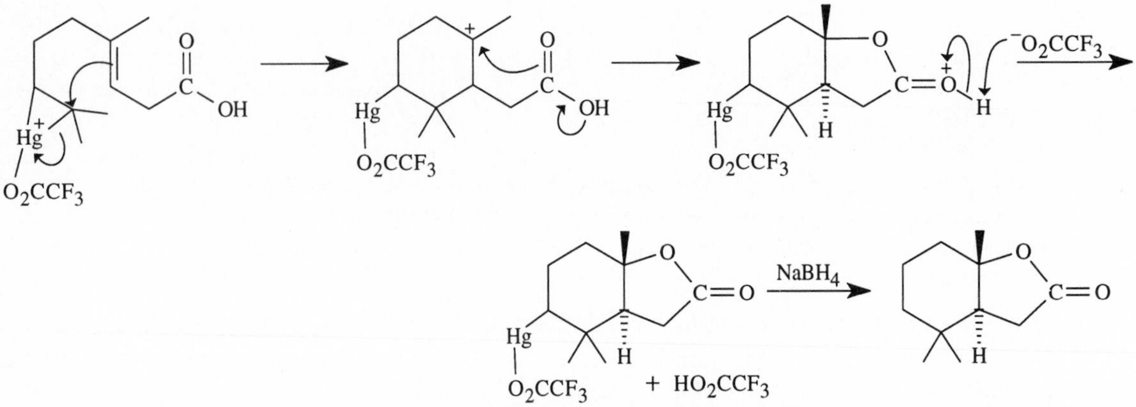

The stereochemistry results from the fact that the attack of the $\pi$ electrons on the mercurinium ion and the attack of the carboxy oxygen occur at opposite faces of the double bond. (The first two steps in the above mechanism may be concerted, that is, may occur essentially at the same time.) As the mechanism is written above, the mercurinium ion sits *above* the $\pi$ bond and the carboxy oxygen attacks from *below* this bond. Of course, the enantiomeric possibility is equally likely.

21.58 This is another example of *neighboring-group participation*; see Sec. 11.6 on text p. 518. In this case, the carboxylate ion displaces bromide intramolecularly to give an $\alpha$-lactone (a three-membered lactone). Because of its strain, this compound reacts rapidly with hydroxide ion in a second displacement that re-inverts the configuration at the asymmetric carbon.

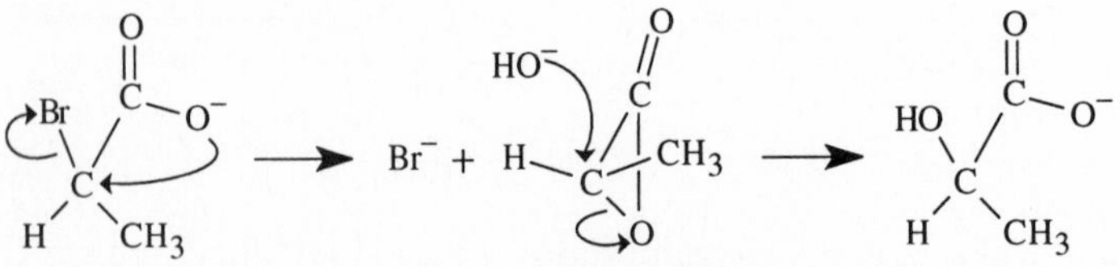

an $\alpha$-lactone

21.59 (a) Because the carbonyl group has trigonal-planar geometry, it must be turned so that its plane is perpendicular to the plane of the benzene ring in order to avoid significant van der Waals repulsions with the *ortho* methyl groups. This occurs even though conjugation is lost.

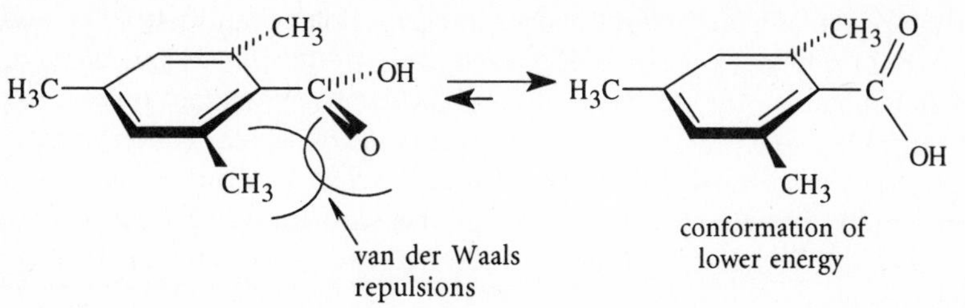

conformation of
lower energy

van der Waals
repulsions

(b) Remember that nucleophiles attack a carbonyl carbon above or below the plane of the carbonyl group. In the "perpendicular" conformation shown on the right, the path of a nucleophile to the carbonyl carbon is blocked on both sides by the *ortho* methyl groups. Because ester hydrolysis occurs by attack of a water molecule at the carbonyl carbon, this mechanism, and thus the hydrolysis reaction, cannot occur.

(c) For the same reason that ester *hydrolysis* cannot occur, ester *formation* also cannot occur by any mechanism that involves attack of a nucleophile at the carbonyl carbon. Because acid-catalyzed esterification involves attack of an alcohol molecule at the carbonyl carbon, this reaction does not take place. However, esterification with diazomethane can occur because the protonated diazomethane molecule is attacked by the carboxylate oxygen (see Eq. 20.21b on text p. 950). The reaction is less susceptible to van der Waals repulsions because it takes place at a site that is much farther removed from the *ortho* methyl groups.

21.60   (a) Phthalic anhydride can be formed by attack of a carboxy group at the carbonyl carbon of the protonated amide. (Note that the amide carbonyl oxygen is about $10^5$ times as basic as the carbonyl oxygen of the carboxylic acid; see text p. 987.)

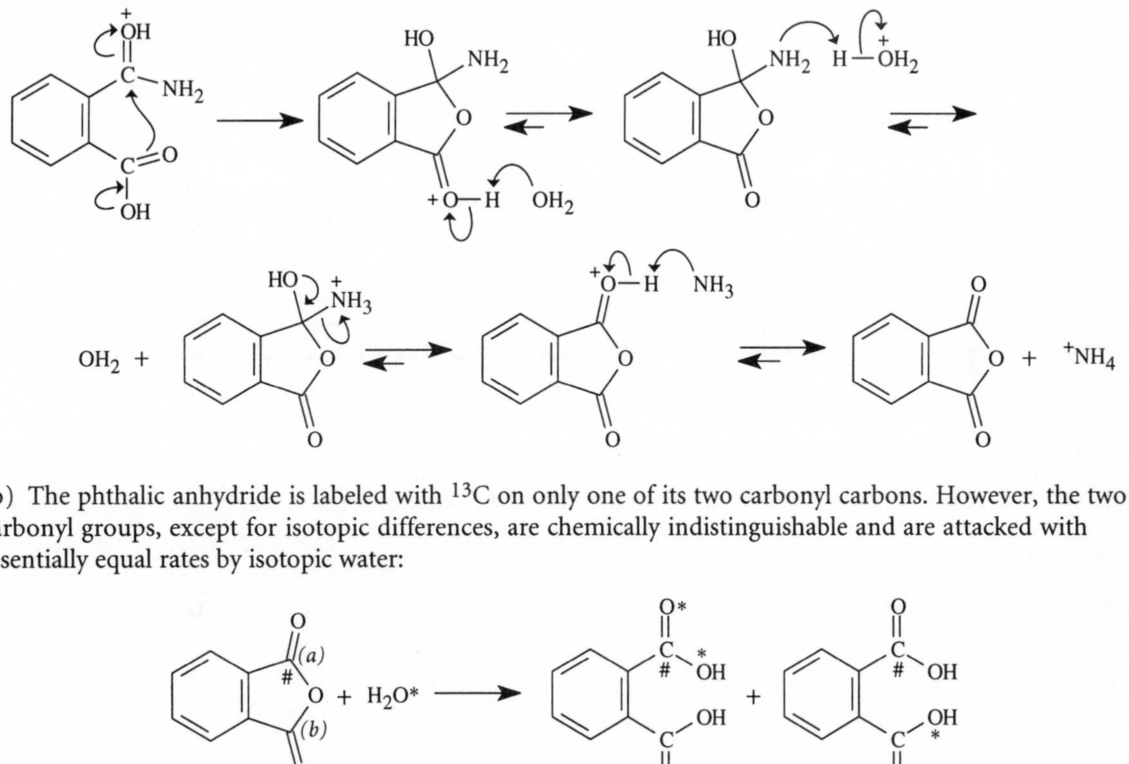

(b) The phthalic anhydride is labeled with $^{13}C$ on only one of its two carbonyl carbons. However, the two carbonyl groups, except for isotopic differences, are chemically indistinguishable and are attacked with essentially equal rates by isotopic water:

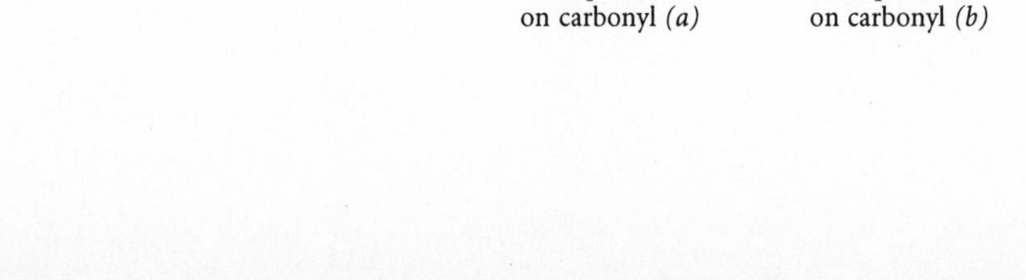

from attack of
isotopic water
on carbonyl *(a)*

from attack of
isotopic water
on carbonyl *(b)*

(c) Formation of the anhydride intermediate must be faster than hydrolysis by the usual mechanism or otherwise it would not occur. Anhydride formation is an *intramolecular* reaction, whereas the ordinary hydrolysis mechanism is *intermolecular*. Intramolecular reactions that involve the formation of five-membered rings are usually much faster than their intermolecular conterparts. (The reasons for the advantage of intramolecular reactions are discussed in Sec. 11.6 on text p. 518.) Although the anhydride is formed very rapidly, it is also hydrolyzed rapidly because anhydrides are generally very reactive toward hydrolysis. Thus, the overall hydrolysis is accelerated because an anhydride intermediate is formed rapidly and then destroyed rapidly.

# 22

# Chemistry of Enolate Ions, Enols, and $\alpha,\beta$-Unsaturated Carbonyl Compounds

## Terms

## Concepts

### I. Acidity of Carbonyl Compounds; Enolate Ions

#### A. FORMATION OF ENOLATE ANIONS

1. The $\alpha$-hydrogens of many carbonyl compounds, as well as those of nitriles, are weakly acidic—much more acidic than other types of hydrogens bound to carbon.

2. Ionization of an $\alpha$-hydrogen gives the conjugate-base anion, called an enolate ion.

    a. Enolate ions are resonance-stabilized.

    b. The anionic carbon of an enolate ion is $sp^2$-hybridized.

       *i.* This hybridization allows the electron pair of an enolate anion to occupy a *p* orbital, which overlaps with the *π* orbital of a carbonyl group.

       *ii.* This additional overlap provides additional bonding and hence additional stabilization.

       *iii.* The negative charge in an enolate ion is delocalized onto oxygen, an electronegative atom.

       *iv.* The polar effect of the carbonyl group stabilizes the enolate anion.

    c. Esters are less acidic than aldehydes or ketones.

       *i.* Esters are stabilized by overlap of the unshared electrons of the carboxylate oxygen with the carbonyl group; ketones lack this type of stabilization.

       *ii.* This stabilization of esters means that more energy is required to form their conjugate-base enolate ions.

       *iii.* The additional resonance stabilization of esters is more important than the polar effect of the carboxylate oxygen, which, in the absence of resonance, would increase the acidity of esters relative to that of aldehydes and ketones.

  3. Amide N—H hydrogens are conceptually *α*-hydrogens and are the most acidic hydrogens in primary and secondary amides; carboxylic acid O—H hydrogens are also *α*-hydrogens.

  4. The acidity order of carbonyl compounds corresponds to the reactive electronegativities of the atoms to which the acidic hydrogens are bound:

        *acidities*:    carboxylic acids  >  amides  >  aldehydes, ketones

   *electronegativities*:     oxygen    > nitrogen  >    carbon

## B. INTRODUCTION TO REACTIONS OF ENOLATE IONS

  1. Enolate ions are key reactive intermediates in many important reactions of carbonyl compounds:

    a. They are Brønsted bases and react with Brønsted acids.

    b. They are Lewis bases also; consequently, they react as nucleophiles.

       *i.* Enolate ions attack the carbon of carbonyl groups. This attack is the first step of a variety of carbonyl addition reactions and nucleophilic acyl substitution reactions of enolate ions.

       *ii.* Enolate ions, like other nucleophiles, also react with alkyl halides and sulfonate esters.

  2. The formation of enolate ions and their reactions with Brønsted acids have two simple but important consequences:

    a. The *α*-hydrogens, and only the *α*-hydrogens, of aldehydes and ketones can be exchanged for deuterium by treating the carbonyl compound with acid or base in $D_2O$.

$$—CH_2—\overset{\displaystyle O}{\overset{\|}{C}}—CH \xrightarrow[D_2O]{D_3O^+ \text{ or } {}^-OD} —CD_2—\overset{\displaystyle O}{\overset{\|}{C}}—CD$$

    b. If an optically active aldehyde or ketone owes its chirality solely to an asymmetric *α*-carbon, and if this carbon bears a hydrogen, the compound will be racemized by base.

       *i.* The enolate ion, which forms in base, is achiral because of the $sp^2$ hybridization at its anionic carbon.

       *ii.* The ionic *α*-carbon and its attached groups lie in one plane.

       *iii.* The anion can be reprotonated at either face to give either enantiomer with equal probability.

chiral            planar and achiral            racemate

3. $\alpha$-Hydrogen exchange and racemization reactions of aldehydes and ketones occur much more readily than those of esters, because aldehydes and ketones are more acidic, and therefore form enolate ions more rapidly and under milder conditions.

## C. ENOLIZATION OF CARBONYL COMPOUNDS

1. Carbonyl compounds with $\alpha$-hydrogens are in equilibrium with vinylic alcohol isomers called enols.

   a. Enols and their parent carbonyl compounds are tautomers of each other, that is, structural isomers that are conceptually related by the shift of a hydrogen and one or more $\pi$ bonds.

a ketone       tautomers       an enol

   b. Most carbonyl compounds are considerably more stable than their corresponding enols, primarily because the C=O bond of a carbonyl group is a stronger bond than the C=C bond of an enol.

   c. Phenol is conceptually an enol, but the enol form of phenol is more stable than its keto tautomers because phenol is aromatic.

   d. The enols of $\beta$-dicarbonyl compounds (compounds in which two carbonyl groups are separated by one carbon) are also relatively stable.

       *i.* These enol forms are conjugated, but their parent carbonyl compounds are not; the resonance stabilization ($\pi$ electron overlap) associated with conjugation provides additional bonding that stabilizes the enol.

      *ii.* The intramolecular hydrogen bond present in such enols provides increased bonding and increased stabilization.

2. The formation of enols and the reverse reaction, conversion of enols into carbonyl compounds, are catalyzed by both acids and bases.

   a. Base-catalyzed enolization involves the intermediacy of the enolate ion, and is thus a consequence of the acidity of the $\alpha$-hydrogen.

       *i.* Protonation of the enolate anion on the $\alpha$-carbon gives back the carbonyl compound; protonation on oxygen gives the enol.

      *ii.* The enolate ion is the conjugate base of not only the carbonyl compound, but also the enol.

   b. Acid-catalyzed enolization involves the conjugate acid of the carbonyl compound.

       *i.* Loss of the proton from the $\alpha$-carbon of a protonated ketone gives the enol.

      *ii.* An enol and its carbonyl tautomer have the same conjugate acid.

3. Exchange of $\alpha$-hydrogens for deuterium as well as racemization at the $\alpha$-carbon are catalyzed not only by bases but also by acids. Acid-catalyzed exhange can be explained by the intermediacy of enols.

   a. Formation of a carbonyl compound from an enol introduces hydrogen from solvent at the $\alpha$-carbon; this fact accounts for the observed isotope exchange.

   b. The $\alpha$-carbon of an enol, like that of an enolate ion, is not an asymmetric carbon.

   c. The absence of chirality in the enol accounts for the racemization observed when an aldehyde or ketone with a hydrogen at an asymmetric $\alpha$-carbon is treated with acid.

## II. *Approaches to Organic Synthesis*

### A. SYNTHETIC SHORT-HAND NOTATION

1. Chemists often use a compact abbreviation for several commonly occurring organic groups.
2. These abbreviations not only save space, but also make the structures of large molecules less cluttered and more easily read.

| | | | | | | | |
|---|---|---|---|---|---|---|---|
| methyl | $CH_3-$ | Me | ethyl | $CH_3CH_2-$ | Et | | |
| propyl | $CH_3CH_2CH_2-$ Pr | | isopropyl | $(CH_3)_2CH-$ | $i$-Pr | | |
| butyl | $CH_3(CH_2)_3-$ Bu | | isobutyl | $(CH_3)_2CHCH_2-$ | $i$-Bu | *tert*-butyl $(CH_3)_3C-$ | $t$-Bu |

acetyl $\underset{\displaystyle CH_3-\overset{\displaystyle O}{\overset{\|}{C}}-}{}$ Ac    acetate or acetoxy $\underset{\displaystyle CH_3-\overset{\displaystyle O}{\overset{\|}{C}}-O-}{}$ AcO

### A. SYNTHESIS WITH THE ALDOL CONDENSATION

1. The aldol condensation:
   a. is an addition reaction involving two aldehyde or ketone molecules (aldol addition) followed by a dehydration of the resulting $\beta$-hydroxy aldehyde
   c. can be applied to the synthesis of a wide variety of $\alpha,\beta$-unsaturated aldehydes and ketones
   b. can be carried out in acidic or basic solution
   d. is a method for the formation of carbon-carbon bonds.

$$R-CH_2-\overset{O}{\overset{\|}{C}}-R' \xrightarrow[\text{}^-OH]{H_3O^+ \text{ or}} R-CH_2-\underset{R'}{\overset{OH}{\underset{|}{\overset{|}{C}}}}-\underset{R}{\overset{}{\underset{|}{CH}}}-\overset{O}{\overset{\|}{C}}-R' \xrightarrow[\text{}^-OH]{H_3O^+ \text{ or}} R-CH_2-\underset{R'}{\overset{}{\underset{|}{C}}}=\underset{R}{\overset{}{\underset{|}{CH}}}-\overset{O}{\overset{\|}{C}}-R'$$

(aldol addition)             (dehydration)

2. To prepare a particular $\alpha,\beta$-unsaturated aldehyde or ketone by the aldol condensation, ask two questions:
   a. What starting materials are required in the aldol condensation?
      i. The starting materials for an aldol condensation can be determined by mentally "splitting" the $\alpha,\beta$-unsaturated carbonyl compound at the double bond.
      ii. Replace the double bond on the carbonyl side by two hydrogens and on the other side by a carbonyl oxygen ($=O$).

   b. With these starting materials, is the aldol condensation of these compounds a feasible one?

### B. SYNTHESIS WITH THE CLAISEN CONDENSATION

1. The Claisen condensation is a nucleophilic acyl substitution reaction.
   a. The nucleophile is an enolate ion derived from an ester.
   b. One full equivalent of base is required.
   c. The ester starting material must have at least two $\alpha$-hydrogens.

2. The Claisen condensation can be used for the synthesis of $\beta$- dicarbonyl compounds: $\beta$-keto esters, $\beta$-diketones, and the like.

a. Mentally reverse the Claisen condensation in two different ways by adding the elements of ethanol (or another alcohol) across either of the carbon-carbon bonds between the carbonyl groups.

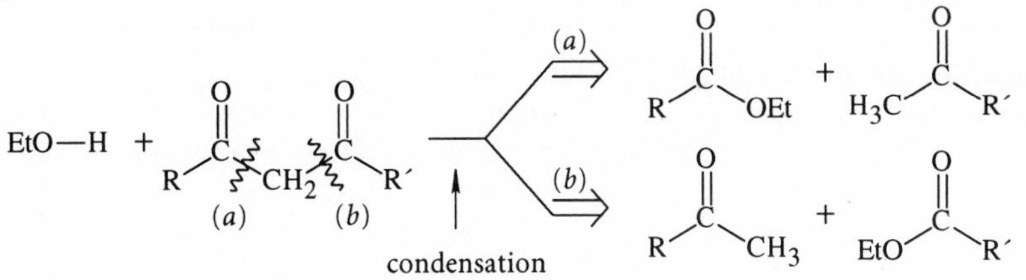

b. Determine whether the Claisen condensation of the required starting materials will give mostly the desired product and not a complex mixture.

## D. SYNTHESIS WITH THE ACETOACETIC ESTER REACTION

1. The acetoacetic ester synthesis involves alkylation of a $\beta$-keto ester. In many cases the alkylated derivative is saponified and decarboxylated to give a branched ketone.

2. Given a target ketone, mentally reverse the acetoacetic ester synthesis.

   a. Replace an $\alpha$-hydrogen of the target ketone with a carboethoxy group to unveil the $\beta$-keto ester required for the synthesis.

   b. The $\beta$-keto ester itself is then analyzed in terms of the alkyl halides required for the alkylation.

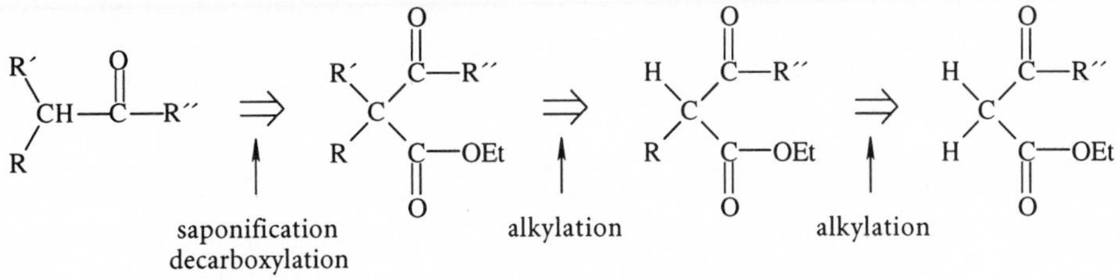

3. Determine whether the $\beta$-keto ester is one that can be made by the Claisen condensation.

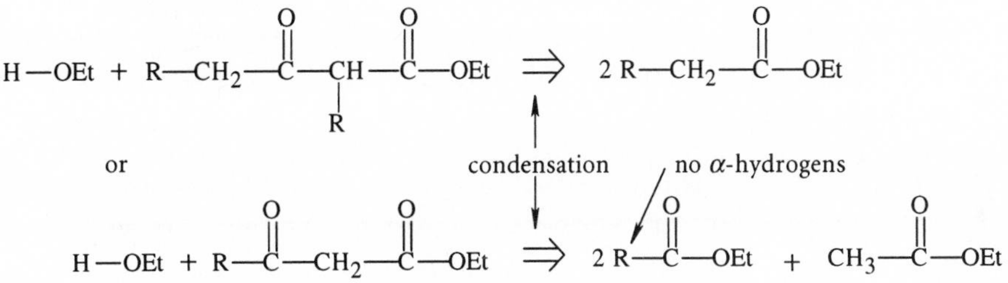

## C. SYNTHESIS WITH THE MALONIC ESTER REACTION

1. The malonic ester synthesis involves alkylation of a malonic ester derivative. In many cases the alkylated derivative is saponified and decarboxylated to give a substituted acetic acid derivative.

2. To determine whether the malonic ester synthesis can be used for the synthesis of a carboxylic acid:

   a. Determine whether the desired carboxylic acid is an acetic acid with one or two alkyl substituents on its $\alpha$-carbon.

   b. Mentally reverse the decarboxylation, hydrolysis, and alkylation steps to arrive at the structures of the alkyl halides (or sulfonate esters) that must be used.

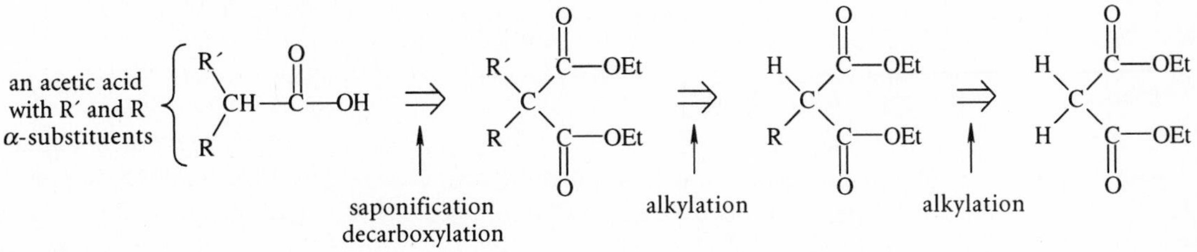

c. If the alkyl halides used in the alkylation step can undergo the $S_N2$ reaction, then the target carboxylic acid can in principle be prepared by the malonic ester synthesis.

## E. Synthesis with Conjugate-Addition Reactions

1. Any group at the β-position of a carbonyl compound (or nitrile) can in principle be delivered as a nucleophile in a conjugate addition.

2. A conjugate addition can be mentally reversed by subtracting a nucleophilic group from the β-position of the target molecule, and a positive fragment (usually a proton) from the α-position.

conjugate addition

## F. Conjugate-Addition vs. Carbonyl-Group Reactions

1. A conjugate-addition reaction usually competes with a carbonyl group reaction.
   a. Conjugate addition retains a carbonyl group at the expense of a carbon-carbon double bond.
   b. Carbonyl addition retains a carbon-carbon double bond at the expense of a carbonyl group.

2. In the case of aldehydes and ketones, conjugate addition competes with addition to the carbonyl group.
   a. Relatively weak bases (cyanide ion, amines, thiolate ions, and enolate ions derived from β-dicarbonyl compounds) that give reversible carbonyl-addition reactions with ordinary aldehydes and ketones tend to give conjugate addition with α,β-unsaturated aldehydes and ketones.
      *i.* Conjugate addition is observed with these nucleophiles primarily because, in most cases, it is irreversible.
      *ii.* Conjugate-addition products are more stable than carbonyl-addition products primarily because a C=O bond is considerably stronger than a C=C bond. Thus, the conjugate-addition product is the thermodynamic (more stable) product of the reaction.
      *iii.* This is another case of kinetic *vs.* thermodynamic control of a reaction.
   b. When nucleophiles are used that undergo irreversible carbonyl additions, then the carbonyl-addition product is observed rather than the conjugate-addition product.
      *i.* Powerful nucleophiles such as $LiAlH_4$ and organolithium reagents add irreversibly to carbonyl groups and form carbonyl-addition products whether the reactant carbonyl compound is α,β-unsaturated or not.

3. Many of the same nucleophiles that undergo conjugate addition with aldehydes and ketones also undergo conjugate addition with esters; in contrast, stronger bases that react irreversibly at the carbonyl carbon react with esters to give nucleophilic acyl substitution products.

## III. Biosynthesis of Compounds Derived from Acetate

### A. Biosynthesis of Fatty Acids

1. The starting material for the biosynthesis of fatty acids is a thiol ester of acetic acid called acetyl-coenzyme-A (abbreviated acetyl-CoA).
   a. Acetyl-CoA is first converted into malonyl-CoA by carboxylation of the α-carbon.
      *i.* The —SCoA group in both acetyl- and malonyl-CoA is then replaced by a different group (abbreviated —SR), called the acyl carrier protein.
      *ii.* The resulting compounds react in an enzyme-catalyzed reaction to give an acetoacetyl thiol ester.
   b. The nucleophilic electron pair is made available not by proton abstraction, but by loss of $CO_2$ from malonyl-CoA; the loss of $CO_2$ as a gaseous by-product also serves to drive the Claisen condensation to completion.
   c. The product, an acetoacetyl thiol ester, then undergoes successively a carbonyl reduction, a dehydration, and a double-bond reduction, each catalyzed by an enzyme; the net result is that the acetyl thiol ester is converted into a thiol ester with two additional carbons.

$$CH_3-\overset{\overset{O}{\|}}{C}-SR \; + \; {}^-O-\overset{\overset{O}{\|}}{C}-CH_2-\overset{\overset{O}{\|}}{C}-SR \xrightarrow{\;\text{condensation}\;} CH_3-\overset{\overset{O}{\|}}{C}-CH_2-\overset{\overset{O}{\|}}{C}-SR \xrightarrow{\;\text{reduction}\;}$$

$$CH_3-\overset{\overset{OH}{|}}{CH}-CH_2-\overset{\overset{O}{\|}}{C}-SR \xrightarrow{\;\text{dehydration}\;} CH_3-CH=CH-\overset{\overset{O}{\|}}{C}-SR \xrightarrow{\;\text{reduction}\;} CH_3-CH_2-CH_2-\overset{\overset{O}{\|}}{C}-SR$$

or

$$CH_3-\overset{\overset{O}{\|}}{C}-SR \; + \; {}^-O-\overset{\overset{O}{\|}}{C}-CH_2-\overset{\overset{O}{\|}}{C}-SR \xrightarrow[\substack{\text{3) dehydration}\\\text{4) reduction}}]{\substack{\text{1) condensation}\\\text{2) reduction}}} CH_3-CH_2-CH_2-\overset{\overset{O}{\|}}{C}-SR$$

2. This sequence of reactions is then repeated until the proper chain length is obtained.
   a. At each cycle another two carbons are added to the chain.
   b. The fatty-acid thiol ester is then transesterified by glycerol to form fats and phospholipids.
3. The common fatty acids have an even number of carbon atoms; those with an odd number of carbon atoms are relatively rare.

### B. BIOSYNTHESIS OF ISOPRENOIDS, STEROIDS, AND SOME AROMATIC COMPOUNDS

1. Isopentenyl pyrophosphate (the basic building block of isopreneoids and steroids) as well as a number of aromatic compounds found in nature are ultimately derived from acetyl-CoA.

## Reactions

### I. α-Halogenation of Carbonyl Compounds

#### A. ACID-CATALYZED α-HALOGENATION OF ALDEHYDES AND KETONES

1. Halogenation of an aldehyde or ketone in acidic solution usually results in the replacement of one α-hydrogen by halogen.
   a. Enols are reactive intermediates in these reactions.
   b. Enols add only one halogen atom.

$$-\overset{\overset{O}{\|}}{C}-CH- \; \underset{\substack{\uparrow\\\text{rate-limiting step}}}{\overset{H_3O^+}{\rightleftharpoons}} \; -\underset{\substack{|\\ \\\text{enol}}}{\overset{\overset{OH}{|}}{C}}=C- \; \xrightarrow{\;Br_2\;} \; -\overset{\overset{O}{\|}}{C}-\underset{|}{\overset{\overset{Br}{|}}{C}}- \; + \; HBr$$

2. The rate law implies that even though the reaction is a halogenation, the rate is independent of the halogen concentration.

$$\text{rate} = k[\text{ketone}][H_3O^+]$$

   a. Halogens are not involved in the transition state for the rate-limiting step of the reaction.
   b. Enol formation is the rate-limiting process in acid-catalyzed halogenation of aldehydes and ketones.
3. Introduction of a second halogen is much slower than introduction of the first halogen.
   a. This may be related to the stability of the cationic intermediate that is formed by attack of the halogen on the halogenated enol.
   b. This cation is destabilized by the electron-attracting polar effect of two halogens.

#### B. HALOGENATION OF ALDEHYDES AND KETONES IN BASE; THE HALOFORM REACTION

1. Halogenation of aldehydes and ketones with α-hydrogens also occurs in base; all α-hydrogens are substituted by halogen.

$$\text{—CH}_2\text{—}\underset{\underset{\|}{O}}{C}\text{—CH—} \xrightarrow[\text{}^-OH]{Br_2} \underset{\underset{Br}{|}}{\overset{Br}{\underset{\|}{C}}}\text{—}\underset{\underset{\|}{O}}{C}\text{—}\overset{Br}{C}\text{—}$$

2. When the aldehyde or ketone starting material is either acetaldehyde or a methyl ketone, the trihalomethyl carbonyl compound is not stable under the reaction conditions, and it reacts further to give, after acidification of the reaction mixture, a carboxylic acid and a haloform.
   a. This reaction is called the haloform reaction.
   b. A carbon-carbon bond is broken.
3. The mechanism of the haloform reaction involves the formation of an enolate ion as a reactive intermediate.
   a. The enolate ion reacts as a nucleophile with halogen to give an α-halo carbonyl compound.
   b. A dihalo and trihalo carbonyl compound are formed more rapidly in successive halogenations.
   c. A carbon-carbon bond is broken when the trihalo carbonyl compound undergoes a nucleophilic acyl substitution reaction.
      i. The leaving group in this reaction is a trihalomethyl anion, which is much less basic than ordinary carbanions.
      ii. The trihalomethyl anion reacts irreversibly with the carboxylic acid by-product to drive the overall haloform reaction to completion.
      iii. The carboxylic acid itself can be isolated by acidifying the reaction mixture.

$$\underset{R}{\overset{O}{\underset{\|}{C}}}\text{CH}_3 \xrightarrow{HO^-} \underset{R}{\overset{O}{\underset{\|}{C}}}\underset{\text{enolate ion}}{\bar{C}H_2} \xrightarrow{Br_2} \underset{R}{\overset{O}{\underset{\|}{C}}}\text{CH}_2\text{—Br} \xrightarrow[\substack{\text{twice} \\ \text{more}}]{Br_2/OH^-} \underset{R}{\overset{O}{\underset{\|}{C}}}\text{CBr}_3 \xrightarrow[\substack{\text{C—C bond} \\ \text{cleavage}}]{Br_2/OH^-}$$

$$\underset{R}{\overset{O}{\underset{\|}{C}}}\text{OH} + {}^-\text{CBr}_3 \underset{\longleftarrow}{\longrightarrow} \underset{R}{\overset{O}{\underset{\|}{C}}}\text{O}^- + \text{HCBr}_3$$

4. The haloform reaction can be used to prepare carboxylic acids from readily available methyl ketones.
5. The haloform reaction is used as a qualitative test for methyl ketones, called the iodoform test.
   a. A compound of unknown structure is mixed with alkaline $I_2$.
   b. A yellow precipitate of iodoform ($HCI_3$) is taken as evidence for a methyl ketone (or acetaldehyde).
   c. α-Substituted ethanol derivatives also give a positive iodoform test because they are oxidized to methyl ketones (or to acetaldehyde, in the case of ethanol) by the basic iodine solution.

## C. α-BROMINATION OF CARBOXYLIC ACIDS; THE HELL-VOLHARD-ZELINSKY REACTION

1. A bromine is substituted for an α-hydrogen when a carboxylic acid is treated with $Br_2$ and a catalytic amount of red phosphorus or $PBr_3$; this reaction is called the Hell-Volhard-Zelinsky (HVZ) reaction.
2. The first stage in the mechanism of this reaction is the conversion of the carboxylic acid into a small amount of acid bromide by the catalyst $PBr_3$.
   a. The acid bromide enolizes in the presence of acid; this reaction is similar to the acid-catalyzed bromination of ketones.
   b. The enol of the acid bromide is the species that actually brominates.

$$\text{—CH—}\underset{\overset{\|}{O}}{C}\text{—OH} \xrightarrow[\text{catalyst}]{PBr_3} \text{—CH—}\underset{\overset{\|}{O}}{C}\text{—Br} \longleftrightarrow \underset{\text{enol}}{\overset{OH}{\underset{Br}{C=C}}} \xrightarrow{Br_2}$$

$$\underset{\underset{|}{}}{\overset{Br}{\underset{|}{C}}}\text{—}\underset{\overset{\|}{O}}{C}\text{—Br} \xrightarrow{RCO_2H} \underset{\underset{|}{}}{\overset{Br}{\underset{|}{C}}}\text{—}\underset{\overset{\|}{O}}{C}\text{—OH} + \underset{R}{\overset{O}{\underset{\|}{C}}}\text{—Br}$$

       *i.* When a small amount of PBr$_3$ catalyst is used, the $\alpha$-bromo acid bromide reacts with the carboxylic acid to form more acid bromide, which is then brominated.

       *ii.* If one full equivalent of PBr$_3$ catalyst is used, the $\alpha$-bromo acid bromide is the reaction product.

       *iii.* The reaction mixture can be treated with an alcohol to give an $\alpha$-bromo ester.

## D. Reactions of $\alpha$-Halo Carbonyl Compounds

1. Most $\alpha$-halo carbonyl compounds are very reactive in S$_N$2 reactions, and can be used to prepare other $\alpha$-substituted carbonyl compounds.

2. In the case of $\alpha$-halo ketones, nucleophiles used in these reactions must not be too basic.
   a. Stronger bases promote enolate-ion formation; and the enolate ions of $\alpha$-halo ketones undergo other reactions.
   b. More basic nucleophiles can be used with $\alpha$-halo acids because, under basic conditions, $\alpha$-halo acids are ionized to form their carboxylate conjugate-base anions, and these do not undergo the side-reactions that $\alpha$-halo ketones undergo.

3. $\alpha$-Halo carbonyl compounds react so slowly by the S$_N$1 mechanism that this reaction is not useful; reactions that require the formation of carbocations alpha to carbonyl groups generally do not occur.

## II. *Condensations Involving Enolate Ions and Enols of Aldehydes and Ketones*

## A. Base-Catalyzed Aldol Addition and Condensation Reactions

1. In aqueous base, acetaldehyde undergoes a reaction called the aldol addition.
   a. The term aldol is both a trivial name for 3-hydroxybutanal and a generic name for $\beta$-hydroxy aldehydes.
   b. The aldol addition is a very important and general reaction of aldehydes and ketones that have $\alpha$-hydrogens.
   c. The aldol addition is another nucleophilic addition to a carbonyl group.

2. The base-catalyzed aldol addition involves an enolate ion as an intermediate; the enolate ion adds to a second molecule of aldehyde or ketone.

3. The aldol addition is reversible.
   a. The equilibrium for the aldol addition is much more favorable for aldehydes than for ketones.
   b. For acetone, the product is obtained only if an apparatus is used that allows the product to be removed from the base catalyst as it is formed.

4. Under severe conditions (higher base concentration and/or heat), the product of aldol addition undergoes a dehydration reaction. (The sequence of reactions consisting of the aldol addition followed by dehydration is called the aldol condensation; a condensation is a reaction in which two molecules combine to form a larger molecule with the elimination of a small molecule, in many cases water.)

   a. The dehydration part of the aldol condensation is catalyzed by base, and occurs in two distinct steps through a carbanion intermediate.

    b. Base-catalyzed dehydration reactions of alcohols are unusual; $\beta$-hydroxy aldehydes and $\beta$-hydroxy ketones undergo base-catalyzed dehydration because:

      *i.* The $\alpha$-hydrogen is relatively acidic.

      *ii.* The product is conjugated and therefore particularly stable.

  5. The product of an aldol addition-dehydration sequence is an $\alpha,\beta$-unsaturated carbonyl compound.

  6. The aldol condensation is an important method for the preparation of certain $\alpha,\beta$-unsaturated carbonyl compounds. Whether the aldol addition product or the condensation product is formed depends on the reaction conditions, which must be determined on a case-by-case basis.

## B. Acid-Catalyzed Aldol Condensation

  1. Aldol condensations are also catalyzed by acid.

    a. Acid-catalyzed aldol condensations generally give $\alpha,\beta$-unsaturated carbonyl compounds as products.

    b. The addition products usually cannot be isolated.

  2. The conjugate acid of the aldehyde or ketone is a key reactive intermediate:

    a. It is transformed into a small amount of the enol.

    b. It is the electrophilic species in the reaction.

    c. It is attacked by the $\pi$ electrons of the enol to give an $\alpha$-hydroxy carbocation, which is the conjugate acid of the addition product. (See Eqs. 22.44a–b on text pp. 1058–1059 for a detailed mechanism.)

  3. The $\alpha$-hydroxy carbocation loses a proton to give the $\beta$-hydroxy carbonyl compound.

    a. Under the acidic conditions, this material spontaneously undergoes acid-catalyzed dehydration to give an $\alpha,\beta$-unsaturated carbonyl compound.

    b. The dehydration drives the aldol condensation to completion.

## C. Special Types of Aldol Condensations

  1. When two different carbonyl compounds are used, a crossed aldol condensation occurs; in many cases, the result is a difficult-to-separate mixture.

  2. Under the usual conditions (aqueous or alcoholic acid or base), useful crossed aldol condensations as a practical matter are limited to situations in which a ketone with $\alpha$-hydrogens is condensed with an aldehyde without $\alpha$-hydrogens.

  3. An important example of this type of crossed aldol condensation is the Claisen-Schmidt condensation.

    a. A ketone with $\alpha$-hydrogens is condensed with an aromatic aldehyde that has no $\alpha$-hydrogens.

    b. There are three reasons why the Claisen-Schmidt condensation occurs instead of other competing processes:

      *i.* Because the aldehyde has no $\alpha$-hydrogens, it cannot act as the enolate or enol component of the aldol condensation.

      *ii.* Addition of the ketone enolate or enol to a ketone occurs more slowly than addition to an aldehyde; aldehydes are more reactive than ketones.

      *iii.* The aldol addition reaction is reversible; thus, to the extent that this side reaction occurs, it is reversed under the reaction conditions.

    c. The Claisen-Schmidt condensation, like other aldol condensations, can be catalyzed by either acid or base.

4. When a molecule contains more than one aldehyde or ketone group, an intramolecular reaction (a reaction within the same molecule) is possible; the resulting formation of a ring is particularly favorable when five- and six-membered rings can be formed.

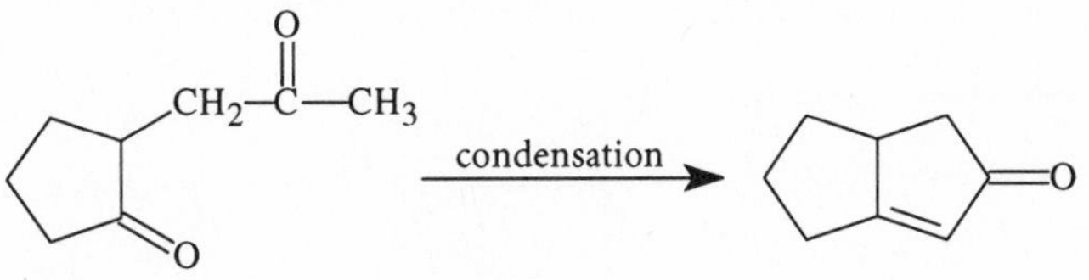

## III. Condensation Reactions Involving Ester Enolate Ions

### A. CLAISEN CONDENSATION

1. The Claisen condensation is used for the formation of $\beta$-dicarbonyl compounds (compounds with two carbonyl groups in a $\beta$-relationship), particularly:
   a. $\beta$-keto esters (a compound with a ketone carbonyl group $\beta$ to an ester carbonyl group)
   b. $\beta$-diketones
2. The mechanism of the Claisen condensation:
   a. An enolate ion is formed by the reaction of an ester with ethoxide as the base. (Although the ester enolate ion is formed in low concentration, it is a strong base and good nucleophile.)

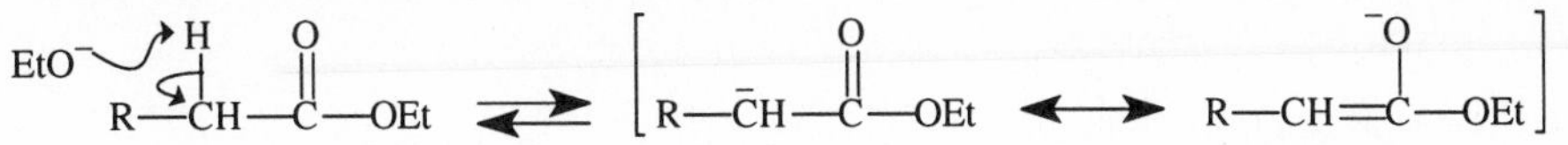

   b. The ester enolate undergoes a nucleophilic acyl substitution reaction with a second molecule of ester.

   c. The overall equilibrium favors the reactants; that is, all $\beta$-keto esters are less stable than the esters from which they are derived.
   d. The Claisen condensation has to be driven to completion by applying LeChatelier's principle.
      i. The most common technique is to use one equivalent of alkoxide catalyst.
      ii. The hydrogens adjacent to both carbonyl groups in the $\beta$-keto ester product are especially acidic; the alkoxide removes one of these hydrogens to form quantitatively the conjugate base of the product.

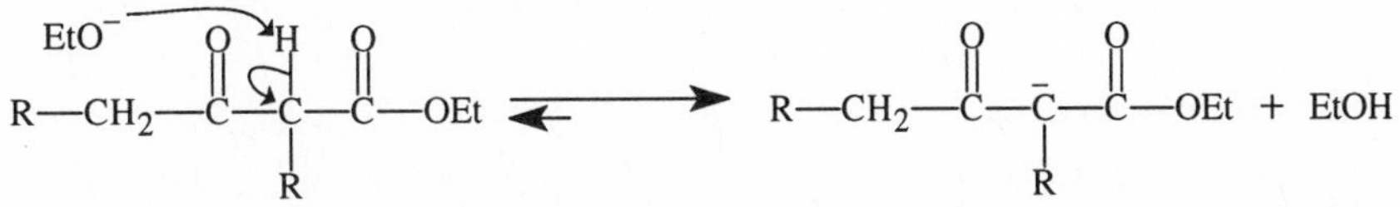

   e. The un-ionized $\beta$-keto ester product is formed when acid is added subsequently to the reaction mixture.

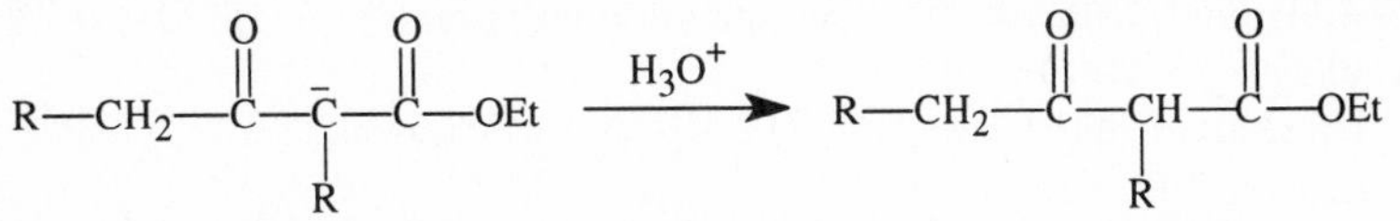

3. Attempts to condense an ester that has only one $\alpha$-hydrogen result in little or no product.
   a. The desired condensation product has no $\alpha$-hydrogens acidic enough to react completely with the alkoxide.
   b. If the desired product (prepared by a different method) is subjected to the conditions of the Claisen condensation, it readily decomposes back to starting materials because of the reversibility of the Claisen condensation.

## B. Dieckmann Condensation

1. Intramolecular Claisen condensations take place readily when five- or six-membered rings can be formed.

2. The intramolecular Claisen condensation reaction is called the Dieckmann condensation.
3. As with the Claisen condensation, one equivalent of base is necessary; the α-proton must be removed from the initially formed product in order for the reaction to be driven to completion.

## C. Crossed Claisen Condensation

1. The Claisen condensation of two different esters is called a crossed Claisen condensation; the crossed Claisen condensation of two esters that both have α-hydrogens gives a mixture of compounds that are typically difficult to separate.
2. Crossed Claisen condensations are useful if one ester is especially reactive or has no α-hydrogens. Formate esters and diethyl carbonate (a less reactive ester that is used in excess) are frequently used in this type of reaction.

3. Another type of crossed Claisen condensation is the reaction of ketones with esters.

   a. In this type of reaction the enolate ion of a ketone attacks the carbonyl group of an ester.
      i. The equilibrium for the aldol addition of two ketones favors the reactants, whereas the Claisen condensation is irreversible because one equivalent of base is used to form the enolate ion of the product.
      ii. Ketones are far more acidic than esters; the enolate ion of the ketone is formed in much greater concentration than the enolate ion of the ester.
   b. The product is a β-dicarbonyl compound, which is especially acidic and is ionized completely by the one equivalent of base; ionization makes this reaction irreversible.

## IV. Alkylation of Ester Enolate Ions

## A. Direct Alkylation of Enolate Ions Derived from Monoesters

1. A family of very strong, highly branched nitrogen bases can be used to form stable enolate ions rapidly at −78° from esters.
   a. These amide bases are strong enough to convert esters completely into their conjugate-base enolate ions.
      i. The ester is added to the base; the reaction of esters with strong amide bases is much faster, even at −78°, than the Claisen condensation; the enolate is formed rapidly.
      ii. Attack of the amide base on the carbonyl carbon of the ester, a reaction that competes with proton removal, is retarded by van der Waals repulsions between groups on the carbonyl compound and the large branched groups on the amide base.
      iii. These van der Waals repulsions have been aptly termed F-strain, or "front-strain."
   b. The ester enolate anions formed with these bases can be alkylated with alkyl halides.
   c. Esters with quaternary α-carbon atoms can be prepared with this method.

d. The nitrogen bases themselves are generated from the corresponding amines and butyllithium at −78° in tetrahydrofuran (THF).

$$i\text{-Pr}_2\text{NH} \quad \xrightarrow[-78°]{\text{BuLi}} \quad i\text{-Pr}_2\text{N}^- \text{ Li}^+ \;+\; \text{CH}_4$$

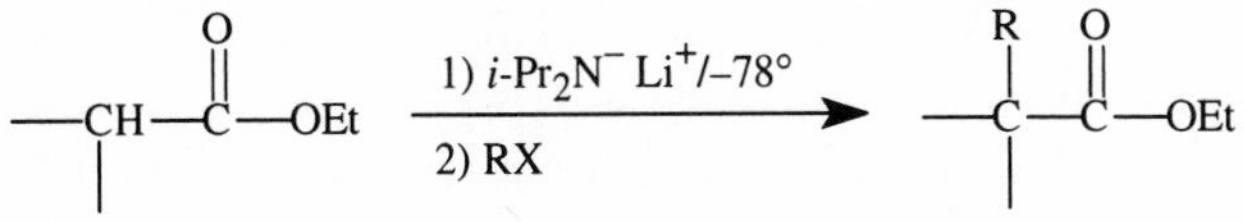

2. This method of ester alkylation is considerably more expensive than the malonic ester synthesis and requires special inert-atmosphere techniques.

## B. ACETOACETIC ESTER SYNTHESIS

1. β-Keto esters, which are substantially more acidic than ordinary esters, are completely ionized by alkoxide bases.
   a. The resulting enolate ions can be alkylated by alkyl halides or sulfonate esters.
   b. Dialkylation of β-keto esters is also possible.

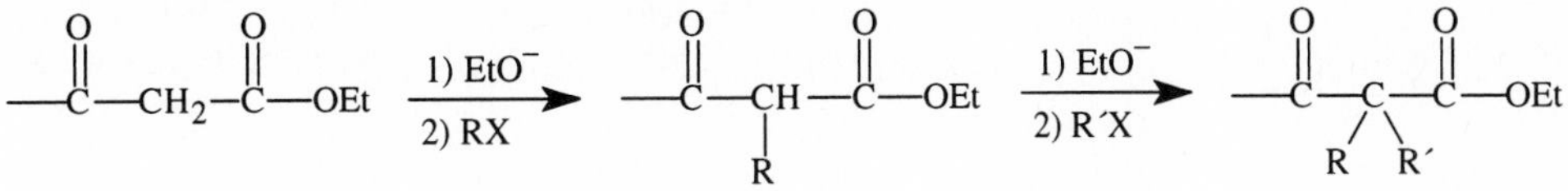

   c. The alkylated derivatives of ethyl acetoacetate can be hydrolyzed and decarboxylated to give ketones.

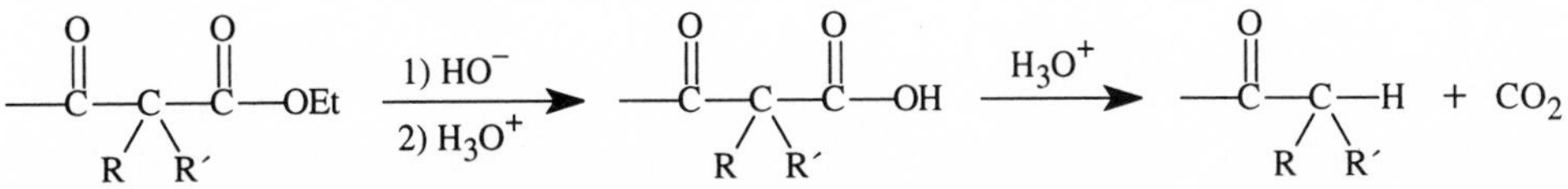

2. The alkylation of ethyl acetoacetate followed by saponification, protonation, and decarboxylation to give a ketone is called the acetoacetic ester synthesis and involves the construction of one or two new carbon-carbon bonds.
3. Whether a target ketone can be prepared by the acetoacetic ester synthesis can be determined mentally by reversing the synthesis:
   a. Replace an α-hydrogen on the target ketone with a —CO₂Et group.
   b. Determine whether the resulting β-keto ester can be prepared by a Claisen condensation or from other β-keto esters by alkylation or di-alkylation with appropriate alkyl halides.

## C. MALONIC ESTER SYNTHESIS

1. Diethyl malonate (malonic ester), like many other β-dicarbonyl compounds, has unusually acidic α-hydrogens.
   a. Its conjugate-base enolate ion can be formed completely with alkoxide bases such as sodium ethoxide.
   b. The conjugate-base anion of diethyl malonate is nucleophilic, and it reacts with alkyl halides and sulfonate esters in typical $S_N2$ reactions.
   c. This reaction can be used to introduce alkyl groups at the α-position of malonic ester.

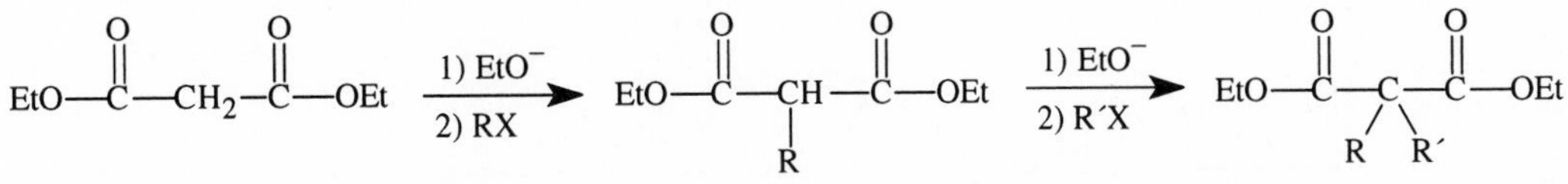

2. The malonic ester synthesis can be extended to the preparation of carboxylic acids.
   a. Saponification of the diester and acidification of the resulting solution give a substituted malonic acid derivative.
   b. Heating any malonic acid derivative causes it to decarboxylate.

$$\text{EtO-}\overset{\overset{\displaystyle O}{\|}}{C}-\overset{\overset{\displaystyle}{\underset{R}{\overset{|}{C}}}}{\underset{R'}{\overset{|}{C}}}-\overset{\overset{\displaystyle O}{\|}}{C}-\text{OEt} \xrightarrow[\text{2) H}_3\text{O}^+]{\text{1) HO}^-} \text{HO-}\overset{\overset{\displaystyle O}{\|}}{C}-\overset{\overset{\displaystyle}{\underset{R}{\overset{|}{C}}}}{\underset{R'}{\overset{|}{C}}}-\overset{\overset{\displaystyle O}{\|}}{C}-\text{OH} \longrightarrow \text{H-}\overset{\overset{\displaystyle}{\underset{R'}{\overset{|}{C}}}}{\underset{R}{\overset{|}{C}}}-\overset{\overset{\displaystyle O}{\|}}{C}-\text{OH} + \text{CO}_2$$

    c. The result of alkylation, saponification, and decarboxylation is a carboxylic acid that conceptually is a disubstituted acetic acid—an acetic acid molecule with two alkyl groups on its α-carbon.

3. The overall sequence of ionization, alkylation, saponification, and decarboxylation starting from diethyl malonate is called the malonic ester synthesis.

    a. The alkylation step of the malonic ester synthesis results in the formation of one or two new carbon-carbon bonds.

    b. The anion of malonic acid can be alkylated twice in two successive reactions with different alkyl halides (if desired) to give, after hydrolysis and decarboxylation, a disubstituted acetic acid.

## V. Conjugate-Addition Reactions

### A. CONJUGATE ADDITION TO α,β-UNSATURATED CARBONYL COMPOUNDS

1. An addition to the double bond of an α,β-unsaturated carbonyl compound is an example of conjugate addition (1,4-addition); nucleophilic conjugate addition has no parallel in the reactions of simple conjugated dienes.

conjugate addition    protonation    tautomerization

2. Nucleophilic addition to the carbon-carbon double bonds of α,β-unsaturated aldehydes, ketones, esters, and nitriles is a rather general reaction that can be observed with a variety of nucleophiles.

    a. The addition of a nucleophile to acrylonitrile ($H_2C=CH-C\equiv N$) is a useful reaction called cyanoethylation.

    b. Quinones are α,β-unsaturated carbonyl compounds, and they also undergo similar addition reactions.

    c. When cyanide is the nucleophile, a new carbon-carbon bond is formed; the nitrile group can then be converted into a carboxylic acid group by acid hydrolysis.

3. Nucleophilic addition to the double bond in an α,β-unsaturated carbonyl compound occurs because it gives a resonance-stabilized enolate ion intermediate.

4. Acid-catalyzed additions to the carbon-carbon double bonds of α,β-unsaturated carbonyl compounds are also known.

### B. CONJUGATE ADDITION OF ENOLATE IONS; MICHAEL ADDITION

1. Enolate ions derived from malonic ester derivatives, β-keto esters, and the like undergo conjugate-addition reactions with α,β-unsaturated carbonyl compounds.

    a. Conjugate additions of carbanions to α,β-unsaturated carbonyl compounds are called Michael additions.

    b. Many Michael additions can originate from either of two pairs of reactants.

    c. To maximize conjugate addition, choose the pair of reactants with the less basic enolate ion.

2. In one useful variation of the Michael addition, called the Robinson annulation, the immediate product of the reaction can be subjected to an aldol condensation that closes a ring. (An annulation is a ring-forming reaction.)

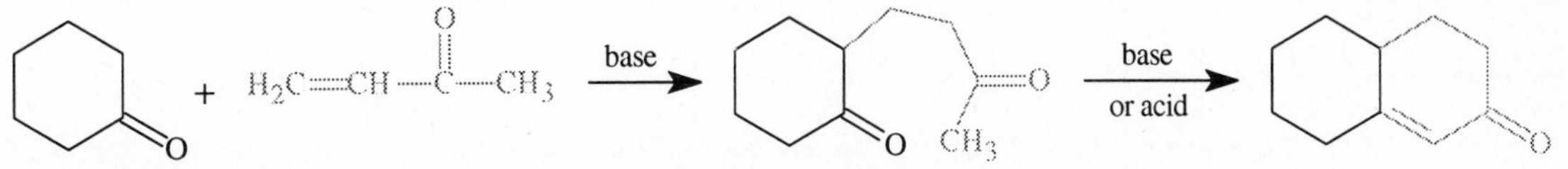

## VI. Reactions of α,β-Unsaturated Carbonyl Compounds with Other Nucleophiles

### A. REDUCTION OF α,β-UNSATURATED CARBONYL COMPOUNDS

1. The carbonyl group of an α,β-unsaturated aldehyde or ketone is reduced to an alcohol with lithium aluminum hydride.

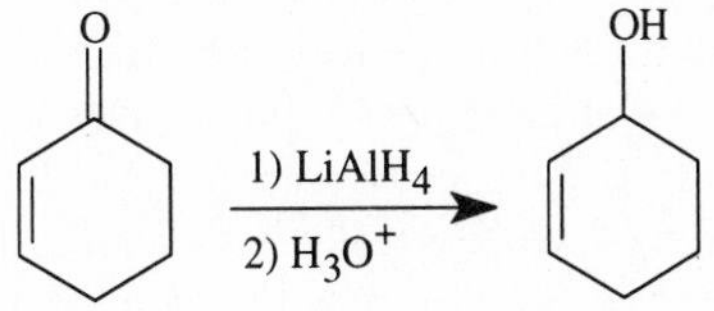

   a. This reaction involves the attack of hydride at the carbonyl carbon and is therefore a carbonyl addition.
   b. Carbonyl addition is not only faster than conjugate addition but, in this case, also irreversible; reduction of the carbonyl group with LiAlH$_4$ is a kinetically controlled reaction.
2. Many α,β-unsaturated carbonyl compounds are reduced by NaBH$_4$ to give mixtures of both carbonyl-addition products and conjugate-addition products.
3. The carbon-carbon double bond of an α,β-unsaturated carbonyl compound can in most cases be reduced selectively by catalytic hydrogenation.

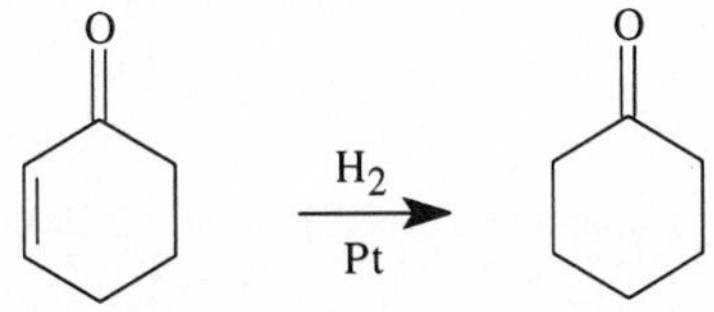

### B. ADDITION OF ORGANOLITHIUM REAGENTS TO THE CARBONYL GROUP

1. Organolithium reagents react with α,β-unsaturated carbonyl compounds to yield products of carbonyl addition.
2. The product is the result of kinetic control, since carbonyl addition is more rapid than conjugate addition and it is also irreversible.

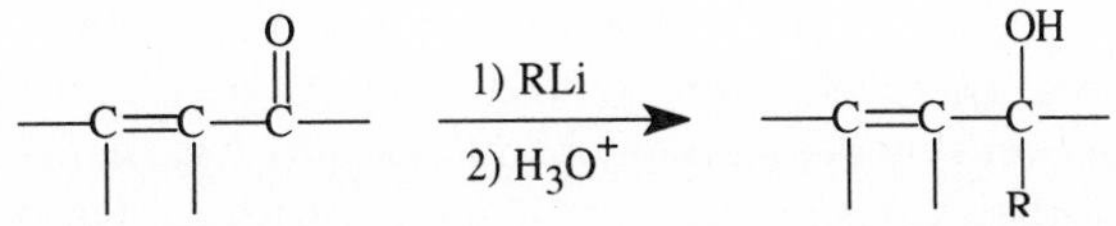

### C. CONJUGATE ADDITION OF LITHIUM DIALKYLCUPRATE REAGENTS

1. Lithium dialkylcuprate reagents give exclusively products of conjugate addition when they react with α,β-unsaturated esters and ketones.

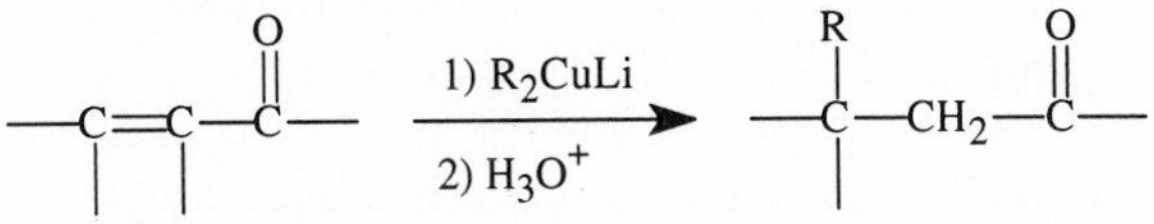

2. The conjugate addition of a lithium dialkylcuprate reagent can be envisioned mechanistically to be similar to other conjugate additions.
   a. Attack of an "alkyl anion" of the dialkylcuprate on the double bond gives a resonance-stabilized enolate ion.
   b. Protonation of the enolate ion by the addition of water gives the conjugate-addition product.

## Study Guide Links

·································································································

## √22.1    Ionization *vs.* Nucleophilic Attack on the Carbonyl Carbon

·································································································

One of the things you learned about Brønsted bases is that they can usually act as nucleophiles, and that nucleophiles attack the carbon atoms of carbonyl groups. Yet in Eqs. 22.1 and 22.2 of the text, bases are shown removing protons. You might be wondering, "What is the major reaction: nucleophilic attack on the carbonyl carbon, or removal of an α-hydrogen to give an enolate ion?" The answer is that both occur simultaneously. This section of the text focuses on the proton removal, but later sections of this chapter show the interplay between both types of reactions. Nevertheless, it might help you to see just one example here.

Suppose an ester is treated with aqueous NaOH. Recall (Sec. 21.7A, text p. 989) that NaOH saponifies esters. Saponification results from nucleophilic attack of hydroxide ion at the carbonyl carbon of an ester. However, while saponification is taking place, hydroxide ion removes the α-protons of a few ester molecules to form some enolate ions. This reaction has no consequences for saponification because saponification occurs much more rapidly than the reactions of enolate ions (many of which are discussed in this chapter). For this reason, it was not important to consider enolate-ion formation in the discussion of saponification.

The point of this chapter is to focus on situations in which enolate ion formation is significant, and ultimately on what happens when *both* proton removal and nucleophilic attack on a carbonyl carbon can occur.

## √22.2    Kinetic *vs.* Thermodynamic Stability of Enols

·································································································

When we think of unstable organic compounds we often think of molecules containing strained rings or molecules containing van der Waals repulsions. The structures of enols have no such destabilizing features. Yet enols in most cases cannot exist for significant lengths of time, while other less stable molecules can be put in bottles and stored for years. Why this seeming contradiction?

The essence of this problem is the meaning of the word *stability* as applied to chemical substances. *Thermodynamic stability* is a term used to describe the relative energy contents of molecules. In general, thermodynamic stability has very little to do with whether a compound can be isolated. *Kinetic stability* is a term used to describe how rapidly molecules revert to more stable molecules. In order for a molecule to be *kinetically unstable*, a mechanism must exist for its conversion into a more stable molecule. For a compound to be kinetically unstable, it must be thermodynamically unstable, but the converse need not be true. Some *thermodynamically unstable compounds* are easy to isolate because they are *kinetically stable*. For example, almost all hydrocarbons in the presence of oxygen are thermodynamically unstable relative to $CO_2$ and $H_2O$ by hundreds of kJ/mol; yet hydrocarbons do not rapidly revert to $CO_2$ and $H_2O$ because there is no readily available mechanism by which this conversion can be accomplished. If we supply a flame, however, such a conversion mechanism becomes available—we call it *combustion*—and we see very graphically how unstable hydrocarbons really are!

Examples of kinetic stability and instability in carbonyl chemistry are provided by acetals (Sec. 19.10, text p. 898) and carbonyl hydrates (Sec. 19.7, text p. 884). The hydrate of cyclohexanone is both thermodynamically and kinetically unstable with respect to the ketone and water:

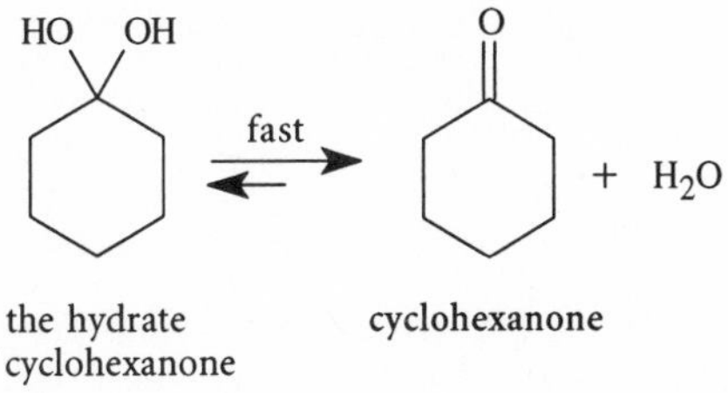
HO
OH
fast
O
+ H2O
the hydrate
of cyclohexanone
cyclohexanone

Yet replacing the O—H hydrogens with alkyl groups gives compounds called *acetals* that are thermodynamically unstable, but *kinetically stable,* in neutral and basic solution.

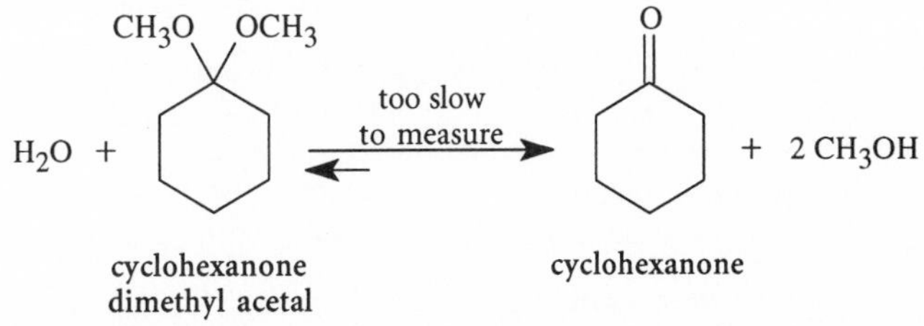
CH3O
OCH3
H2O +
too slow
to measure
O
+ 2 CH3OH
cyclohexanone
dimethyl acetal
cyclohexanone

The base-catalyzed mechanism of hydrate decomposition requires the presence of an O—H hydrogen, which acetals lack. Hence, the base-catalyzed pathway of hydrate decomposition cannot operate with acetals. Therefore, acetals, although thermodynamically unstable relative to their corresponding carbonyl compounds, are kinetically stable in basic solution. This is why acetals are good protecting groups for carbonyl groups under basic or neutral conditions. Under acidic conditions, however, acetals are kinetically unstable because an acid-catalyzed mechanism exists for their conversion into ketones.

The thermodynamic stabilities of most enols are not much less than the stabilities of their corresponding carbonyl isomers—typically, about 30–40 kJ/mol (7–10 kcal/mol). What makes enols difficult to isolate is their *kinetic* instabilities. Simple paths, or mechanisms, exist for their rapid transformation into carbonyl compounds in the presence of acids or bases; these mechanisms are shown in Eqs. 22.15a–b in the text. Enols are kinetically unstable molecules because the conditions under which they are converted into carbonyl compounds—the presence of acids and/or bases—are so prevalent. It's hard to generate an enol without an acid or base around. In fact, enols themselves are weak acids, and can catalyze their own slow conversion into ketones. Enols simply don't stand much of a chance for survival.

Not many years ago chemists figured out ways to generate simple enols in solution in the absence of other acids and bases and, indeed, found that they could be isolated and observed for significant periods of time.

## ✓22.3 Dehydration of $\beta$-Hydroxy Carbonyl Compounds

At first sight, the dehydration of $\beta$-hydroxy carbonyl compounds would appear to be an ordinary acid-catalyzed dehydration as discussed in Sec. 10.1, text p. 443. Recall that in the mechanism of such a dehydration, protonation of the hydroxy group is followed by carbocation formation:

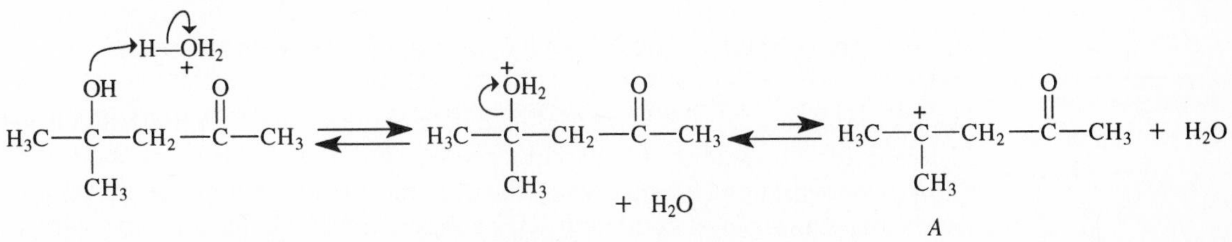
H—OH2
+
OH
O
H3C—C—CH2—C—CH3
CH3
+
OH2
O
H3C—C—CH2—C—CH3
CH3
+ H2O
O
+
H3C—C—CH2—C—CH3 + H2O
CH3
A

Loss of a C—H proton from carbocation *A* gives the $\alpha,\beta$-unsaturated ketone. However, a different mechanism—and, as it turns out, the correct one—is possible. This mechanism involves dehydration not of the ketone, but instead of the *enol.*

$$H_3C-\underset{\underset{CH_3}{|}}{\overset{\overset{OH}{|}}{C}}-CH_2-\overset{\overset{O}{||}}{C}-CH_3 \quad \xleftarrow{\hspace{1cm}} \quad H_3C-\underset{\underset{CH_3}{|}}{\overset{\overset{OH}{|}}{C}}-CH=\overset{\overset{OH}{|}}{C}-CH_3$$

an enol form

The dehydration involves protonation of the —OH group, formation of carbocation *B*, and loss of a proton from this carbocation:

$$H_3C-\underset{\underset{CH_3}{|}}{\overset{\overset{+OH_2}{|}}{C}}-CH=\overset{\overset{OH}{|}}{C}-CH_3 \quad \xleftarrow{\hspace{1cm}} \quad H_3C-\underset{\underset{CH_3}{|}}{\overset{+}{C}}-CH=\overset{\overset{O-H}{|}}{C}-CH_3 + H_2O \quad \xrightarrow{\hspace{1cm}} \quad H_3C-\underset{\underset{CH_3}{|}}{C}=CH-\overset{\overset{O}{||}}{C}-CH_3$$

*B*

+ H_3\overset{+}{O}

The reason the enol mechanism is favored is that *carbocation B is much more stable than carbocation A.* Can you see why? Draw resonance structure(s) for carbocation *B*. Is carbocation *A* resonance-stabilized? What is the *polar effect* of the carbonyl group on the stability of carbocation *A*?

## ✓22.4   Understanding Condensation Reactions

Condensations such as the aldol condensation and its variants, as well as the other reactions you'll encounter in Chapter 22, present a particular challenge to the beginning student, because the products are somewhat more complex than the products of many typical organic reactions, and because more than one functional-group transformation is involved (an addition followed by a dehydration). The way to develop an understanding of these reactions is to write the mechanism of each new reaction you see. First, identify the sources of the atoms in the product. (See Study Guide Link 4.7 on p. 64, Vol. 1, of this manual; the discussion in Sec. 22.4D on text p. 1062 should also be useful.) Use the base-catalyzed reaction of acetaldehyde (Eqs. 22.38a–b and 22.42 of the text) and the acid-catalyzed reaction of acetone (Eqs. 22.44a–c of the text) as models.  If you still have difficulty, *seek assistance!* You should have taken this approach with the crossed-aldol products in Study Problem 22.1. Be sure you understand the reactions shown in Eqs. 22.45 and 22.46. If you work through these examples, the reaction in Eq. 22.47 should be clear. The only thing different in this case is that the condensation occurs internally. Which carbons of the starting material are joined in the product? What is the structure of the enol intermediate involved? Use Eq. 22.44b to help you understand the formation of the carbon-carbon bond.

Just as in a sport, or in music, or in any other worthwhile endeavor, improving your skill requires practice. If you are concerned or discouraged that these transformations are not immediately obvious, don't be! They are more likely to become easier to understand, however, if you'll take the approach described here.

## ✓22.5   Variants of the Aldol and Claisen Condensations

Many variations of the aldol and Claisen condensations exist. The text has considered two: the Claisen-Schmidt variation of the aldol condensation, and the Dieckmann variation of the Claisen condensation. Possibly you'll hear of the Perkin condensation, the Stobbe condensation, the Knoevenagel condensation, or other variations that have no name. To cover all these in a text of reasonable (or even unreasonable) size is neither possible nor even necessary. For if you master the principles of reactivity contained in Chapters 21 and 22, understanding variations of these condensation reactions will only be a matter of re-applying the principles.

In fact, mastering these principles might even allow you to invent your own condensation reaction!

## 22.6    Malonic Ester Alkylation

Notice in Eq. 22.61b of the text that a *secondary* alkyl halide is used to alkylate the enolate ion of malonic ester. When you studied the $S_N2$ reaction, you learned that a side reaction is E2 elimination, and that this side reaction is particularly prevalent with secondary and tertiary alkyl halides. Because the product of Eq. 22.61b is formed in high yield, elimination is evidently not a problem in this case, and one might ask why.

Recall (Sec. 9.5F on text pp. 411–415) that, although most strong bases are good nucleophiles, a number of relatively weak bases are also good nucleophiles, and that branched or secondary alkyl halides undergo $S_N2$ reactions with such nucleophiles with little interference from E2 reactions. (Note particularly the example in Eq. 9.46 on text p. 414.) The enolate ion of diethyl malonate is in this category. Chemists have found that delocalized carbanions such as enolate ions are particularly good nucleophiles (for reasons that are not completely understood). Consequently, the $S_N2$ reaction is particularly rapid for these anions, and the E2 reaction does not interfere unless highly branched secondary or tertiary alkyl halides are used.

## 22.7    Alkylation of Enolate Ions

The acetoacetic ester synthesis requires the presence of the ester group of a $\beta$-keto ester; this group is then "thrown away" by hydrolysis and decarboxylation at the end of the synthesis and replaced by a hydrogen. For the same reasons that carboxylic acids with quaternary $\alpha$-carbons cannot be prepared by the malonic ester synthesis, ketones with quaternary $\alpha$-carbons cannot be prepared by the acetoacetic ester synthesis. For example, suppose we want to prepare 2,2-dimethylcyclohexanone:

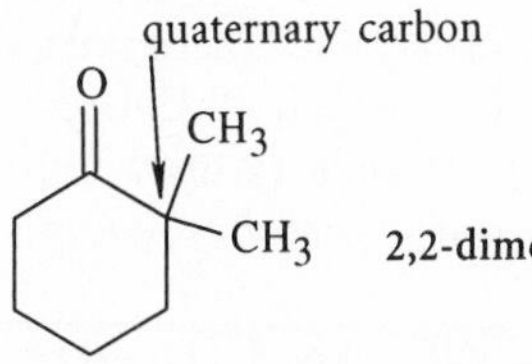

Although 2-methylcyclohexanone is easily prepared by a Dieckman condensation-alkylation sequence, introduction of the second methyl group is not possible in such a reaction scheme.

Why not, then, resort to the strategy used in Sec. 22.6B on p. 1076 of the text? Methylcyclohexanone could be alkylated *directly* by forming its enolate ion with a strong base such as LCHIA and treatment of that enolate with methyl iodide:

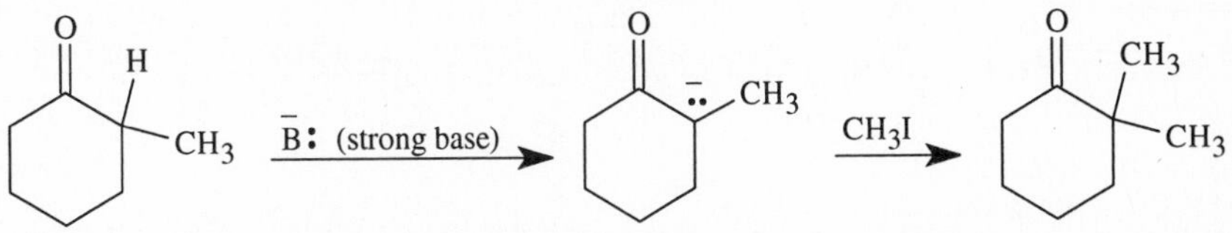

Can you see the problem with this approach? With an ester, only one enolate ion is possible. But with a ketone, in general *two* enolate ions are possible:

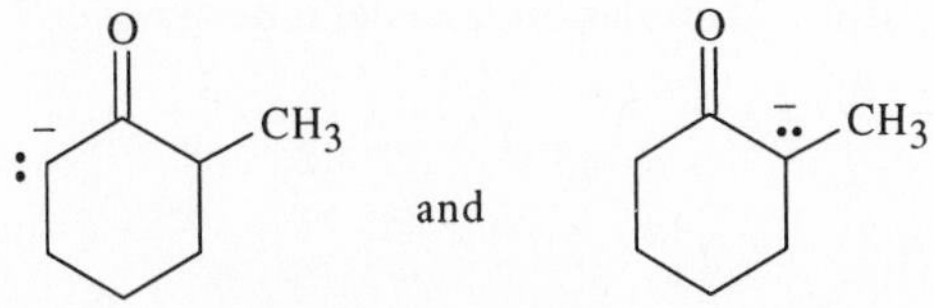

The problem, then, is how to form one enolate and not the other, and to keep the two enolate ions from equilibrating once the desired one is formed.

But the problems don't stop here. Suppose a way could be found to form the desired enolate ion. An additional problem is that enolate ions can react in more than one way. The enolate ions of ketones are not only nucleophilic at carbon; they are also nucleophilic at *oxygen*:

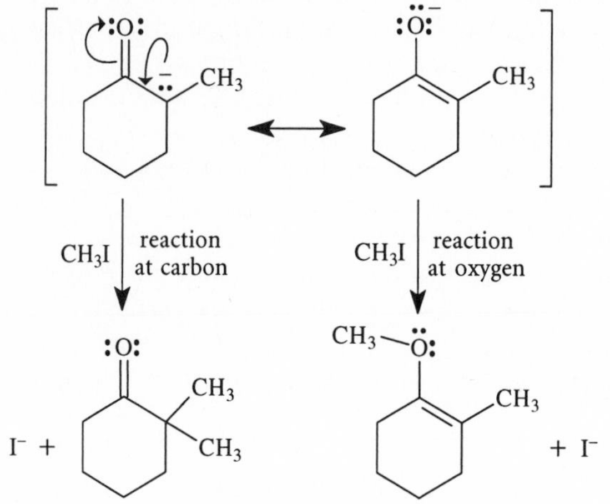

a vinyl ether

Alkylation at oxygen gives a vinyl ether—sometimes called an *enol ether,* because conceptually it is derived by replacement of the —OH hydrogen of an enol with an alkyl group. In general, both products are formed. (Alkylation of oxygen is not usually observed with esters because enols of esters—and enol ethers of esters—are *much* less stable relative to their parent esters than enols of ketones are; see Eq. 22.11 on text p. 1045.)

Direct alkylation of ketones, then, is a much more complex problem than alkylation of esters. Nevertheless, chemists have discovered ways to control, in many cases, which enolate ion is formed, and to control which alkylation product is produced. These solutions, although beyond the scope of an introductory text, are conceptually not difficult to understand if you understand the principles of enolate-ion chemistry.

## ✓22.8  Further Analysis of the Claisen Condensation

If you imagine the Claisen condensation on an ester of general structure *A*, you'll see that saponification and decarboxylation *must* yield a *symmetrical ketone* of general structure *C*. (A symmetrical ketone is a ketone in which the two groups attached to the carbonyl carbon are the same.)

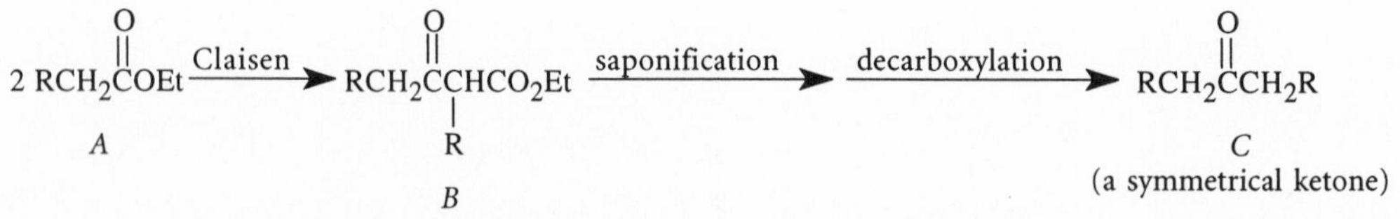

(Be sure to go through this case in detail to verify the foregoing statement.) Alkylation of the Claisen product *B* can lead to a variety of *substituted derivatives* of *C*. For example, treatment of *B* with NaOEt, then a general alkyl halide R′—I, would give, after saponification, acidification, and decarboxylation, a general derivative *D*.

$$RCH_2—\overset{\overset{\displaystyle O}{\|}}{C}—\underset{\underset{\displaystyle R'}{|}}{CHR} \quad \longleftarrow \quad \text{symmetrical ketone structure}$$

$$D$$

Notice the symmetrical ketone structure "lurking" as the circled part of *D*.

When you are asked to prepare a ketone by the Claisen condensation, look for such symmetrical-ketone "substructures" within the target molecule. Problem 22.34 of the text gives you practice in applying this strategy.

## ✓22.9  Synthetic Equivalents in Conjugate Addition

Study Guide Links 17.3 (p. 397 of this manual) and 20.7 (p. 503 of this manual) discussed the notion of *synthetic equivalence*, that is, thinking of one group in terms of other groups from which it might be formed. This notion is particularly useful in planning conjugate additions. In Study Problem 22.6, for example, the di(ethoxycarbonyl)methyl anion (the conjugate base of diethyl malonate, $^-:CH(CO_2Et)_2$) is used as the *synthetic equivalent* of a carboxymethyl anion ($^-:CH_2CO_2H$), which itself cannot exist. (Why can't this anion exist?).

What are some other synthetic equivalents that are useful in conjugate-addition reactions?

$$:\overset{-}{C}N \quad \text{is equivalent to} \quad :\overset{-}{\underset{}{C}}\!\!\overset{\overset{\displaystyle O}{\|}}{}\!\!—OH \quad \text{(see Eq. 22.77 on text p. 1086)}$$

$$:\overset{-}{C}N \quad \text{is also equivalent to} \quad :\overset{-}{C}H_2—NH_2 \quad \text{(see Sec. 21.9C on text p. 1010)}$$

$$:\underset{\underset{\displaystyle CO_2Et}{|}}{\overset{-}{C}H}—\overset{\overset{\displaystyle O}{\|}}{C}—R \quad \text{is equivalent to} \quad :\overset{-}{C}H_2—\overset{\overset{\displaystyle O}{\|}}{C}—R \quad \text{(see Eq. 22.72 on text pp. 1080)}$$

(You should show how each group can be converted into the other, for example, how a —C≡N group can be converted into a —CH₂NH₂ group.)

When you can start thinking in terms of synthetic equivalents you really have started to master the art of organic synthesis.

## 22.10  Conjugate Addition of Organocuprate Reagents

Although the mechanism presented in the text is conceptually useful, it does not explain the effect of copper in promoting conjugate addition. Because carbonyl addition rather than conjugate addition is observed in the *absence* of copper, it is evident that copper has a special effect. A substantial body of evidence has shown that conjugate-addition reactions of lithium organocuprate reagents involve free-radical intermediates, and that conjugate addition is a consequence of a free-radical mechanism.

Mixtures of conjugate-addition and carbonyl-addition products are often observed when Grignard reagents react with $\alpha,\beta$-unsaturated carbonyl compounds. The conjugate-addition reactions may be due to traces of transition-metal impurities in the magnesium used to form the Grignard reagent. Many transition metals, for example, copper, also promote conjugate addition. Thus, if copper(I) chloride, CuCl, is intentionally added to a Grignard reagent, conjugate addition is the *only* reaction observed.

## Solutions

## Solutions to In-Text Problems

**22.1**   The acidic hydrogens are the α-hydrogens on the carbons between the two carbonyl groups. (—OEt = ethoxy group = —OCH₂CH₃.)

$$\text{acidic hydrogens}$$

EtOC—CH₂—COEt        CH₃C—CH₂—COEt

diethyl malonate            ethyl acetoacetate

The reason that these hydrogens are particularly acidic is that their conjugate-base enolate ions are stabilized by the polar effects and resonance effects of two carbonyl groups, whereas the conjugate-base enolate ion of an ordinary ester is stabilized by the corresponding effects of only one carbonyl group. The resonance structures of the conjugate-base enolate ion of diethyl malonate are as follows; the resonance structures of the conjugate-base enolate ion of ethyl acetoacetate are similar.

$$\left[ \; \text{EtO—C—CH—C—OEt} \longleftrightarrow \text{EtO—C—CH=C—OEt} \longleftrightarrow \text{EtO—C=CH—C—OEt} \; \right]$$

resonance structures for the conjugate-base enolate ion of diethyl malonate

**22.3**   A mechanism for replacement of one hydrogen in the reaction of Eq. 22.4 is shown in the following equation. (The mechanisms for replacement of the others are identical.) Only the α-hydrogens are replaced because the enolate ion is the only carbanion stable enough to be formed. The carbanion intermediates required in a similar mechanism for the replacement of the hydrogens other than the α-hydrogens are not resonance-stabilized.

$$\text{Et}_3\text{N} \quad \text{D—OD} \rightleftharpoons \text{Et}_3\overset{+}{\text{N}}\text{D} \;\; ^-\text{OD}$$

$$\text{H—OD} + \qquad \qquad \qquad + \; ^-\text{OD}$$

resonance-stabilized
enolate ion

**22.5**   Exchange does not occur because the orbital containing the unshared electron pair on the α-carbon of the enolate ion is perpendicular to the π-electron system of the carbonyl group; therefore, the enolate ion is not stabilized by resonance. Another way to view this situation is suggested by the hint: the resonance structure of the enolate ion violates Bredt's rule. This situation is discussed in Guideline 5, text pp. 711–712.

**22.6**   (a) All α-hydrogens as well as the O—H hydrogen are exchanged for deuterium:

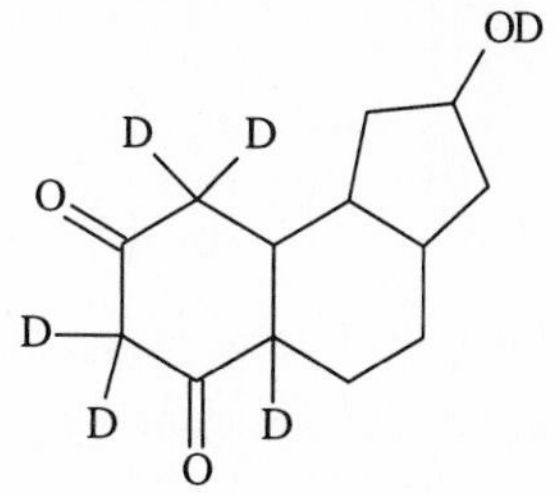

**22.7** (a)

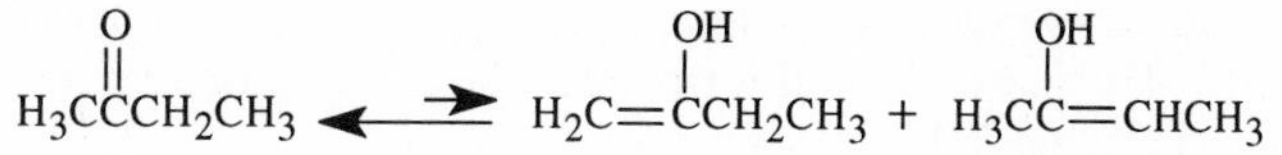

2-butanone — enol forms of 2-butanone (*E*) and (*Z*)

(c)

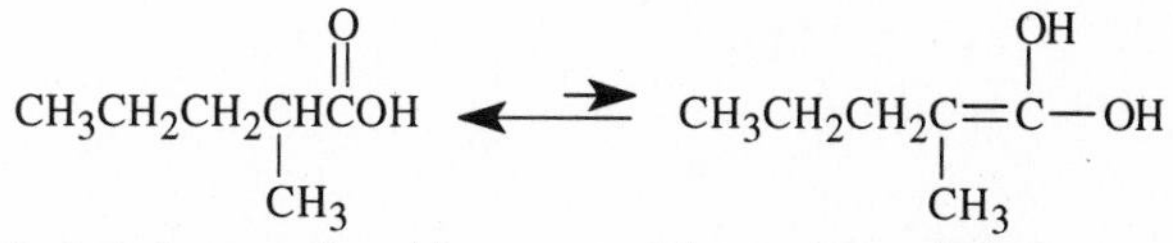

2-methylpentanoic acid — enol form of 2-methylpentanoic acid

(e)

*N,N*-dimethylacetamide — enol form of *N,N*-dimethylacetamide

**22.8** (a)  (c)

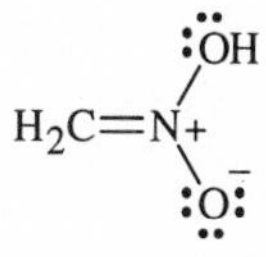

**22.9** (a) One source of stabilization of the enols of β-diketones is their intramolecular hydrogen bonds; see the structure preceding Problem 22.7 on text p. 1046. In aqueous solution, the solvent can serve as a hydrogen-bond donor to the carbonyl oxygen; hence the energetic advantage of the enol that results from intramolecular hydrogen bonding is reduced in aqueous solution. In a solvent that cannot donate hydrogen bonds, such as hexane, there are no external hydrogen bonds, and the internal hydrogen bond thus provides an energetic advantage to the enol.

**22.10** (a) The enol is formed by the mechanism shown in Eq. 22.15b on text p. 1047. The enol is achiral, and can be protonated at either face of the double bond at the same rate to give the racemate of the protonated ketone.

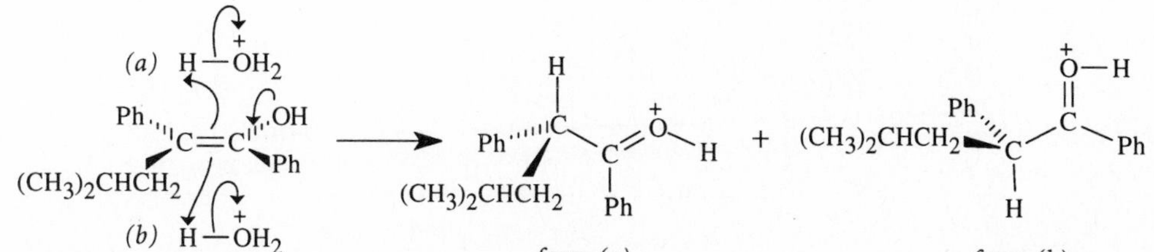

from (a)    from (b)

The ketone itself is formed when a proton is removed by the solvent. (This is shown below for enantiomer *a*.)

Ph⟨⟩C—C=O⁺—H   OH₂ ⟶ Ph⟨⟩C—C=O   + H—OH₂⁺
(CH₃)₂CHCH₂   Ph          (CH₃)₂CHCH₂   Ph

22.11   (a)–(b)   A reaction-free energy diagram for halogenation of a ketone is shown in Fig. SG22.1. The first two steps are the reaction-free energy diagram for enol formation. (The addition of halogen is treated as a single step.)

22.12   (a)   Because enolization is the process responsible for acid-catalyzed racemization, and because the rate-limiting process in acid-catalyzed ketone halogenation is enol formation, the rates of the two processes are the same.

22.13   (a)                                        (c)

O=C—O⁻ Na⁺   + HCBr₃                    Ph—C(=O)—O⁻ Na⁺ + HCBr₃

In part (c), the alcohol is first oxidized to the the corresponding ketone, acetophenone, which then undergoes the haloform reaction.

22.14   2,4-Hexanedione gives succinic acid and iodoform when subjected to the iodoform reaction.

$$CH_3CCH_2CH_2CCH_3 \xrightarrow[\text{2) } H_3O^+]{\text{1) } I_2, \text{ base}} HOCCH_2CH_2COH + 2HCI_3$$

2,4-hexanedione                              succinic acid

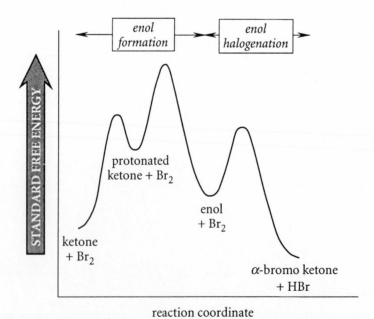

Figure SG22.1   *Reaction free-energy diagram to accompany the solution to Problem 22.11. Notice that the rate-limiting steps for enol formation and overall halogenation are the same.*

**22.15** (a) The $\alpha$-bromo acid bromide is formed, and it reacts with ammonia to give the amide of the $\alpha$-amino acid.

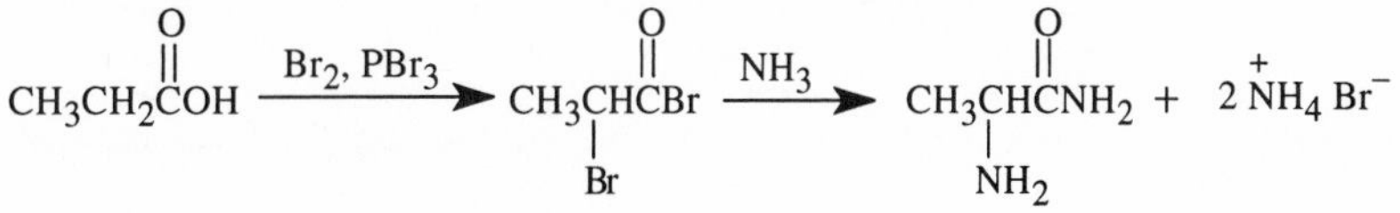

$$CH_3CH_2COH \xrightarrow{Br_2, PBr_3} CH_3CHCBr \xrightarrow{NH_3} CH_3CHCNH_2 + 2\overset{+}{NH_4}Br^-$$

**22.16** (a)

$$CH_3CH_2CCH_2—\overset{+}{N}\langle\rangle \quad Br^-$$

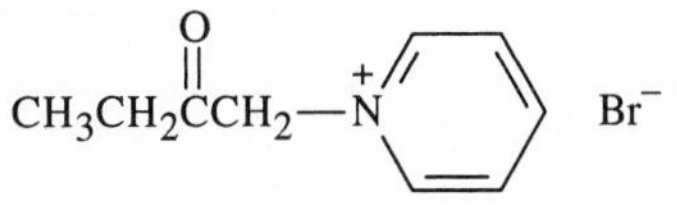

**22.17** One equivalent of NaOH is consumed by its reaction with the carboxylic acid. The second equivalent forms the 2,4-dichlorophenolate ion, which is the nucleophile in the ensuing $S_N2$ reaction.

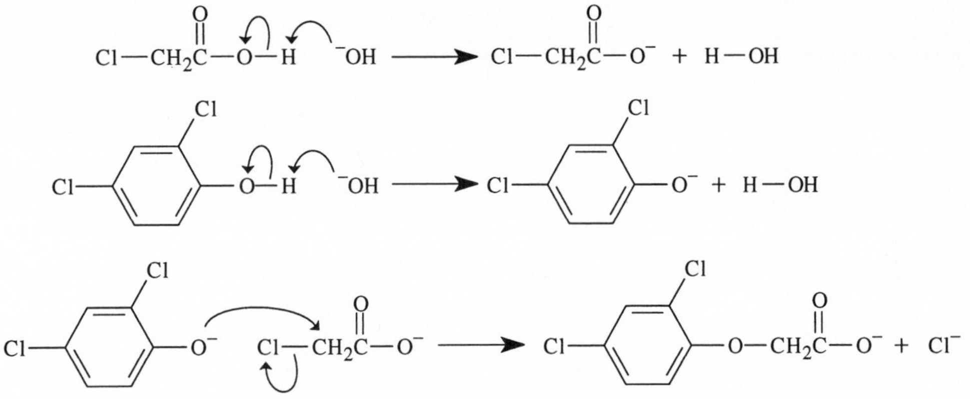

$$Cl—CH_2C—O—H \quad ^-OH \longrightarrow Cl—CH_2C—O^- + H—OH$$

Addition of acid ($H_3O^+$) converts the product into the corresponding carboxylic acid.

**22.18** (a) The mechanism of the aldol addition reaction of $\alpha$-phenylacetaldehyde:

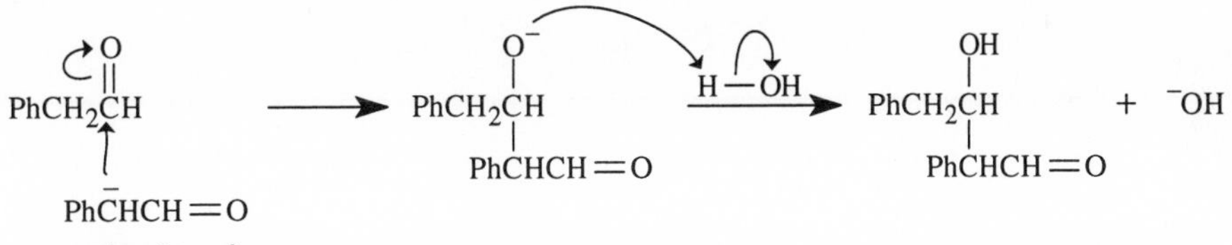

PhCH$_2$CH
PhCHCH=O
enolate ion of
$\alpha$-phenylacetaldehyde

PhCH$_2$CH
PhCHCH=O
H—OH

PhCH$_2$CH
PhCHCH=O
+ $^-$OH

**22.19** (a) Assume a reaction of one equivalent of each component. The enolate ion of the ketone reacts with the aldehyde because aldehyde carbonyl groups are more reactive in additions than ketone carbonyl groups.

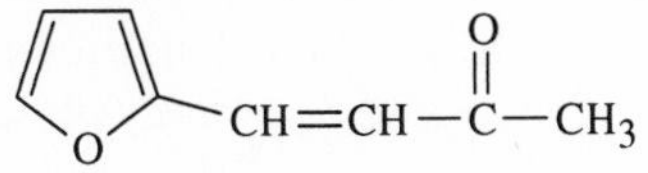

$$\langle\rangle\text{—CH=CH—}\overset{O}{\overset{\|}{C}}\text{—CH}_3$$

(c) This is an intramolecular aldol condensation.

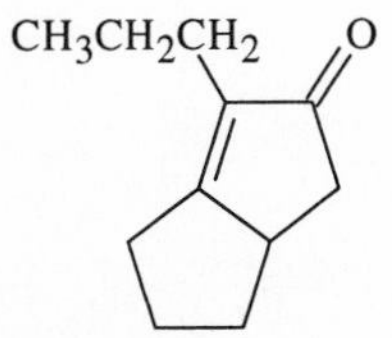

CH$_3$CH$_2$CH$_2$

22.20 (a) In this case, the enolate ion of a symmetrical ketone, 3-pentanone, would have to react with *p*-methoxybenzaldehyde. There is only one possible enolate ion, and the aldehyde carbonyl group is much more reactive than the ketone carbonyl group. Hence, the reaction is reasonable.

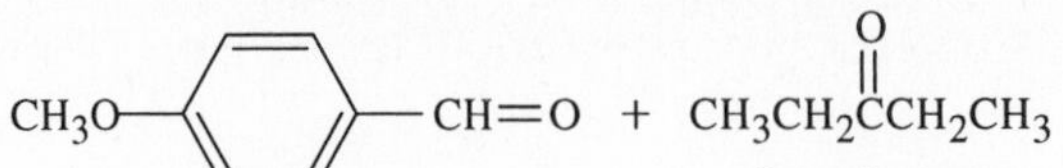

$$CH_3O-\underset{\text{\emph{p}-methoxybenzaldehyde}}{\boxed{\phantom{xx}}}-CH=O \;+\; CH_3CH_2\overset{O}{\overset{\|}{C}}CH_2CH_3$$

3-pentanone

(c) In this case, the enolate ion of an aldehyde would have to react with a ketone carbonyl group. Because the aldehyde carbonyl group is more reactive, and because its intramolecular reaction with an enolate ion of the ketone would also give a six-membered ring, synthesis of the compound shown in the problem is not reasonable.

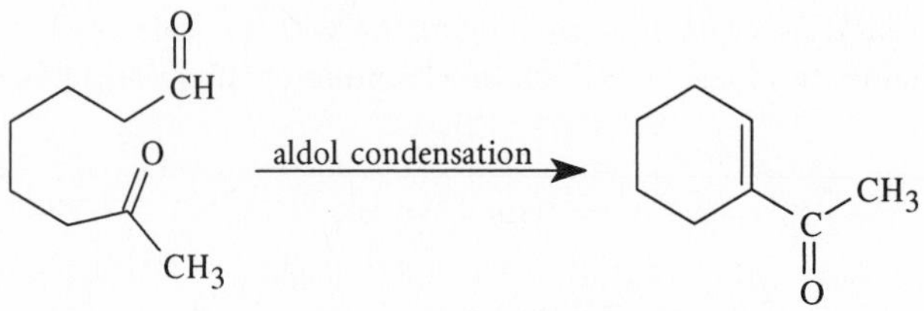

(e) A double aldol condensation of the following starting materials would give the desired product. The reaction is particularly favorable because of the highly conjugated nature of the product.

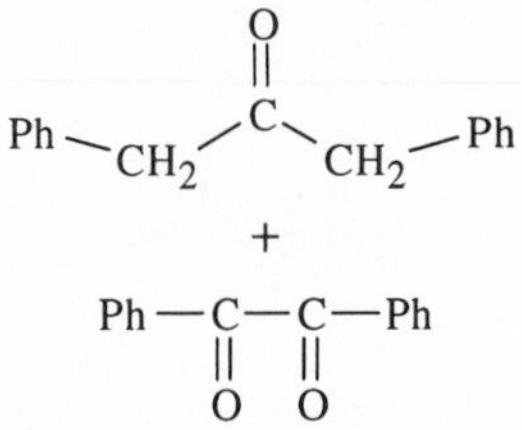

$$Ph-CH_2\overset{O}{\overset{\|}{C}}CH_2-Ph$$

$$+$$

$$Ph-\overset{\|}{\underset{O}{C}}-\overset{\|}{\underset{O}{C}}-Ph$$

(g) Because the diketone starting material is symmetrical, the product shown is the only possible one, and thus the synthesis is reasonable.

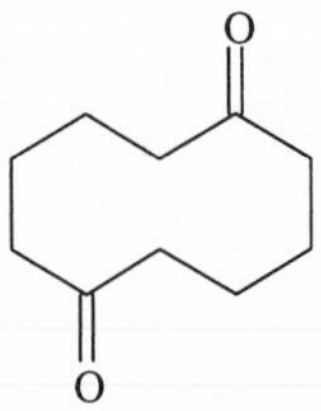

(h) This product would require that the carbonyl group of acetone, $(CH_3)_2C=O$, react with the conjugate-base enolate ion of acetaldehyde, $CH_3CH=O$. Because the aldehyde carbonyl group is more reactive than the ketone carbonyl group, the more likely products would be the aldol condensation product of acetaldehyde along with the aldol condensation product derived from the reaction of the conjugate-base enolate ion of acetone with the carbonyl group of acetaldehyde.

22.21 First, break the double bond as shown in Eq. 22.48 on text p. 1062 to reveal the possible starting material.

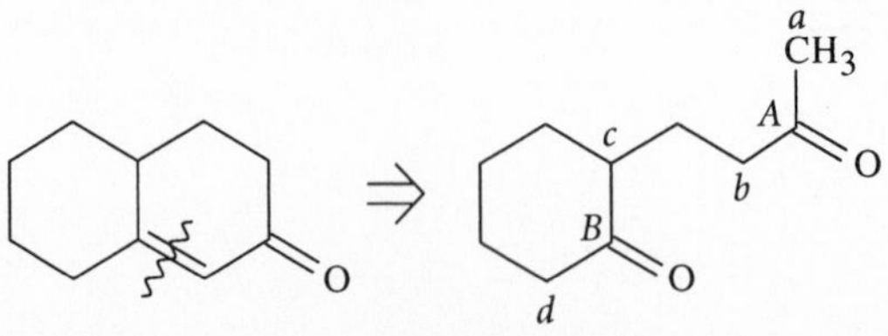

The product shown in the problem results from the reaction of an enolate ion formed at carbon *a* with carbonyl group *B*. Reaction of the enolate ion formed at carbon *b* with carbonyl *B* would require formation of a strained four-membered ring, and is thus not likely. Reaction of an enolate ion formed at carbon *c* with carbonyl *A* also would require the formation of a strained four-membered ring. Reaction of an enolate ion formed at carbon *d* with carbonyl *A* would give the following bicyclic compound:

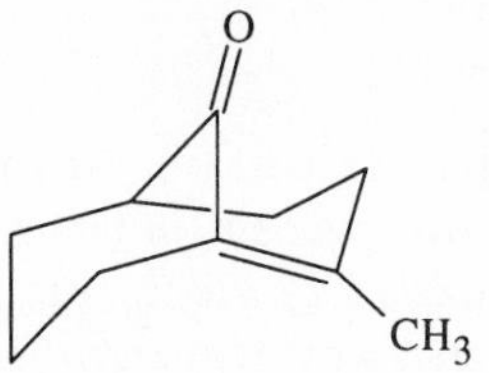

This compound is a modest violation of Bredt's rule (Sec. 7.6C of the text; build a model to verify that it is strained). The aldol *addition* product could form, thus avoiding dehydration to the strained alkene, but aldol additions of ketones are highly reversible. The product shown in the problem is thus the only one that can readily be driven to completion by dehydration.

**22.22** (a)  (c)   (e)

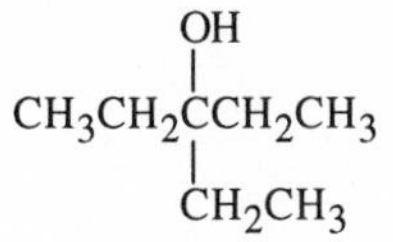

**22.23** (a) The following product is formed after acidification of the reaction mixture:

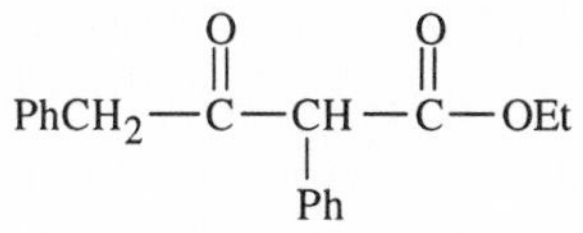

**22.24** Hydroxide ion would *not* be a good base for the Claisen condensation because it would saponify ethyl acetate and convert it into acetate ion. The role of the ethoxy group in ethyl acetate is to serve as a leaving group; acetate ion lacks this leaving group and therefore will not react in the Claisen condensation.

**22.25** (a) An ordinary Claisen condensation is driven to completion by ionization of an $\alpha$-hydrogen more acidic than ethanol, the conjugate acid of the base ethoxide. Compound *A* lacks such an acidic $\alpha$-hydrogen. Consequently, ethoxide attacks the ketone carbonyl group and converts it into *B*. This reaction is driven to completion by ionization of the acidic $\alpha$-hydrogen of compound *B*. Compound *B* itself is formed on acidification of the reaction mixture.

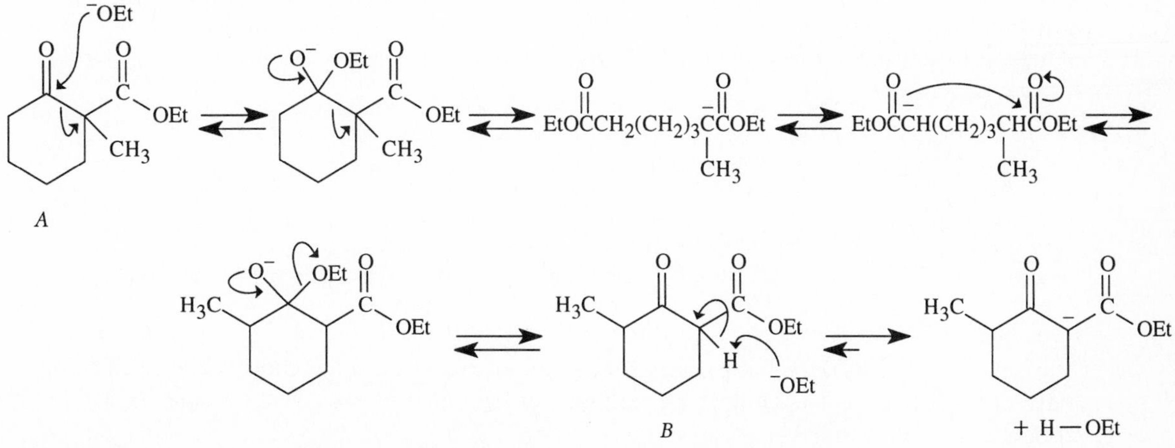

22.26 (a)

$$\mathrm{EtO\overset{O}{\overset{\|}{C}}CH_2\overset{O}{\overset{\|}{C}}CMe_3}$$

(c)

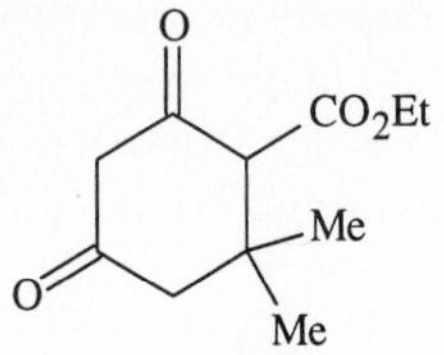

In both parts, ethanol is a by-product. In part (c), the carbanion formed at carbon *a* leads to the product. Hydrogens *b* are more acidic than hydrogens *a*, but a carbanion formed at carbon *b* cannot lead to a product containing a ring of reasonable size. Hence, the two anions are in an equilibrium that is driven by ionization of the product (LeChatelier's principle).

$$\mathrm{\overset{a}{C}H_3\overset{O}{\overset{\|}{C}}CH_2\overset{Me}{\underset{Me}{C}}\overset{b}{C}H(CO_2Et)_2}$$

22.27 (a) Apply the analysis shown in Eq. 22.60 on text p. 1072. The two possible sets of starting materials are as follows:

$$\underbrace{\mathrm{CH_3\overset{O}{\overset{\|}{C}}OEt + CH_3CH_2\overset{O}{\overset{\|}{C}}CH_2CH_3}}_{A} \quad \text{and} \quad \underbrace{\mathrm{CH_3\overset{O}{\overset{\|}{C}}CH_2CH_3 + EtO\overset{O}{\overset{\|}{C}}CH_2CH_3}}_{B}$$

Set *A* could work provided that an excess of the ester is used. The analysis is similar to the analysis of Eq. 22.58 on text p. 1070. Because the ketone is symmetrical, only one enolate ion is possible. Set *B* would give a mixture of products because the ketone could form two different conjugate-base enolate ions, and there is no reason for one to be strongly preferred over the other.

(c) Apply the analysis shown in Eq. 22.59 on text p. 1071. The two possible sets of starting materials are as follows:

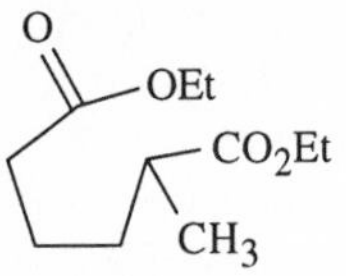

*A*

and

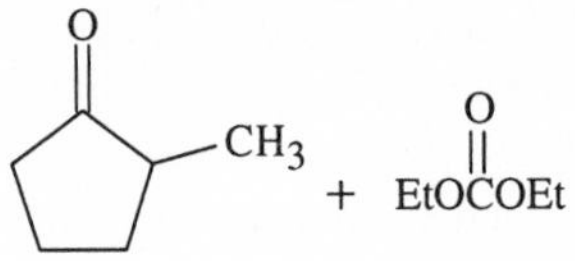

*B*

Neither set of starting materials is a good choice. Because the desired product has no α-hydrogen more acidic than ethanol, it cannot be formed in a Claisen condensation because the equilibrium would be unfavorable. In fact, *A could* react in a Claisen condensation to give a different product. (Can you give its structure?) Reactants *B* are also not a good choice for the same reason: the desired product is not acidic enough to be ionized by ethoxide. Furthermore, in this case a different product would be formed.

22.28 (a)

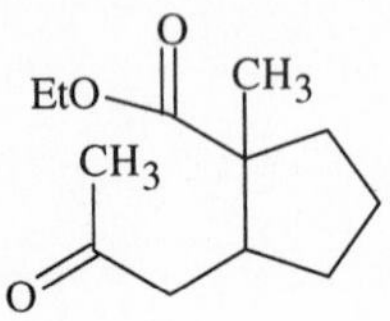

*A*

or

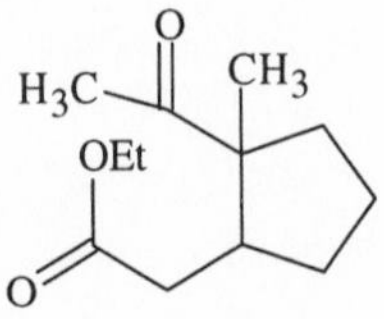

*B*

Either answer is satisfactory, but perhaps compound *B* would be a better choice because, in compound *A*, attack of the ketone enolate on the ester carbonyl carbon might be somewhat retarded by van der Waals repulsions with the α-methyl group.

**22.29** (a)

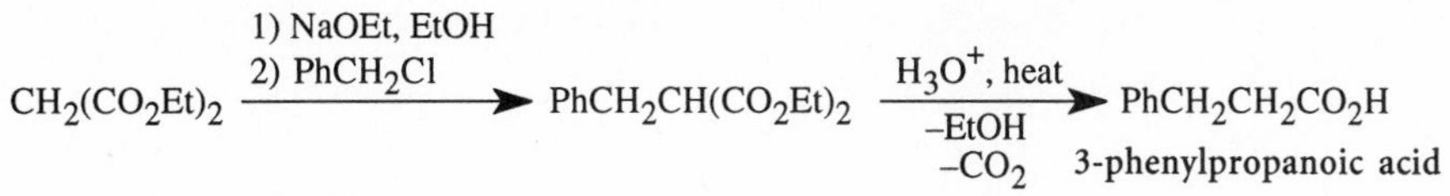

$$CH_2(CO_2Et)_2 \xrightarrow[\text{2) PhCH}_2\text{Cl}]{\text{1) NaOEt, EtOH}} PhCH_2CH(CO_2Et)_2 \xrightarrow[\substack{-\text{EtOH} \\ -\text{CO}_2}]{\text{H}_3\text{O}^+, \text{heat}} PhCH_2CH_2CO_2H \quad \text{3-phenylpropanoic acid}$$

Recall that benzyl chloride, $PhCH_2Cl$, is exceptionally reactive in $S_N2$ reactions. (See Eq. 17.31a, text p. 803.)

(c)  3,3-Dimethylbutanoic acid cannot be prepared by a malonic ester synthesis because the requisite alkyl halide, neopentyl bromide, is virtually unreactive in $S_N2$ reactions. (See Fig. 9.3(b) on text p. 399 and the associated discussion.)

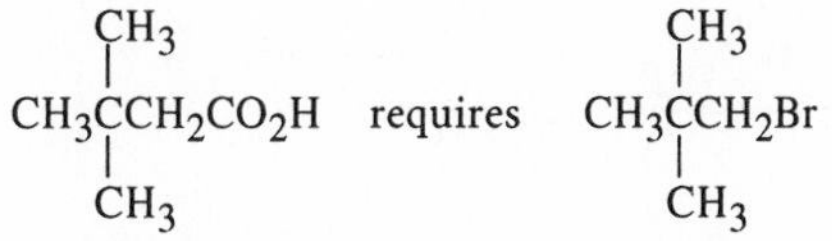

|  |  |  |
|---|---|---|
| $\underset{|}{CH_3}$ | | $\underset{|}{CH_3}$ |
| $CH_3\overset{|}{\underset{|}{C}}CH_2CO_2H$ requires | | $CH_3\overset{|}{\underset{|}{C}}CH_2Br$ |
| $CH_3$ | | $CH_3$ |
| 3,3-dimethylbutanoic acid | | neopentyl bromide |

**22.30**  On the assumption that hydrolysis of the ester groups and decarboxylation of one carboxy group have occurred, the unsaturation number (2) of the product calculated from the molecular formula indicates that a ring has been formed. The formation of cyclobutanecarboxylic acid by the following reaction sequence accounts for the results:

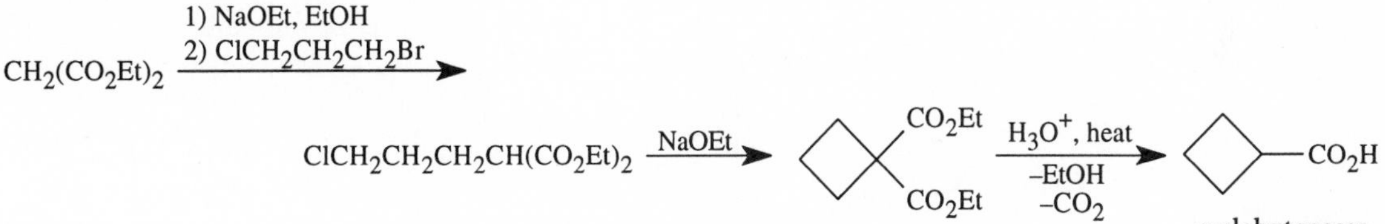

$$CH_2(CO_2Et)_2 \xrightarrow[\text{2) ClCH}_2\text{CH}_2\text{CH}_2\text{Br}]{\text{1) NaOEt, EtOH}}$$

$$ClCH_2CH_2CH_2CH(CO_2Et)_2 \xrightarrow{\text{NaOEt}} \underset{CO_2Et}{\overset{CO_2Et}{\square}} \xrightarrow[\substack{-\text{EtOH} \\ -\text{CO}_2}]{\text{H}_3\text{O}^+, \text{heat}} \square\text{—}CO_2H$$

cyclobutanecarboxylic acid

**22.31**  Because carboxylic acids with quaternary $\alpha$-carbons cannot be prepared by the malonic ester synthesis, the amide-base method will have to be used for the compound in part (c).

(a)  Note that allylic halides such as allyl bromide ($H_2C{=}CHCH_2Br$) are very reactive in $S_N2$ reactions. (See Eq. 17.30a on text p. 803.)

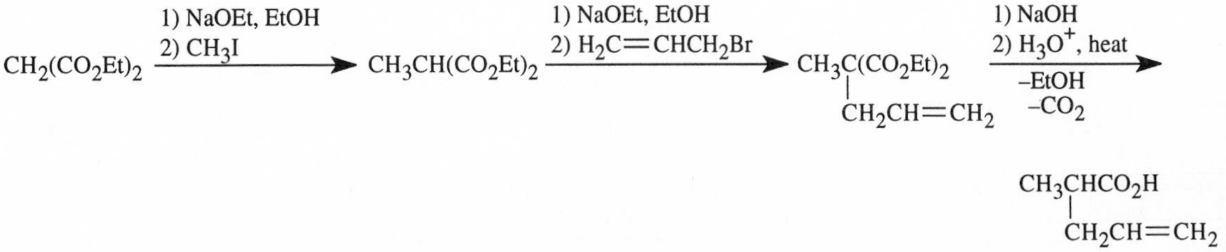

$$CH_2(CO_2Et)_2 \xrightarrow[\text{2) CH}_3\text{I}]{\text{1) NaOEt, EtOH}} CH_3CH(CO_2Et)_2 \xrightarrow[\text{2) H}_2\text{C}{=}\text{CHCH}_2\text{Br}]{\text{1) NaOEt, EtOH}} \underset{CH_2CH=CH_2}{CH_3\overset{|}{C}(CO_2Et)_2} \xrightarrow[\substack{-\text{EtOH} \\ -\text{CO}_2}]{\text{1) NaOH} \\ \text{2) H}_3\text{O}^+, \text{heat}}$$

$$\underset{CH_2CH=CH_2}{CH_3\overset{|}{C}HCO_2H}$$

(c)

$$CH_2(CO_2Et)_2 \xrightarrow[\text{2) EtBr}]{\text{1) NaOEt, EtOH}} EtCH(CO_2Et)_2 \xrightarrow[\text{2) EtBr}]{\text{1) NaOEt, EtOH}} Et_2C(CO_2Et)_2 \xrightarrow[\substack{-\text{EtOH} \\ -\text{CO}_2}]{\text{H}_3\text{O}^+, \text{heat}}$$

$$Et_2CHCO_2H \xrightarrow[\text{H}_2\text{SO}_4 \text{ (catalyst)}]{\text{EtOH (solvent)}} Et_2CHCO_2Et \xrightarrow[\text{2) H}_2\text{C}{=}\text{CHCH}_2\text{Br}]{\text{1) LDA}} \underset{CH_2CH=CH_2}{Et_2\overset{|}{C}CO_2Et}$$

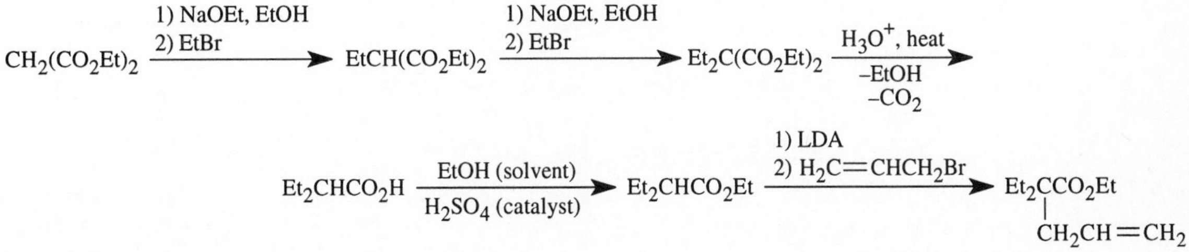

**22.32**  (a)  The enolate ion of *tert*-butyl acetate, formed by reaction of the ester with LCHIA, adds to the carbonyl group of acetone to give the lithium alkoxide, which is protonated when acid is added to give the β-hydroxy ester.

$$CH_3\overset{O}{\overset{\|}{C}}CH_3 \;+\; Li^+ \;{}^-CH_2\overset{O}{\overset{\|}{C}}OC(CH_3)_3 \longrightarrow CH_3\underset{\underset{CH_3}{|}}{\overset{\overset{Li^+}{O^-}}{C}}CH_2\overset{O}{\overset{\|}{C}}OC(CH_3)_3 \xrightarrow{H_3O^+} CH_3\underset{\underset{CH_3}{|}}{\overset{\overset{OH}{|}}{C}}CH_2\overset{O}{\overset{\|}{C}}OC(CH_3)_3$$

**22.33**  (a)

$$CH_3\overset{O}{\overset{\|}{C}}CH_2CO_2Et \xrightarrow[\text{2) BrCH}_2CH(CH_3)_2]{\text{1) NaOEt, EtOH}} CH_3\overset{O}{\overset{\|}{C}}\underset{\underset{CH_2CH(CH_3)_2}{|}}{C}HCO_2Et \xrightarrow[\text{2) H}_3O^+,\ -CO_2]{\text{1) NaOH}} CH_3\overset{O}{\overset{\|}{C}}CH_2CH_2CH(CH_3)_2$$

ethyl acetoacetate                                                                                                                            5-methyl-2-hexanone

**22.34**  (a)  Follow the hint in Study Guide Link 22.8.

$$\underset{CH_3CH}{\overset{CO_2Et}{|}}\!-\!\overset{O}{\overset{\|}{C}}\!-\!CH_2CH_3 \xrightarrow[\text{2) PhCH}_2Br]{\text{1) NaOEt, EtOH}} CH_3\underset{\underset{CH_2Ph}{|}}{\overset{\overset{CO_2Et}{|}}{C}}\!-\!\overset{O}{\overset{\|}{C}}\!-\!CH_2CH_3 \xrightarrow[\text{2) H}_3O^+,\ -CO_2]{\text{1) NaOH}} CH_3\underset{\underset{CH_2Ph}{|}}{C}H\overset{O}{\overset{\|}{C}}CH_2CH_3$$

The starting β-keto ester is prepared by a Claisen condensation of ethyl propionate:

$$2\ EtO\overset{O}{\overset{\|}{C}}CH_2CH_3 \xrightarrow[\text{2) H}_3O^+]{\text{1) NaOEt, EtOH}} \underset{CH_3CH}{\overset{CO_2Et}{|}}\!-\!\overset{O}{\overset{\|}{C}}CH_2CH_3$$

ethyl propionate

**22.35**  *A* is the enolate ion of diethyl malonate. Attack of the enolate ion on the methylene carbon of the epoxide gives the alkoxide *B*, which, in turn, undergoes an intramolecular transesterification to give a lactone.

**22.36**  (a)  Follow the pattern in Eqs. 22.73b–c on text p. 1084.

$$CH_3(CH_2)_4\overset{O}{\overset{\|}{C}}SR \;+\; {}^-\overset{O}{\overset{\|}{O}}CCH_2\overset{O}{\overset{\|}{C}}SR \xrightarrow[\text{-HSR, -CO}_2]{H_3O^+,} CH_3(CH_2)_4\overset{O}{\overset{\|}{C}}CH_2\overset{O}{\overset{\|}{C}}SR \xrightarrow{\text{reduction}}$$

$$CH_3(CH_2)_4\underset{\underset{}{\overset{OH}{|}}}{C}HCH_2\overset{O}{\overset{\|}{C}}SR \xrightarrow{\text{dehydration}} CH_3(CH_2)_4CH\!=\!CH\!-\!\overset{O}{\overset{\|}{C}}SR \xrightarrow{\text{reduction}} CH_3(CH_2)_6\overset{O}{\overset{\|}{C}}SR$$

**22.37**  (a)                                                                                               (c)

$$\underset{\underset{CH_3}{|}}{H_2C\!=\!C}\!-\!CO_2CH_3 \xrightarrow{HCN} N\!\equiv\!CCH_2\!-\!\underset{\underset{CH_3}{|}}{C}H\!-\!CO_2CH_3$$

methyl methacrylate

$$BrCH_2\!-\!\underset{\underset{CH_3}{|}}{C}H\!-\!CO_2CH_3$$

When considering acid-catalyzed additions to $\alpha,\beta$-unsaturated carbonyl compounds, pay careful attention to the mechanism in Eq. 22.84 on text p. 1087. In part (c), it is tempting to add HBr to the double bond to give the tertiary carbocation *A*; however, such a mechanism would lead to a product *B* that is different from the product observed:

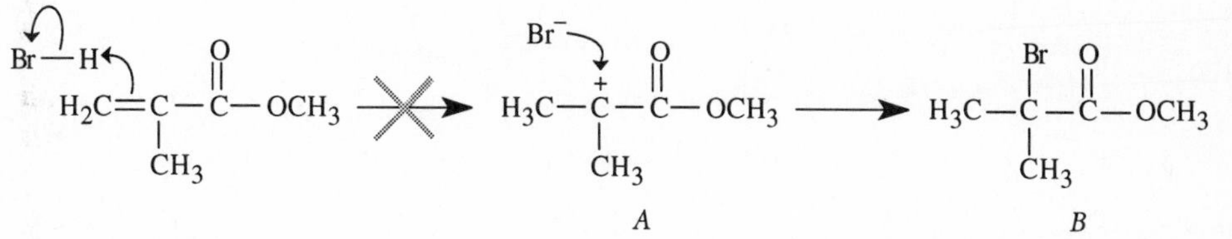

Remember that carbocations $\alpha$ to carbonyl groups are quite unstable and are almost never formed; see text p. 1054. In contrast, the mechanism in Eq. 22.84 on text p. 1087 involves a ketone that is protonated on the carbonyl oxygen; this is a resonance-stabilized species in which positive charge is shared at the terminal carbon. (See the solution to Problem 21.9(a) on page 546 of this manual.) Attack of bromide at this carbon leads to the observed product.

**22.38** (a) The product results from two successive conjugate additions. (Proton transfers are shown for simplicity as intramolecular processes, but they could be mediated by other acids and bases in solution.)

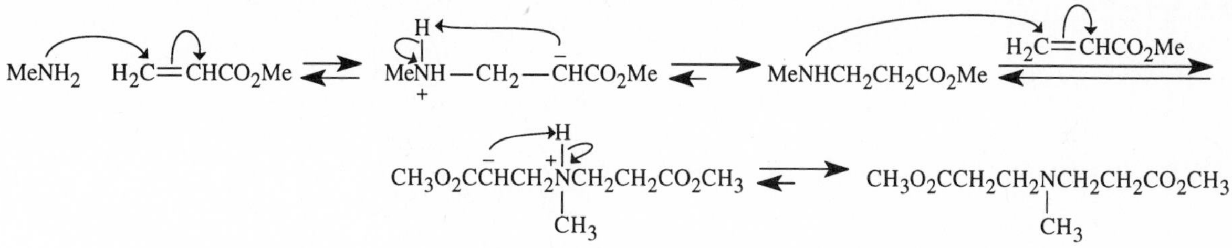

(c) The conjugate-addition product has no carbon–carbon double bond. Internal rotation within this adduct, followed by reversal of the addition, gives the isomerized product.

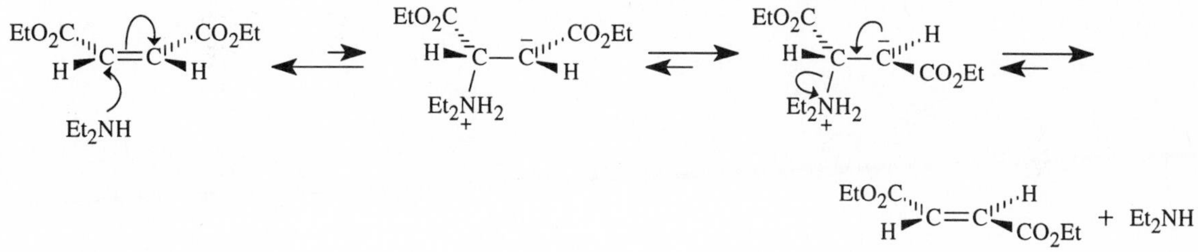

**22.39** (a) The key is to recognize that the desired product results from the addition of *two equivalents* of acrylonitrile, which end up as the two —CH$_2$CH$_2$CO$_2$H groups of the product. The conjugate-base anion of ethyl acetoacetate would be a suitable nucleophile *Y*. Reaction of *Y* with one equivalent of acrylonitrile would give compound *A*. Ionization of *A* gives anion *B*, which reacts with a second equivalent of acrylonitrile to give compound *C*. The nitrile groups are hydrolyzed to carboxy groups and the ester group is hydrolyzed and decarboxylated to give the final product when compound *C* is heated in acid.

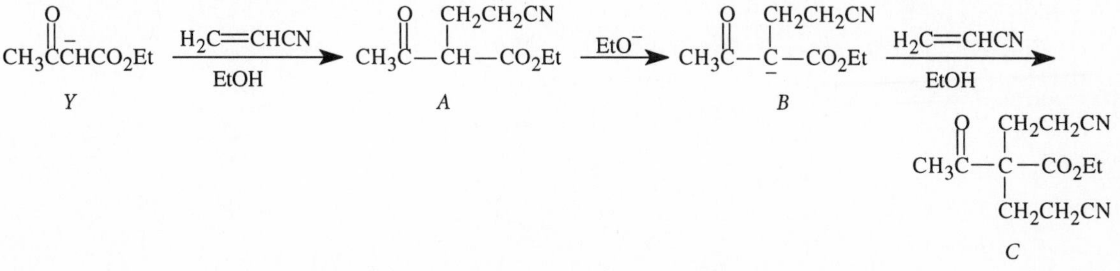

**22.40** (a) The first part of the mechanism is a Michael addition followed by loss of formic acid. Either *A1* or *A2* is a reasonable candidate for "intermediate *A*."

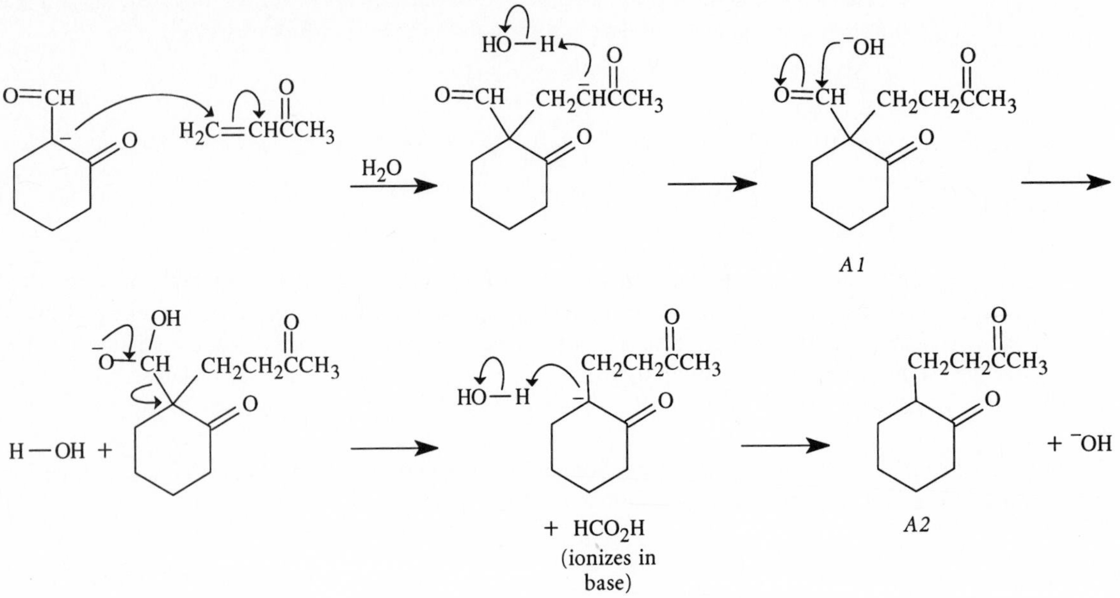

$H_2O$

*A1*

$H-OH +$

$+ HCO_2H$
(ionizes in base)

*A2*

$+ \ ^-OH$

Compound *A2* then undergoes an intramolecular aldol condensation reaction.

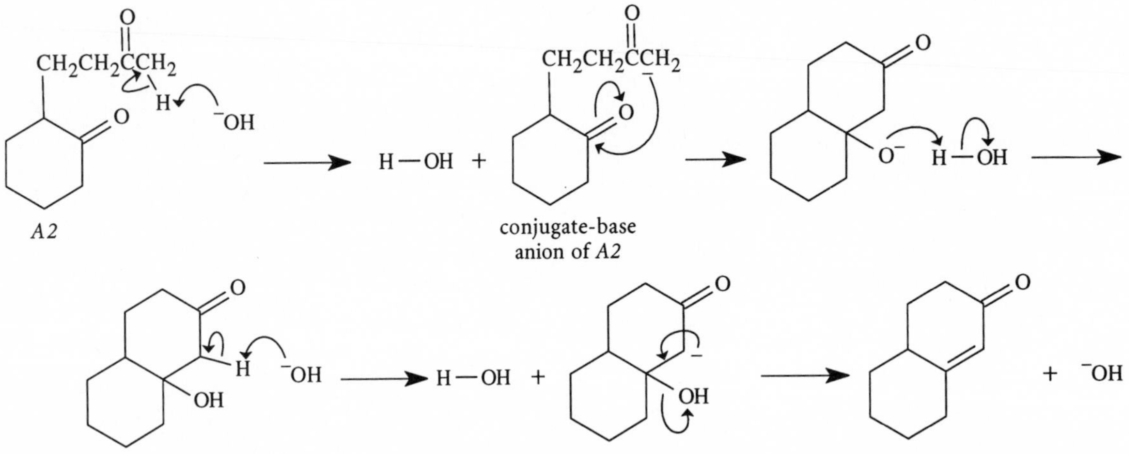

*A2*

$H-OH +$

conjugate-base anion of *A2*

$H-OH +$

$+ \ ^-OH$

In the foregoing mechanism many of the steps are reversible and should rigorously be shown with equilibrium arrows. However, the focus of the problem is on the mechanism of the forward reaction and not on the reversibility of the reaction. In such cases, forward arrows only are shown.

**22.41** (a)

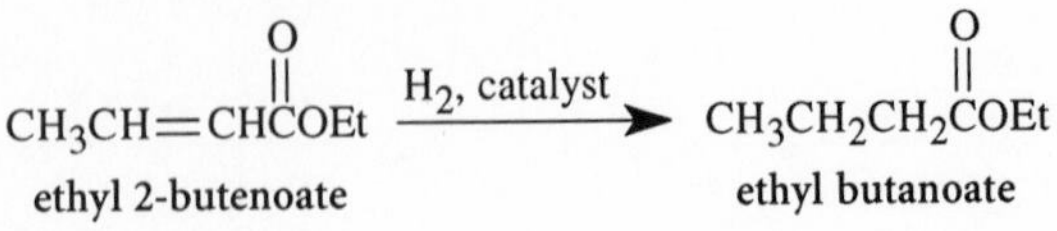

$$CH_3CH=CHCOEt \xrightarrow{H_2,\ catalyst} CH_3CH_2CH_2COEt$$

ethyl 2-butenoate          ethyl butanoate

**22.42** (a)

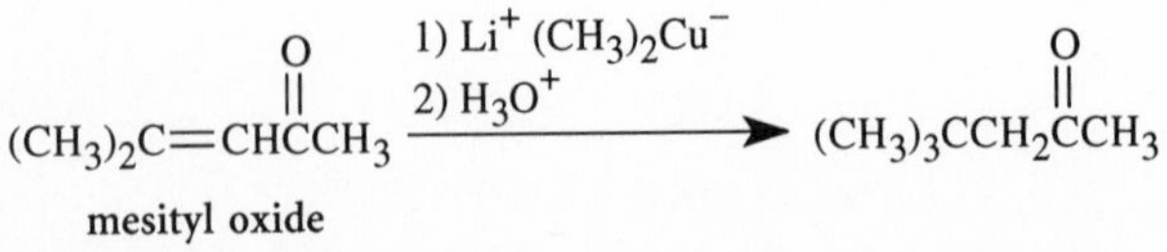

$$(CH_3)_2C=CHCCH_3 \xrightarrow[\ 2)\ H_3O^+\ ]{1)\ Li^+\ (CH_3)_2Cu^-} (CH_3)_3CCH_2CCH_3$$

mesityl oxide

(c)

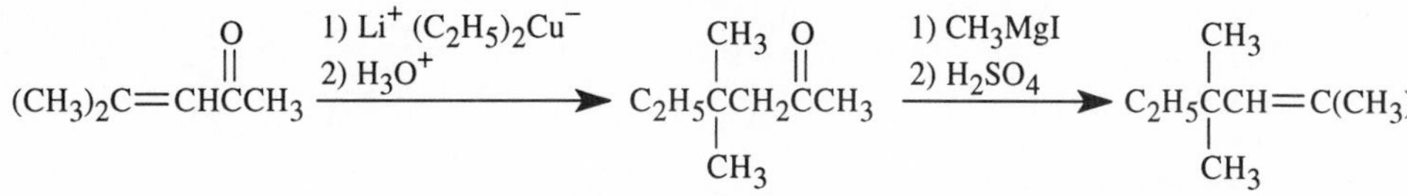

$$(CH_3)_2C=CHCCH_3 \xrightarrow[\text{2) H}_3\text{O}^+]{\text{1) Li}^+ (C_2H_5)_2Cu^-} C_2H_5CCH_2CCH_3 \xrightarrow[\text{2) H}_2\text{SO}_4]{\text{1) CH}_3\text{MgI}} C_2H_5CCH=C(CH_3)_2$$

Notice that the conjugate addition must *precede* the carbonyl addition. (If the carbonyl addition were carried out first, the product would not be an $\alpha,\beta$-unsaturated carbonyl compound, and conjugate addition could not occur.)

**22.43** (a) The question in this problem is what type of conjugate addition occurs (1,4 or 1,6). The 1,6-addition occurs because it gives the enolate ion intermediate with the greater number of resonance structures and hence, the more stable enolate ion.

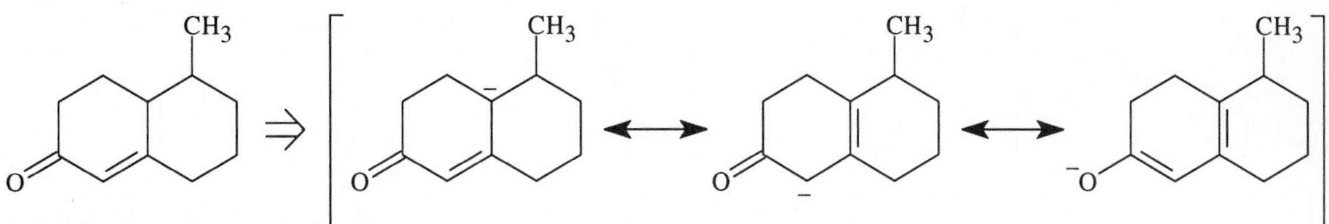

1,6-addition product
(observed)

enolate ion intermediate

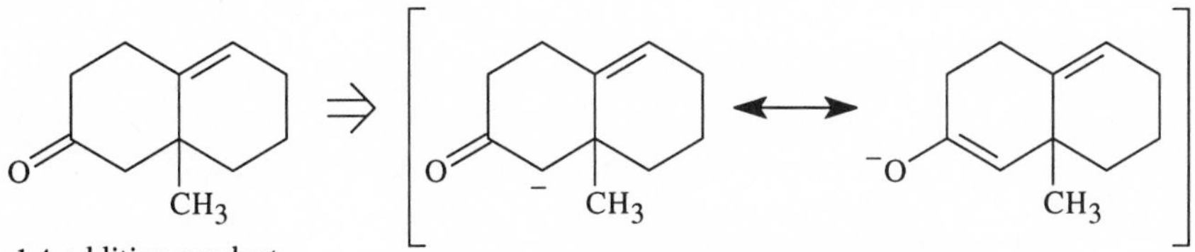

1,4-addition product

enolate ion intermediate

**22.44** (a) An analysis similar to that used in Study Problem 22.7 on text p. 1097 reveals that either a methyl group or an ethyl group can be added in the conjugate addition. (Addition of an ethyl group is illustrated here.)

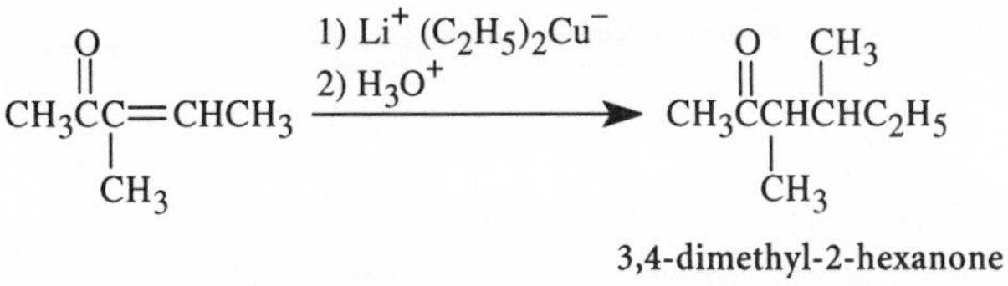

$$CH_3CC=CHCH_3 \xrightarrow[\text{2) H}_3\text{O}^+]{\text{1) Li}^+ (C_2H_5)_2Cu^-} CH_3CCHCHC_2H_5$$

**3,4-dimethyl-2-hexanone**

(c) An analysis along the lines of Study Problem 22.7 reveals the following two possibilities:

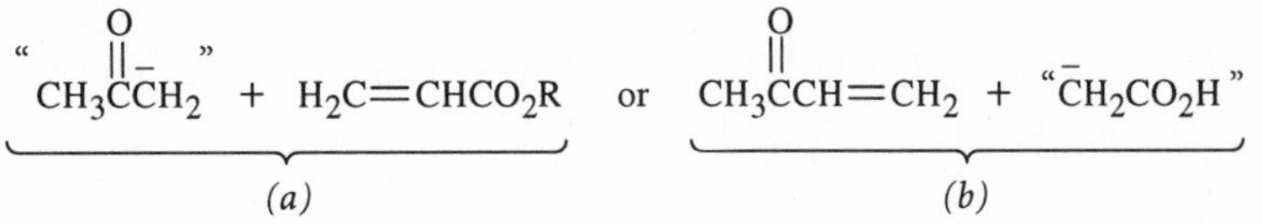

$$\underbrace{\text{``}CH_3CCH_2\text{''} + H_2C=CHCO_2R}_{(a)} \quad \text{or} \quad \underbrace{CH_3CCH=CH_2 + \text{``}^-CH_2CO_2H\text{''}}_{(b)}$$

A practical synthetic equivalent to the anion in (*a*) is the conjugate base of ethyl acetoacetate:

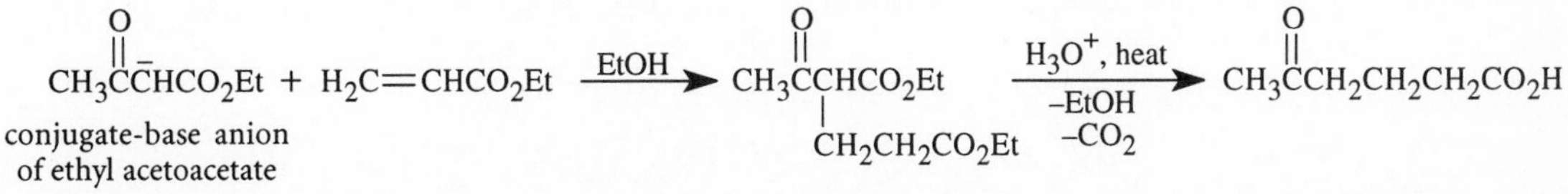

$$CH_3C\bar{C}HCO_2Et + H_2C=CHCO_2Et \xrightarrow{\text{EtOH}} CH_3CCHCO_2Et \xrightarrow[\text{–EtOH} \; -CO_2]{\text{H}_3\text{O}^+, \text{ heat}} CH_3CCH_2CH_2CH_2CO_2H$$

conjugate-base anion
of ethyl acetoacetate

The practical synthetic equivalent to the anion in (*b*) is the conjugate base of diethyl malonate:

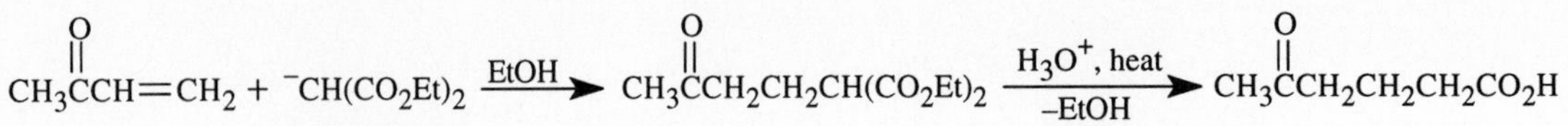

$$CH_3\overset{\overset{\displaystyle O}{\|}}{C}CH=CH_2 + {}^-CH(CO_2Et)_2 \xrightarrow{\text{EtOH}} CH_3\overset{\overset{\displaystyle O}{\|}}{C}CH_2CH_2CH(CO_2Et)_2 \xrightarrow[\substack{-\text{EtOH} \\ -CO_2}]{H_3O^+,\ heat} CH_3\overset{\overset{\displaystyle O}{\|}}{C}CH_2CH_2CH_2CO_2H$$

conjugate-base
anion of
diethyl malonate

## Solutions to Additional Problems

**22.45**

(a)
$$H_2C=CH\overset{\overset{\displaystyle O}{\|}}{C}CH_3$$
3-buten-2-one

(b)
$$BrCH_2CH_2\overset{\overset{\displaystyle O}{\|}}{C}CH_3$$

(c)
$$BrCH_2\underset{\underset{\displaystyle Br}{|}}{CH}\overset{\overset{\displaystyle O}{\|}}{C}CH_3$$

(d)
$$H_2C=CH\underset{\underset{\displaystyle OH}{|}}{CH}CH_3$$

$$N\equiv C CH_2CH_2\overset{\overset{\displaystyle O}{\|}}{C}CH_3$$

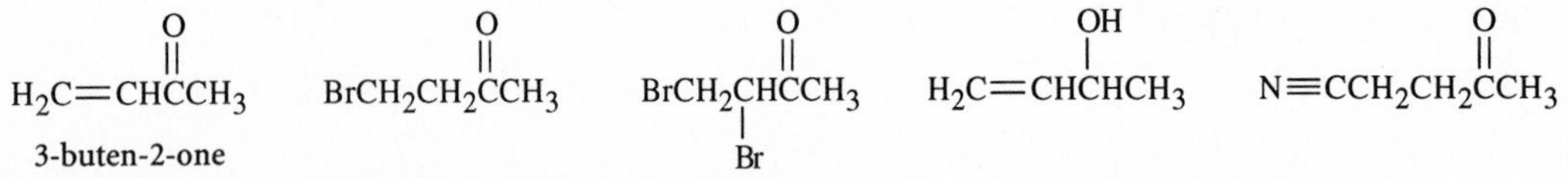

(e)
$$CH_3CH_2CH_2CH_2\overset{\overset{\displaystyle O}{\|}}{C}CH_3$$

(f)
$$(EtO_2C)_2CHCH_2CH_2\overset{\overset{\displaystyle O}{\|}}{C}CH_3$$

(g)

(h)

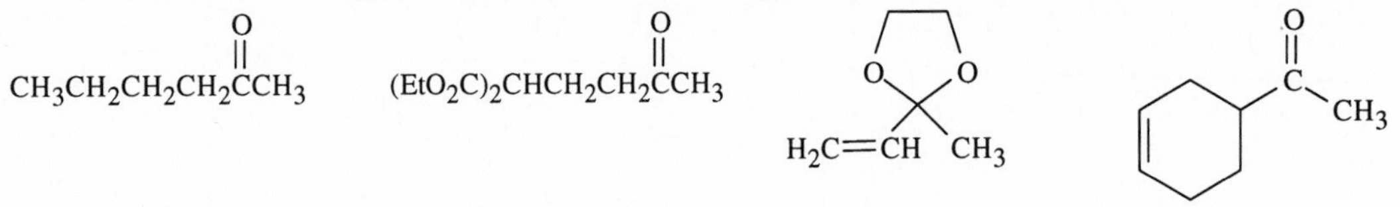

Although the Diels-Alder reaction is covered in Chapter 15, part (h) serves as a reminder that it is a very important reaction of $\alpha,\beta$-unsaturated carbonyl compounds.

**22.47** (a)
$$CH_3CH_2\underset{\underset{\displaystyle CH_3}{|}}{CH}\overset{\overset{\displaystyle O}{\|}}{C}CH_3$$
(either enantiomer)

(c)

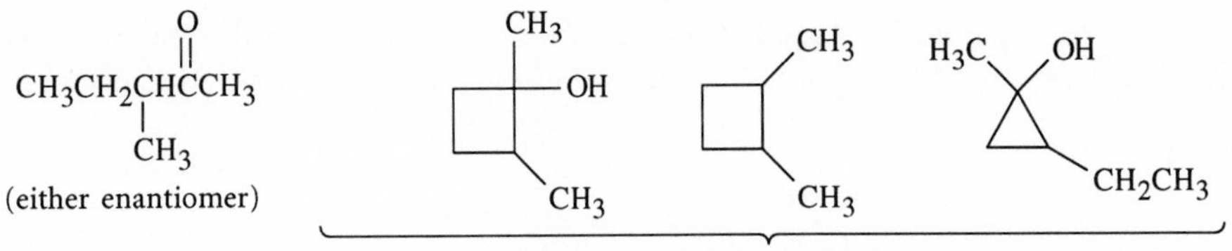

(each could be any one of four stereoisomers)

and many others

**22.48** (a)
$$CH_3\overset{\overset{\displaystyle O}{\|}}{C}CH_2CH_3$$

(c)
$$H_2C=CH\underset{\underset{\displaystyle OH}{|}}{CH}CH_3$$

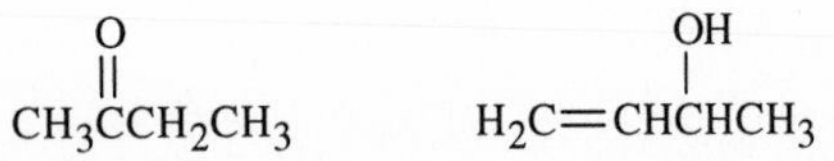

**22.49** (a) This compound is an "ynol," the acetylenic analog of an enol. It is spontaneously converted into its tautomeric ketene derivative, $CH_3CH=C=O$.

(c) This compound is a hemiacetal, and it spontaneously decomposes to acetaldehyde and an enol *A*, which, in turn, spontaneously forms propionaldehyde.

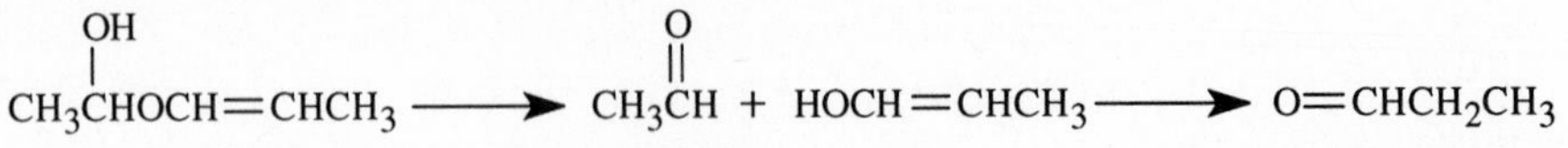

$$CH_3\underset{\underset{\displaystyle OH}{|}}{CH}OCH=CHCH_3 \longrightarrow CH_3\overset{\overset{\displaystyle O}{\|}}{C}H + HOCH=CHCH_3 \longrightarrow O=CHCH_2CH_3$$

acetaldehyde      *A*      propionaldehyde

**22.50** (a) The conjugate bases of the two compounds, because they are resonance structures, are identical.

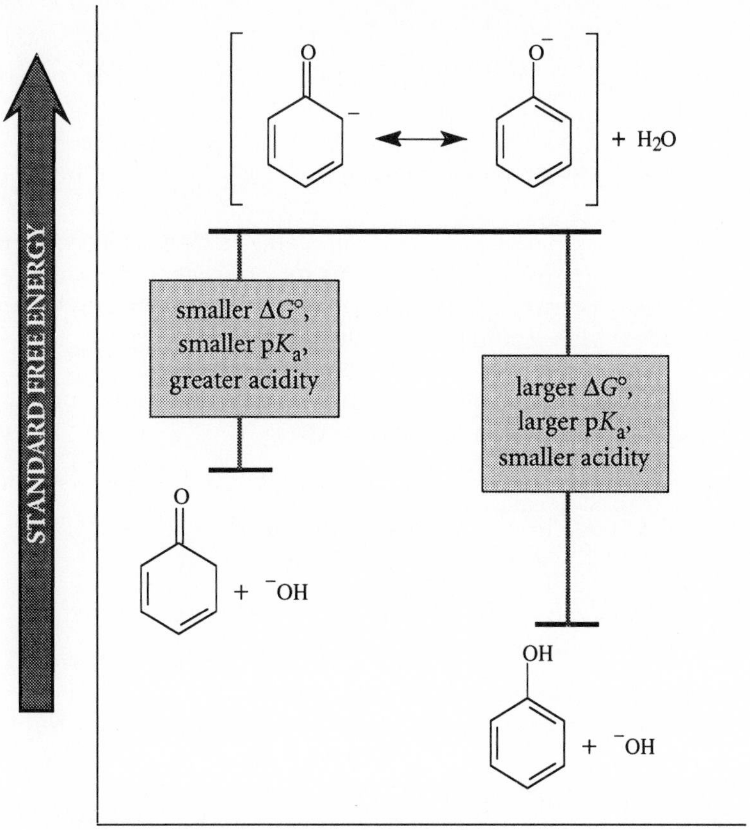

**Figure SG22.2**  *Diagram for the solution to Problem 22.50, showing the effect of relative stability on the acidity of two tautomers. Because the conjugate acids of the two tautomers are identical, the more stable tautomer is less acidic. The phenol tautomer is more stable because it is aromatic.*

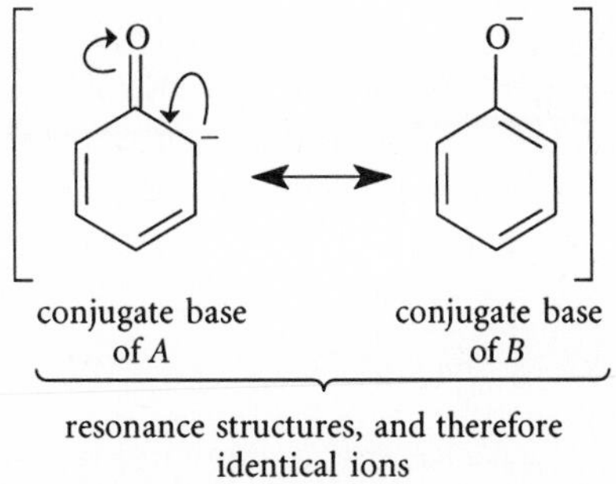

(b)  Because compound *B* (phenol) is aromatic, it is more stable, and thus more energy is required to convert it into the conjugate base shown in part (a). From Eq. 3.27 on text p. 103, $pK_a = \Delta G_a^{\circ}/2.3RT$. Thus, compound *B* has the greater $pK_a$ (it is less acidic). This is demonstrated in Fig. SG22.2 above.

22.52  (a)  The second compound is most acidic because the conjugate-base anion (structure below) has the greatest number of resonance structures and is therefore most stable. (Draw these structures!) This anion is stabilized by resonance interaction with two carbonyl groups and a benzene ring. In the first compound, the conjugate-base anion lacks the resonance interaction with a benzene ring; and in the third compound, the conjugate-base anion at the central carbon also lacks the resonance interaction with a benzene ring, and the conjugate-base anion at the benzylic carbon lacks the resonance interaction with a second carbonyl group.

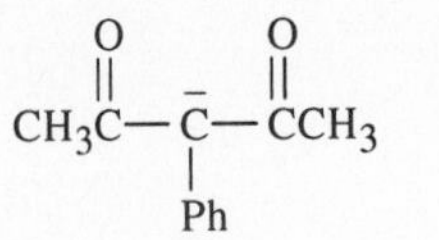

$$CH_3C(=O)-\overset{-}{\underset{|}{C}}(Ph)-C(=O)CH_3 \qquad H_2C=CH\overset{-}{C}HC(=O)CH_3$$

conjugate-base anion
of the most acidic compound
in part (a)

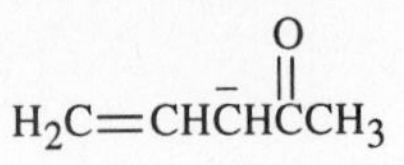

conjugate-base anion
of the most acidic compound
in part (b)

(b) The first compound is more acidic because the conjugate-base anion (see structure above) has the greater number of resonance structures. It is stabilized by resonance interaction with both the carbon-carbon double bond and the carbonyl group. The two possible enolate ions of the second compound lack the resonance interaction with a carbon-carbon double bond.

**22.54** (a) Bromination can occur at either of the two α-carbons; compounds *B* and *C* are diastereomers. (Each of the three compounds can also exist as enantiomers.)

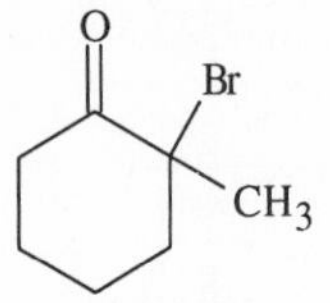

*A* 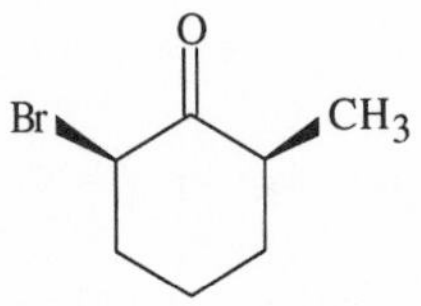                *B* 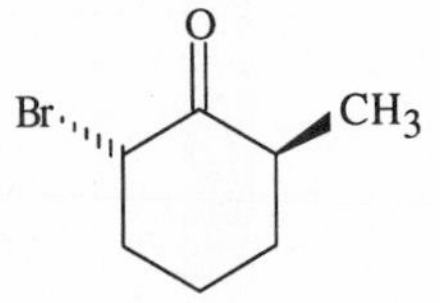                *C*

(b) The enol leading to *A* is the more stable one because it has the greater number of alkyl branches on the double bond. Hence, compound *A* is the α-bromo derivative that is formed.

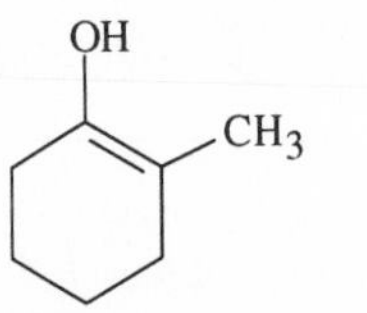

enol intermediate
involved in the
formation of *A*
(more stable enol)

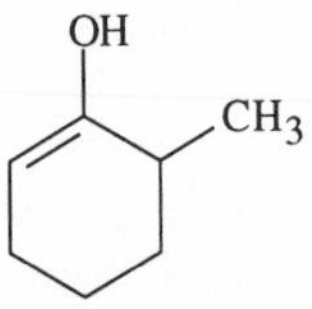

enol intermediate
involved in the
formation of
compounds *B* and *C*

**22.56** The discussion on text pp. 1048–1049 explains that enols are intermediates in the acid-catalyzed bromination of aldehydes and ketones, and that enol formation is the rate-limiting process. As explained on text p. 958, an enol intermediate is also involved in the decarboxylation of a β-keto acid. Thus, the bromination reaction described in the problem evidently involves an enol intermediate, and formation of the enol is the rate-limiting process. Hence, the rate of the bromination reaction is zero-order in bromine.

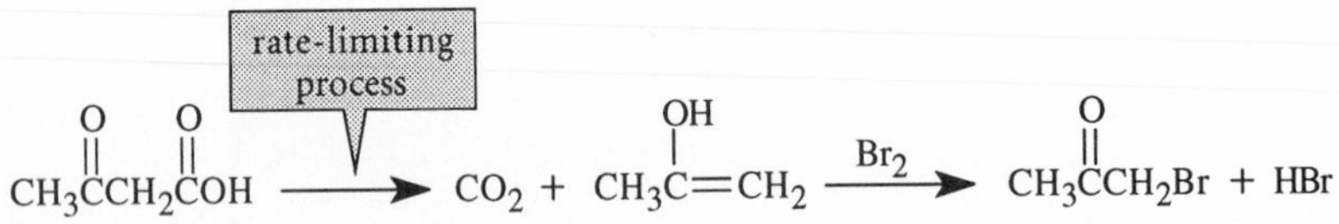

$$CH_3CCH_2COH \xrightarrow{\text{rate-limiting process}} CO_2 + CH_3C(OH)=CH_2 \xrightarrow{Br_2} CH_3CCH_2Br + HBr$$

**22.57** It is reasonable to hypothesize that the carbon atom of the iodoform must originate from the central carbon of the β-diketone:

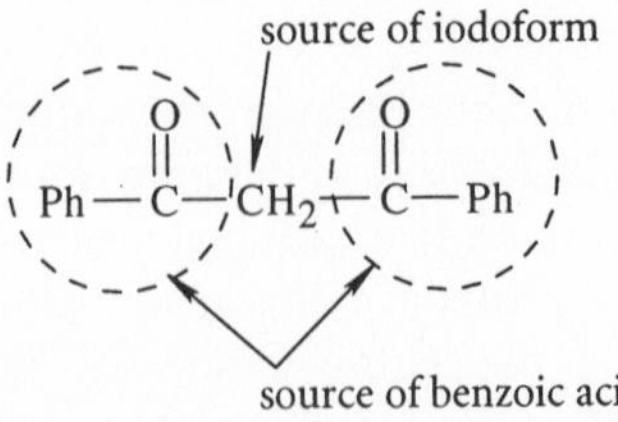

source of iodoform

$$Ph-C(=O)-CH_2-C(=O)-Ph$$

source of benzoic acid

Base-promoted iodination occurs at the central carbon by the usual mechanism involving enolate ions to give the diiodo ketone *A*. Hydroxide then displaces an *α*-diiodo enolate anion *B* in a nucleophilic acyl substitution reaction; this resonance-stabilized anion is a fairly good leaving group. Anion *B* is then iodinated, and the resulting triiodo ketone *C* undergoes the final steps of the haloform reaction illustrated in Eq. 22.25e on text p. 1051 to give iodoform and benzoate anion. The two equivalents of benzoate anion produced in this sequence give benzoic acid when the reaction mixture is acidified.

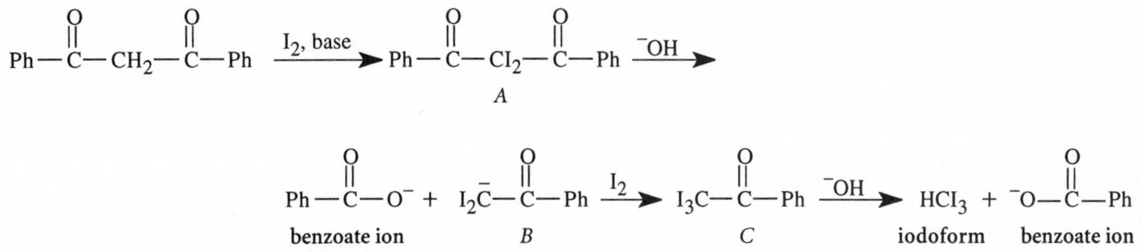

**22.58** (a) The isomerization favors compound *B* because *trans*-decalin derivatives are more stable than *cis*-decalin derivatives. (See Sec. 7.6B on text p. 297.) The isomerization mechanism involves formation of an enolate ion intermediate, which can be protonated on either face to give *A* or *B*, respectively.

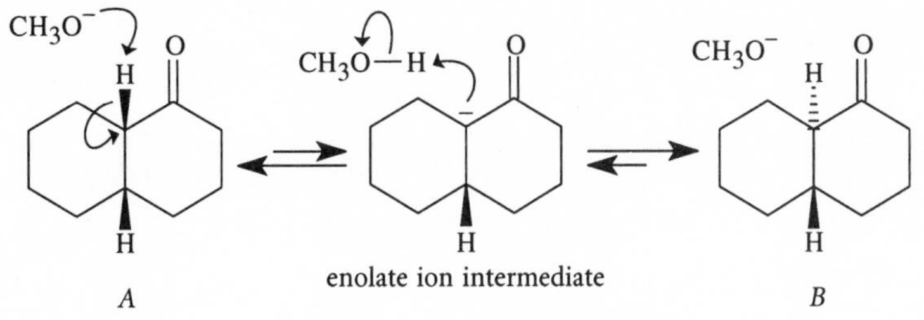

**22.59** (a) The *α*-hydrogens and the N—H hydrogen are acidic enough to be replaced. These hydrogens are shown as deuteriums in the following structure.

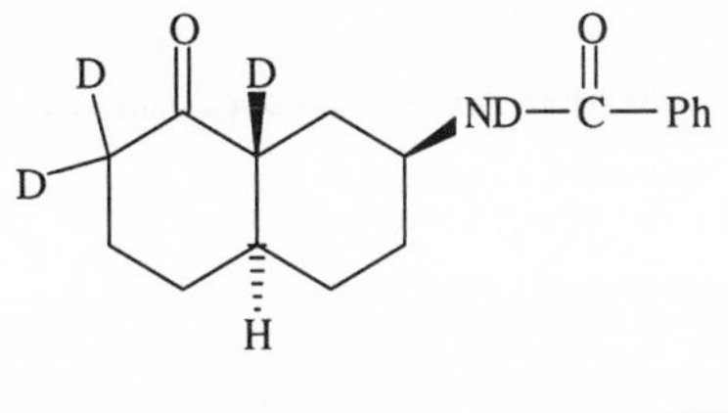

The stereochemistry of the deuterium at the ring junction could be a mixture of "up" and "down" (that is, *trans* and *cis*), because the planar carbanion intermediate could be protonated from either face. (See the solution to Problem 22.10(a) on p. 587 of this manual.) However, because the *trans*-decalin derivatives are more stable than *cis*-decalin derivatives, the stereoisomer shown above should be the predominant one.

**22.60** (a) The equilibrium favors the *α,β*-unsaturated isomer because it is conjugated. Conjugation is a stabilizing effect because of the additional *π*-electron overlap that is possible in conjugated compounds. (See Sec. 15.1A on text p. 680.)

(b) The mechanism in aqueous base involves a resonance-stabilized enolate ion intermediate, which can be protonated on either the *α*-carbon or the *γ*-carbon to give the respective unsaturated ketones.

*(solution continues)*

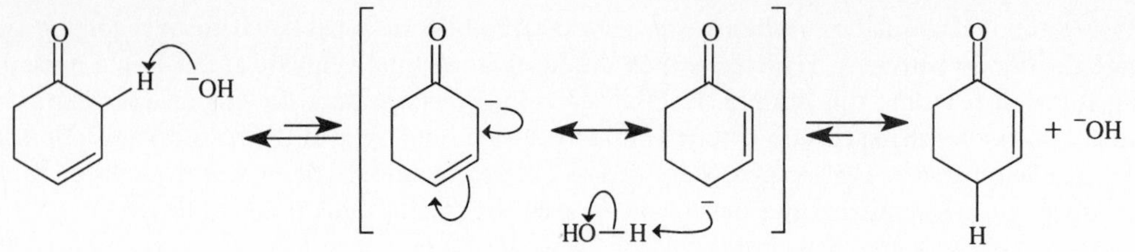

(c)  The mechanism below commences with the dienol 1,3-cyclohexadienol, which is formed by the usual enolization mechanism (Eq. 22.15b on text p. 1047). Protonation of this dienol gives a resonance-stabilized carbocation, which is the same as the protonated α,β-unsaturated ketone. Loss of the O—H proton gives the α,β-unsaturated ketone itself.

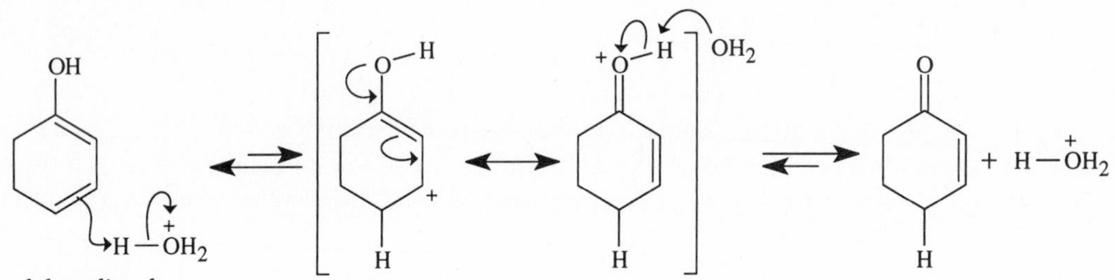

1,3-cyclohexadienol

(d)  The equilibrium constant for conversion of 4-methyl-3-cyclohexenone into 4-methyl-2-cyclohexenone should be smaller, because the additional alkyl branch on the β,γ-double bond tends to offset the stabilizing effect of conjugation in the α,β-unsaturated derivative.

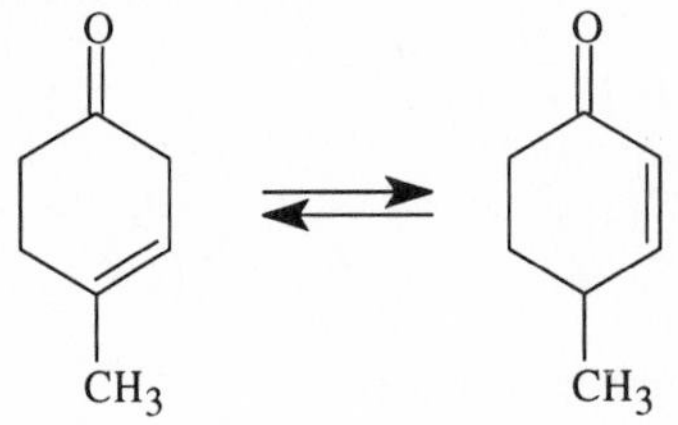

4-methyl-3-cyclohexenone  4-methyl-2-cyclohexenone

**22.61**  (a)  The exchange of $H^a$ occurs by the usual enolate-ion mechanism.

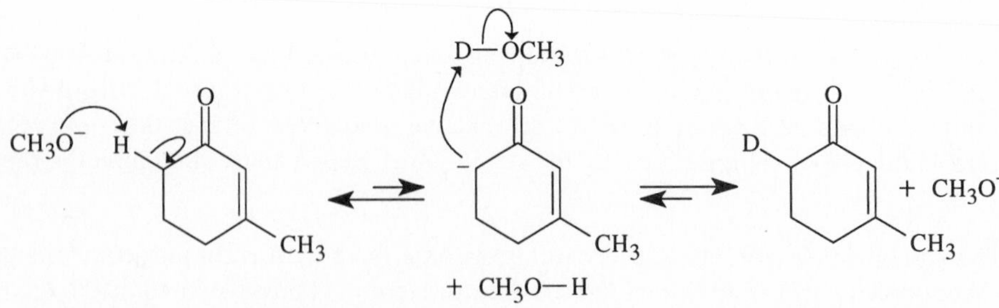

Removal of either hydrogen $H^c$ or $H^d$ gives an anion that has resonance structures which show that it is an enolate ion. Hence, these hydrogens are acidic enough to undergo exchange. The exchange of $H^c$ is shown explicitly below; you should show the exchange of $H^d$, which occurs by essentially the same mechanism.

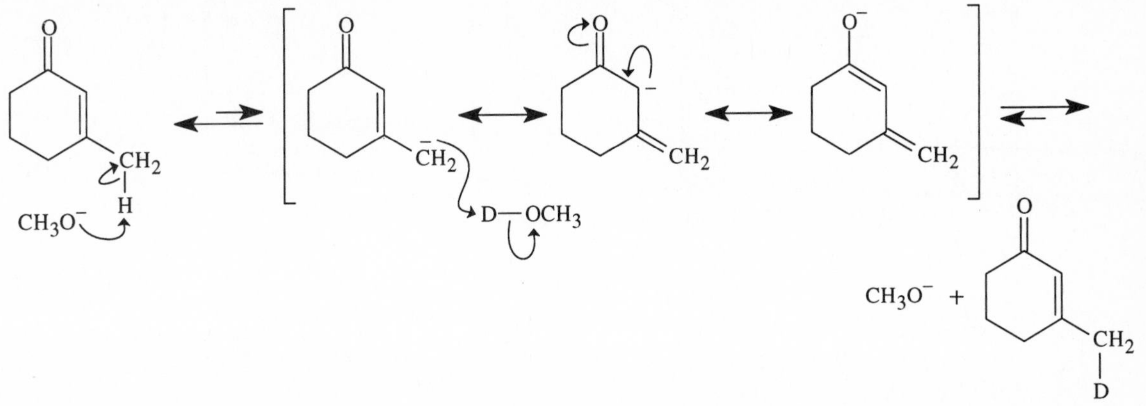

(b) Although $H^b$ is an $\alpha$-proton, it is not acidic because the anion that would result from its removal is not resonance-stabilized. The reason is that the orbital containing the unshared electron pair in this anion is perpendicular to the $\pi$-electron system of the carbonyl group and cannot overlap with it. An equivalent resonance argument is that if the overlap which lies at the basis of resonance were to occur, the nuclei would have to move and a cumulated double bond and the attendant large amount of strain would be introduced into the six-membered ring.

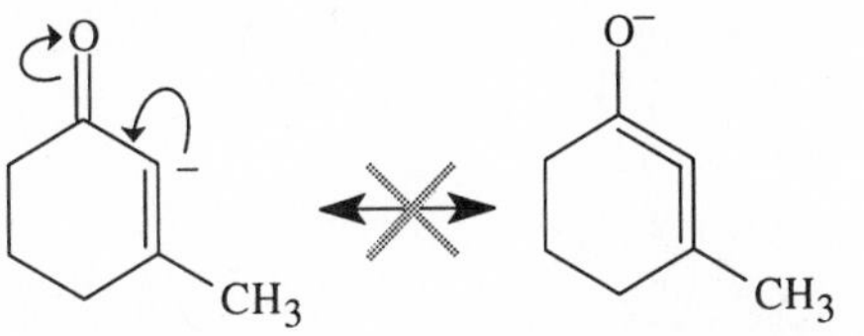

overlap introduces
strained cumulated
double bond

(c) Hydrogen $H^b$ is not acidic in the $\alpha,\beta$-unsaturated ketone. However, as shown in Problem 22.60, the $\alpha,\beta$-unsaturated ketone readily isomerizes to a $\beta,\gamma$-unsaturated ketone. Ionization of $H^b$ in this compound gives a resonance-stabilized anion, as shown in the solution to Problem 22.60(b). Hence, $H^b$ exchanges, and the $\beta,\gamma$-unsaturated ketone isomerizes back to the $\alpha,\beta$-unsaturated ketone.

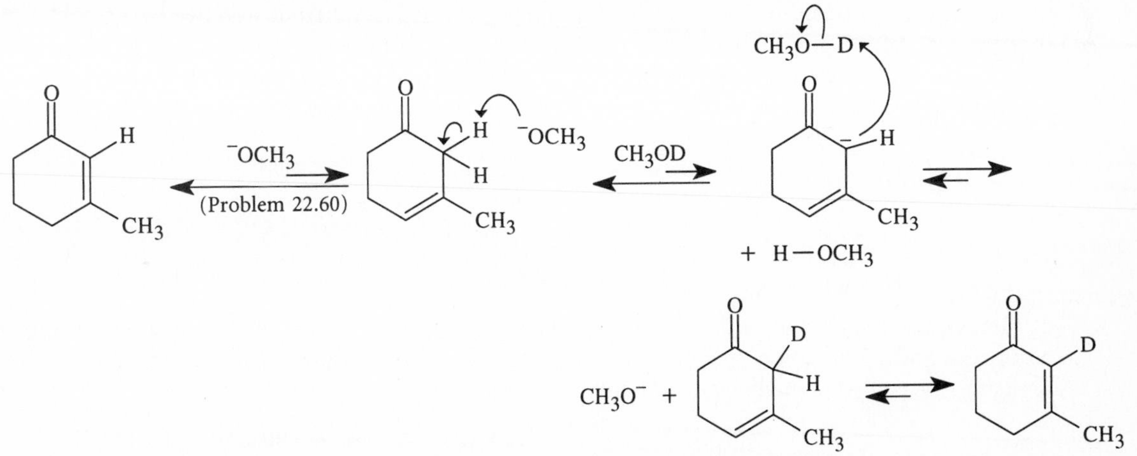

22.63 (a) Although ethers are usually more basic than carbonyl oxygens, in compound A the carbonyl oxygen is protonated because the resulting carbocation has $4n + 2$ $\pi$ electrons; that is, it is aromatic. The aromaticity of its conjugate acid explains why compound A is more basic than compound B. Compound B is protonated on the ether oxygen and has the basicity typical of an ether.

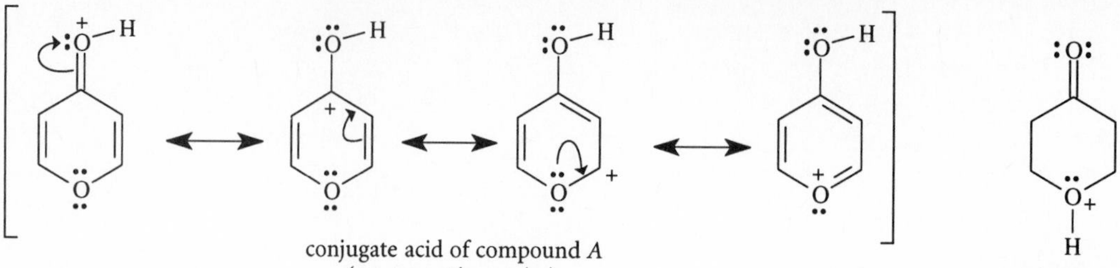

conjugate acid of compound *A*
(an aromatic species)

conjugate acid
of compound *B*

**22.64** (a) As shown in Fig. SG22.2 on page 599 of this manual, the more stable a species is relative to its conjugate base the less acidic it is. Consequently, resonance stabilization of an $\alpha,\beta$-unsaturated carboxylic acid decreases its acidity.

(b) The p$K_a$ of the last compound, butanoic acid, is the reference value—the p$K_a$ to be expected for a four-carbon carboxylic acid containing no carbon-carbon double bonds. In the first compound, 3-butenoic acid, the carbon-carbon double bond is not conjugated with the carbonyl group. Hence, the difference between its p$K_a$ value and that of butanoic acid is due to the electron-withdrawing polar effect of the carbon-carbon double bond. If the polar effect were the only effect operating in the middle compound, *trans*-2-butenoic acid (crotonic acid), it should be the most acidic compound in the series, because the polar effect of any group on acidity increases when the group, in this case a carbon-carbon double bond, is closer to the site of ionization. Because *trans*-2-butenoic acid is in fact *less* acidic than 3-butenoic acid, some acid-weakening effect must also be operating, and this is the resonance effect discussed in part (a).

**22.65** (a) The $\alpha$-hydrogen of any nitro compound is particularly acidic because the conjugate-base anion is stabilized both by resonance and by the polar effect of the nitro group (note the positive charge next to the negative charge of the carbanion).

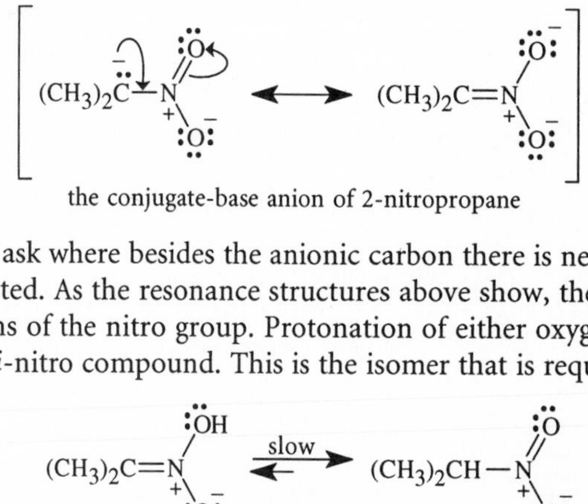

the conjugate-base anion of 2-nitropropane

(b) To answer this question, ask where besides the anionic carbon there is negative charge and a pair of electrons that can be protonated. As the resonance structures above show, the negative charge of the carbanion is shared by the oxygens of the nitro group. Protonation of either oxygen gives the nitro analog of an enol, which is called an *aci*-nitro compound. This is the isomer that is requested in the problem.

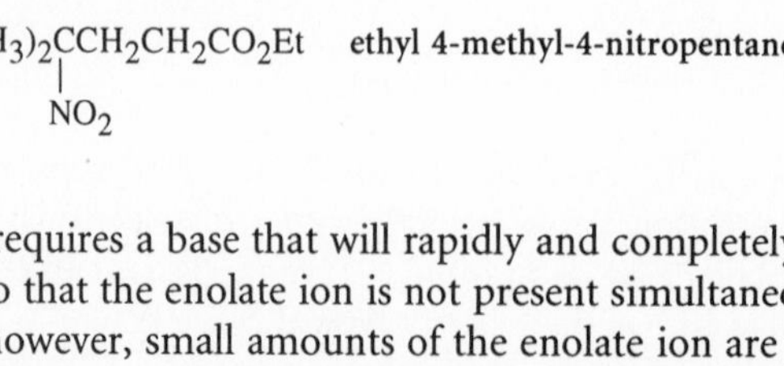

an *aci*-nitro compound

(c) Because 2-nitropropane is much more acidic than ethanol, it is completely converted into its anion by sodium ethoxide. This anion, like many other "enolate ions," undergoes a Michael addition to ethyl acrylate to give ethyl 4-methyl-4-nitropentanoate.

$$(CH_3)_2CCH_2CH_2CO_2Et \quad \text{ethyl 4-methyl-4-nitropentanoate}$$
$$|$$
$$NO_2$$

**22.66** (a) An alkylation of this type requires a base that will rapidly and completely convert the ester into its conjugate-base enolate anion so that the enolate ion is not present simultaneously with the ester. In the presence of sodium ethoxide, however, small amounts of the enolate ion are present together with large amounts of the un-ionized ester; consequently, the Claisen condensation will occur as the major reaction.

(c) Alcohol dehydration involves a carbocation intermediate, and carbocations $\alpha$ to carbonyl groups are particularly unstable. The desired dehydration will not occur. The more likely reaction would be polymerization of the ester by reaction of the hydroxy group of one molecule with the carbonyl of the other.

(e) In addition to bromination of the benzene ring, the acidic conditions will also promote $\alpha$-bromination at the methyl group. Lewis acids can catalyze this reaction just as Brønsted acids can. In addition, HBr is a by-product of ring halogenation, and this can also catalyze $\alpha$-halogenation.

**22.67** (a) The conjugate-base enolate ion of diethyl malonate reacts at the carbonyl group of acetone. The addition product dehydrates to give the following product.

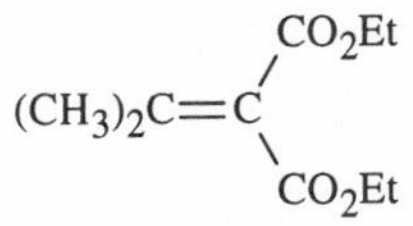

(c)

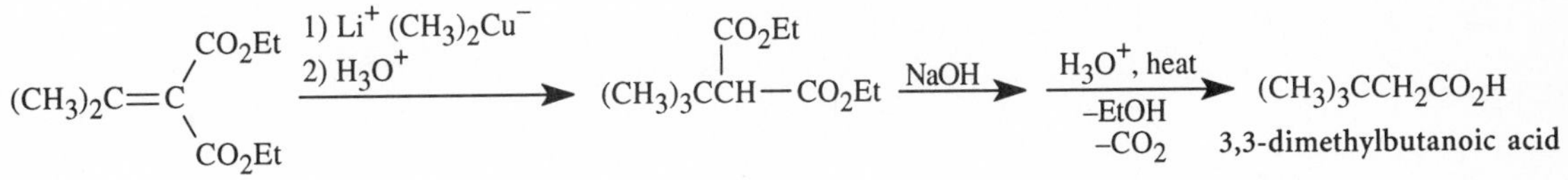

Note that this product cannot be prepared by a conventional malonic ester synthesis involving alkylation with *tert*-butyl bromide. (Why? See Sec. 9.4C on text pp. 397–399.)

**22.68** Intermediate *A* is the anionic product of a crossed Claisen condensation. This anion undergoes a Michael addition to the $\alpha,\beta$-unsaturated ketone, and the product of that reaction undergoes an intramolecular aldol condensation to give compound *B*. Hydroxide ion effects a reverse Claisen condensation that removes the formyl group. The steps are outlined below; you should provide the mechanistic details using the curved-arrow formalism.

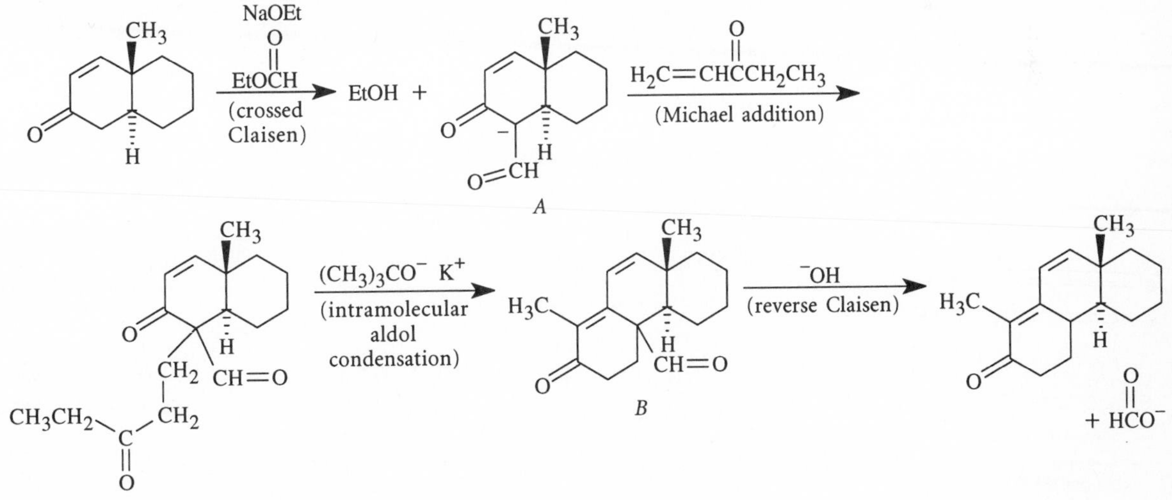

**22.70** (a) The gas is $H_2$ and the species *A* is the sodium salt of the conjugate-base enolate ion of 2,4-pentanedione. As the resonance structures indicate, the negative charge, and therefore the nucleophilic character, of this anion is shared by both the anionic carbon and the oxygens.

*(solution continues)*

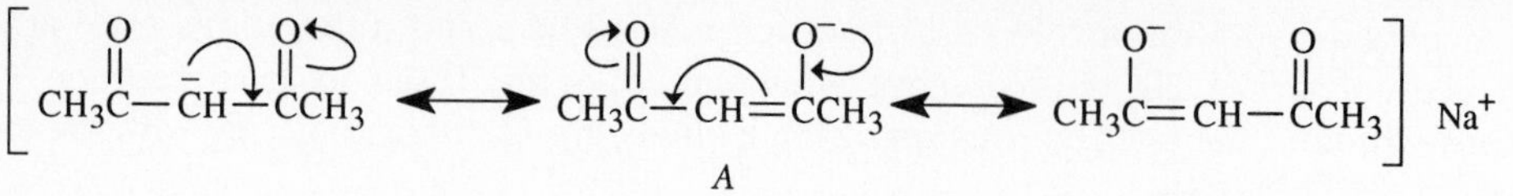

$$\left[ \ \underset{A}{CH_3\overset{O}{\overset{||}{C}}-\bar{C}H-\overset{O}{\overset{||}{C}}CH_3} \longleftrightarrow CH_3\overset{O}{\overset{||}{C}}-CH=\overset{O^-}{\overset{|}{C}}CH_3 \longleftrightarrow CH_3\overset{O^-}{\overset{|}{C}}=CH-\overset{O}{\overset{||}{C}}CH_3 \ \right] Na^+$$

(b) The three species that are formed all result from reaction with a nucleophilic atom with methyl iodide in an S$_N$2 reaction. The products are the two stereoisomeric ethers *B* and *C* and the alkylated β-diketone *D*.

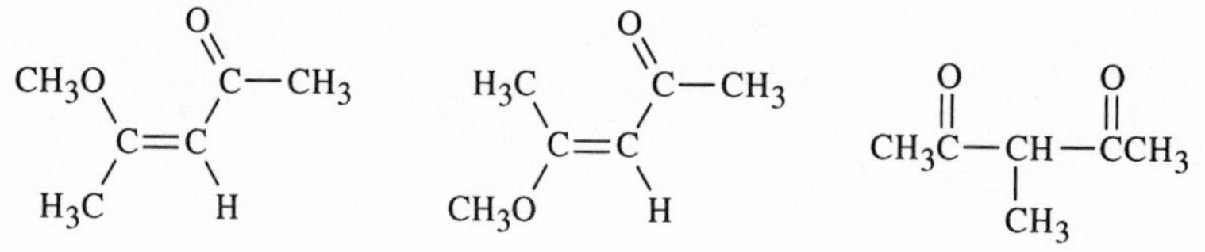

*B*          *C*          *D*

**22.71** (a)

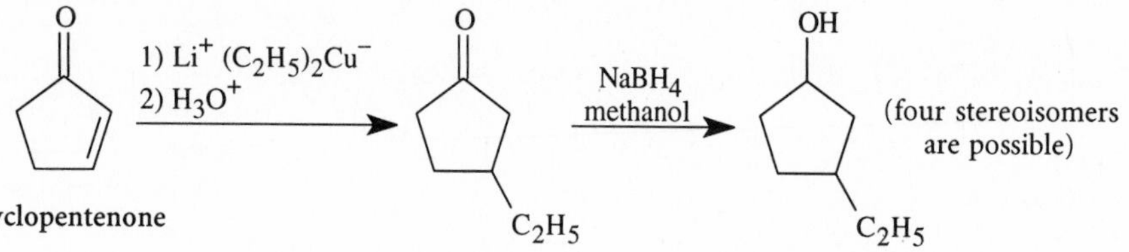

2-cyclopentenone                                          3-ethylcyclopentanol

(c)

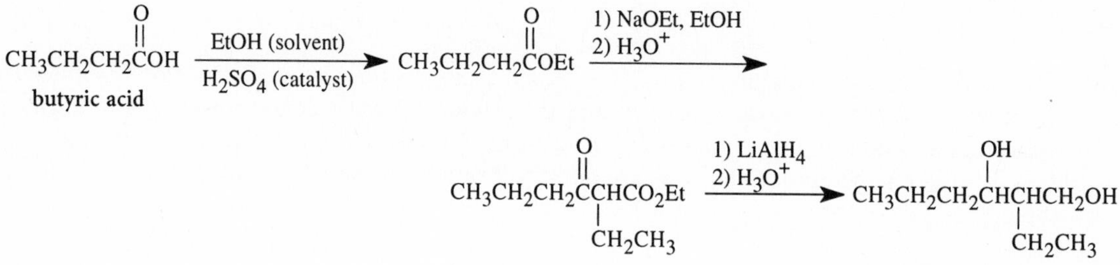

butyric acid

2-ethyl-1,3-hexanediol

(e)

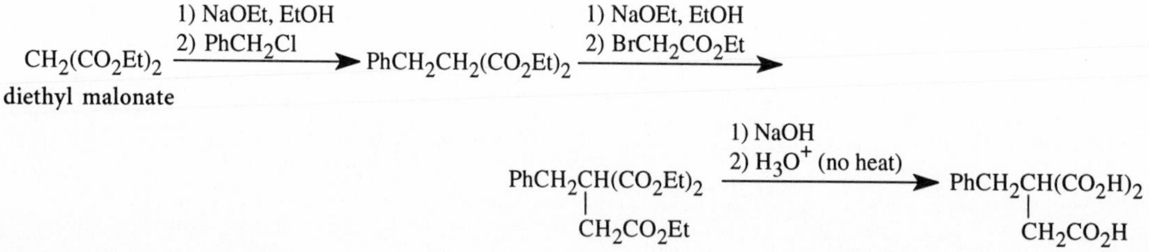

diethyl malonate

Note in the second step that α-bromo esters react rapidly in S$_N$2 reactions. Note also that heat is avoided after acid is added in the last step so that decarboxylation of the product does not occur. (The product is a malonic acid derivative.)

(g)

$$H_2C=CHCO_2Et \xrightarrow{HCN} N\equiv CCH_2CH_2CO_2Et \xrightarrow[2)\ H_2O]{1)\ LiAlH_4} H_2NCH_2CH_2CH_2CH_2OH$$

ethyl acrylate

(i)

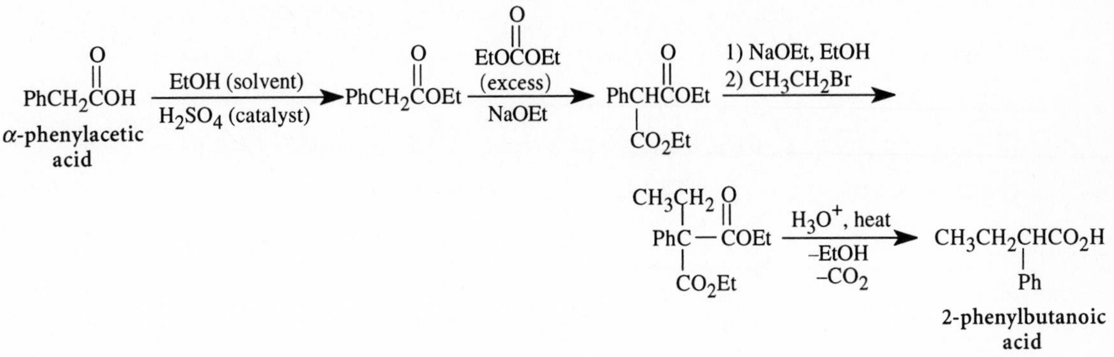

$$PhCH_2\overset{O}{\overset{\|}{C}}OH \xrightarrow[\text{H}_2\text{SO}_4 \text{ (catalyst)}]{\text{EtOH (solvent)}} PhCH_2\overset{O}{\overset{\|}{C}}OEt \xrightarrow[\text{NaOEt}]{\underset{\text{(excess)}}{EtO\overset{O}{\overset{\|}{C}}OEt}} PhCH\overset{O}{\overset{\|}{C}}OEt \xrightarrow[\text{2) CH}_3\text{CH}_2\text{Br}]{\text{1) NaOEt, EtOH}}$$

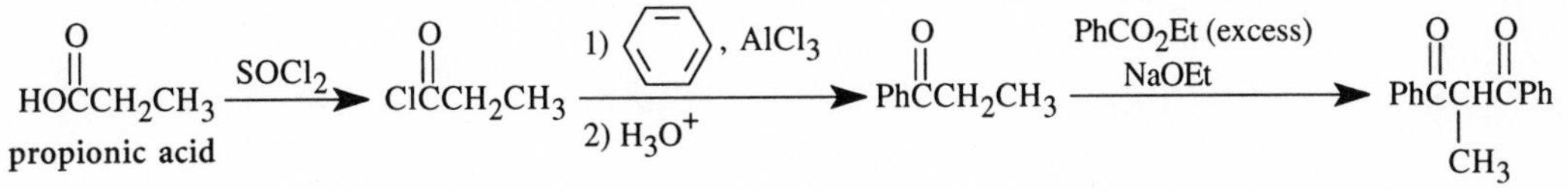

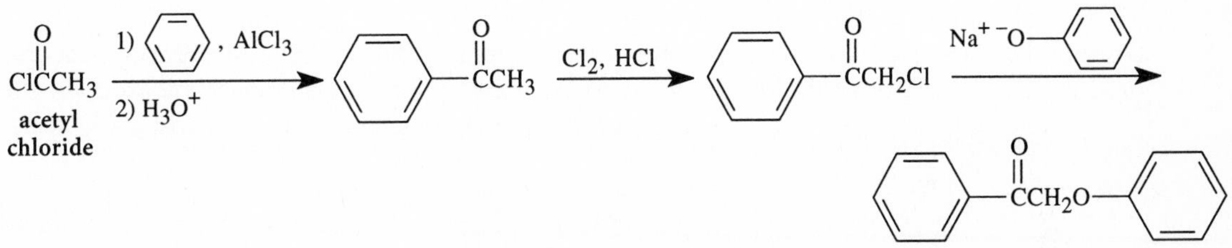

α-phenylacetic acid

(below, CO₂Et under PhCHCOEt)

$$\underset{\overset{|}{CO_2Et}}{\underset{|}{PhC}}\overset{CH_3CH_2\ O}{\overset{|\qquad\|}{\underset{}{C}}}OEt \xrightarrow[\text{−EtOH}\ \text{−CO}_2]{H_3O^+,\ \text{heat}} CH_3CH_2\underset{\overset{|}{Ph}}{CH}CO_2H$$

CO₂Et

2-phenylbutanoic acid

(k)

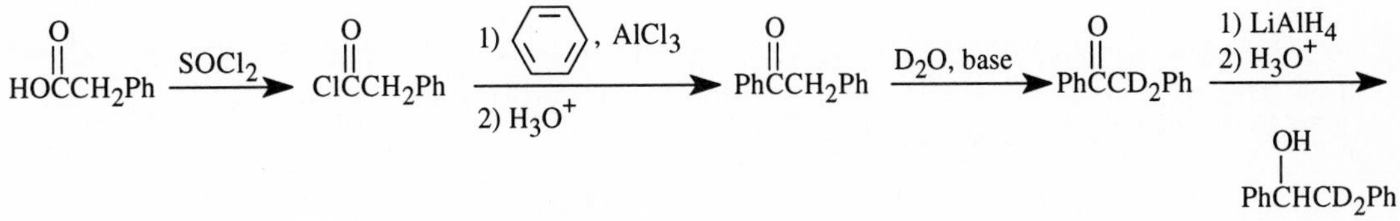

$$HO\overset{O}{\overset{\|}{C}}CH_2Ph \xrightarrow{SOCl_2} Cl\overset{O}{\overset{\|}{C}}CH_2Ph \xrightarrow[\text{2) H}_3\text{O}^+]{\text{1) } \langle\bigcirc\rangle, \text{ AlCl}_3} Ph\overset{O}{\overset{\|}{C}}CH_2Ph \xrightarrow{D_2O,\ \text{base}} Ph\overset{O}{\overset{\|}{C}}CD_2Ph \xrightarrow[\text{2) H}_3\text{O}^+]{\text{1) LiAlH}_4}$$

$$\underset{\overset{|}{PhCHCD_2Ph}}{\overset{OH}{|}}$$

(m)

$$HO\overset{O}{\overset{\|}{C}}CH_2CH_3 \xrightarrow{SOCl_2} Cl\overset{O}{\overset{\|}{C}}CH_2CH_3 \xrightarrow[\text{2) H}_3\text{O}^+]{\text{1) }\langle\bigcirc\rangle,\ \text{AlCl}_3} Ph\overset{O}{\overset{\|}{C}}CH_2CH_3 \xrightarrow[\text{NaOEt}]{PhCO_2Et \text{ (excess)}} Ph\overset{O}{\overset{\|}{C}}\underset{\overset{|}{CH_3}}{CH}\overset{O}{\overset{\|}{C}}Ph$$

propionic acid

(n)

$$Cl\overset{O}{\overset{\|}{C}}CH_3 \xrightarrow[\text{2) H}_3\text{O}^+]{\text{1) }\langle\bigcirc\rangle,\ \text{AlCl}_3} \langle\bigcirc\rangle\overset{O}{\overset{\|}{C}}CH_3 \xrightarrow{Cl_2,\ HCl} \langle\bigcirc\rangle\overset{O}{\overset{\|}{C}}CH_2Cl \xrightarrow{Na^{+-}O-\langle\bigcirc\rangle}$$

acetyl chloride

$$\langle\bigcirc\rangle\overset{O}{\overset{\|}{C}}CH_2O-\langle\bigcirc\rangle$$

(o)

$$(CH_3)_2C=CH\overset{O}{\overset{\|}{C}}CH_3 + CH_2(CO_2Et)_2 \xrightarrow{\text{NaOEt, EtOH}} \left( (CH_3)_2\underset{\overset{|}{CH(CO_2Et)_2}}{C}-CH_2\overset{O}{\overset{\|}{C}}CH_3 \right) \xrightarrow[\text{condensation}]{\text{Dieckmann}} \xrightarrow{H_3O^+}$$

[cyclohexanone ring structure: H₃C, H₃C substituents, CO₂Et, two ketone O groups]

$$\xrightarrow[\text{2) H}_3\text{O}^+,\ \text{−CO}_2]{\text{1) NaOH}}$$

[dimedone structure: H₃C, H₃C substituents, two ketone groups]

dimedone

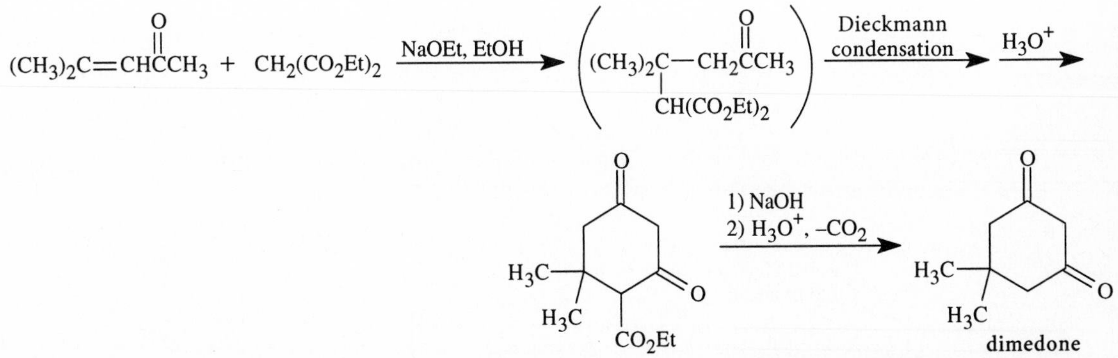

The first step is a Michael addition; the compound in parentheses under the basic reaction conditions undergoes an intramolecular crossed Claisen condensation (that is, a Dieckmann condensation) to form the ring.

(p)

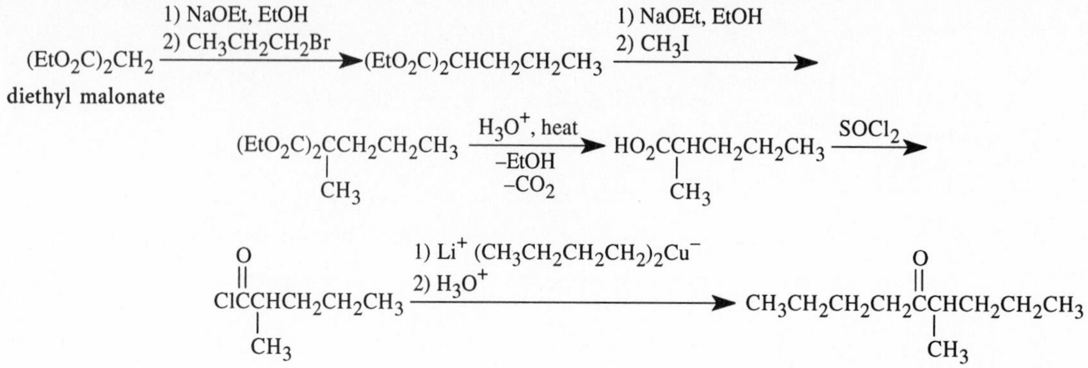

$$(EtO_2C)_2CH_2 \xrightarrow[\text{2) } CH_3CH_2CH_2Br]{\text{1) NaOEt, EtOH}} (EtO_2C)_2CHCH_2CH_2CH_3 \xrightarrow[\text{2) } CH_3I]{\text{1) NaOEt, EtOH}}$$

diethyl malonate

$$(EtO_2C)_2\underset{\underset{CH_3}{|}}{C}CH_2CH_2CH_3 \xrightarrow[\substack{-EtOH \\ -CO_2}]{H_3O^+, \text{ heat}} HO_2C\underset{\underset{CH_3}{|}}{C}HCH_2CH_2CH_3 \xrightarrow{SOCl_2}$$

$$\underset{\underset{CH_3}{|}}{ClC(O)CH}CH_2CH_2CH_3 \xrightarrow[\text{2) } H_3O^+]{\text{1) } Li^+ (CH_3CH_2CH_2CH_2)_2Cu^-} CH_3CH_2CH_2CH_2\underset{\underset{}{||}}{C(O)}\underset{\underset{CH_3}{|}}{C}HCH_2CH_2CH_3$$

**22.72** (a) This is an aldol addition; dehydration cannot occur because the resulting alkene would violate Bredt's rule. The mechanism below begins with the enolate ion of the methyl ketone, which is formed when a proton is removed by potassium carbonate.

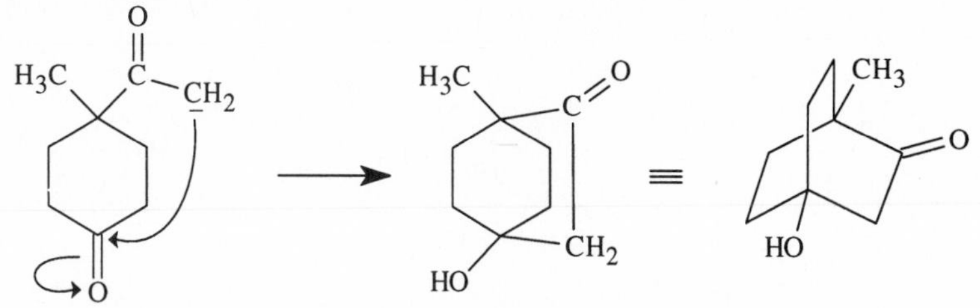

(c) The mechanism shown below begins with an α-bromo enolate ion, formed by the removal of an α-proton by NaH. This ion displaces a bromine in an internal nucleophilic substitution reaction to form the ring. An E2 reaction of the resulting compound with NaH as the base gives the product. (Recall that E2 reactions are particularly facile when the proton undergoing elimination is particularly acidic. See Sec. 17.3B on text p. 802.)

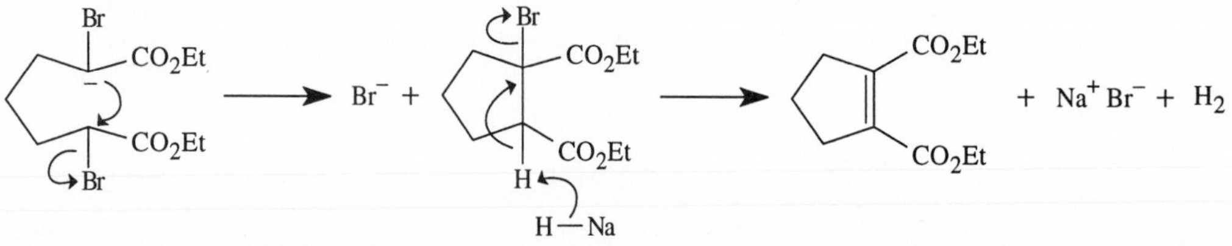

(e) A carbon-carbon double bond of a dienol on one ring attacks the protonated carbonyl group on the other. The mechanism below begins with a protonated ketone.

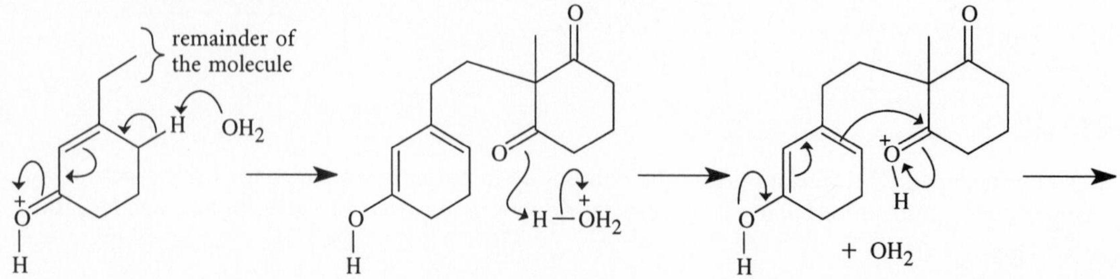

*(equation continues)*

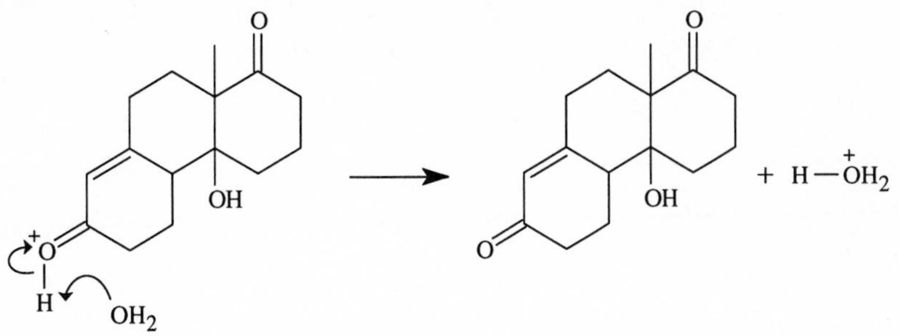

**(f)** This reaction is a double crossed-Claisen condensation followed by an intramolecular transesterification reaction in which the anionic oxygen of an enolate ion serves as the nucleophile. (The oxygen rather than the anion carbon acts as the nucleophile because a six-membered ring is formed; attack of the anionic carbon would give a strained four-membered ring.) The mechanism that follows begins with the conjugate-base enolate ion of acetone. Acid-catalyzed hydrolysis of the diester by the usual mechanism gives chelidonic acid.

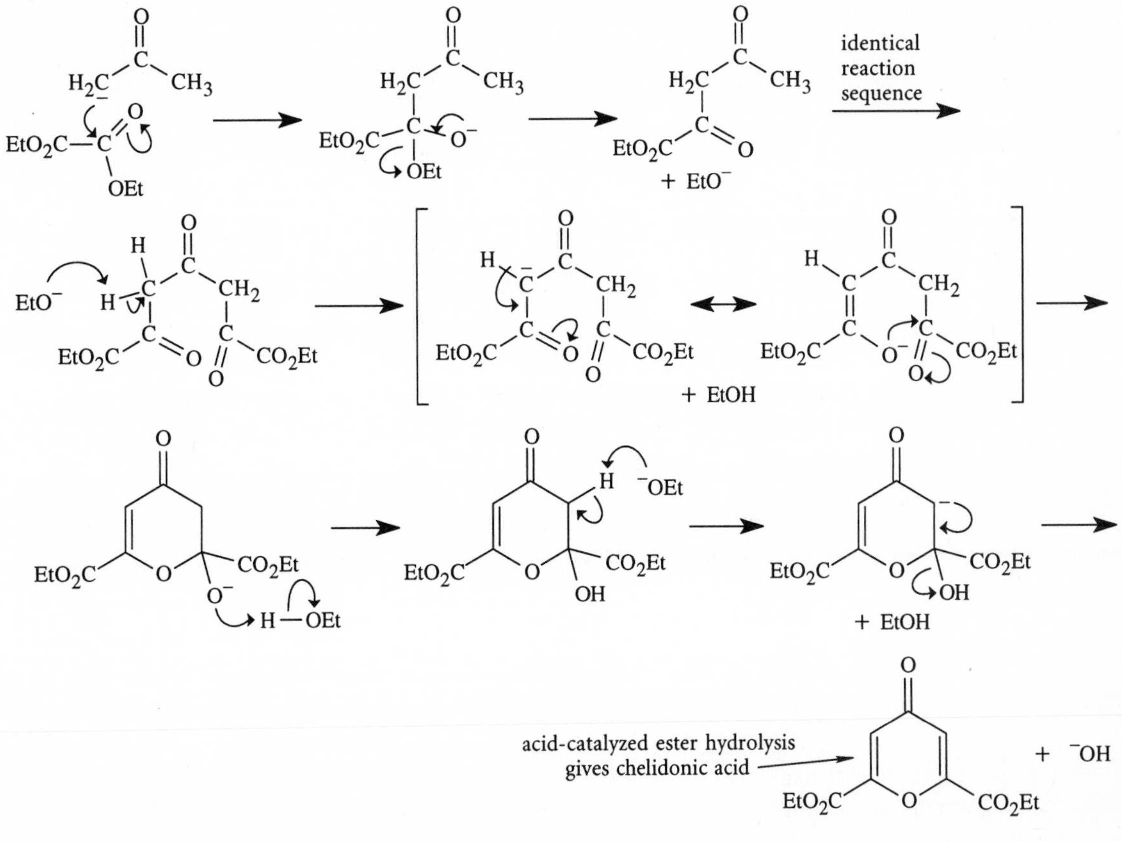

**22.73** This reaction is essentially a type of crossed Claisen condensation in which the "enolate ion" is the conjugate-base anion of urea. A second reaction of the same type closes the ring. The following mechanism starts with the attack of the ionized urea on one of the ester groups.

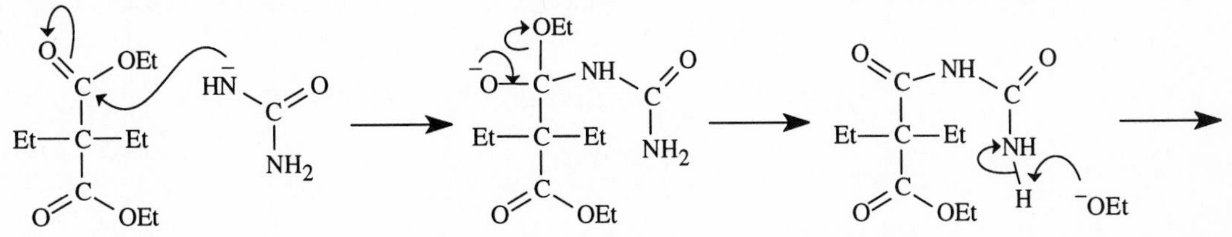

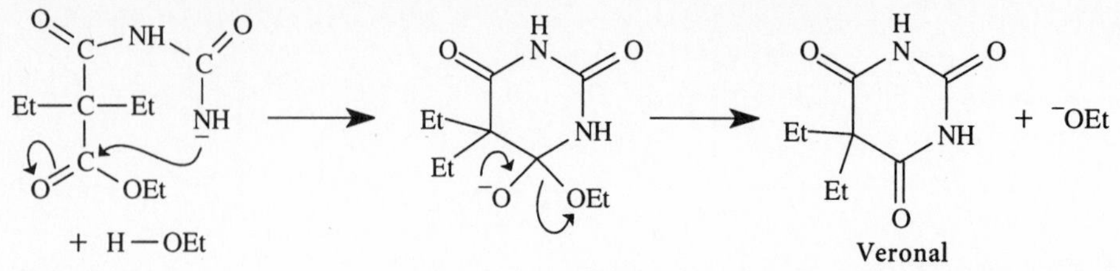

Veronal

As in other Claisen-type condensation reactions, the product Veronal is ionized by the ethoxide base, and Veronal is regenerated when acid is added. (These final steps are not shown in the foregoing mechanism.)

**22.75** (a) Use a conjugate addition of cyanide to an α,β-unsaturated ester, and then hydrolyze. (Either stereoisomer of the starting ester can be used.)

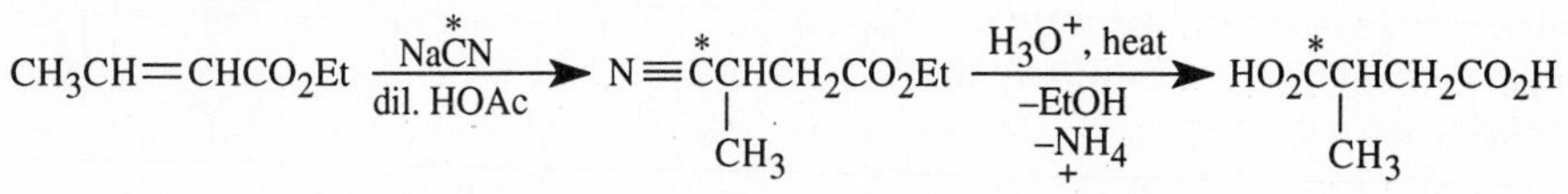

**22.76** (a) A conjugate addition of hydroxide to pulegone gives a β-hydroxy ketone, which then undergoes the reverse of an aldol addition to give acetone and 3-methylcyclohexanone.

3-methylcyclohexanone

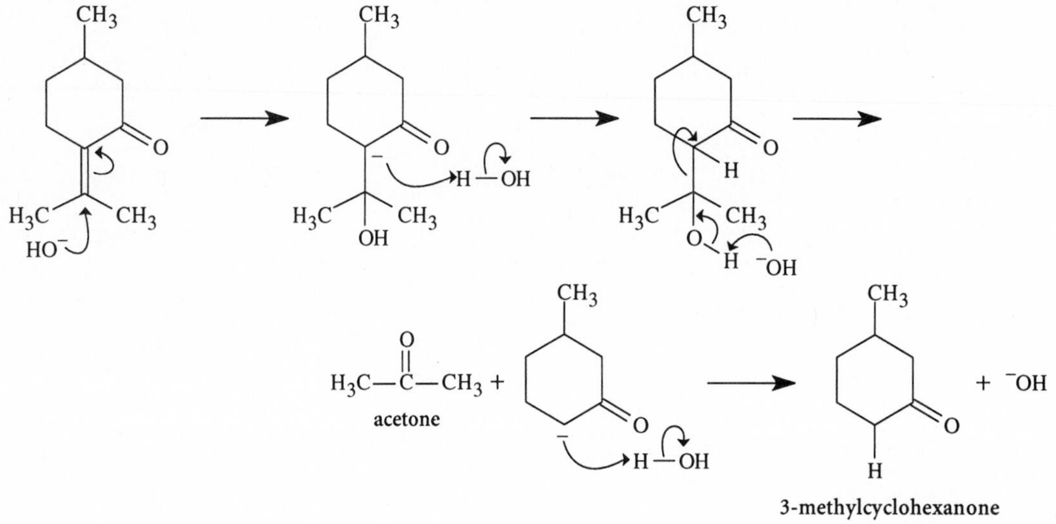

**22.77** Compound *B* is an aldehyde (it gives a positive Tollens' test); therefore, compound *A* is a primary alcohol. Compound *B* is conjugated (it has a strong UV spectrum), and it has the same carbon skeleton as octanoic acid (an unbranched chain of eight carbons). Therefore, compound *B* is 2-octenal, and compound *A* is 2-octenol. (The data do not differentiate between the *E* and *Z* stereoisomers.) In order to form 2-octenol, lithium dibutylcuprate must undergo a *conjugate addition* to the epoxide:

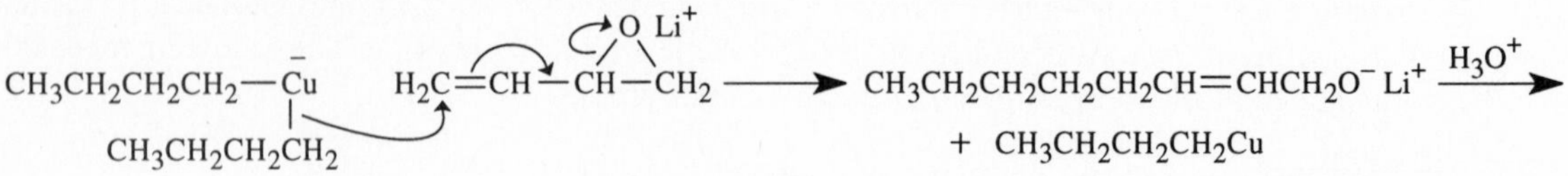

2-octenol (compound *A*)

2-octenal (compound *B*)

**22.79** (a) The conjugate-base enolate ion of the α-bromo ester, formed by reaction of the α-bromo ester with potassium *tert*-butoxide, attacks the carbonyl carbon of the aldehyde to give β-bromo alkoxide *A*, the conjugate base of a bromohydrin. As shown in Eq. 11.19 on text p. 505, an alkoxide of this type readily cyclizes to an epoxide by an internal nucleophilic substitution mechanism.

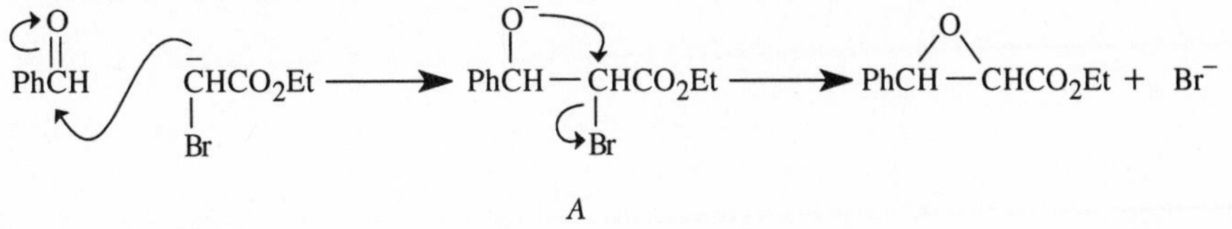

*A*

**22.80** The deuterium incorporation results show that the carbon-carbon double bond is protonated. This is exactly the way that enols are protonated (see the reverse of Eq. 22.15b on text p. 1047). As a result, a relatively stable α-alkoxy carbocation is formed, and this is attacked by water to give a hemiacetal. The hemiacetal breaks down by the usual mechanism (see text p. 899) to give acetaldehyde and the alcohol. Hydrolysis of vinyl ethers is much faster than hydrolysis of ordinary ethers because vinyl ether hydrolysis involves a relatively stable carbocation intermediate, an α-alkoxy carbocation.

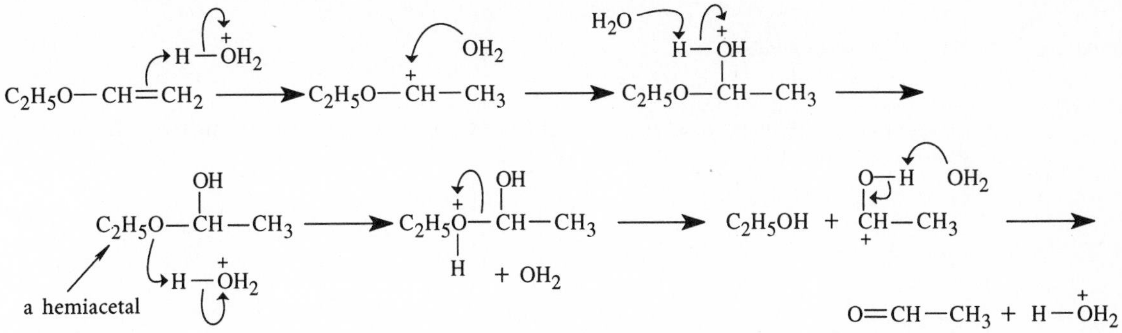

**22.82** (a) The conjugate-base enolate ion is alkylated by the alkyl halide. Then the conjugate-base enolate ion of the alkylation product is formed and reacts intramolecularly with the second alkyl halide group to form a ring. Hydrolysis and decarboxylation of the ester group give the following product.

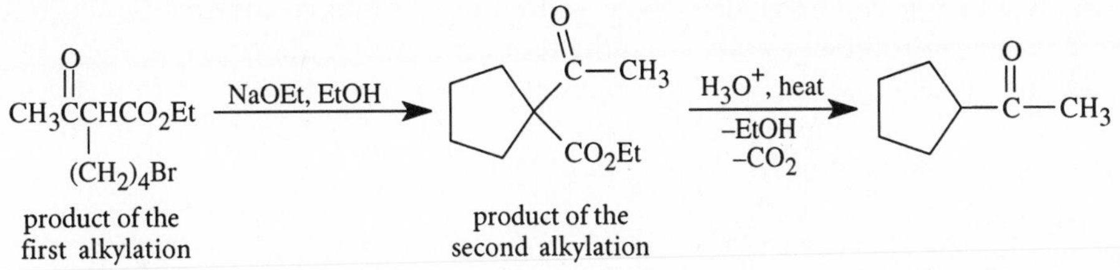

product of the first alkylation

product of the second alkylation

(c) This is a nucleophilic aromatic substitution reaction in which the nucleophile is an enolate ion.

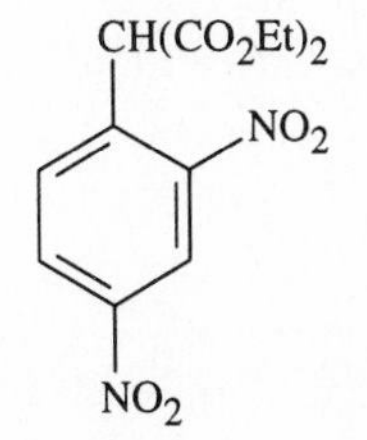

(e) This is a Michael addition in which the conjugate-base enolate ion of the substituted malonic ester is the nucleophile. (Note that four diastereomers of this compound are possible.)

*(solution continues)*

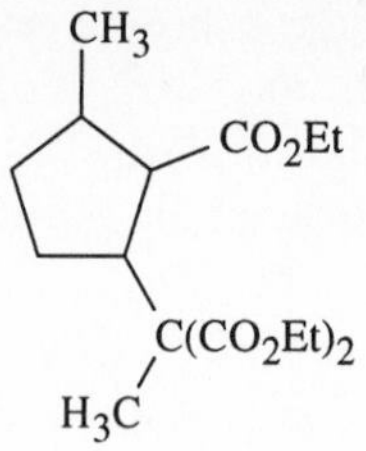

(g) The magnesium organocuprate reagent undergoes conjugate addition; the acetal group hydrolyzes; and an acid-catalyzed intramolecular aldol condensation ensues.

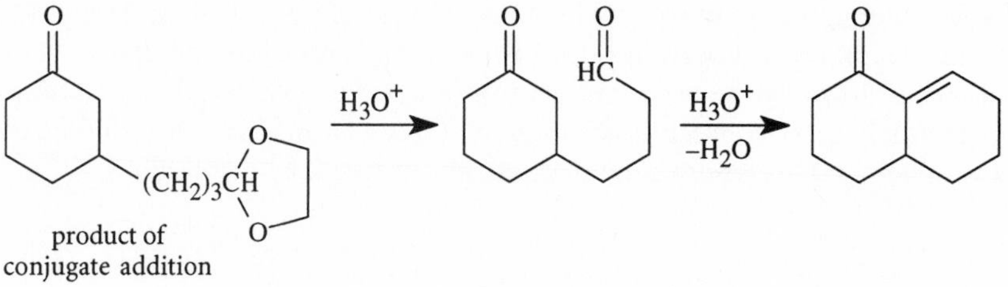

product of
conjugate addition

(i) The benzene $\pi$ electrons serve as the nucleophile in an acid-catalyzed conjugate addition, which can also be viewed as a variation of Friedel-Crafts alkylation.

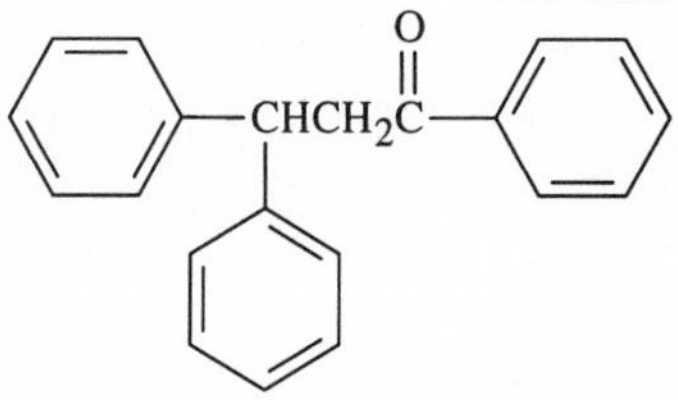

22.83 (a) A crossed Claisen condensation is followed by a reverse Claisen condensation, driven by expulsion of the more volatile ester, ethyl acetate, from the reaction mixture. The tertiary proton in compound *A* is very acidic and is removed to give a large amount of a conjugate-base anion (not shown) in equilibrium with a small amount of compound *A*; however, by LeChatelier's principle, the reaction of *A* shown in the following mechanism eventually depletes this anion.

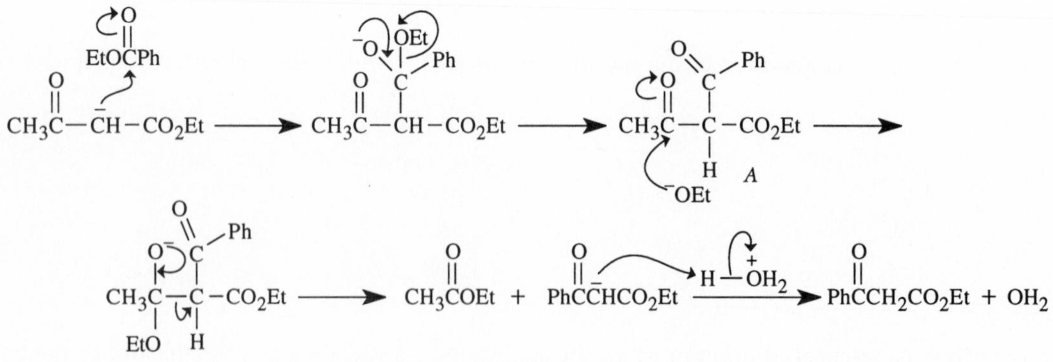

(c) This is an acid-catalyzed conjugate addition of water followed by elimination of phenol. The resulting enol is then spontaneously converted into the corresponding aldehyde.

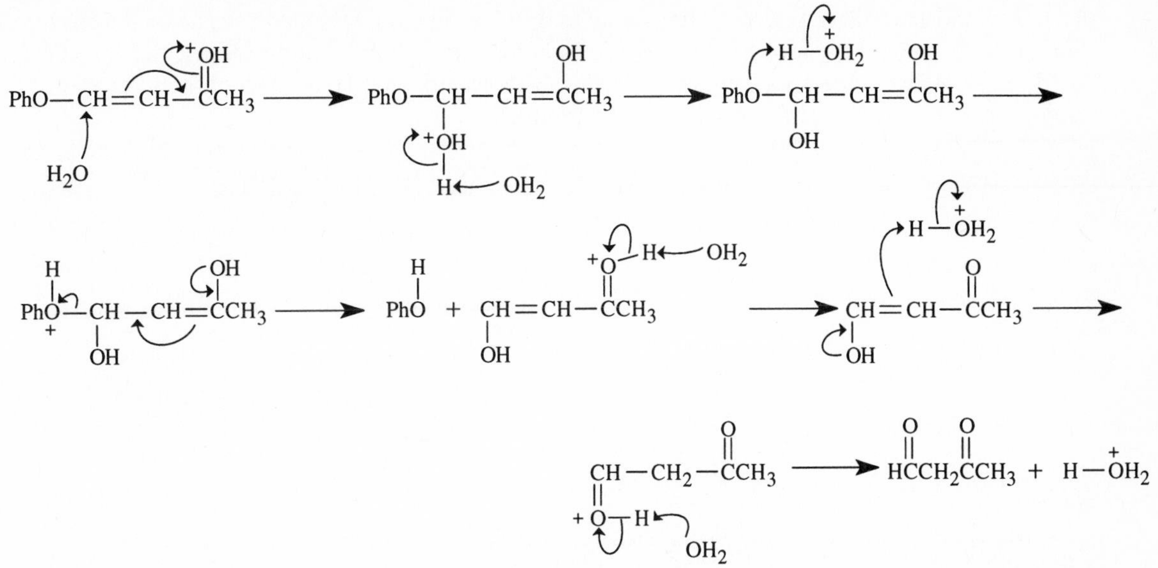

(e)  Ionization of the benzylic hydrogen gives a carbanion that not only is allylic and benzylic but also is stabilized by the polar effect of the α-oxygen. The O—H hydrogen is also acidic, and removal of it occurs, but the resulting anion does not lead to a reaction. The equilibrium in the reaction is pulled to the right by formation of the ketone from the enol.

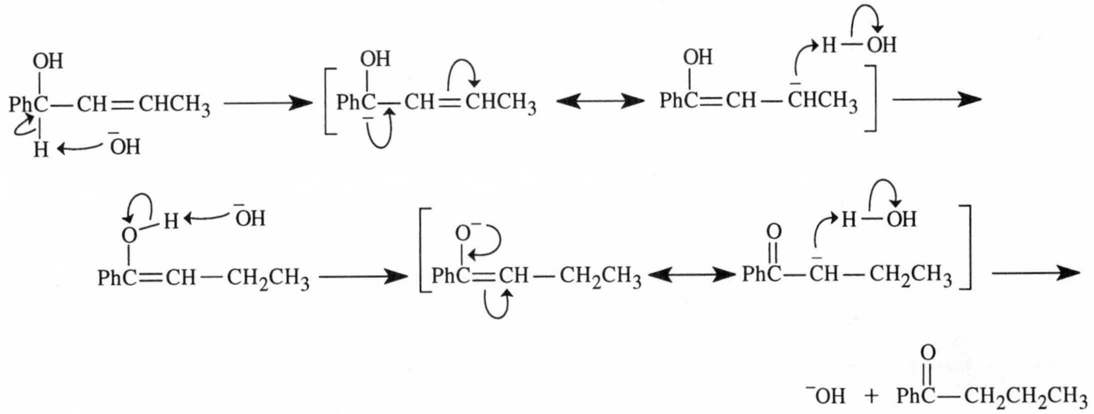

(g)  Formation of the α,β-unsaturated ester releases HCl, which undergoes conjugate addition.

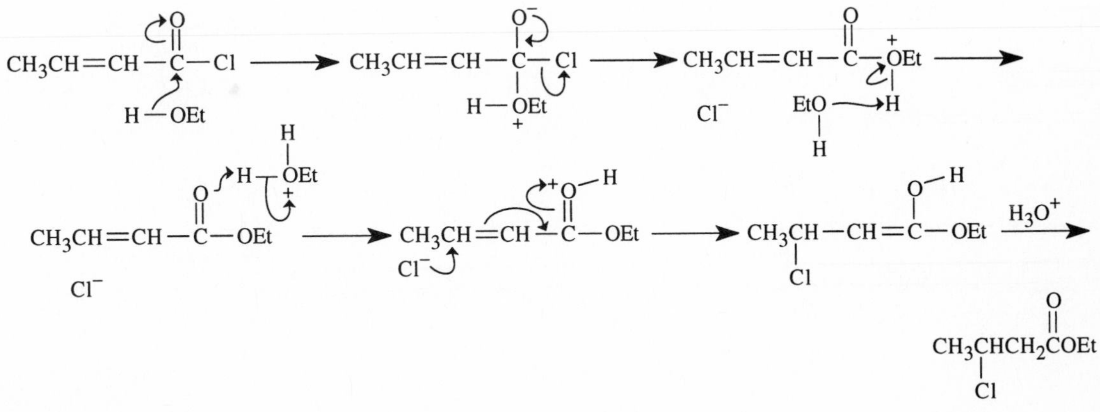

The last step, formation of the ketone from the enol, occurs by the reverse of the mechanism shown in Eq. 22.15b, text p. 1047.

(h) α-Cyano esters have acidities comparable to those of β-keto esters. (Why?) The conjugate-base enolate ion of the α-cyano ester adds to the aldehyde carbonyl group. Elimination of water gives an α,β-unsaturated nitrile, which undergoes conjugate addition with cyanide ion. Hydrolysis of the ester and decarboxylation of the resulting acid give the product.

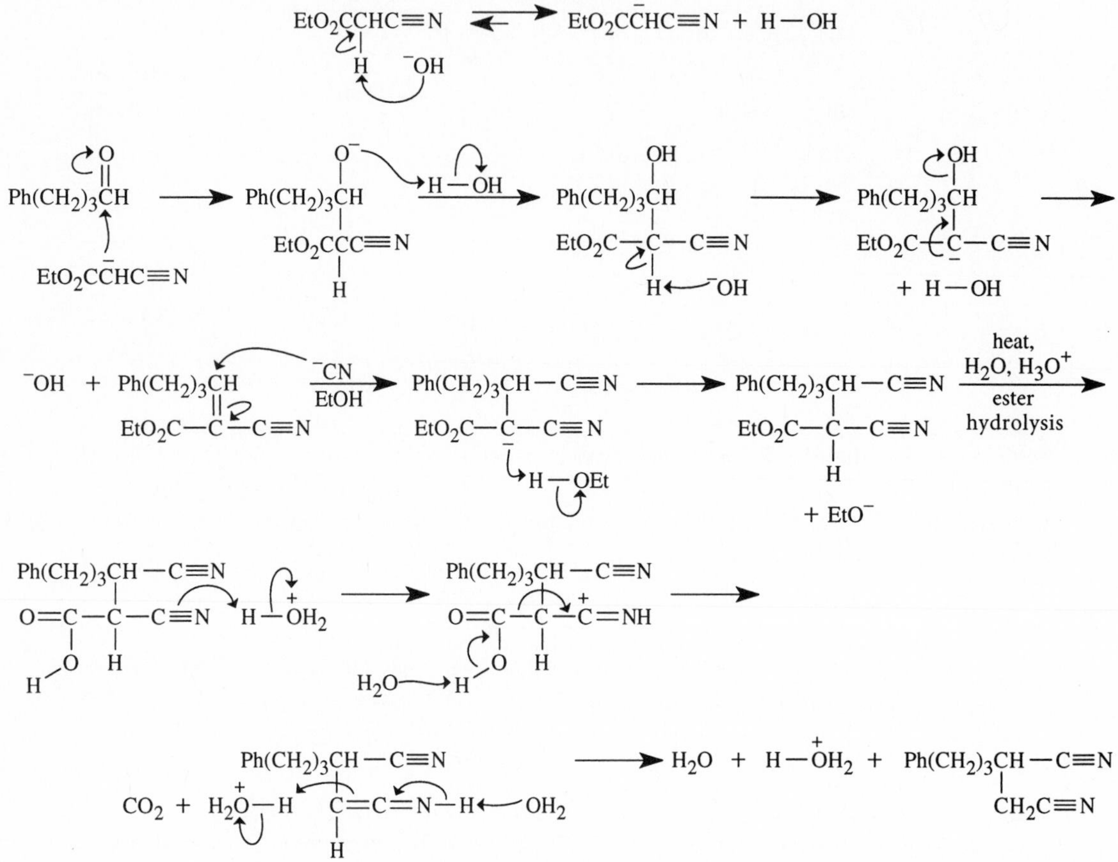

(j) Conjugate addition of triphenylphosphine to the anhydride starting material followed by proton transfer to the initially formed anion gives an ylid *A*, which undergoes a Wittig reaction with the aldehyde butanal. Hydrolysis of the resulting anhydride gives the dicarboxylic acid.

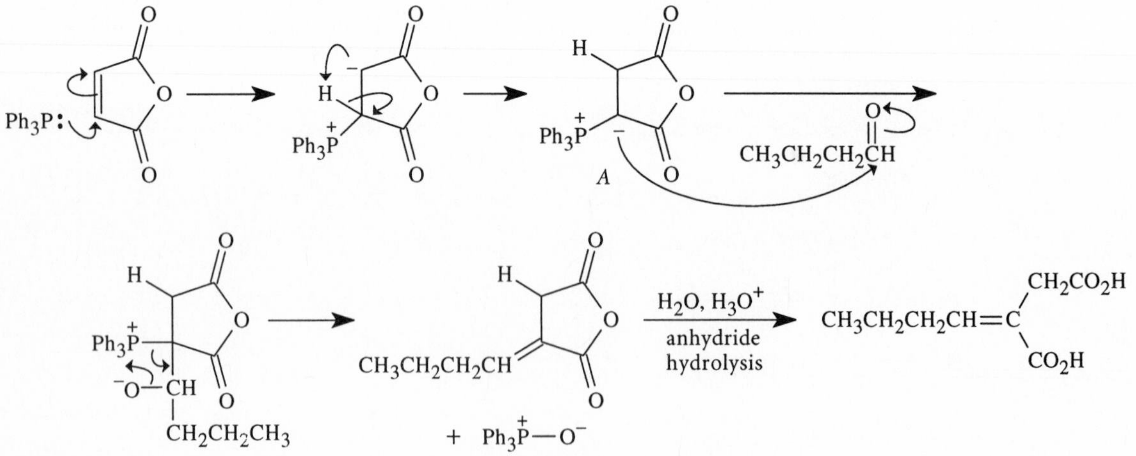

# 23

# Chemistry of Amines

## Terms

| | |
|---|---|
| acyl azide | 23.11C |
| acyl hydrazide | 23.11C |
| alkaloid | 23.12B |
| amide base | 23.5D |
| amine | Introduction |
| amino group | 23.1B |
| ammonium salt | 23.5A |
| azo dye | 23.10B |
| azobenzene | 23.10B |
| Curtius rearrangement | 23.11C |
| diazonium salt | 23.10A |
| diazotization | 23.10A |
| exhaustive methylation | 23.7A |
| Gabriel synthesis | 23.11A |
| Hofmann elimination | 23.8 |
| Hofmann hypobromite reaction | 23.11C |
| Hofmann rearrangement | 23.11C |
| imminium ion | 23.7B |
| isocyanate | 23.11C |
| neurochemistry | 23.12B |
| neurotransmitter | 23.12B |
| nitrosamine | 23.10C |
| primary amine | Introduction |
| quaternary ammonium salt | 23.6 |
| quaternization | 23.7A |
| reductive amination | 23.7B |
| Sandmeyer reaction | 23.10A |
| secondary amine | Introduction |
| tertiary amine | Introduction |

## Concepts

### I. Introduction to Amines

#### A. General

1. Amines are organic derivatives of ammonia in which the ammonia hydrogens are replaced by alkyl or aryl groups.
2. Amines are classified by the number of alkyl or aryl substituents (R groups) on the amine nitrogen.
   a. A primary amine has one substituent.
   b. A secondary amine has two substituents.
   c. A tertiary amine has three substituents.

$CH_3—NH_2$

a primary amine

a secondary amine

a tertiary amine

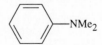

### B.  COMMON NOMENCLATURE

1. In common nomenclature an amine is named by appending the suffix *amine* to the name of the alkyl group; the name of the amine is written as one word.
2. When two or more alkyl groups in a secondary or tertiary amine are different, the compound is named as an *N*-substituted derivative of the larger group.  (An *N* or <u>N</u> designates that the substituent is on the amine nitrogen.)
3. Aromatic amines are named as derivatives of aniline.

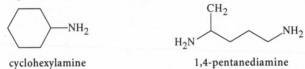

|  |  |  |
|:--:|:--:|:--:|
| (CH₃)₂CHNH₂ | | |
| isopropylamine | *N*-ethylbutylamine | *p*-chloro-*N*,*N*-diethylaniline |
| ⎵ | (a secondary amine with | (an aromatic amine) |
| name of the | two different alkyl groups) | |
| alkyl group | | |

### C.  SYSTEMATIC NOMENCLATURE

1. The most widely used system of substitutive amine nomenclature is that of *Chemical Abstracts*.
   a. In this system, an amine is named in the same way as the analogous alcohol, except that the suffix *amine* is used.
   b. In diamine nomenclature, the final *e* of the hydrocarbon name is retained; an *N′* is used when one of the groups is on a different nitrogen.

|  |  |
|:--:|:--:|
| cyclohexylamine | 1,4-pentanediamine |

2. The priority for citation of amine groups as principal groups is just below that of alcohols.  (A complete list of group priorities is given in Appendix I, text page A-1.)

$$—CO_2H \text{ and derivatives} \ > \ —CH{=}O, \ \text{\Large)}C{=}O \ > \ —OH > —NR_2$$

3. When cited as a substituent, the —NH₂ group is called the amino group.

| H₂NCH₂CH₂OH | CH₃CH₂NH——⟨benzene⟩——CO₂H |
|:--:|:--:|
| 2-aminoethanol | 4-(ethylamino)benzoic acid |

4. Although *Chemical Abstracts* calls aniline *benzenamine*, the more common practice is to use the common name *aniline* in substitutive nomenclature.
5. Many important nitrogen-containing heterocyclic compounds are known by specific names that should be learned (numbering generally begins with the heteroatom).

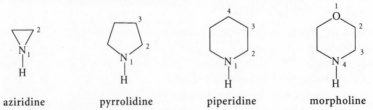

| aziridine | pyrrolidine | piperidine | morpholine |
|:--:|:--:|:--:|:--:|

### D.  STRUCTURE OF AMINES

1. The C—N bonds of aliphatic amines are longer than the C—O bonds of alcohols, but shorter than the C—C bonds of alkanes.
2. Aliphatic amines have a pyramidal shape (or approximately tetrahedral shape, if the electron pair is considered to be a "group").

a. Most amines undergo rapid inversion at nitrogen; this occurs through a planar transition state and converts an amine into its mirror image.
b. Because of this inversion, amines in which the only asymmetric atom is the amine nitrogen cannot be resolved into enantiomers.

3. The C—N bond in aniline is shorter than that in aliphatic amines because of:
   a. $sp^2$ hybridization of the adjacent carbon.
   b. overlap of the unshared electrons on nitrogen with the $\pi$-electron system of the ring; this overlap gives double-bond character to the C—N bond.

E. PHYSICAL PROPERTIES OF AMINES
   1. Most amines are somewhat polar liquids with unpleasant odors.
   2. The boiling points of amines depend on whether they are primary, secondary, or tertiary.
      a. Primary and secondary amines, which can both donate and accept hydrogen bonds, have higher boiling points than isomeric tertiary amines, which cannot donate hydrogen bonds.
      b. Primary and secondary amines have higher boiling points than ethers but lower boiling points than alcohols (alcohols are better hydrogen-bond donors).
   3. Most primary and secondary amines with four or fewer carbons, as well as trimethylamine, are miscible with water; amines with large carbon groups have little or no water solubility.

F. USES AND OCCURRENCE OF AMINES
   1. Among the relatively few industrially important amines is hexamethylenediamine, which is used in the synthesis of nylon-6,6.
   2. Among the many types of naturally occurring amines are the alkaloids (nitrogen-containing bases that occur naturally in plants).
      a. Most alkaloids are amines, and many are heterocyclic amines.
      b. Many alkaloids have biological activity; others have no known activity, and their functions within the plants from which they come are, in many cases, obscure.
   3. Some hormones, such as neurotransmitters, are amines.
      a. Hormones are compounds that regulate the biochemistry of multicellular organisms, particularly vertebrates.
      b. Neurotransmitters are molecules that are important in the communication between nerve cells, or between nerve cells and their target organs.

II. *Spectroscopy of Amines*

A. IR SPECTROSCOPY
   1. The most important absorptions in the infrared spectra of primary amines are the N—H stretching absorptions, which usually occur as two or more peaks at 3200–3375 cm$^{-1}$.
   2. Primary amines show an NH$_2$ scissoring absorption near 1600 cm$^{-1}$.
   3. Most secondary amines show a single N—H stretching absorption rather than the multiple peaks observed for primary amines.
   4. Tertiary amines show no N—H absorptions.

B. NMR SPECTROSCOPY
   1. The characteristic resonances in the NMR spectra of amines are those of the protons adjacent to the nitrogen (the $\alpha$-protons) and the N—H protons.
      a. In alkylamines, the $\alpha$-protons are observed in the $\delta$2.5–3.0 region of the spectrum.
      b. In aromatic amines, the $\alpha$-protons of *N*-alkyl groups are somewhat further downfield near $\delta$3.
   2. The chemical shift of the N—H proton depends on the concentration of the amine and on the conditions of the NMR experiment.
      a. In alkylamines, this resonance typically occurs at rather high field, typically around $\delta$1.

      b. In aromatic amines, this resonance is at considerably lower field.

  3. The N—H protons of amines under most conditions undergo rapid exchange.

      a. The N—H absorption can be obliterated from the spectrum by exchange with $D_2O$.

      b. In some amine samples the N—H resonance is broadened.

      c. Splitting between the amine N—H and adjacent C—H groups is usually not observed.

  4. The characteristic CMR absorptions of amines are those of the $\alpha$-carbon; these absorptions occur in the $\delta$ 30–50 chemical shift range.

## C. Mass Spectrometry

  1. $\alpha$-Cleavage is a particularly important fragmentation mode of aliphatic amines.

  2. The molecular ion occurs at an odd mass if the amine contains an odd number of nitrogens because compounds containing an odd number of nitrogens have odd molecular weights.

      a. Odd-electron ions containing an odd number of nitrogens are observed at odd mass.

      b. Even-electron ions containing an odd number of nitrogens are observed at even mass.

## III. Acidity and Basicity of Amines

### A. Acidity of Amines

  1. Primary and secondary amines are very weakly acidic.

  2. The conjugate base of an amine is called an amide base. Amide bases are very strong.

      a. The amide conjugate base of ammonia is usually prepared by dissolving an alkali metal such as sodium in liquid ammonia in the presence of a trace of ferric ion.

      b. The conjugate bases of alkylamines are prepared by treating the amine with butyllithium in an ether solvent such as THF.

  3. The p$K_a$ of a typical amine is about 35. (This p$K_a$ refers to the process $R_2\ddot{N}$—H + B$^- \rightarrow$ $R_2\ddot{N}$:$^-$ + BH, where B is a base, and not to $R_3\overset{+}{N}$—H + B$^- \rightarrow$ $R_3N$: + BH.)

### B. Basicity of Amines

  1. Amines are strong enough bases that they are completely protonated in dilute acid solutions.

  2. The salts of protonated amines are called ammonium salts.

      a. The ammonium salts of simple alkylamines are named as substituted derivatives of the ammonium ion, *e. g.* methylammonium chloride.

      b. Other ammonium salts are named by replacing the final *e* in the name of the amine with the suffix *ium*, *e. g.* pyridinium chloride.

  3. Ammonium salts are fully ionic compounds; the N—H bonds are covalent, but there is no covalent bond between the nitrogen and the counter-ion.

  4. The basicity of an amine is expressed in terms of the p$K_a$ of its conjugate-acid ammonium salt; the higher the p$K_a$ of an ammonium ion, the more basic is its conjugate-base amine.

### C. Factors Affecting Amine Basicity

  1. Three effects influence the basicity of amines (the same effects that govern the acidity and basicity of other compounds):

      a. the effect of alkyl substitution

      b. the polar effect

      c. the resonance effect

  2. In alkyl substitution two opposing factors are involved:

      a. The first is the tendency of alkyl groups to stabilize charge through a polarization effect.

        *i.* The polarization of alkyl groups can act to stabilize either positive or negative charge.

        *ii.* This effect governs the acidity of ammonium ions in the gas phase, in which the acidity of ammonium ions decreases regularly with increasing alkyl substitution.

      b. The second factor is a solvent effect.

        *i.* Ammonium ions in solution are stabilized by hydrogen-bond donation to the solvent.

        *ii.* Primary ammonium ions are stabilized by hydrogen bonding more than tertiary ones; primary ammonium salts have three hydrogens that can be donated to form hydrogen bonds, but a tertiary ammonium salt has only one.

  3. The polar effect operates largely on the conjugate acid of the amine because this cation is the charged species in the acid-base equilibrium.

    a. An electronegative (electron-withdrawing) group destabilizes an ammonium ion because of a repulsive electrostatic interaction between the positive charge on the ammonium ion and the positive end of the substituent bond dipole.

    b. The base-weakening effect of electron-withdrawing substituents decreases rapidly with distance between the substituent and the charged nitrogen.

    c. The electron-withdrawing polar effect of the aromatic ring also contributes significantly to the reduced basicity of aromatic amines.

4. The basicity of aniline derivatives is affected by resonance:

    a. Aniline is stabilized by resonance interaction of the unshared electron pair on nitrogen with the aromatic ring.

       *i.* When aniline is protonated, this resonance stabilization is no longer present, because the unshared electron pair is bound to a proton.

       *ii.* The resonance stabilization of aniline relative to its conjugate acid reduces its basicity.

    b. The resonance stabilization of aniline lowers the energy required for its formation from its conjugate acid, and thus lowers its basicity relative to that of cyclohexylamine.

5. The $pK_a$ of an ammonium ion is directly related to the standard free-energy difference $\Delta G°$ between it and its conjugate base.

$$\Delta G° = 2.3RT\,pK_a$$

    a. If a substituent stabilizes an amine ion more than it stabilizes the conjugate-acid ammonium ion, the standard free energy of the amine is lowered, $\Delta G°$ is decreased, and the $pK_a$ of the ammonium ion is reduced.

    b. If a substituent stabilizes the ammonium ion more than its conjugate-base amine, the opposite effect is observed: the $pK_a$ is increased, and the amine basicity is also increased.

## D. QUATERNARY AMMONIUM SALTS

1. Closely related to ammonium salts are compounds in which all four hydrogens of $^+NH_4$ are replaced by alkyl or aryl groups; such compounds are called quaternary ammonium salts.

2. Quaternary ammonium salts are ionic compounds.

3. Many quaternary ammonium salts containing large organic groups are soluble in nonaqueous solvents.

$$CH_3CH_2CH_2CH_2CH_2CH_2CH_2CH_2CH_2CH_2-\overset{\overset{\displaystyle CH_3}{|}}{\underset{\underset{\displaystyle CH_3}{|}}{N^{\pm}}}-CH_2Ph \qquad Cl^-$$

benzyldecyldimethylammonium  chloride

## E. SEPARATIONS USING AMINE BASICITY

1. Because ammonium salts are ionic compounds, many have appreciable water solubilities.

    a. When a water-insoluble amine is treated with dilute aqueous acid, the amine dissolves as its ammonium salt.

    b. Upon treatment with base, an ammonium salt is converted back into the corresponding amine.

2. This property of ammonium salts can be used to design separations of amines from other compounds.

3. Amine basicities can play a key role in the design of enantiomeric resolutions.

# Reactions

......................................................................................................

## I. Synthesis of Amines; Alkylation Reactions

### A. GENERAL

1. Amines are good nucleophiles (Lewis bases).
2. Three reactions of nucleophiles have been studied in previous sections:
   a. $S_N2$ reaction with alkyl halides, sulfonate esters, or epoxides (Sec. 9.1A, text pp. 385–387; Sec. 9.4, text pp. 394–404; Secs. 10.3A–C, text pp. 450–456; and Sec. 11.4, text pp. 509–516).
   b. Addition to aldehydes, ketones, and $\alpha,\beta$-unsaturated carbonyl compounds (Sec. 19.7, text pp. 884–890; Sec. 19.11, text pp. 904–909; and Sec. 22.8A, text pp. 1085–1088).
   c. Nucleophilic acyl substitution at the carbonyl groups of carboxylic acid derivatives (Sec. 21.8, text pp. 1000–1007).

### B. DIRECT ALKYLATION OF AMINES

1. Treatment of ammonia or an amine with an alkyl halide or other alkylating agent results in alkylation of the nitrogen. This is an example of an $S_N2$ reaction in which the amine acts as the nucleophile.
   a. The immediate product of the reaction is an alkylammonium ion.
   b. If this ammonium ion has N—H bonds, further alkylations can take place to give a complex product mixture.

$$NH_3 \;+\; RX \quad\longrightarrow\quad RNH_2 \;+\; R_2NH \;+\; R_3N \;+\; R_4N^+$$

   c. A mixture of products is formed because the alkylammonium ion produced initially can be deprotonated by the amine starting material.
2. Epoxides, as well as $\alpha,\beta$-unsaturated carbonyl compounds and $\alpha,\beta$-unsaturated nitriles, also react with amines and ammonia; multiple alkylation can occur with these alkylating agents as well.
3. In an alkylation reaction, the exact amount of each product obtained depends on the precise reaction conditions and on relative amounts of starting amine and alkyl halide; the utility of alkylation as a preparative method for amines is limited.

### C. QUATERNIZATION OF AMINES

1. Amines can be converted into quaternary ammonium salts with excess alkyl halide under forcing conditions.
   a. This process, called quaternization, is one of the most important synthetic applications of amine alkylation.
   b. The reaction is particularly useful when especially reactive alkyl halides are used.
2. Conversion of an amine into a quaternary ammonium salt with excess methyl iodide is called exhaustive methylation.

### D. REDUCTIVE AMINATION

1. Primary and secondary amines form imines and enamines, respectively, when they react with aldehydes and ketones; in the presence of a reducing agent, imines and enamines are reduced to amines.
   a. Reduction of the C=N double bond is analogous to reduction of the C=O double bond.
   b. The imine or enamine does not have to be isolated; because imines and enamines are reduced much more rapidly than carbonyl compounds, reduction of the carbonyl compound is not a competing reaction.
2. The formation of an amine from the reaction of an aldehyde or ketone with another amine and a reducing agent, usually $NaBH_4$ or sodium cyanoborohydride ($NaBH_3CN$), is called reductive amination.

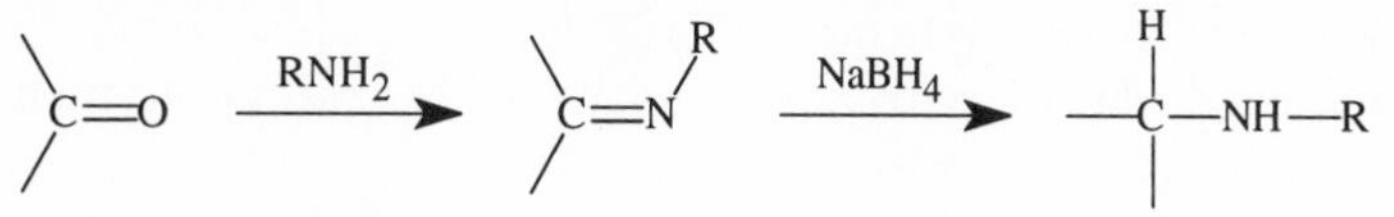

a. An imminium ion is formed in solution by protonation and dehydration of the carbinolamine intermediate in the reductive amination reaction of a secondary amine and formaldehyde; this ion is rapidly and irreversibly reduced.

$$H_2C{=}O \xrightleftharpoons{R_2NH} \underset{\text{a carbinolamine}}{R_2N{-}CH_2{-}OH} \xrightleftharpoons{H_3O^+} \underset{\text{an iminium ion}}{R_2\overset{+}{N}{=}CH_2} + H_2O$$

b. The reaction of an amine with an excess of formaldehyde is a useful way to introduce methyl groups to the level of a tertiary amine; quaternization does not occur in this reaction.

3. To determine the required starting materials:
   a. starting with the desired compound, mentally reverse the reductive amination process by breaking one of the C—N bonds:
      *i.* on the nitrogen side, replace it with an N—H bond
      *ii.* on the carbon side, remove a hydrogen from the carbon and add a carbonyl oxygen.
   b. repeat this process for each C—N bond and evaluate the best pathway.

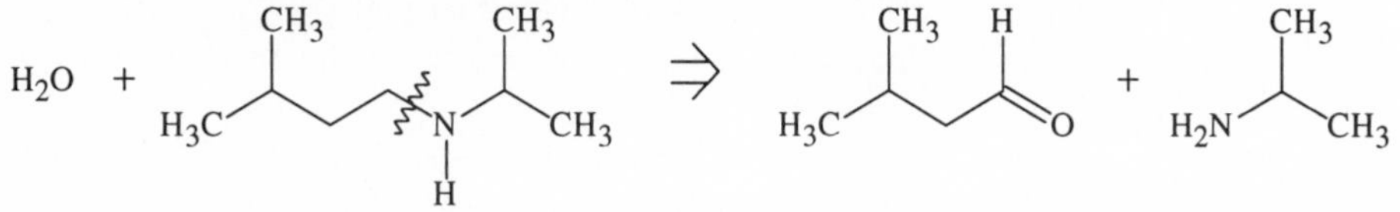

## *II. Other Syntheses of Amines*

### A. Gabriel Synthesis of Primary Amines

1. The alkylation of phthalimide anion followed by hydrolysis of the alkylated derivative to the primary amine is called the Gabriel synthesis.

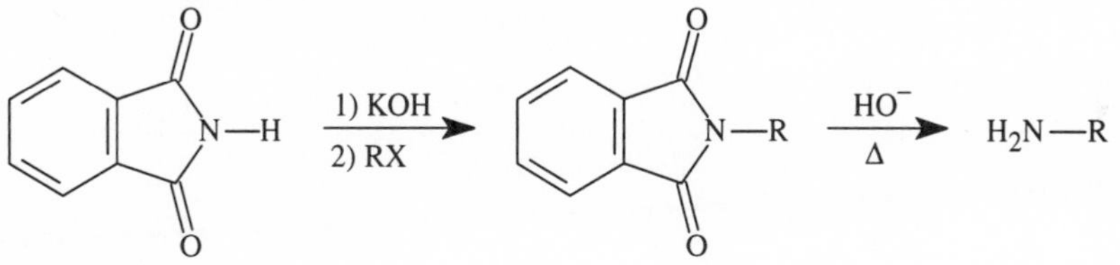

phthalimide

   a. The nitrogen in phthalimide has only one acidic hydrogen, and thus it can be alkylated only once.
      *i.* The conjugate base of phthalimide is easily formed with KOH or NaOH.
      *ii.* The phthalimide anion is a good nucleophile, and is alkylated by alkyl halides or sulfonate esters (primary or unbranched secondary) in $S_N2$ reactions.
      *iii.* The *N*-alkylated phthalimide formed is converted into the free amine by hydrolysis in either strong acid or base.
   b. Although *N*-alkylphthalimides also have a pair of unshared electrons on nitrogen, they do not alkylate further, because neutral imides are much less basic, and therefore less nucleophilic, than the phthalimide anion.
2. Multiple alkylation, which occurs in the direct alkylation of ammonia, does not occur in the Gabriel synthesis.

### B. Reduction of Nitro Compounds

1. Nitro compounds can be reduced to amines under a variety of conditions.
   a. The nitro group is usually reduced very easily by catalytic hydrogenation.
   b. The reduction of aromatic nitro compounds to primary amines can be accomplished with finely divided metal powders and HCl.

    *i.* Iron or tin powder is frequently used.

    *ii.* In this reaction the nitro compound is reduced at nitrogen, and the metal, which is oxidized to a metal salt, is the reducing agent.

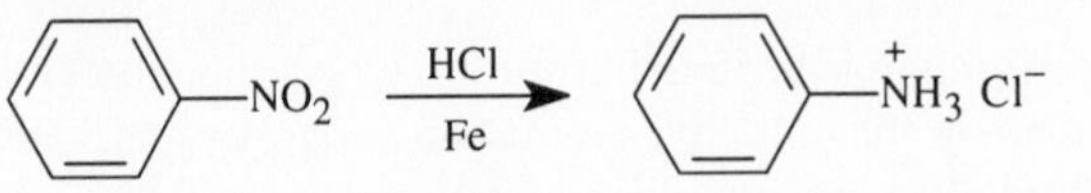

2. Aromatic nitro compounds react with $LiAlH_4$, but the reduction products are azobenzenes, not amines.

3. Nitro groups do not react at all with $NaBH_4$ under the usual conditions.

## C. HOFMANN REARRANGEMENT

1. Treatment of a primary amide with bromine in base gives rise to a rearrangement called the Hofmann rearrangement or Hofmann hypobromite reaction.

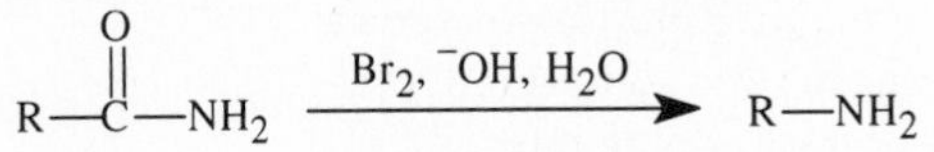

2. The first step in the mechanism is ionization of the amide N—H; the resulting anion is then brominated.

    a. The *N*-bromoamide product is more acidic than the amide starting material and it ionizes.

       *i.* The *N*-bromo anion rearranges rapidly to an isocyanate.

       *ii.* Because the Hofmann rearrangement is carried out in aqueous base, the isocyanate cannot be isolated.

    b. The isocyanate formed spontaneously hydrates to a carbamate ion, which then decarboxylates to the amine product under the strongly basic reaction conditions.

    c. When the reaction mixture is acidified, the carbamate decarboxylates to give the amine.

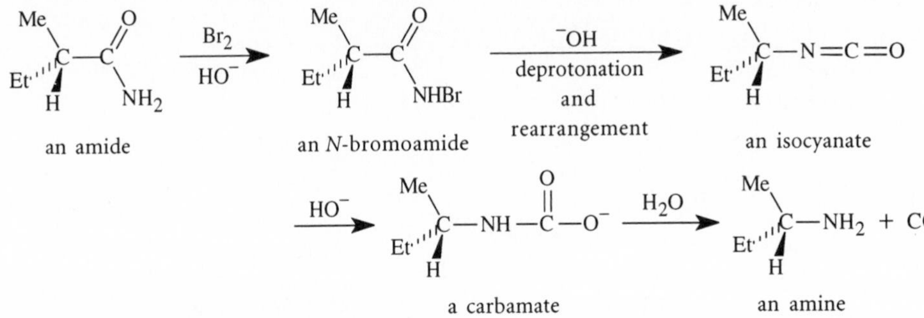

3. The Hofmann rearrangement takes place with retention of stereochemical configuration in the migrating alkyl group.

## D. CURTIUS REARRANGEMENT

1. When an acyl azide is heated in an inert solvent such as benzene or toluene, it is transformed with loss of nitrogen into an isocyanate; this concerted reaction is called the Curtius rearrangement.

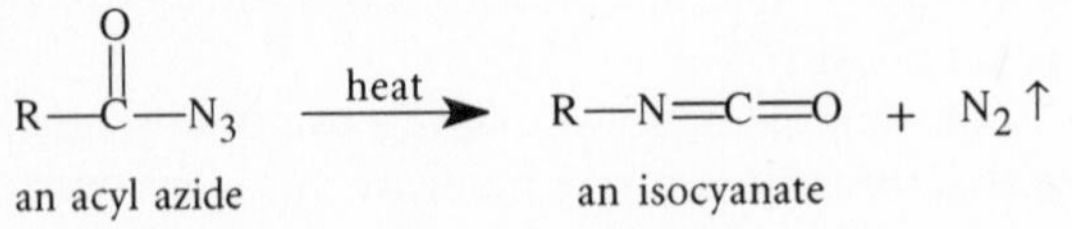

2. The overall transformation that occurs as a result of the Curtius rearrangement followed by hydration is the removal of the carbonyl carbon of the acyl azide as $CO_2$.

    a. The isocyanate product of the Curtius rearrangement can be transformed into a carbamic acid by hydration in either acid or base.

    b. Spontaneous decarboxylation of the carbamic acid gives the amine, which is protonated under the acidic conditions of the reaction.

    c. The free amine is obtained by neutralization.

an acyl azide      an isocyanate

a carbamic acid      an amine

3. The Curtius rearrangement takes place with retention of stereochemical configuration in the migrating alkyl group.
4. The Curtius reaction can be run under mild, neutral conditions.
   a. The isocyanate can be isolated if desired.
   b. Some acyl azides in the pure state can detonate without warning, and extreme caution is required in handling them.
5. An important use of the Curtius rearrangement is for the preparation of carbamic acid derivatives.
   a. Reaction of isocyanates with alcohols or phenols yields carbamate esters.
   b. Reaction of isocyanates with amines yields ureas.
6. There are two ways to prepare acyl azides:
   a. The most straightforward preparation is the reaction of an acid chloride with sodium azide.

an acyl azide

   b. Another widely used method is to convert an ester into an acyl derivative of hydrazine by aminolysis; the resulting amide, an acyl hydrazide, is then diazotized with nitrous acid to give the acyl azide.

an acyl hydrazide      an acyl azide

## E. AROMATIC SUBSTITUTION REACTIONS OF ANILINE DERIVATIVES

1. Aromatic amines can undergo electrophilic aromatic substitution reactions on the aromatic ring.
   a. The amino group is one of the most powerful *ortho*, *para*-directing groups in electrophilic aromatic substitution.
   b. A protonated amino group does not have the unshared electron pair on nitrogen that gives rise through resonance to the activating, *ortho*, *para*-directing effect of a free amino group.
   c. Ammonium salts are largely *meta*-directing groups.
2. Aniline can be nitrated regioselectively at the *para* position if the amino group is first protected from protonation by conversion into an acetamide derivative by acetylation.

## F. SYNTHESIS OF AMINES; SUMMARY

1. Methods of amine synthesis:
   a. Reduction of amides with LiAlH$_4$ (Sec. 21.9B, text pp. 1008–1010).
   b. Reduction of nitriles with LiAlH$_4$ (Sec. 21.9C, text pp. 1010–1011).
   c. Direct alkylation of amines (Sec. 23.7A, text pp. 1128–1130). This reaction is of limited utility, but it is useful for preparing quaternary ammonium salts.
   d. Reductive amination (Sec. 23.7B, text pp. 1130–1131).
   e. Aromatic substitution reactions of anilines (Sec. 23.9, text pp. 1137–1138).
   f. Gabriel synthesis of primary amines (Sec. 23.11A, text pp. 1146–1147).
   g. Reduction of nitro compounds (Sec. 23.11B, text pp. 1147–1148).

h. Hofmann and Curtius rearrangements (Sec. 23.11C, text pp. 1148–1152).
2. Methods a), c), d), and e) represent methods of preparing amines from other amines.
3. Methods b), f), g), and h) are limited to the preparation of primary amines.
4. Methods a), b), d), g), and h) can be used to obtain amines from other functional groups.

## III. *Syntheses of Organic Compounds with Amines*

### A. ACYLATION OF AMINES

1. Amines can be converted into amides by acylation with acid chlorides, anhydrides or esters.
2. The reaction of an amine with an acid chloride or anhydride requires either two equivalents of the amine or one equivalent of the amine and an additional equivalent of another base such as a tertiary amine or hydroxide ion.

### B. HOFMANN ELIMINATION OF QUATERNARY AMMONIUM HYDROXIDES

1. Alkenes can be formed from amines by a three-step process:
   a. exhaustive methylation
   b. conversion of the ammonium salt to the hydroxide
   c. Hofmann elimination.

2. The Hofmann elimination involves a quaternary ammonium hydroxide as the starting material; an amine acts as the leaving group.
   a. When a quaternary ammonium hydroxide is heated, a $\beta$-elimination reaction takes place to give an alkene, which distills from the reaction mixture.
   b. A quaternary ammonium hydroxide is formed by treating a quaternary ammonium salt with silver hydroxide (AgOH, formed from water and silver oxide, $AgO_2$).
   c. The Hofmann elimination is conceptually analogous to the E2 reaction of alkyl halides, in which a proton and a halide ion are eliminated; in the Hofmann elimination, a proton and a tertiary amine are eliminated.
   d. Hofmann elimination generally occurs as an *anti* elimination.
   e. The conditions of the Hofmann elimination are typically harsh.
3. The elimination reactions of alkyl halides and those of quaternary ammonium salts show distinct differences in regiochemistry.
   a. E2 elimination of most alkyl halides gives a predominance of the alkene with the greatest amount of branching at the double bond.
   b. In contrast, Hofmann elimination of the corresponding trialkylammonium salt generally occurs so that the base abstracts a proton from the $\beta$-carbon with the least branching.

c. This behavior is accounted for by the preference for *anti*-elimination and the minimization of van der Waals repulsions in the transition state of the reaction.
4. Especially acidic $\beta$-hydrogens tend to be eliminated even if they are on a more highly branched carbon.

### C. Diazotization; Reactions of Diazonium Ions

1. Oxidation of amines generally occurs at the amino nitrogen; an important oxidation reaction of amines is called diazotization.

   a. Diazotization is the reaction of a primary amine with nitrous acid ($HNO_2$) to form a diazonium salt.

   b. A diazonium salt is a compound of the form $R-\overset{+}{N}\equiv N:\ X^-$, in which $X^-$ is a typical anion.

   c. Both aliphatic and aromatic primary amines are readily diazotized.

2. Diazonium ions incorporate one of the best leaving groups, molecular nitrogen.

3. Aliphatic diazonium ions react immediately as they are formed by $S_N1$, $E1$, and/or $S_N2$ mechanisms to give substitution and elimination products along with nitrogen gas. (The rapid liberation of nitrogen gas on treatment with nitrous acid is a qualitative test for primary alkylamines.)

4. Aromatic diazonium salts may be isolated and used in a variety of reactions; however, they are usually prepared in solution at 0–5 °C and used without isolation, because they lose nitrogen on heating and they are explosive in the dry state.

### D. Aromatic Substitution with Diazonium Ions

1. Aryldiazonium ions react with aromatic compounds containing strongly activating substituent groups, such as amines and phenols, to give substituted azobenzenes. This is an electrophilic aromatic substitution reaction in which the terminal nitrogen of the diazonium ion is the electrophile.

an azobenzene

2. The mechanism follows the usual pattern of electrophilic aromatic substitution.

   a. First, the electrophile is attacked by the $\pi$ electrons of the aromatic compound to give a resonance-stabilized carbocation.

   b. This carbocation then loses a proton to give the substitution product.

3. The azobenzene derivatives formed in these reactions have extensive conjugated $\pi$-electron systems, and most of them are colored.

4. Some of these compounds are used as dyes and indicators; as a class they are known as azo dyes. (An azo dye is a colored derivative of azobenzene.)

### E. Substitution Reactions of Aryldiazonium Salts

1. Among the most important reactions of aryldiazonium salts are substitution reactions with cuprous halides; in these reactions the diazonium group is replaced by a halogen.

   a. An analogous reaction occurs with CuCN.

      i. This reaction is another way of forming a carbon-carbon bond, in this case to an aromatic ring.

      ii. The resulting nitrile can be converted by hydrolysis into a carboxylic acid.

   b. The reaction of an aryldiazonium ion with a cuprous salt is called the Sandmeyer reaction.

   c. This reaction is an important method for the synthesis of aryl halides and nitriles.

2. Aryl iodides can also be made by the reaction of diazonium salts with potassium iodide; a cuprous salt is not required.

3. Diazonium salts can be hydrolyzed to phenols by heating them with water or by treating them with cuprous oxide ($Cu_2O$) and an excess of aqueous cupric nitrate [$Cu(NO_3)_2$] at room temperature.

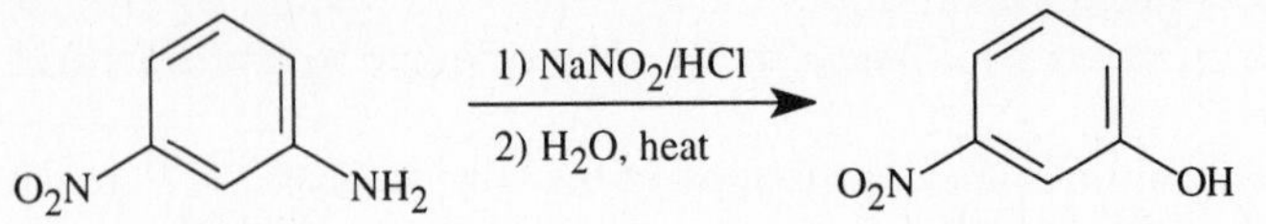

4. The diazonium group is replaced by hydrogen when the diazonium salt is treated with hypophosphorous acid, $H_3PO_2$.

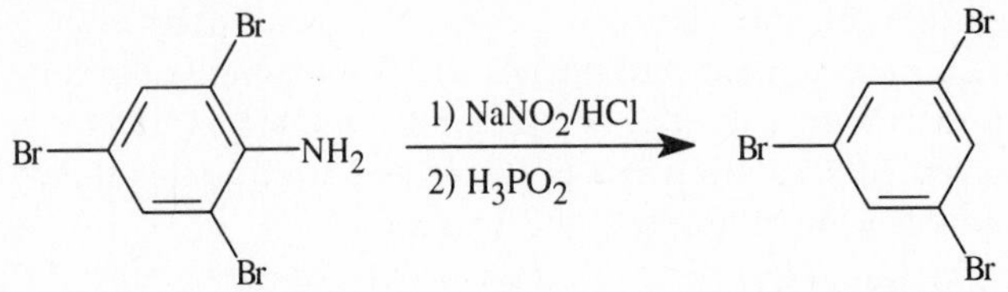

5. All of the diazonium salt reactions listed above are substitution reactions but none are $S_N2$ or $S_N1$ reactions because aromatic rings do not undergo substitution by these mechanisms.
   a. The Sandmeyer and related reactions occur by radical-like mechanisms mediated by the copper.
   b. The reaction of diazonium salts with KI probably occurs by a similar mechanism.
   c. The reaction of diazonium salts with $H_3PO_2$ has been shown definitely to be a free-radical chain reaction.
6. The substitution reactions of diazonium salts achieve ring-substitution patterns that cannot be obtained in other ways.

## F.  REACTIONS OF SECONDARY AND TERTIARY AMINES WITH NITROUS ACID

1. Secondary amines react with nitrous acid to yield *N*-nitrosoamines, usually called simply nitrosamines. Nitrosamines are known to be potent carcinogens.

$$R_2NH \ + \ HNO_2 \longrightarrow R_2N-N{=}O \qquad \text{a nitrosamine}$$

2. The nitrogen of tertiary amines does not react under the strongly acidic conditions used in diazotization reactions.
3. *N,N*-Disubstituted aromatic amines undergo electrophilic aromatic substitution on the benzene ring; the electrophile is the nitrosyl cation, which is generated from nitrous acid under acidic conditions.

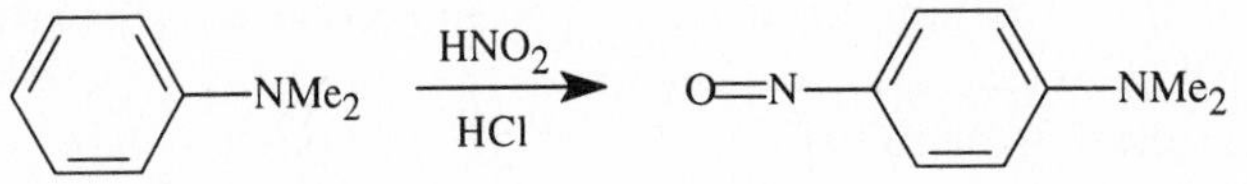

## Study Guide Links

### 23.1 Structures of Amide Bases

Although the structures of lithium amide bases are written as ionic compounds for conceptual simplicity, recent research has shown that the structures of these species are considerably more complex. For example, lithium diisopropylamide in tetrahydrofuran (THF) has a symmetrical dimer structure (that is, it is a symmetrical aggregate of two units):

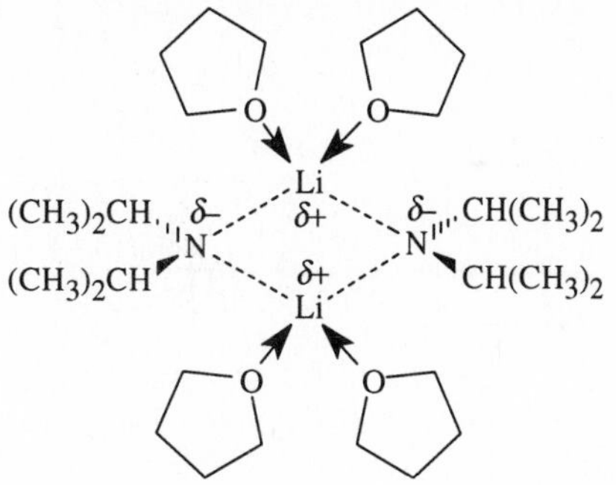

Each lithium is partially bonded to two nitrogens with "half bonds" (dashed lines in the structure above). The nitrogen-lithium bonds are covalent, although they have significant ionic character. Each lithium is strongly solvated by two solvent molecules (THF in this case). The degree to which these amide bases are aggregated into larger structures varies with their concentration and with the nature of the solvent.

### ✓23.2 Nitration of Aniline

Remember that rates of reactions are affected by both intrinsic reactivity (reflected in the magnitude of the rate constant) and concentrations of the species involved (Sec. 9.3B, text p. 391). Under the acidic conditions of nitration, the highly reactive free aniline is present in very small concentration; the much less reactive protonated aniline is present in much greater concentration.

$$\text{rate of formation of } \textit{para}\text{-nitroaniline} = k_{para}[\text{PhNH}_2]$$

very large    very small

$$\text{rate of formation of } \textit{meta}\text{-nitroaniline} = k_{meta}[\text{Ph}\overset{+}{\text{N}}\text{H}_3]$$

very small    very large

These two competing rates must be about the same, because comparable amounts of *meta*- and *para*-substituted products are formed in text Eq. 23.41.

### 23.3 Mechanism of Diazotization

Diazotization at first sight may seem to be a very strange reaction, but once you understand the reactivity of nitrous acid, diazotization should make more sense.

Nitrous acid in solution is in equilibrium with its anhydride, dinitrogen trioxide.

$$2\,HNO_2 \;\rightleftharpoons\; O\!\!=\!\!N\!-\!O\!-\!N\!\!=\!\!O \;+\; H_2O$$

dinitrogen trioxide

Dinitrogen trioxide is the actual diazotizing agent under conditions of dilute acid; under conditions of very strong acid, the diazotizing agent is the nitrosyl cation, $^+N\!\!=\!\!\ddot{O}\!:$, formed by dehydration of nitrous acid. The dinitrogen trioxide mechanism is explored here.

The N=O double bonds of dinitrogen trioxide have a reactivity somewhat analogous to that of a carbonyl group. They react at nitrogen with nucleophiles such as amines with loss of nitrite.

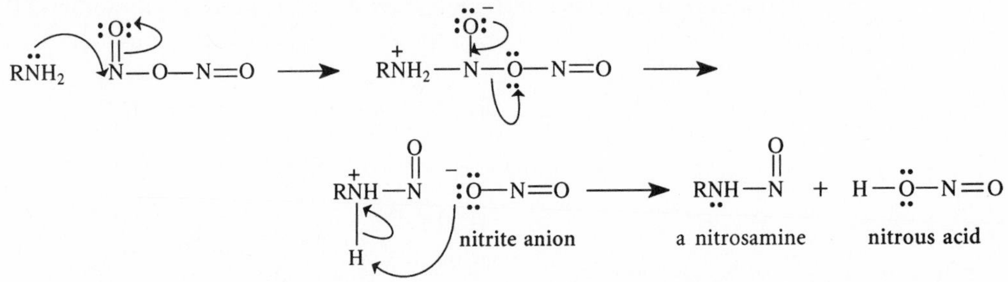

(The p$K_a$ of nitrous acid is 3.2; hence, nitrite is a weak base, and a good leaving group.) The product of this sequence is a *nitrosamine.* Because the amine nitrogen loses only one hydrogen in forming a nitrosamine, secondary amines as well as primary amines form nitrosamines when they react with nitrous acid (Sec. 23.10C, text p. 1144).

In the case of primary amines, nitrosamines react further under the acidic conditions. First, they undergo tautomerization to give another intermediate called a diazoic acid. (You should draw the arrow formalism for this step, which is analogous to enol formation from a ketone.)

$$RNH\!-\!N\!\!=\!\!O \;\xrightleftharpoons{H_2O,\,H_3O^+}\; R\!-\!N\!\!=\!\!N\!-\!OH$$

a nitrosamine          a diazoic acid

Under the acidic conditions of diazotization, diazoic acids dehydrate to give diazonium ions:

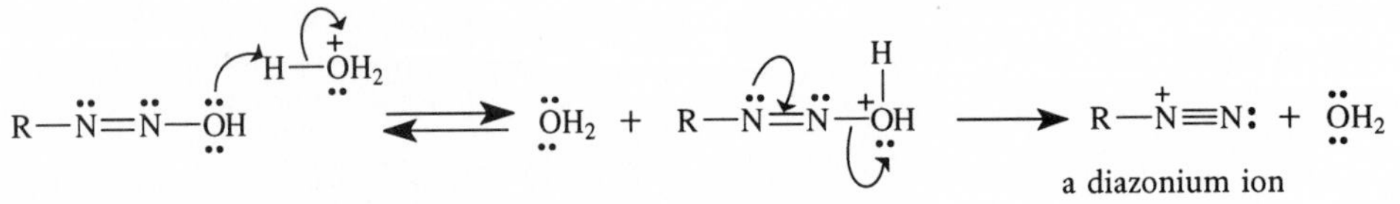

## ✓23.4    Mechanism of the Curtius Rearrangement

Although the Curtius rearrangement is concerted, as shown in the text, it is helpful to *think* of the reaction in a stepwise fashion to see why the rearrangement occurs. There is a strong driving force for loss of dinitrogen, a very stable molecule:

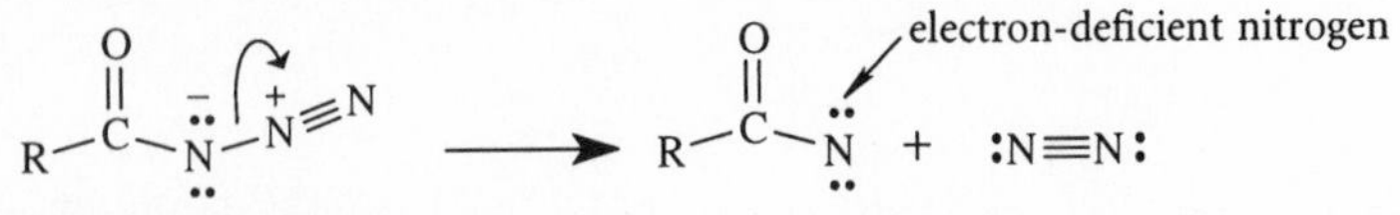

Loss of dinitrogen leaves behind an *electron-deficient* nitrogen; it is electron-deficient because it is short one electron pair of an octet.

Just as rearrangements of alkyl groups occur to the electron-deficient *carbons* of carbocations, so they also occur to electron-deficient nitrogens.

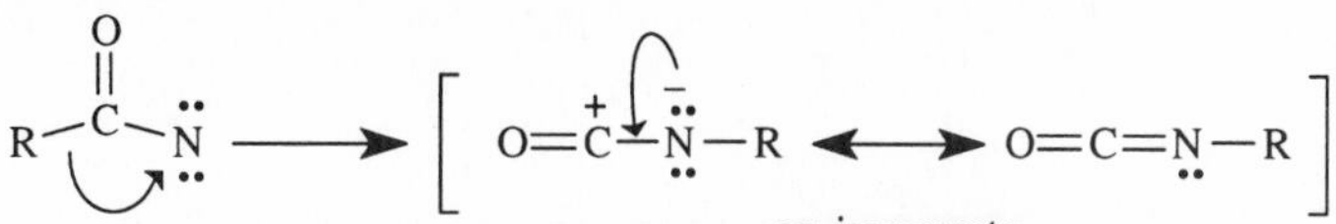

an isocyanate

In other words, the Curtius rearrangement is driven by the very strong tendency of nitrogen to have a complete octet.

## ✓23.5  Formation and Decarboxylation of Carbamic Acids

The formation of carbamic acids by hydration of isocyanates is an acid-catalyzed addition to the C=N double bond:

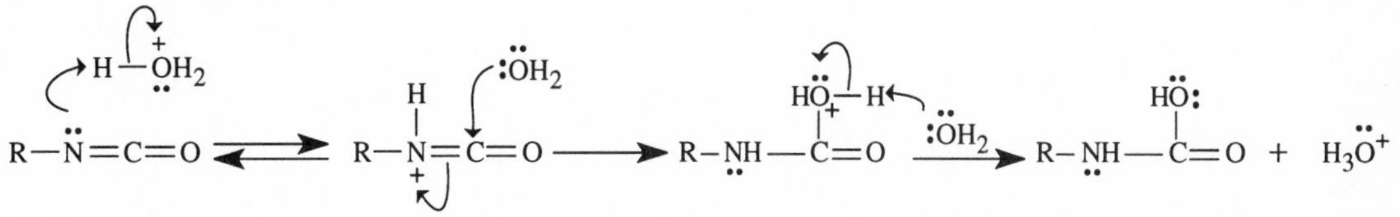

Hydration can also occur in base by attack of $^-OH$ on the carbonyl carbon. (Write the mechanism of this reaction.) The hydration in basic solution gives the carbamate ion, the conjugate base of a carbamic acid. The hydration of isocyanates under basic conditions is important in the Hofmann rearrangement, which is also discussed in Sec. 23.11C of the text.

Decarboxylation is a concerted reaction in which the same water molecule acts as both an acid and base catalyst at the same time.

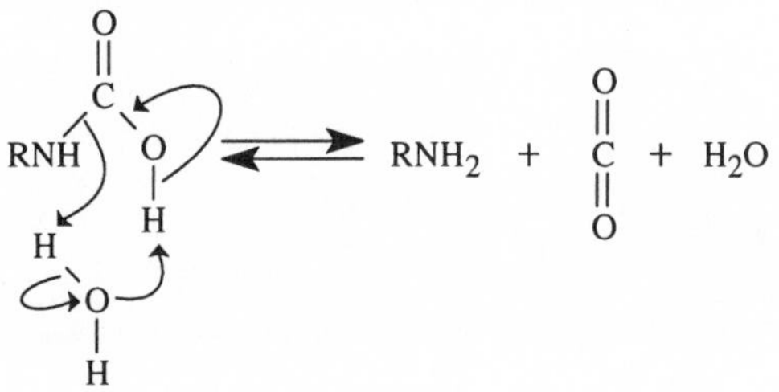

The decarboxylation reaction is reversible; that is, amines react readily with $CO_2$ to give carbamic acids. However, the decarboxylation reaction in acidic solution is driven to completion by two factors: (1) protonation of the product amine, and (2) loss of $CO_2$ from solution as a gas.

Decarboxylation is fastest in acidic solution, but also occurs in base:

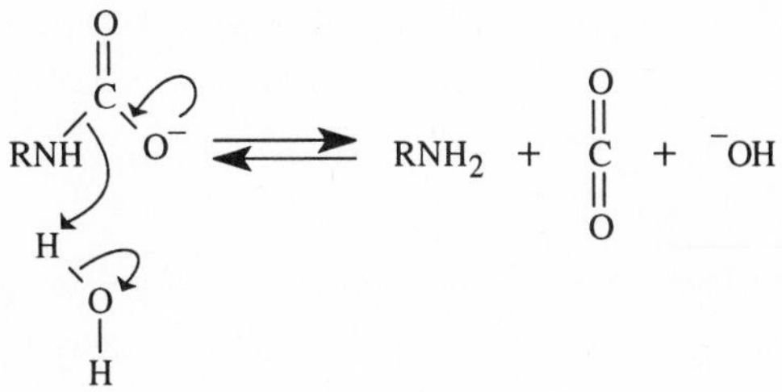

This reaction is driven to completion by reaction of the hydroxide ion with $CO_2$ to form bicarbonate ion:

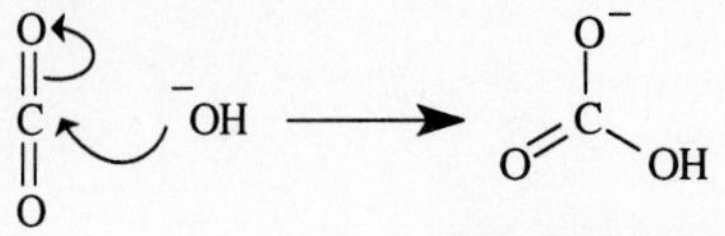

bicarbonate ion

If the solution is strongly basic, bicarbonate ion itself ionizes further to carbonate ion.

Decarboxylation of carbamates in basic solution is important in the Hofmann rearrangement.

# Solutions

## Solutions to In-Text Problems

**23.1** (a)

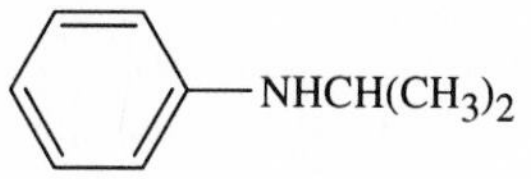

*N*-isopropylaniline

(c)

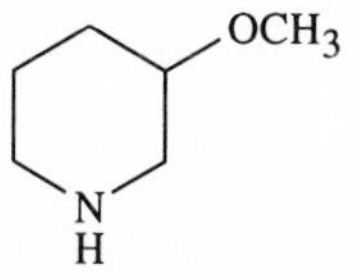

3-methoxypiperidine

(e)

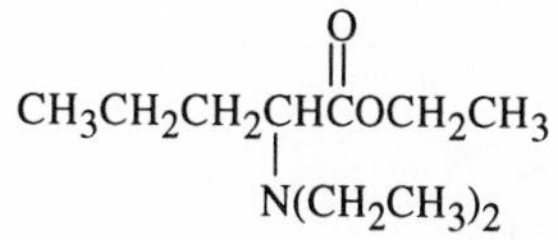

ethyl 2-(diethylamino)pentanoate

**23.2** (a) *N*-Ethyl-*N*-methylisobutylamine (common), or *N*-ethyl-*N,2*-dimethyl-1-propanamine
(c) Dicyclohexylamine (common), or *N*-cyclohexylcyclohexanamine
(e) 1-(2-Chloroethyl)-3-propylpyrrolidine

**23.3** (a) The principle to apply is that the more double-bond character the carbon-nitrogen bond has, the shorter it is. The order of decreasing double-bond character, and hence increasing carbon-nitrogen bond length, is

*p*-nitroaniline < aniline < cyclohexylamine

Because of resonance interaction of their nitrogen lone pair with the benzene ring, aniline and *p*-nitroaniline have the most double-bond character; cyclohexylamine lacks this interaction and therefore has the least carbon-nitrogen double-bond character. In *p*-nitroaniline, the unshared electron pair on the amine nitrogen can be delocalized into the *p*-nitro group:

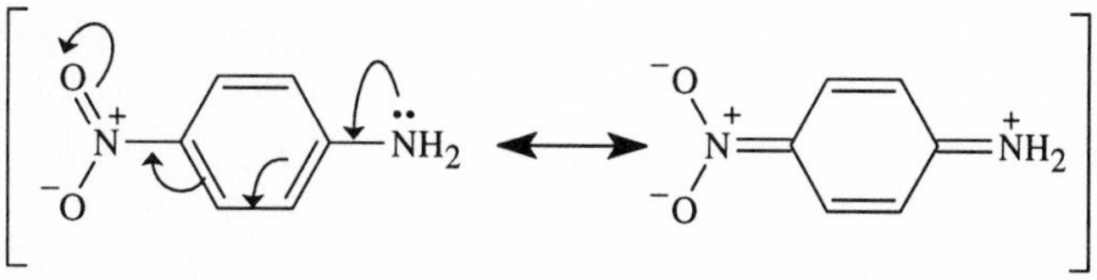

This interaction is in addition to the resonance interaction of the unshared pair with the ring of the type shown in Eq. 23.3 of the text. This additional delocalization increases the carbon-nitrogen double-bond character and decreases the carbon-nitrogen bond length.

**23.4** The five-proton resonance at $\delta$ 7.18 in the NMR spectrum indicates the presence of a monosubstituted benzene ring (which accounts for 77 mass units), and the mutually split triplet-quartet pattern at $\delta$ 1.07 and $\delta$ 2.60 indicates an ethyl group (which accounts for 29 mass units). The IR spectrum indicates the presence of an amine; if it is a secondary amine, as suggested by the integral of the $\delta$ 0.91 resonance in the NMR spectrum, the NH accounts for 15 mass units, leaving 14 mass units unaccounted for. A $CH_2$ group could account for this remaining mass, and the two-proton singlet at $\delta$ 3.70 could correspond to such a resonance. On the assumption that the N—H proton is undergoing rapid exchange and does not show splitting with neighboring protons, *N*-ethylbenzylamine fits the data.

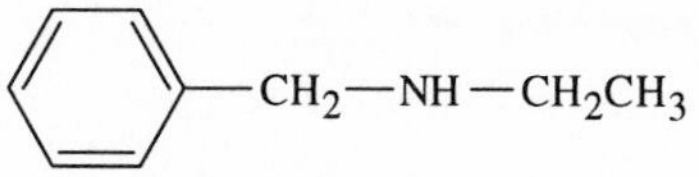

$CH_2$—NH—$CH_2CH_3$ *N*-ethylbenzylamine

The predicted chemical shift (Eq. 13.3 on text p. 588) for the benzylic $CH_2$ group is $\delta$ 3.6. The $m/z = 120$ peak in the mass spectrum corresponds to loss of a methyl radical by $\alpha$-cleavage; and the $m/z = 91$ peak corresponds to a benzyl cation, which is formed by inductive cleavage. (See text p. 564 for a discussion of these cleavage mechanisms.)

**23.6**   The peaks arise from $\alpha$-cleavage of the molecular ions derived from the respective amines. 2-Methyl-2-heptanamine can form a cation with $m/z = 58$ by this mechanism, and is therefore compound *B*. *N*-Ethyl-4-methyl-2-pentanamine can form a cation of $m/z = 72$ by this mechanism, and is therefore compound *A*. The $\alpha$-cleavage mechanism is shown below for *B*; you should show it for compound *A*.

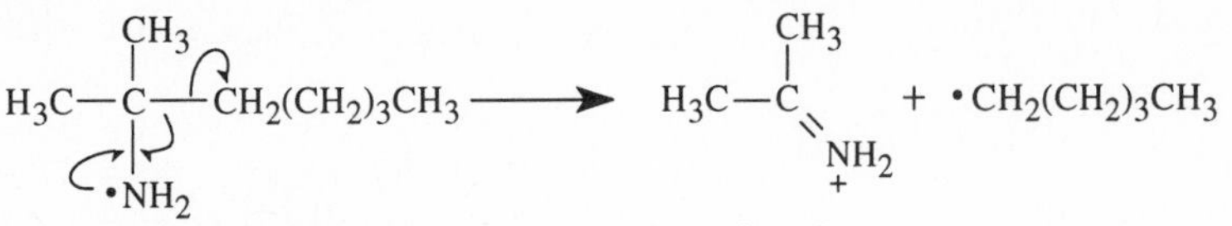

molecular ion from
2-methyl-2-heptanamine
(compound *B*)

$m/z = 58$

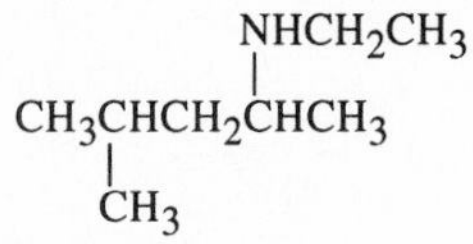

*N*-ethyl-4-methyl-2-pentanamine
(compound *A*)

**23.7**   (a)  2,2-Dimethyl-1-propanamine (neopentylamine, $(CH_3)_3CCH_2NH_2$) has a maximum of three resonances in its CMR spectrum. 2-Methyl-2-butanamine, $CH_3CH_2C(CH_3)_2NH_2$, should have four resonances in its CMR spectrum.

**23.8**   (a)  The order of increasing basicity is $NH_3$(ammonia) < $PrNH_2$ (propylamine) < $Pr_2NH_2$ (dipropyl-amine). As the text discussion on p. 1123 shows, within a series of amines of increasing alkyl substitution basicity increases from ammonia to the secondary amine.

(b)  The order of increasing basicity is,

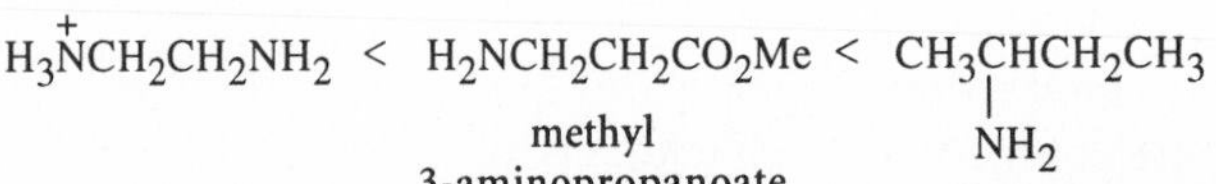

methyl
3-aminopropanoate

*sec*-butylamine

The electron-withdrawing polar effect of the partial positive charge on the carbonyl carbon of the ester reduces the basicity of its $\beta$-amino group, and the polar effect of the full positive charge in the ammonium ion reduces the basicity of its $\beta$-amino group even more. Note that the polar effect of a substituent containing a fully charged atom is generally greater than the polar effect of a substituent containing the same atom without a full charge.

(c)  The order of increasing basicity is

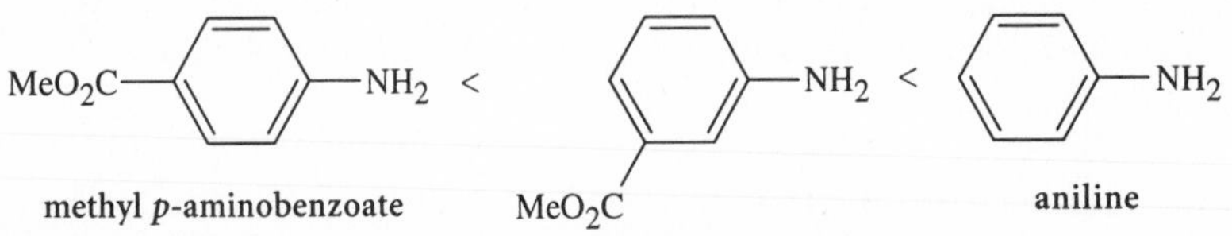

methyl *p*-aminobenzoate

methyl *m*-aminobenzoate

aniline

The methoxycarbonyl group (that is, the methyl ester group) decreases basicity in the first two derivatives by a polar effect. Furthermore, methyl *p*-aminobenzoate is stabilized by resonance interaction of the nitrogen unshared pair with the carbonyl group:

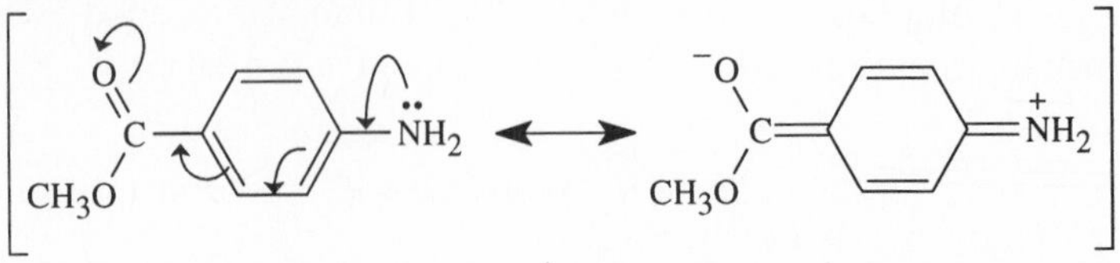

Because there is no unshared pair on nitrogen in the conjugate-acid ammonium ion, such resonance stabilization of the conjugate acid is not possible. As Fig. 23.2 on text p. 1122 shows, anything that stabilizes an amine relative to its conjugate-acid ammonium ion lowers its basicity. Note that resonance interaction between the carbonyl group and a *meta*-amino group is not possible.

(d) The order of increasing basicity is

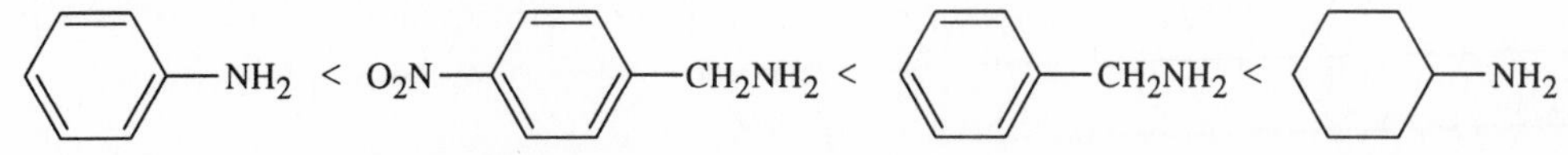


aniline — *p*-nitrobenzylamine — benzylamine — cyclohexylamine

The resonance and electron-withdrawing polar effect of the benzene ring severely reduces the nitrogen basicity of aniline relative to that of an alkylamine. The electron-withdrawing polar effect of the nitro group of *p*-nitrobenzylamine reduces its basicity relative to that of benzylamine. And the electron-withdrawing polar effect of the benzene rings reduces the basicity of both benzylamines relative to cyclohexylamine, an ordinary alkylamine.

**23.10** Dissolve the mixture in a low-boiling, water-insoluble organic solvent such as methylene chloride. Extract this solution with 5% sodium bicarbonate or 5% sodium hydroxide solution; *p*-chlorobenzoic acid will dissolve in the aqueous layer as its sodium salt. It can be recovered as the free acid by acidifying the aqueous extracts. Then extract the methylene chloride solution with 5% HCl solution. *p*-Chloroaniline will dissolve in the aqueous layer as its hydrochloride salt. It can be recovered as the free amine by neutralizing the aqueous layer with dilute NaOH. *p*-Chlorotoluene remains in the methylene chloride layer; it can be isolated by drying the solution and evaporating the methylene chloride. (Note that the order of the extractions could be interchanged.)

**23.12** (a) $(CH_3CH_2)_4N^+ F^-$ tetraethylammonium fluoride

**23.13** Compound *A* has an asymmetric nitrogen stereocenter; therefore, it is chiral and can exist as two enantiomers. Compound *B* also contains an asymmetric nitrogen; however, compound *B* undergoes an acid-base equilibrium by which it is in equilibrium with a very small amount of the free amine:

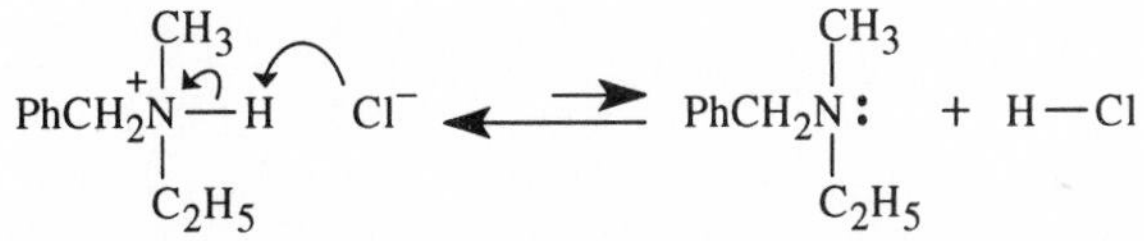


This equilibrium takes place so rapidly, and the minuscule amount of free amine undergoes nitrogen inversion so quickly, that the ammonium salt is rapidly racemized. Because compound *A* does not have a hydrogen that can be involved in an equilibrium of this type, it is not racemized.

**23.14** Either the ethyl group can originate from acetaldehyde, or the cyclohexyl group can originate from cyclohexanone.

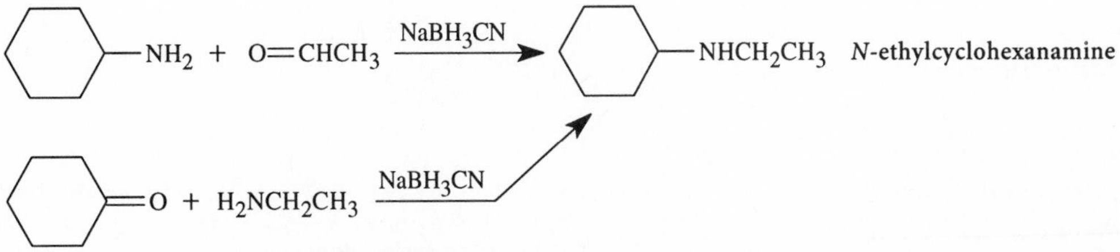


**23.16** (a) Prepare the tertiary amine by reductive amination and alkylate it to give the ammonium salt.

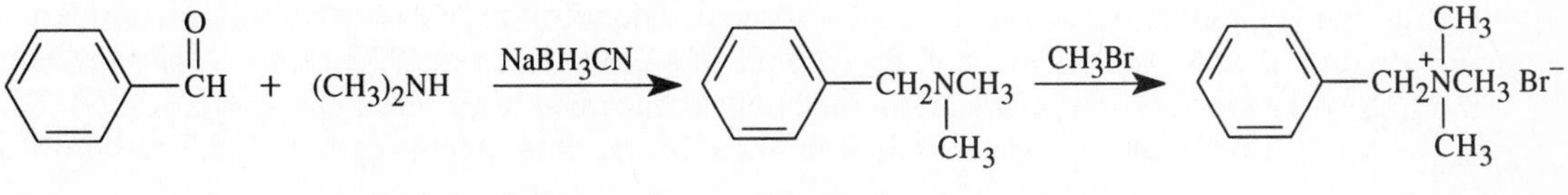

An alternative synthesis would be to reductively aminate the same amine with formaldehyde to give trimethylamine and then alkylate this tertiary amine with benzyl bromide ($PhCH_2Br$).

23.17   Caleb has forgotten, if he ever knew, that aryl halides such as bromobenzene do not undergo $S_N2$ reactions (Sec. 18.1 on text p. 824).

23.19   (a)

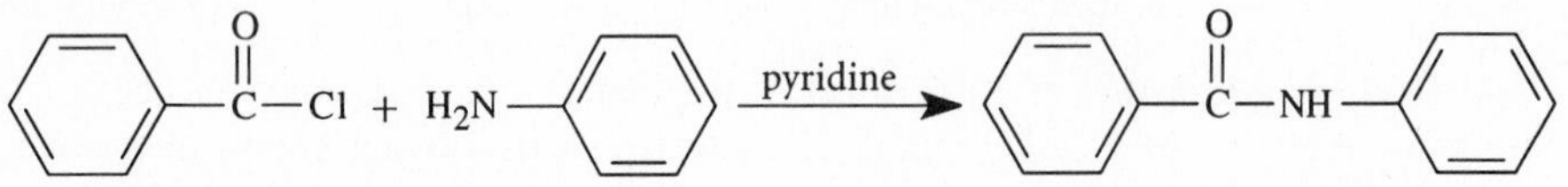

*N*-phenylbenzamide

23.20   (a)  Draw the molecule in a conformation in which the trimethylammonium group is *anti* to the β-hydrogen, and examine the relative positions of the phenyl groups. In this part, the alkene product must have the *Z* configuration.

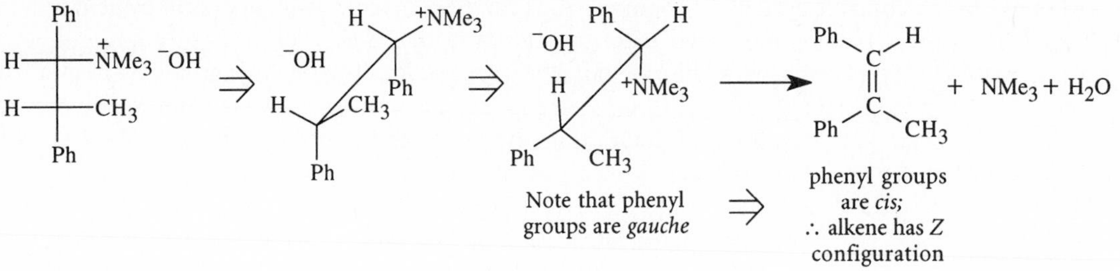

23.21   (a)  The hydroxide ion removes the proton from the β-carbon with the least amount of branching, that is, from the methyl carbon.

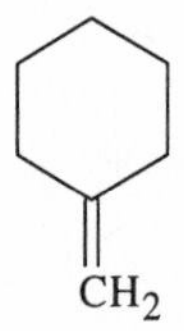

(b)  The hydrogen α to the carbonyl group is removed because it is considerably more acidic than the other β-hydrogens.

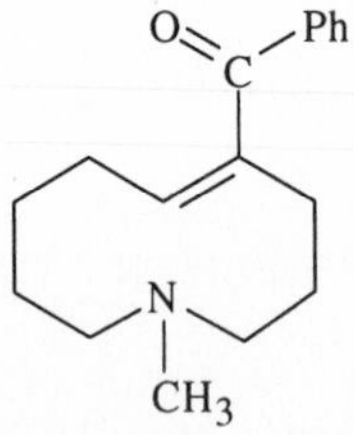

23.23   Compound *A* is evidently one of the alkenes produced by application of an exhaustive methylation-Hofmann elimination sequence to coniine. The alkene double bond accounts for the only degree of unsaturation of *A*; hence, the nitrogen is part of an amine group. These deductions are reasonable, because it is given that coniine is a piperidine, a cyclic amine (structure on text p. 1115); when a cyclic amine is taken through exhaustive methylation and Hofmann elimination, if a carbon-nitrogen bond is broken, the ring is thus opened and an unsaturated amine is formed. The second cycle of methylation/elimination forms the mixture of octadienes. There are five carbons in a piperidine ring. The carbons bearing the nitrogen in the piperidine are the ones to which double bonds are formed in the elimination steps. From the structures of the octadienes, it follows that the nitrogen of coniine is connected to carbon-1 and carbon-5. This is the only connection that could give *both* alkenes from the same starting material.

N was connected here                                    N was connected here

H₂C=CHCH₂CH=CHCH₂CH₂CH₃               H₂C=CHCH₂CH₂CH=CHCH₂CH₃

1,4-octadiene                                              1,5-octadiene

The only structure for coniine consistent with this analysis is that of 2-propylpiperidine:

2-propylpiperidine
(coniine)

CH₂CH₂CH₃

  In a number of problems, the text uses Ag₂O (silver(I) oxide) for the conversion of quaternary ammonium halides into quaternary ammonium hydroxides. Attempts to balance these reactions will show that a hydrogen is unaccounted for. The resolution of this problem is that silver(I) oxide is typically used in a *hydrated* form; hydrated Ag₂O is equivalent to AgOH. Consequently, you should assume in all problems involving Hofmann elimination that Ag₂O is operationally the same thing as AgOH.

**23.24**  Use an acetyl protecting group as outlined in Study Problem 23.4.

H₂N—⟨⟩  →(CH₃CCl, pyridine)→  CH₃CNH—⟨⟩  →(ClSO₃H)→  CH₃CNH—⟨⟩—S—Cl  →(NH₃ (excess))→

*p*-acetamidobenzenesulfonyl
chloride

CH₃CNH—⟨⟩—S—NH₂  →(1) H₃O⁺, heat  2) dil. ⁻OH)→  H₂N—⟨⟩—S—NH₂

sulfanilamide

Note that if the acetyl protecting group were not used, the amine group on one molecule would react with a chlorosulfonyl group on the other and a polymer would be formed.

**23.25**  (a)  Begin with *p*-nitroacetanilide, prepared as shown in Study Problem 22.4. Then convert it into 2,4-dinitroaniline. The acetyl protecting group is used to avoid protonation of the nitrogen, which would result in a significant amount of unwanted *meta* substitution. (See Eq. 23.41 on text p. 1137.)

NHCCH₃ ... →(HNO₃, H₂SO₄)→ ... NHCCH₃, NO₂ ... →(1) H₃O⁺, heat  2) dil. ⁻OH)→ ... NH₂, NO₂

NO₂                                       NO₂                                       NO₂

*p*-nitroacetanilide                                                              2,4-dinitroaniline

**23.26** (a)

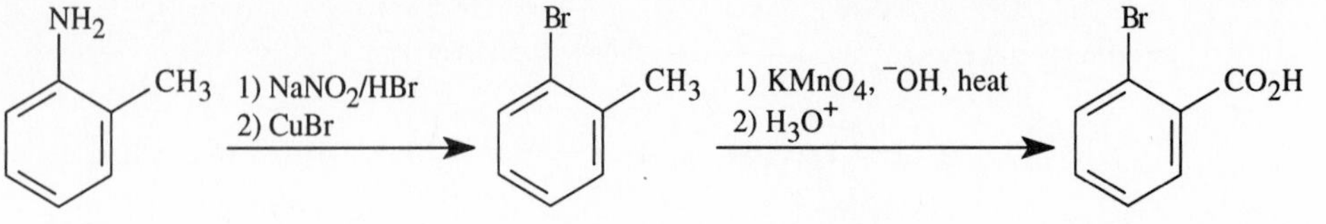

o-toluidine → 1) NaNO$_2$/HBr  2) CuBr → 1) KMnO$_4$, $^-$OH, heat  2) H$_3$O$^+$ → 2-bromobenzoic acid

**23.27** (a) Because diazotization does not break the carbon-nitrogen bond, the diazonium ion intermediate, like the starting amine, has the *R* configuration. Hence, inversion of stereochemical configuration occurs in the reaction with water.

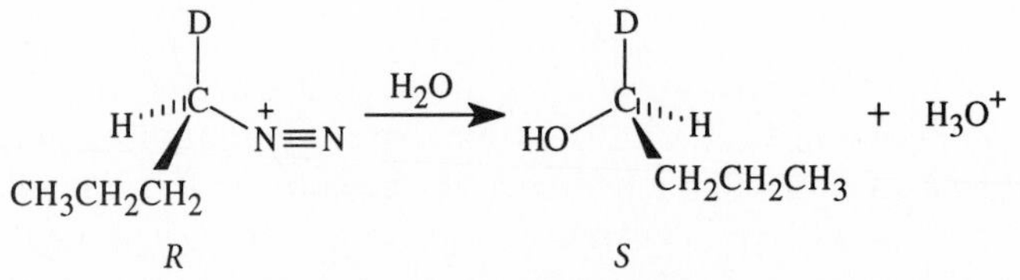

$$R \xrightarrow{H_2O} S + H_3O^+$$

(b) Because inversion of stereochemical configuration is observed, an S$_N$2 reaction in which water is the nucleophile and N$_2$ is the leaving group is likely to be involved. An S$_N$1 mechanism involving a carbocation intermediate would be expected to result in some racemization. (Why?)

**23.28** *p*-Aminobenzenesulfonic acid is prepared and diazotized to give the diazonium ion *A*. This synthesis begins with *p*-acetamidobenzenesulfonyl chloride, prepared as shown in the solution to Problem 23.24 on p. 635 of this manual. Formaldehyde is reductively aminated with aniline to give *N,N*-dimethylaniline, *B*. Compounds *A* and *B* are coupled to give methyl orange.

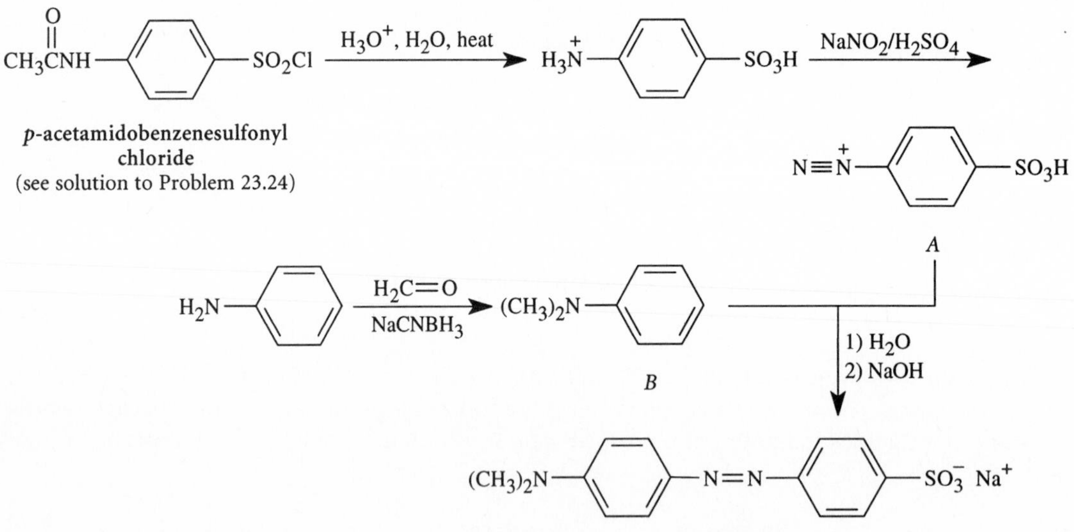

*p*-acetamidobenzenesulfonyl chloride
(see solution to Problem 23.24) → H$_3$O$^+$, H$_2$O, heat → NaNO$_2$/H$_2$SO$_4$ →

*A*

H$_2$N— → H$_2$C=O / NaCNBH$_3$ → (CH$_3$)$_2$N—

*B*

1) H$_2$O
2) NaOH

methyl orange

Early printings of the text have an error in the structure of FD & C #6 on p. 1144 caused by a typesetter error. The correct structure is as follows:

*(correction continues)*

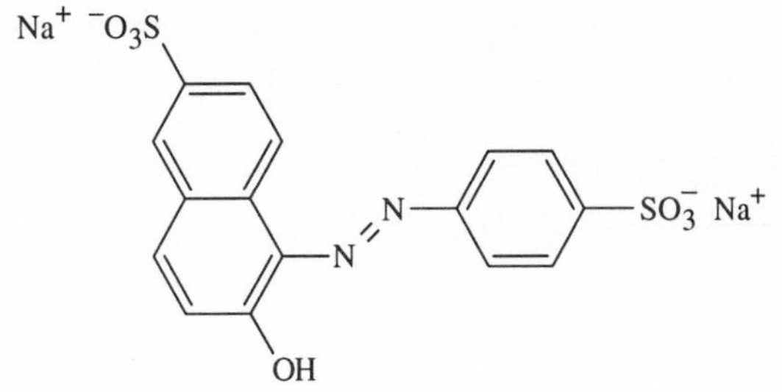

FD & C #6

**23.30** (a) Protonation of the —OH oxygen and loss of water give the nitrosyl cation:

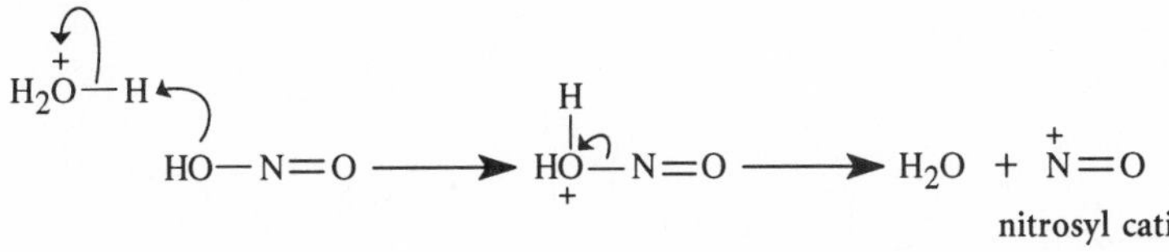

nitrosyl cation

**23.31** (a)

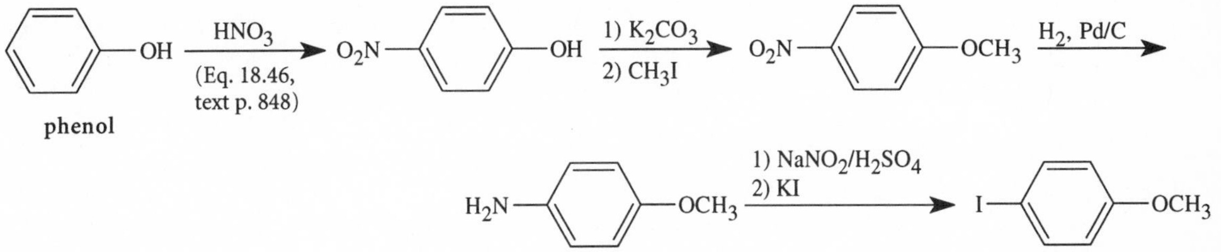

phenol

*p*-iodoanisole

Note that ether formation is carried out before reduction of the nitro group because otherwise the amino group could also be alkylated with methyl iodide.

**23.32** (a) *Tert*-butylamine cannot be prepared by the Gabriel synthesis because it would require that the nitrogen of phthalimide be alkylated with *tert*-butyl bromide in an $S_N2$ reaction. Tertiary alkyl halides generally do not undergo $S_N2$ reactions but, under basic conditions, undergo E2 reactions instead.

**23.33** (a)

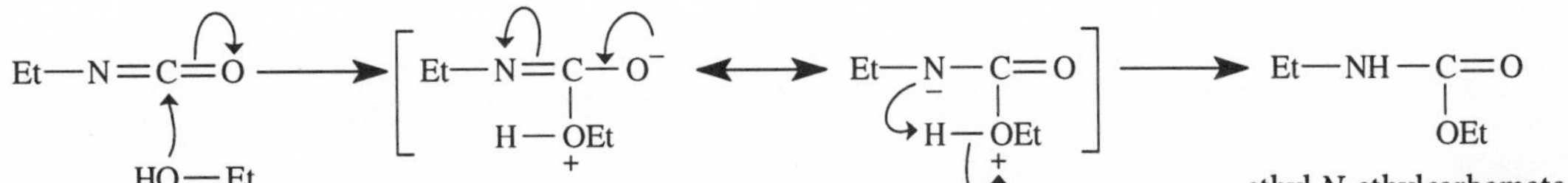

ethyl *N*-ethylcarbamate

Note that the proton transfer, which is shown as an intramolecular process for brevity, could also be mediated by the solvent.

**23.34** (a) In ethanol solvent, ethanol (or its conjugate base ethoxide) is the most abundant nucleophile present, and thus an ethyl carbamate is obtained as the product.

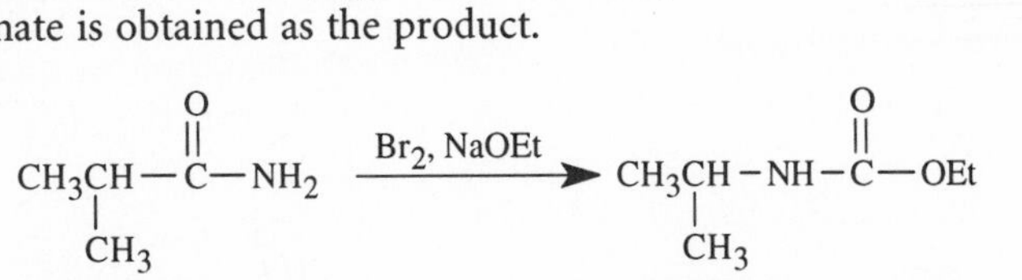

2-methylpropanamide

ethyl *N*-isopropylcarbamate

**23.35** Pentyl isocyanate is an intermediate in the reaction. Because it is largely insoluble in water, it forms a separate phase. A small amount of pentyl isocyanate reacts with the aqueous NaOH to form pentylamine, which is then extracted into the isocyanate, with which it reacts to give the urea.

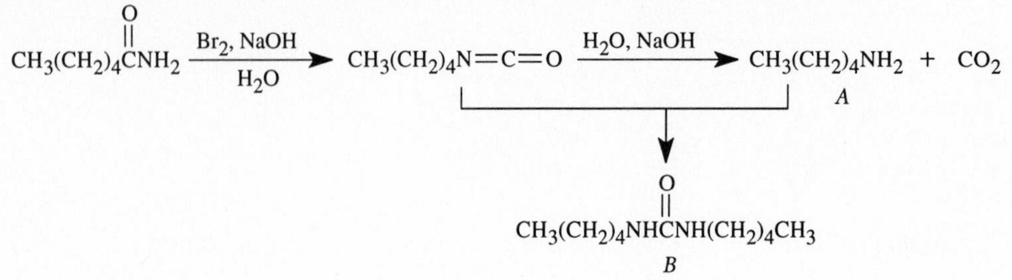

**23.36** (a)

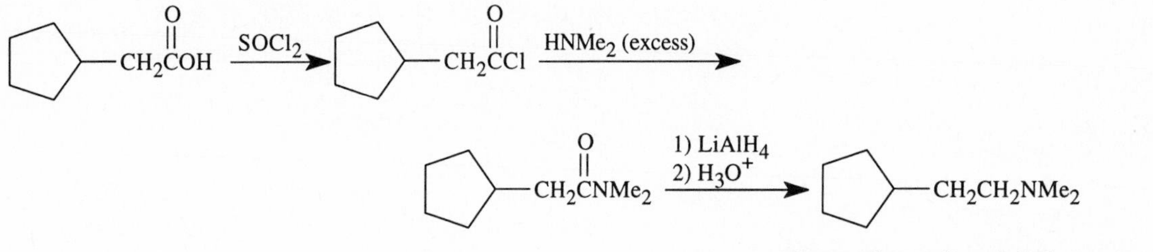

2-cyclopentyl-*N*,*N*-dimethylethanamine

(c)

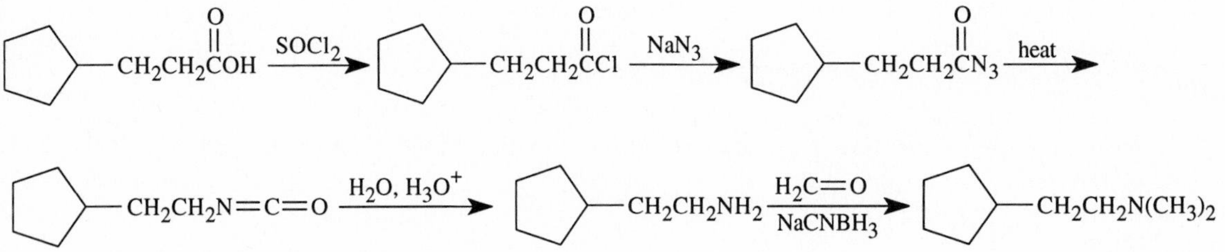

**23.37** (a) In morphine, the nitrogen of the tertiary amine is the most basic atom. Therefore the conjugate acid of morphine has the following structure:

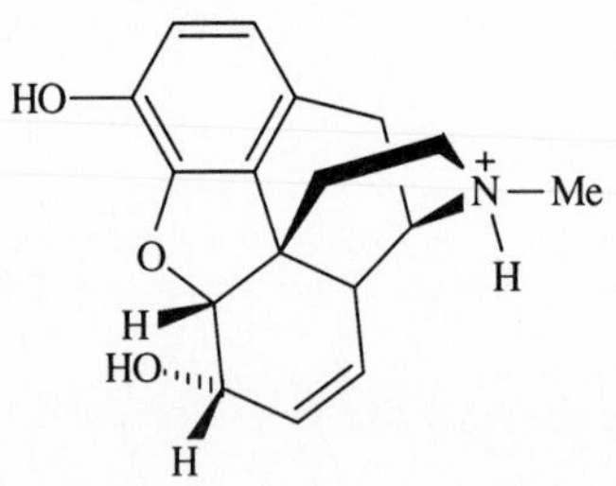

conjugate acid of morphine

## Solutions to Additional Problems

**23.38**  (a) 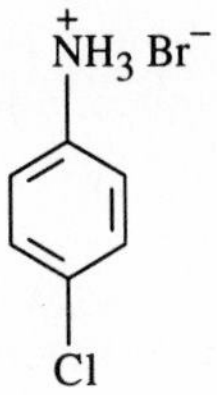   (b) 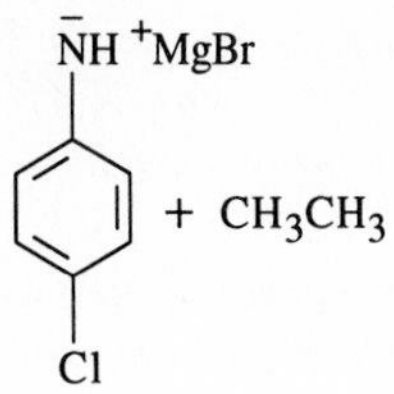   (c) 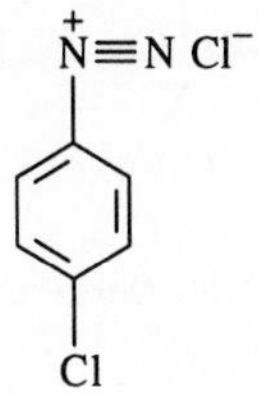   (d) 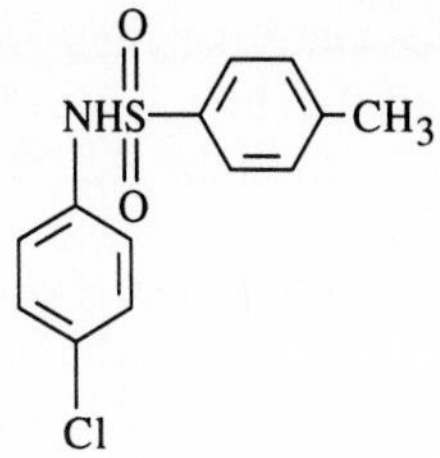   (e)

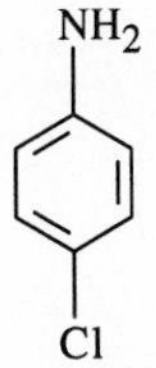

NH$_2$ / Cl — *p*-chloroaniline

$\overset{+}{N}H_3$ Br$^-$ / Cl

$\overset{-}{N}H$ $^+$MgBr / Cl  + CH$_3$CH$_3$

$\overset{+}{N}{\equiv}N$ Cl$^-$ / Cl

NHS(=O)$_2$—CH$_3$ / Cl

OH / Cl

(f) 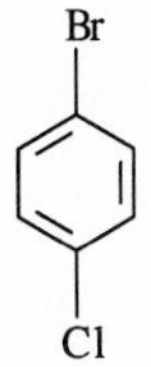   (g) 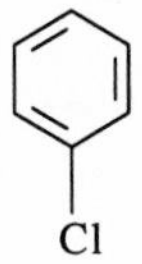   (h)    (i)

Br / Cl

Cl (phenyl)

CN / Cl

Na$^+$ $\overset{-}{N}$–S(=O)$_2$—CH$_3$ / Cl

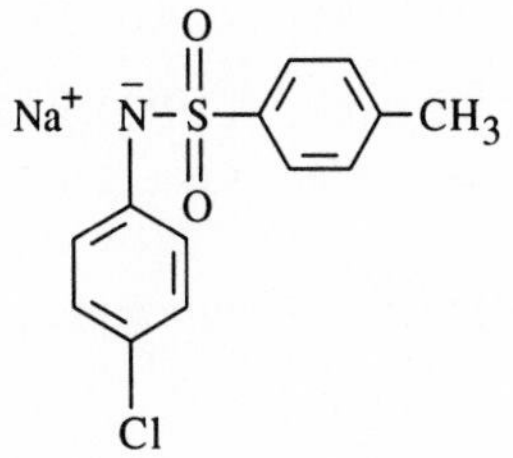

In part (i), sulfonamides are more acidic than carboxamides for the same reasons that sulfonic acids are more acidic than carboxylic acids. In addition, the conjugate-base anion shown in (i) is stabilized by resonance interaction with the phenyl ring.

**23.40**  (a)   (b) no reaction   (c)   (d)

(CH$_3$)$_2$CH$\overset{+}{N}$H$_2$ $^-$OSO$_3$H

(CH$_3$)$_2$CHN$^-$H Li$^+$  + CH$_3$CH$_2$CH$_2$CH$_3$

(CH$_3$)$_2$CHNH$\overset{O}{\overset{\|}{C}}$CH$_3$

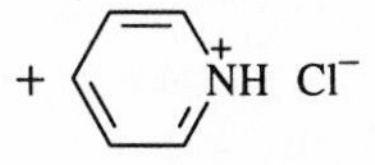

+ pyridinium NH Cl$^-$

(e)   (f)   (g)   (h)

(CH$_3$)$_2$CHBr
+ (CH$_3$)$_2$CHOH
+ H$_2$C=CHCH$_3$

(CH$_3$)$_2$CHNHCH(CH$_3$)$_2$

(CH$_3$)$_2$CH$\overset{+}{N}$(CH$_3$)$_3$ I$^-$

(CH$_3$)$_2$CH$\overset{+}{N}$H$_3$  $^-$OCPh (with O double bond)

(i)   (j)   (k)

(CH$_3$)$_2$CHN(CH$_3$)$_2$

H$_2$C=CHCH$_3$
+ N(CH$_3$)$_3$

(CH$_3$)$_2$CHNHCH$_2$CH$_3$

For part (j) and other problems involving Ag$_2$O, note the icon comment following the solution to Problem 23.23 on p. 635 of this manual.

**23.41**  (a) 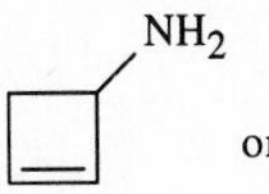   (c)

NH$_2$ (cyclobutene)  or  NH$_2$ (cyclopropene) CH$_3$

2-cyclobutenamine

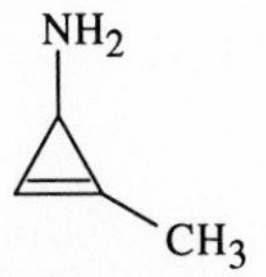

2-methyl-2-cyclo-propenamine

NHCH(CH$_3$)$_2$ (phenyl)

+

NHCH$_2$CH$_2$CH$_3$ (phenyl)

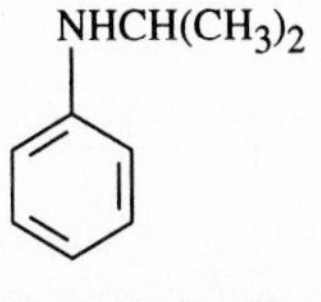

*N*-isopropylaniline

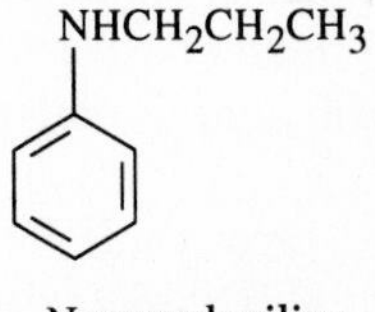

*N*-propylaniline

(d) The required compound could be either $(CH_3)_2C{=}NCH_3$, the imine that results from the reaction of acetone and methylamine, or $(CH_3)_2CHN{=}CH_2$, the imine that results from the reaction of isopropyl-amine and formaldehyde.

23.42  (a)  *N*-Methylhexanamide is an amide, and it is therefore not significantly soluble in dilute HCl; the two amines are soluble as their conjugate-acid ammonium salts in dilute HCl solution. 1-Octanamine, like all primary alkylamines, gives off a gas ($N_2$) when diazotized with $NaNO_2$ in aqueous acid. *N,N*-Dimethyl-1-hexanamine does not give off a gas when treated with nitrous acid.

23.43  (a)  The amine nitrogen (but *not* the amide nitrogen) is the most basic atom in the molecule. Hence, labetalol hydrochloride is the ammonium ion that results from protonation of this nitrogen.

structure of labetalol hydrochloride

(b)  When labetalol hydrochloride is treated with one molar equivalent of NaOH, a proton on the positively charged nitrogen is removed and neutral labetalol is formed.

(c)  Hot NaOH ionizes the phenol group and also brings about hydrolysis of the amide group to a carboxylate.

(d)  Hot 6 *M* aqueous HCl solution results in protonation of the amine nitrogen (as in part (a) and hydrolysis of the amide.

23.45  Dissolve the mixture in a suitable low-boiling solvent such as ether or methylene chloride. Extract with 5% aqueous sodium bicarbonate solution. The carboxylic acid *p*-nitrobenzoic acid will dissolve in the aqueous solution as its sodium salt. Isolate this aqueous solution and acidify it with HCl to precipitate the carboxylic acid. Then extract the remaining organic solution with 5% aqueous NaOH solution. *p*-Chlorophenol will dissolve in the aqueous solution as its conjugate-base phenolate ion. Isolate this aqueous layer and acidify it to obtain the neutral phenol. Then extract the organic solution with 5% aqueous HCl solution. The amine aniline will dissolve in the aqueous layer as its conjugate-acid ammonium ion. Separate the aqueous layer and neutralize it with NaOH. Aniline will form a separate layer that can be separated, dried, and distilled. Finally, dry the remaining organic layer and remove the solvent to obtain nitrobenzene.

23.47  (a)  Nitrous acid forms a diazonium ion from anthranilic acid. This diazonium ion undergoes coupling with *N,N*-dimethylaniline to give methyl red.

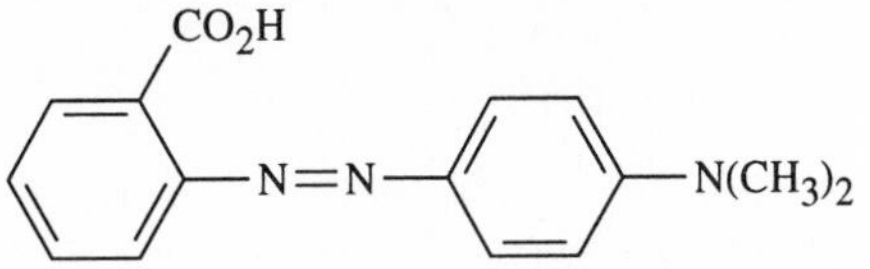

$$CO_2H$$

[structure: methyl red — benzene ring with CO₂H, —N=N— linking to another ring with —N(CH₃)₂]   methyl red

(b) In acidic solution, methyl red is protonated to give its conjugate acid. Protonation occurs on one of the diazo nitrogens because the resulting conjugate acid is resonance-stabilized.

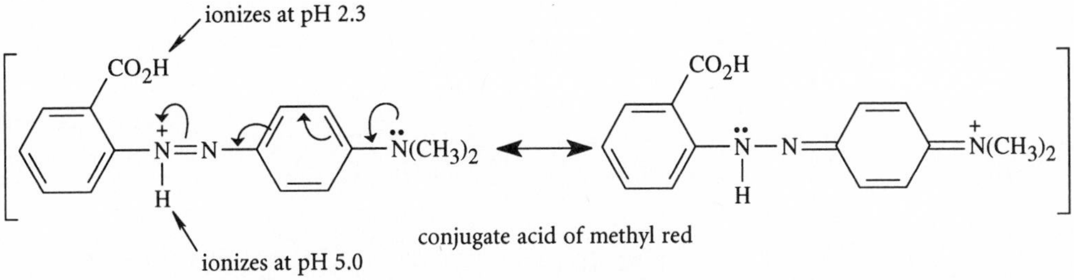

ionizes at pH 2.3

$$CO_2H \qquad\qquad CO_2H$$

[resonance structures of the conjugate acid of methyl red]

conjugate acid of methyl red

ionizes at pH 5.0

There are two $pK_a$ values observed because there are two acidic groups, the carboxy group and the protonated diazo group. Because the diazo group is part of the conjugated system of $\pi$ electrons responsible for the color of methyl red, the protonation-deprotonation equilibrium of this group should have the greatest effect on color. Since the color of methyl red changes most around pH = 5, the $pK_a$ of 5.0 is the $pK_a$ of the protonated diazo group, as shown in the structure above. Ionization of the carboxy group does not affect the conjugated $\pi$-electron system, and therefore does not affect the color of the molecule; thus, the $pK_a$ of 2.3 is the $pK_a$ of the carboxy group.

 You may have noticed that the $pK_a$ of the carboxy group is unusually low for the $pK_a$ of a carboxylic acid. Can you explain why this is reasonable?

**23.49** (a) Amanda is evidently trying to run a Friedel-Crafts acylation on aniline. This cannot work as shown for two reasons. First, the aluminum trichloride, a Lewis acid, will react with the electron pair on the amino group, thus deactivating the ring. (This is much like its effect on phenol acylation; see text pp. 848–849.) Second, a competing reaction is acylation of the amino group to form acetanilide, a reaction that will undoubtedly be catalyzed by the aluminum trichloride.

(c) Amanda is attempting to nitrate the ring and is relying on the activating, *ortho, para*-directing effect of the dimethylamino group to direct the substitution to the *para* position. The fallacy here is that under the very acidic conditions the dimethylamino group is protonated, and the protonated dimethylamino group is expected to be a *meta* director. (See Eq. 23.41 and related discussion on text p. 1137.)

(e) Amanda is attempting to displace water from the protonated alcohol with dimethylamine, which is a good nucleophile. Unfortunately, the acidic conditions required to convert the —OH group of the alcohol into a good leaving group also protonate the amine, which is much more basic than the alcohol. The protonated amine no longer has the unshared electron pair that is responsible for its nucleophilicity.

(g) Amanda is attempting an exhaustive methylation followed by Hofmann elimination. An elimination would occur, but unfortunately, the alkene shown would not be the major product of the reaction. (Why?)

**23.50** (a)

$$PhNH_2 \xrightarrow[\text{2) CuCN}]{\text{1) NaNO}_2/\text{HCl}} PhC\equiv N \xrightarrow[]{H_2,\ cat.} PhCH_2NH_2 \quad \text{benzylamine}$$

(c)

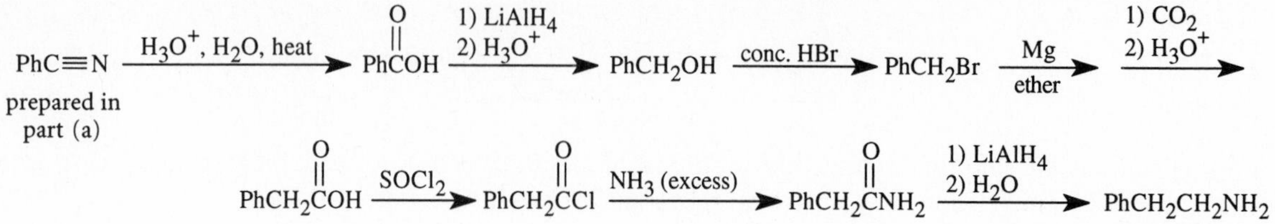

$$PhC\equiv N \xrightarrow{H_3O^+, H_2O, heat} \underset{\underset{PhCOH}{\|}}{\overset{O}{}} \xrightarrow[2) H_3O^+]{1) LiAlH_4} PhCH_2OH \xrightarrow{conc. HBr} PhCH_2Br \xrightarrow[ether]{Mg} \xrightarrow[2) H_3O^+]{1) CO_2}$$

prepared in
part (a)

$$\underset{\underset{PhCH_2COH}{\|}}{\overset{O}{}} \xrightarrow{SOCl_2} \underset{\underset{PhCH_2CCl}{\|}}{\overset{O}{}} \xrightarrow{NH_3 \text{ (excess)}} \underset{\underset{PhCH_2CNH_2}{\|}}{\overset{O}{}} \xrightarrow[2) H_2O]{1) LiAlH_4} PhCH_2CH_2NH_2$$

2-phenylethanamine

(e)

$$\text{aniline} \xrightarrow{Ac_2O} \text{—NHAc} \xrightarrow{Cl_2, FeCl_3} Cl\text{—}\text{—NHAc} \xrightarrow[2) NaOH \text{ (dilute)}]{1) H_3O^+, H_2O, heat}$$

$$Cl\text{—}\text{—}NH_2 \xrightarrow[2) CuCN]{1) NaNO_2/HCl} Cl\text{—}\text{—}C\equiv N \xrightarrow[H_2O]{H_3O^+, heat} Cl\text{—}\text{—}CO_2H$$

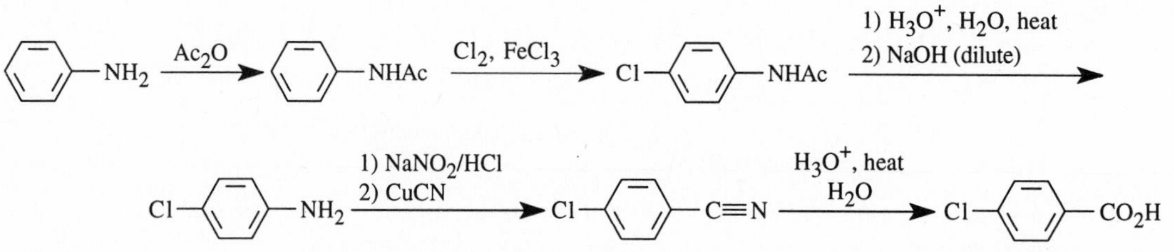

*p*-chlorobenzoic acid

**23.51**   First identify compound *B* by completing the reaction sequence:

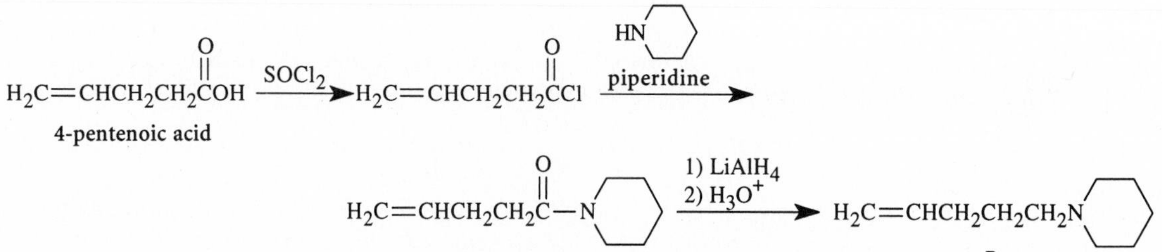

$$\underset{\underset{H_2C=CHCH_2CH_2COH}{\|}}{\overset{O}{}} \xrightarrow{SOCl_2} \underset{\underset{H_2C=CHCH_2CH_2CCl}{\|}}{\overset{O}{}} \xrightarrow[piperidine]{HN\diamond}$$

4-pentenoic acid

$$\underset{\underset{H_2C=CHCH_2CH_2C-N\diamond}{\|}}{\overset{O}{}} \xrightarrow[2) H_3O^+]{1) LiAlH_4} H_2C=CHCH_2CH_2CH_2N\diamond$$

*B*

Compound *A*, because of its reaction with Ag$^+$ to give AgBr, appears to be an ionic compound. The formula of compound *B* shows that it must be derived from two molecules of 1,5-dibromopentane. Because the alkene *B* is obtained by an elimination reaction, it is reasonable to suppose that compound *A* is a quaternary amine salt. A reasonable structure of such a salt that could result from the reaction of two equivalents of 1,5-dibromopentane and one of ammonia is the following:

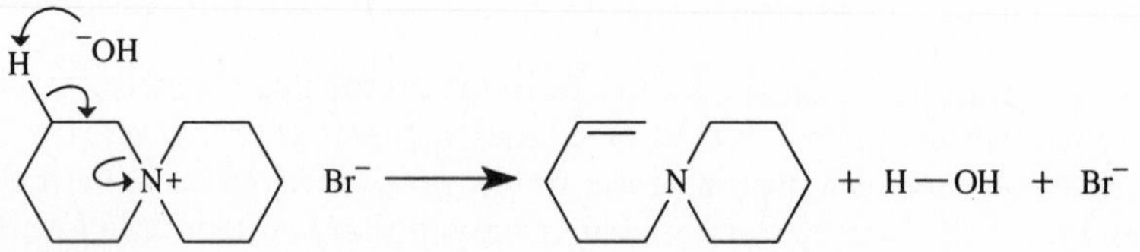

compound *A*                           compound *B*

The formation of compound *A* from 1,5-dibromopentane and ammonia involves the following steps.

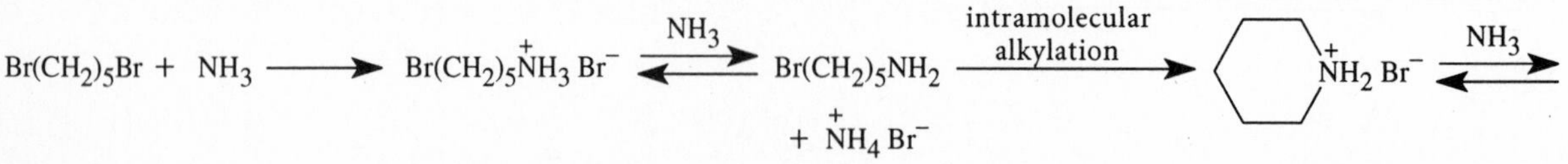

$$Br(CH_2)_5Br + NH_3 \longrightarrow Br(CH_2)_5\overset{+}{N}H_3 \ Br^- \underset{\xrightarrow{}}{\overset{NH_3}{\rightleftharpoons}} Br(CH_2)_5NH_2 \xrightarrow[alkylation]{intramolecular} \overset{+}{N}H_2 \ Br^- \overset{NH_3}{\underset{\xrightarrow{}}{\rightleftharpoons}}$$

$$+ \ \overset{+}{N}H_4 \ Br^-$$

*(equation continues)*

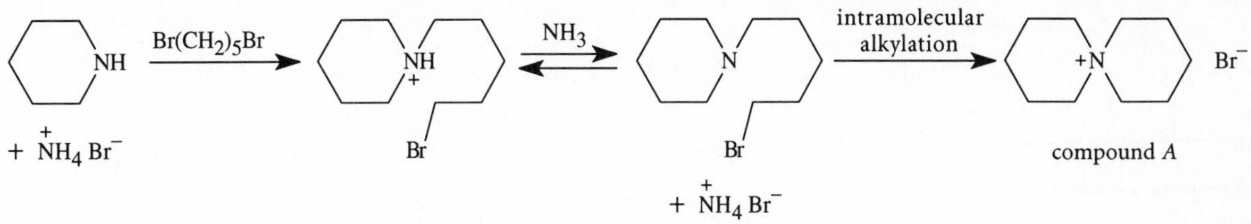

23.53 (a) The barrier to internal rotation about the N-phenyl bond in *N*-methyl-*p*-nitroaniline is greater because there is more double bond character in this bond than there is in *N*-methylaniline. There is more double bond character because of the resonance interaction between the unshared electron pair on the amine nitrogen and the *p*-nitro group.

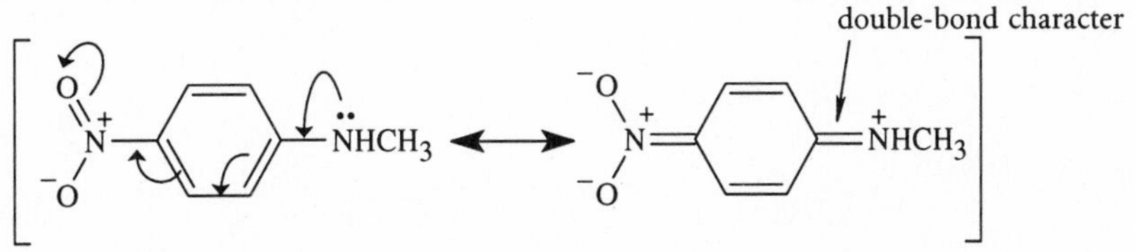

(c) This compound is the nitrogen analog of an acetal and, by a mechanism similar to that for the hydrolysis of an acetal, decomposes to formaldehyde and two equivalents of methylamine in aqueous solution.

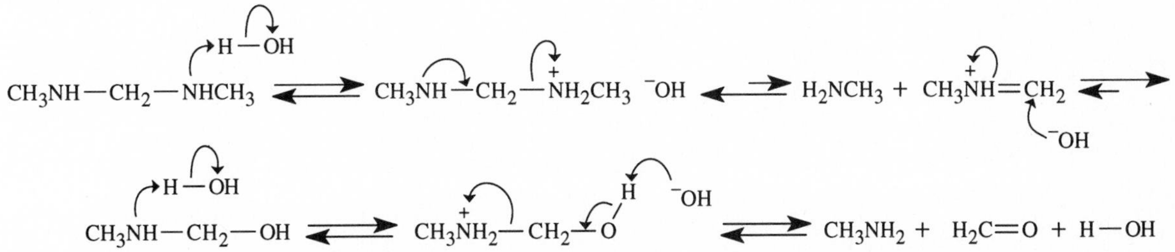

(e) The diazonium salt does not decompose to a carbocation because the carbocation has a $4n$ $\pi$-electron system and is therefore antiaromatic and very unstable.

2,4-cyclopentadien-1-amine

antiaromatic; unstable

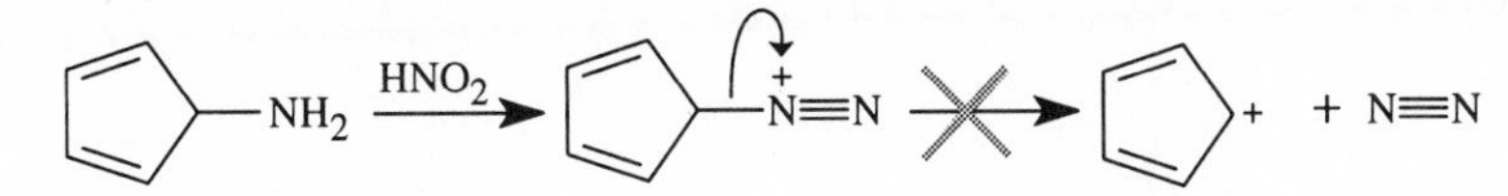

23.54 First, resolve 2-phenylbutanoic acid into enantiomers using an enantiomerically pure amine as the resolving agent. Then allow the appropriate enantiomer of the carboxylic acid to undergo the reactions shown in each part below. The key element in all these syntheses is that the Hofmann and Curtius rearrangements take place with retention of stereochemical configuration.

(a)

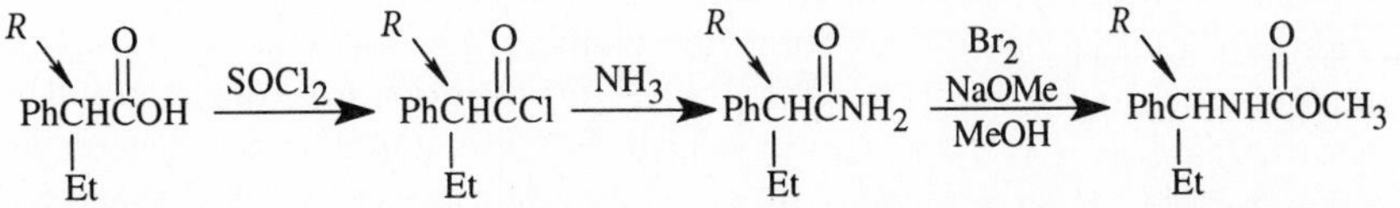

(c) In this part, the isocyanate product from the Curtius rearrangement is divided in half, and part of it is converted into an amine. This amine is then allowed to react with the remaining isocyanate.

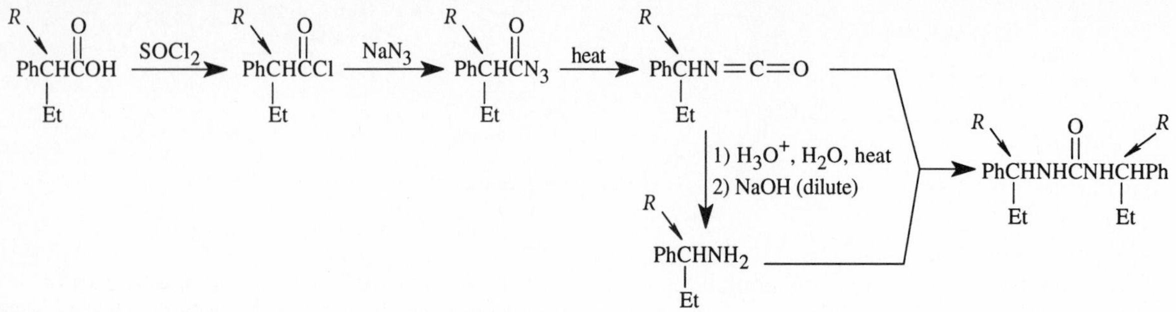

**23.55** In all three parts of this problem the Curtius rearrangement is used. Reasonable routes involving the Hofmann rearrangement can also be devised.

(a)

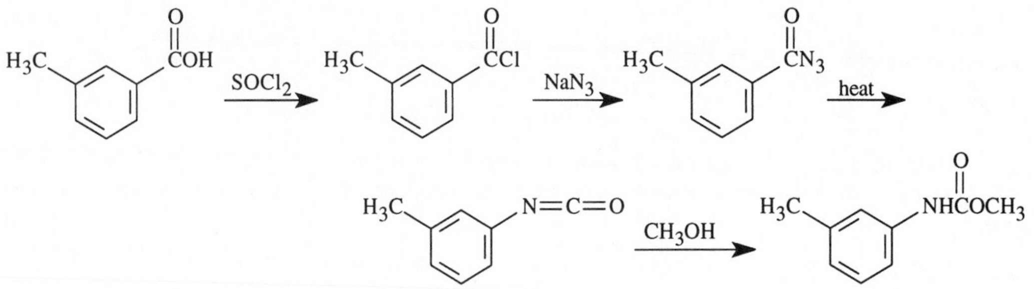

(b)

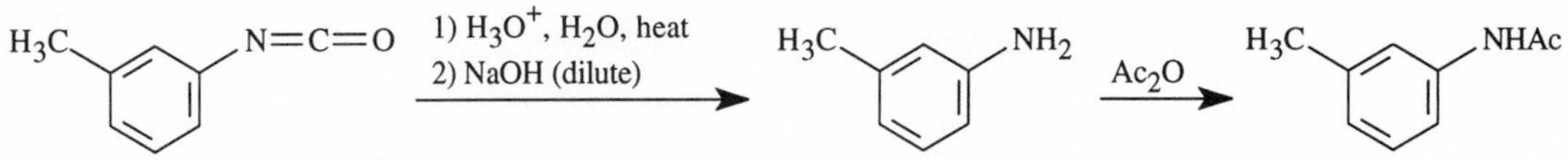

prepared in part (a)

Treatment of the isocyanate with acetic acid also gives the same product plus $CO_2$. Try to write a mechanism for this process.

(c)

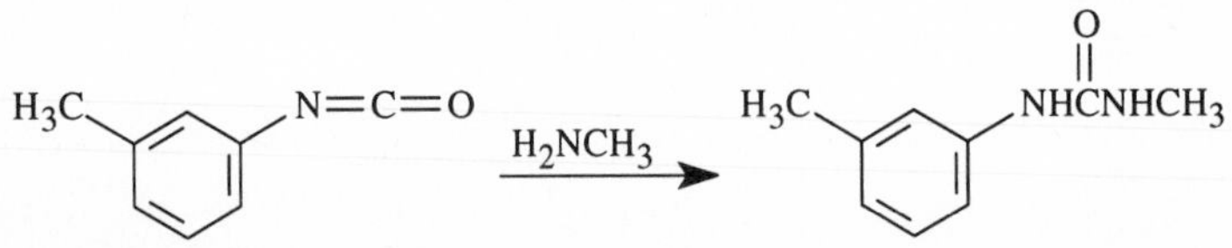

prepared in part (a)

**23.57** Compound *A* appears to be an *N*-benzoyl amide of compound *B* because it is hydrolyzed to benzoic acid, and because it is re-formed when *B* is treated with benzoyl chloride. Compound *A* cannot be an ester because it contains only one oxygen. Compound *B* is therefore an achiral primary or secondary amine. It cannot be a primary amine because it does not liberate nitrogen gas after treatment with nitrous acid; thus, it must be secondary because a tertiary amine cannot form an amide. The formula of *B* is obtained by subtracting the formula of a benzoyl group (Ph—CO—, $C_7H_5O$) from the formula of *A* and adding a hydrogen. Thus, the formula of *B* is $C_{15}H_{23}N$. Compound *B* therefore contains five degrees of unsaturation. One degree of unsaturation is accounted for by a cyclohexane ring, reflected in the degradation product *D*, and the remaining four by a phenyl ring, reflected in the degradation product styrene, which is formed along with product *C*. There are only two structures for compound *B*, shown below as *B1* and *B2*, that would give the indicated degradation products:

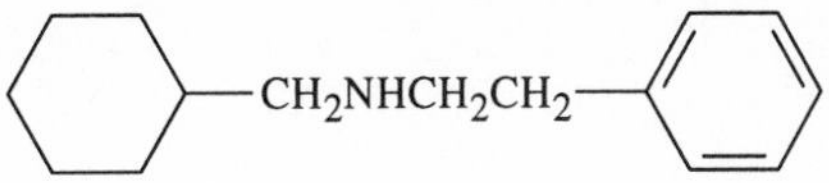

*B1*

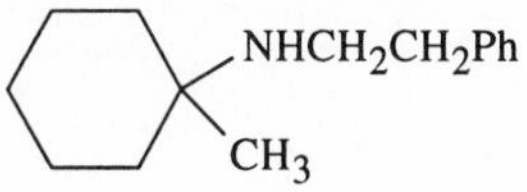

*B2*

Although both are reasonable answers, structure *B1* is better because it would lead to *one* alkene *D*, whereas *B2* would lead to *two* different alkenes of which *D* is the major one. Taking *B1* as the correct structure for *B*, the identities of the unknown compounds are as follows:

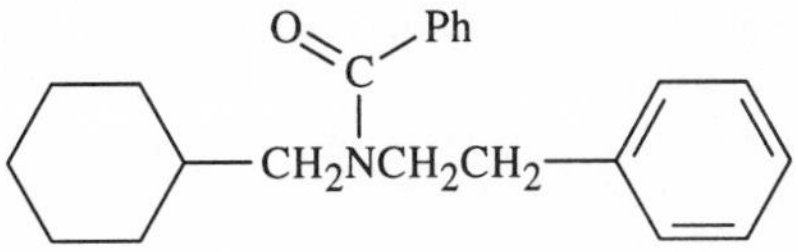

*A*

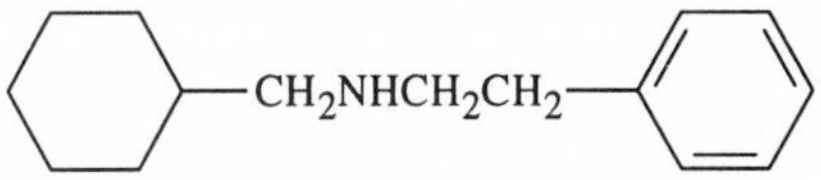

*B*

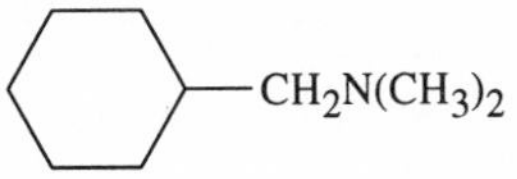

*C*

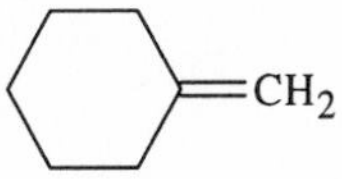

*D*

23.59 Compound *A* is the hydrochloride salt of *B*, and *B* is a primary alkylamine, because it liberates a gas ($N_2$) upon diazotization. The elemental analysis of *B* yields the formula $C_9H_{13}N$; a compound with this formula has an unsaturation number of 4. Compound *D* is propylbenzene, $PhCH_2CH_2CH_3$, because this is the compound that would be obtained by the Wolff-Kishner reduction of propiophenone:

$$\underset{\text{propiophenone}}{\overset{\text{O}}{\overset{\|}{\text{PhCCH}_2\text{CH}_3}}} \xrightarrow[\text{heat}]{\text{NH}_2\text{NH}_2,\ ^-\text{OH}} \underset{\substack{\text{propylbenzene}\\(\text{compound } D)}}{\text{PhCH}_2\text{CH}_2\text{CH}_3}$$

The fact that the degradation product *D* contains a phenyl ring shows that the four degrees of unsaturation of *B* are accounted for by a phenyl ring. Three carbons and an $—NH_2$ group remain to be accounted for. The mixture of alkenes *C* that would hydrogenate to propylbenzene are the following:

$$\underbrace{\underset{C1}{\text{PhCH}{=}\text{CHCH}_3} + \underset{C2}{\text{PhCH}_2\text{CH}{=}\text{CH}_2}}_{\text{the mixture of alkenes } C}$$

The mixture of alkenes might also be only the *cis* and *trans* isomers of alkene *C1*. Two structures for *B* fit the data discussed so far:

$$\underset{B1}{\overset{\text{NH}_2}{\overset{|}{\text{PhCHCH}_2\text{CH}_3}}} \qquad \underset{B2}{\overset{\text{NH}_2}{\overset{|}{\text{PhCH}_2\text{CHCH}_3}}}$$

Compound *B1* would give alkenes *C1* following exhaustive methylation/Hofmann elimination, and compound *B2* would give alkenes *C1* and *C2*. However, *only* compound *B2* would have a doublet at $\delta$ 1.1 in its proton NMR spectrum. Therefore, compound *B* = *B2*, and compound *A* is its HCl salt.

23.60 (a) Since diazotization of $R—NH_2$ gives $N_2$ and products derived from $R^+$, then diazotization of $H—NH_2$ should give $H^+$, that is, $H_3O^+$, and $N_2$.

(b) Exhausive methylation is followed by Hofmann elimination in which a proton is removed from the $\beta$-carbon with the fewest branches.

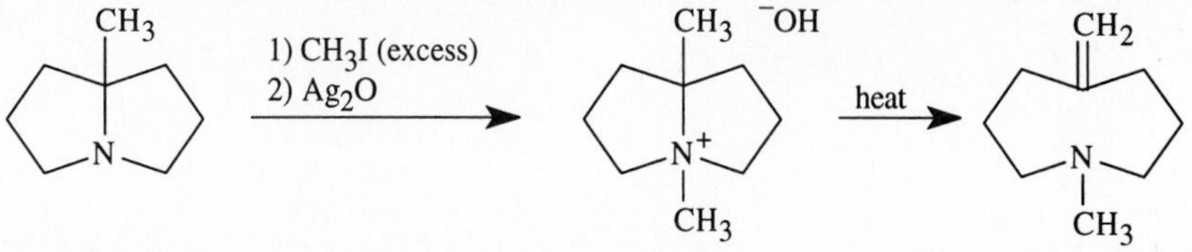

(d) The nitrosyl cation acts as the electrophile in an electrophilic aromatic substitution reaction of phenol.

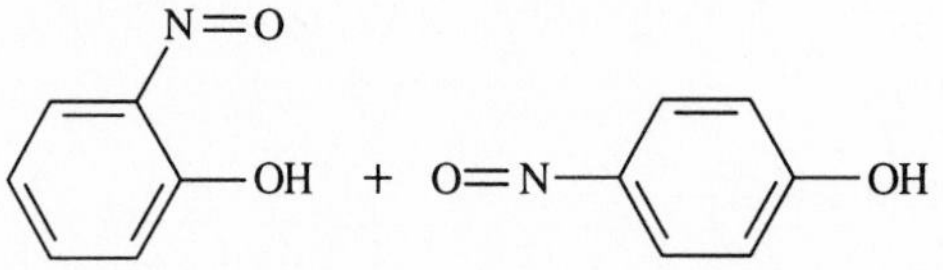

(f) The amino group is alkylated twice by ethylene oxide to give $CH_3CH_2CH_2CH_2N(CH_2CH_2OH)_2$.

(h) The conjugate-base anion of phthalimide reacts at the less branched carbon of the epoxide to give compound *A*; hydrolysis of the amide groups gives the amine product.

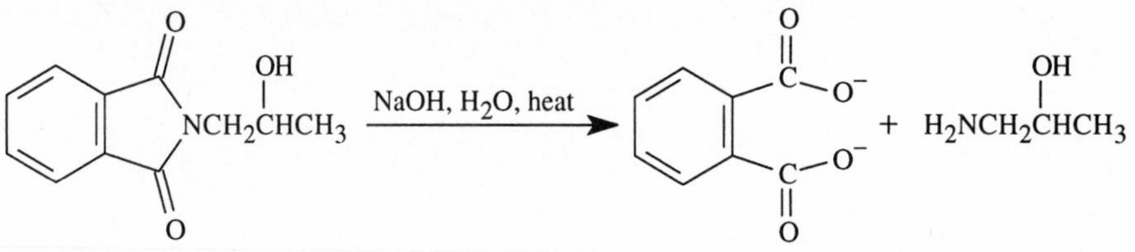

*A*

(j) Vinylic halides are inert in $S_N2$ reactions, and allylic halides are very reactive. Because an excess of the amine is used, only one halide molecule reacts with each amine, and the allylic bromine is displaced.

$$H_2C{=}\underset{\displaystyle Br}{C}{-}CH_2\overset{+}{N}H_2Et\ Br^-$$

(k) Catalytic hydrogenation reduces nitrobenzene to aniline, which then reductively aminates the aldehyde; because excess aldehyde is present, the resulting secondary amine then reductively amin ates a second molecule of aldehyde. The final product is the tertiary amine *N,N*-dibutylaniline, $PhN(CH_2CH_2CH_2CH_3)_2$.

23.61 (a)

$$\underset{\text{pentanoic acid}}{CH_3(CH_2)_3\overset{O}{\overset{\|}{C}}OH} \xrightarrow[]{\text{1) LiAlH}_4\text{; 2) H}_3\text{O}^+} CH_3(CH_2)_3CH_2OH \xrightarrow[H_2SO_4]{\text{conc. HBr}} CH_3(CH_2)_3CH_2Br \xrightarrow[\text{ether}]{\text{Mg}} \xrightarrow{\text{1) CO}_2\text{; 2) H}_3\text{O}^+}$$

$$CH_3(CH_2)_3CH_2\overset{O}{\overset{\|}{C}}OH \xrightarrow{SOCl_2} CH_3(CH_2)_3CH_2\overset{O}{\overset{\|}{C}}Cl \xrightarrow{H_2NCH_3\text{ (excess)}} CH_3(CH_2)_3CH_2\overset{O}{\overset{\|}{C}}NHCH_3 \xrightarrow{\text{1) LiAlH}_4\text{; 2) H}_3\text{O}^+}$$

$$\underset{N\text{-methyl-1-hexanamine}}{CH_3(CH_2)_3CH_2CH_2NHCH_3}$$

(c)

$$CH_3(CH_2)_3\overset{O}{\overset{\|}{C}}OH \xrightarrow{SOCl_2} CH_3(CH_2)_3\overset{O}{\overset{\|}{C}}Cl \xrightarrow{HN(CH_3)_2\text{ (excess)}}$$

*(equation continues)*

$$CH_3(CH_2)_3\overset{\overset{\displaystyle O}{\|}}{C}N(CH_3)_2 \xrightarrow[\text{2) H}_3\text{O}^+]{\text{1) LiAlH}_4} CH_3(CH_2)_3CH_2N(CH_3)_2$$

*N,N*-dimethyl-1-pentanamine

(e)

$$CH_3(CH_2)_3CH_2\overset{\overset{\displaystyle O}{\|}}{C}Cl \xrightarrow{NH_3 \text{ (excess)}} CH_3(CH_2)_3CH_2\overset{\overset{\displaystyle O}{\|}}{C}NH_2 \xrightarrow[\text{2) H}_3\text{O}^+]{\text{1) LiAlH}_4} CH_3(CH_2)_3CH_2CH_2NH_2$$

prepared in part (a)

hexylamine
(1-hexanamine)

(g)

$$PhCH_3 \xrightarrow[\text{peroxides}]{\text{NBS, CCl}_4} PhCH_2Br \xrightarrow[\text{ether}]{\text{Mg}} \xrightarrow[\text{2) H}_3\text{O}^+]{\text{1) }\triangle\text{O}} PhCH_2CH_2CH_2OH \xrightarrow[\text{2) H}_3\text{O}^+]{\text{1) KMnO}_4,\ ^-\text{OH}}$$

$$PhCH_2CH_2\overset{\overset{\displaystyle O}{\|}}{C}OH \xrightarrow{\text{SOCl}_2} PhCH_2CH_2\overset{\overset{\displaystyle O}{\|}}{C}Cl \xrightarrow[\text{(excess)}]{\text{H}_2\text{NEt}} PhCH_2CH_2\overset{\overset{\displaystyle O}{\|}}{C}NHEt \xrightarrow[\text{2) H}_3\text{O}^+]{\text{1) LiAlH}_4} PhCH_2CH_2CH_2NHEt$$

*N*-ethyl-3-phenyl-
1-propanamine

(h)

$$CH_2(CO_2Et)_2 \xrightarrow[\text{2) CH}_3\text{CH}_2\text{CH}_2\text{Br}]{\text{1) NaOEt, EtOH}} CH_3CH_2CH_2CH(CO_2Et)_2 \xrightarrow[\text{2) CH}_3\text{I}]{\text{1) NaOEt, EtOH}}$$

diethyl malonate

$$CH_3CH_2CH_2\underset{\underset{\displaystyle CH_3}{|}}{C}(CO_2Et)_2 \xrightarrow[\text{2) NaOH (dilute)}]{\text{1) H}_3\text{O}^+,\ \text{H}_2\text{O, heat}} CH_3CH_2CH_2\underset{\underset{\displaystyle CH_3}{|}}{\overset{\overset{\displaystyle O}{\|}}{C}H}COH \xrightarrow[\text{2) NH}_3]{\text{1) SOCl}_2}$$

$$CH_3CH_2CH_2\underset{\underset{\displaystyle CH_3}{|}}{\overset{\overset{\displaystyle O}{\|}}{C}H}CNH_2 \xrightarrow[\text{H}_2\text{O}]{\text{Br}_2,\ \text{NaOH}} CH_3CH_2CH_2\underset{\underset{\displaystyle CH_3}{|}}{C}HNH_2$$

2-pentanamine

(i)

$$(CH_3)_2C{=}O \xrightarrow[\text{2) H}_3\text{O}^+]{\text{1) LiAlH}_4} (CH_3)_2CHOH \xrightarrow[\text{H}_2\text{SO}_4]{\text{conc. HBr}} (CH_3)_2CHBr \xrightarrow[\text{ether}]{\text{Mg}} \xrightarrow[\text{2) H}_3\text{O}^+]{\text{1) }\triangle\text{O}}$$

acetone

$$(CH_3)_2CHCH_2CH_2OH \xrightarrow[\text{2) H}_3\text{O}^+]{\text{1) KMnO}_4,\ ^-\text{OH}} (CH_3)_2CHCH_2\overset{\overset{\displaystyle O}{\|}}{C}OH \xrightarrow[\text{2) NH}_3]{\text{1) SOCl}_2}$$

$$(CH_3)_2CHCH_2\overset{\overset{\displaystyle O}{\|}}{C}NH_2 \xrightarrow[\text{H}_2\text{O}]{\text{Br}_2,\ \text{NaOH}} (CH_3)_2CHCH_2NH_2$$

isobutylamine

(k)

nitrobenzene                                    *m*-chlorobromobenzene

(m)

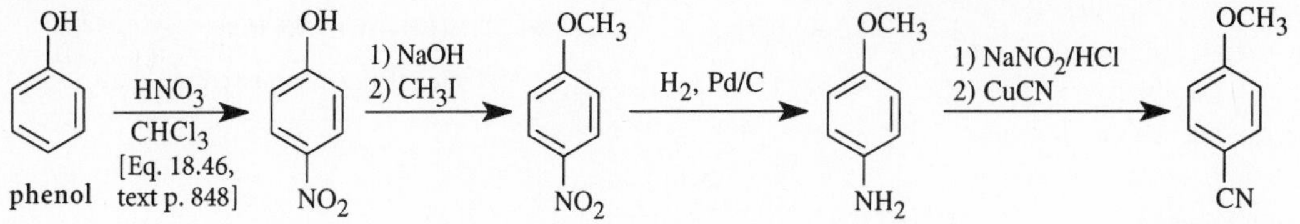

*p*-methoxybenzonitrile

(o)

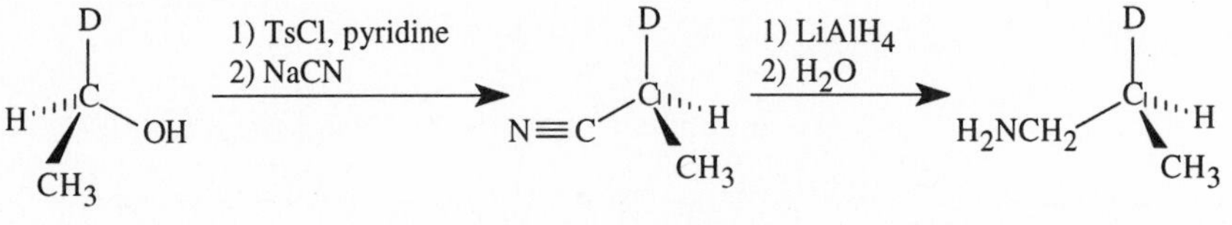

(p)

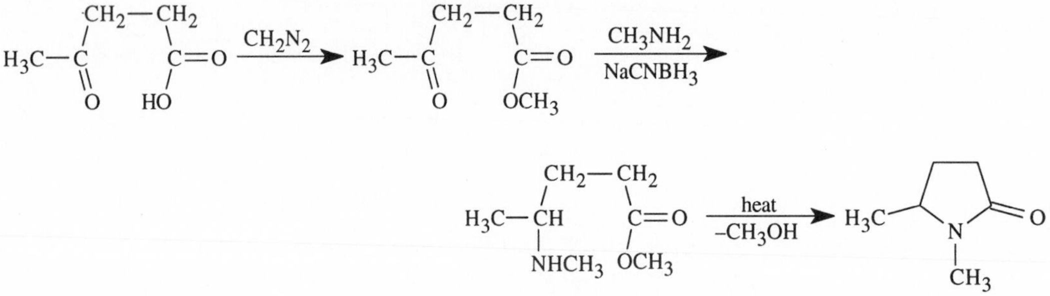

**23.62**  In order for the proton that is lost and the trimethylammonium group to be *anti*, the R-group and the trimethylammonium group must be *gauche* in the transition state for elimination:

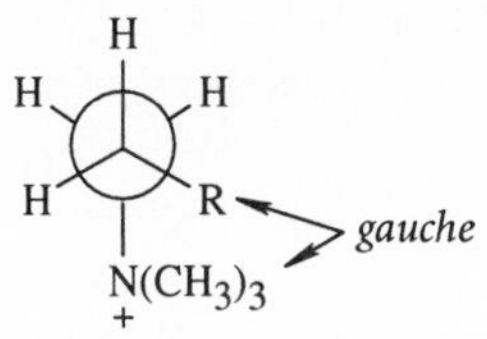

As the size of the R-group is increased, the van der Waals repulsions in this transition state are more severe, the transition state is less stable, and the rate of the Hofmann elimination is slower. Therefore, the rates of Hofmann elimination increase in the order

$$R\!\!-\!\!\ = (CH_3)_3C\!\!-\!\!\ <\ R\!\!-\!\!\ = CH_3\!\!-\!\!\ <\ R\!\!-\!\!\ = H\!\!-\!\!$$

**23.63**  At pH 1, methylamine exists almost completely as the methylammonium ion, $CH_3\overset{+}{N}H_3$. The methyl quartet is expected for the presence of three neighboring nitrogen protons provided that chemical exchange is slow. As the pH is raised, more free methylamine is present and chemical exchange of the protons on the nitrogen can occur rapidly by the following mechanism. (The asterisk is used to differentiate the two nitrogens.)

$$CH_3\overset{+}{\underset{*}{N}}H_3 \ +\ CH_3NH_2 \ \rightleftharpoons\ CH_3NH_2 \ +\ CH_3\overset{+}{\underset{*}{N}}H_3$$

Evidently, this exchange is so fast on the NMR time scale that splitting is obliterated. (See Sec. 13.6D on text p. 616 and Sec. 13.7 on p. 619.)

**23.64**  The $\lambda_{max}$ of aniline occurs at longer wavelength than that of benzene because of the conjugation of the aniline unshared electron pair with the $\pi$-electron system of the benzene ring. Recall that the more atoms are involved in a conjugated $\pi$-electron system, the greater is the intensity of the UV spectrum and the greater is the $\lambda_{max}$ (Sec. 15.2C on text p. 688). When aniline is protonated, there are no unshared pairs on nitrogen, and the conjugated $\pi$-electron system is restricted to the phenyl ring, in which case the UV spectrum looks much as it does in benzene. In other words, protonated aniline, as far as the $\pi$-electron system and the UV spectrum are concerned, is essentially the same as benzene.

23.65 (a) The IR spectrum indicates the presence of an amine. The integral of the entire NMR spectrum is 39.5 spaces, which corresponds to 2.6 spaces per proton. The "two leaning doublet" pattern centered at $\delta$ 7 in the NMR spectrum indicates a *para*-disubstituted benzene ring, and the mutually split triplet-quartet pattern at $\delta$ 1.1 and $\delta$ 2.6 indicates the presence of an ethyl group. The one-proton singlet at $\delta$ 1.6 is consistent with a single NH, and this is consistent with the presence of a secondary amine. The unsaturation number is 4; thus, there is no unsaturation in addition to the benzene ring. The strange-looking "almost-singlet" at $\delta$ 3.8 is consistent with the presence of a methoxy group, except that it integrates for five protons. The only structure consistent with these data is the following; evidently, the benzylic $CH_2$ group is a singlet that nearly overlaps with the methyl singlet of the methoxy group. Indeed, the predicted chemical shift (Table 13.2, text p. 589) for this group is $\delta$ 3.6. No splitting is observed between the carbon and nitrogen protons because the N—H is rapidly exchanging.

$\delta$3.8       $\delta$3.7<br>
$CH_3O$—⟨benzene ring⟩—$CH_2NHCH_2CH_3$

23.66 The unsaturation number of this compound is 0. The instability of this compound in aqueous acid and the presence of two oxygens is consistent with the presence of an acetal. The six-proton singlet at $\delta$ 3.27 could be due to the methyl protons of a dimethyl acetal; the $\delta$ 53.2 resonance in the CMR spectrum could be due to the carbons of these methyl groups, and the $\delta$ 102.4 resonance could be due to the carbon of the acetal group, that is, —O—CH—O— . If so, the resonance in the proton NMR at $\delta$ 4.50 is due to the proton of this CH group, and its splitting indicates two adjacent protons. The resonance for these two protons is found at $\delta$ 2.45. This leaves a six-proton singlet at $\delta$ 2.30 unaccounted for; this could be due to the six methyl protons of a dimethylamino group. The molecule is dimethylaminoacetaldehyde dimethyl acetal, or 2,2-dimethoxy-*N,N*-dimethylethanamine, $(CH_3)_2NCH_2CH(OCH_3)_2$.

23.67 (a) The chemical data indicate the presence of a primary amine. The compound has one degree of unsaturation, and the chemical shifts in the CMR are too small for any carbon to be involved in a double bond. Hence, compound $A$ contains a ring, and only CH and $CH_2$ groups are present. The presence of only three resonances means that two carbons are equivalent. Compound $A$ is cyclobutylamine. Cyclopropylmethanamine is ruled out by the fact that the carbon bound to one hydrogen has the largest chemical shift and therefore must be bound to the nitrogen.

⟨cyclobutyl⟩—$NH_2$        ⟨cyclopropyl⟩—$CH_2NH_2$<br>
cyclobutylamine       cyclopropylmethanamine<br>
(compound *A*)         (ruled out)

23.68 All of these reactions are related to the Hofmann and Curtius rearrangements.

(a) Potassium hydride removes the NH proton to form an anion that rearranges to an isocyanate. This is exactly like the Hofmann rearrangement shown in Eq. 23.75d on text p. 1151, except that the leaving group is benzoate. (This type of reaction is a *Lossen rearrangement*.) The amine is formed by hydration of the isocyanate and decarboxylation of the resulting carbamate. (For brevity, the proton transfer in step *(a)* is shown as an intramolecular concerted process, but it could instead be a two-step process involving $^-OH$ and $H_2O$.)

*(solution continues)*

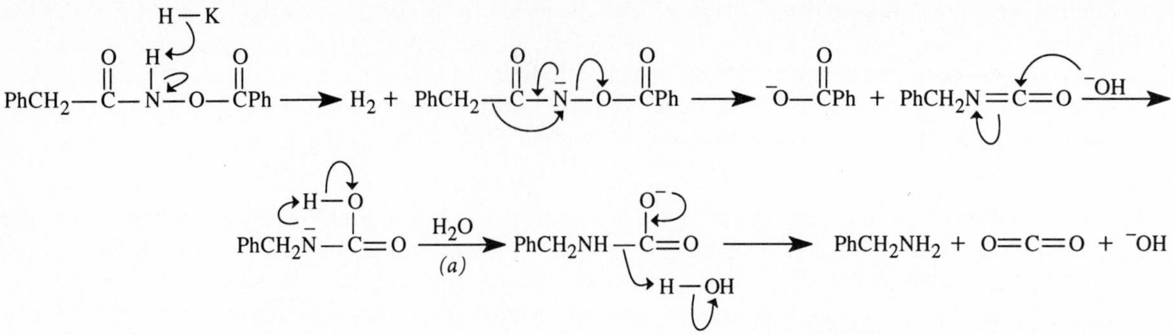

(c) In this case, the amide nitrogen attacks the isocyanate following a Curtius rearrangement. As in part (a), the proton transfer is shown for brevity as an intramolecular process.

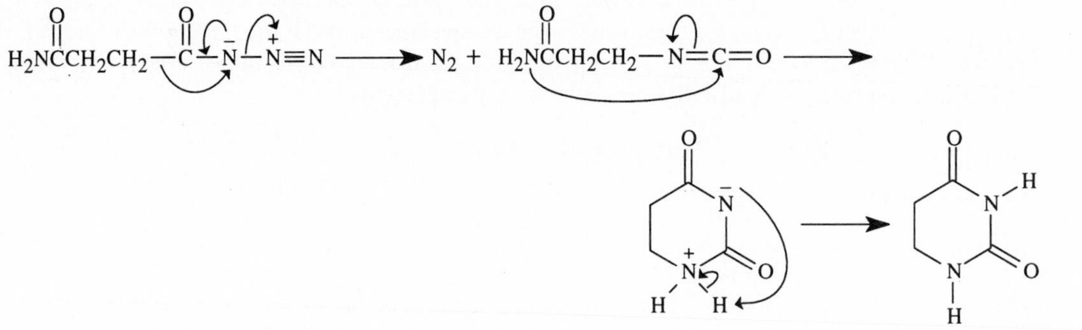

**23.69** (a) Amides, like amines, can be diazotized. The result is an acyldiazonium ion $A$, which rapidly hydrolyzes; remember that $N_2$ is a superb leaving group. This produces carbamic acid, which decomposes to ammonia and $CO_2$. Then ammonia is itself diazotized to give $N_2$ as shown in the solution to Problem 23.60(a) on p. 645 of this manual.

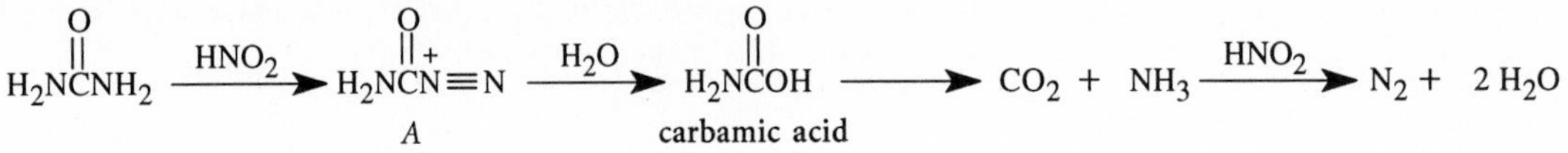

(b) The amine reacts intramolecularly with the carbonyl group to form a cyclic imine $A$, which is reduced by the sodium borohydride.

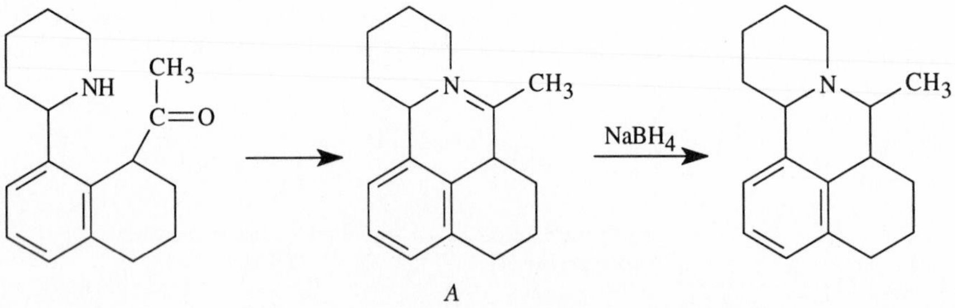

(c) Cyanide ion serves as a base to promote a $\beta$-elimination of trimethylamine and form an $\alpha,\beta$-unsaturated ketone $A$. This undergoes conjugate addition of cyanide ion to give the product.

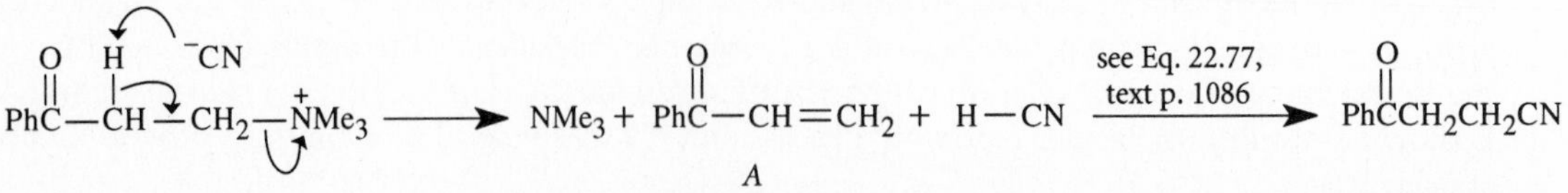

(d) The acyl azide $A$ undergoes Curtius rearrangement to an enamine $B$, which hydrolyzes under the aqueous reaction conditions to the product ketone.

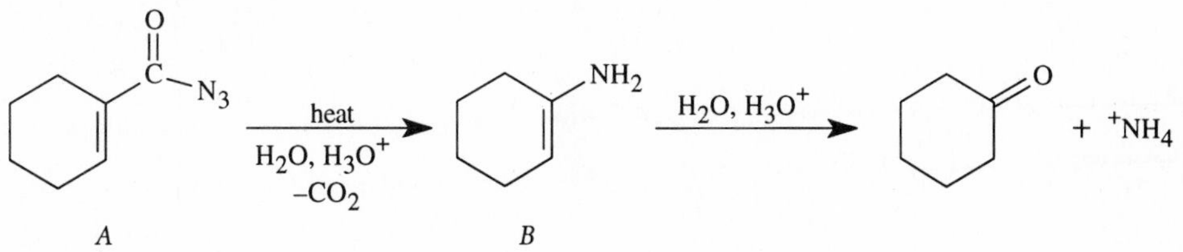

(e) The product shows that a 1,4-addition of benzyne to furan (that is, a Diels-Alder reaction) has occurred. (See the solution to Problem 15.58(a) on p. 356, Vol. 1, of this manual for the approach to this type of problem.) The benzyne is formed by elimination of $CO_2$ and $N_2$ from the diazotization product *A* of the starting material, anthranilic acid.

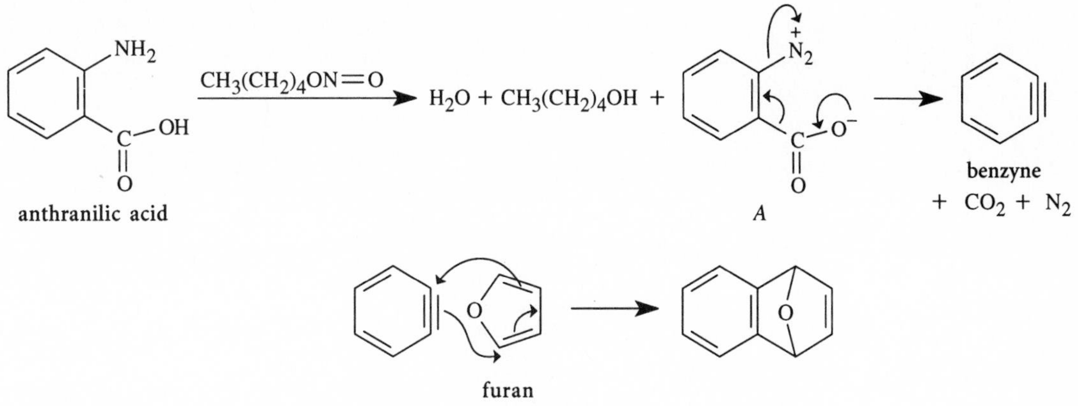

(f) The diazonium ion *A* formed from the starting material loses nitrogen to give a carbocation *B*. This carbocation undergoes a rearrangement to the product. (See Study Problem 19.2 on text p. 882 for a discussion of a closely related reaction.)

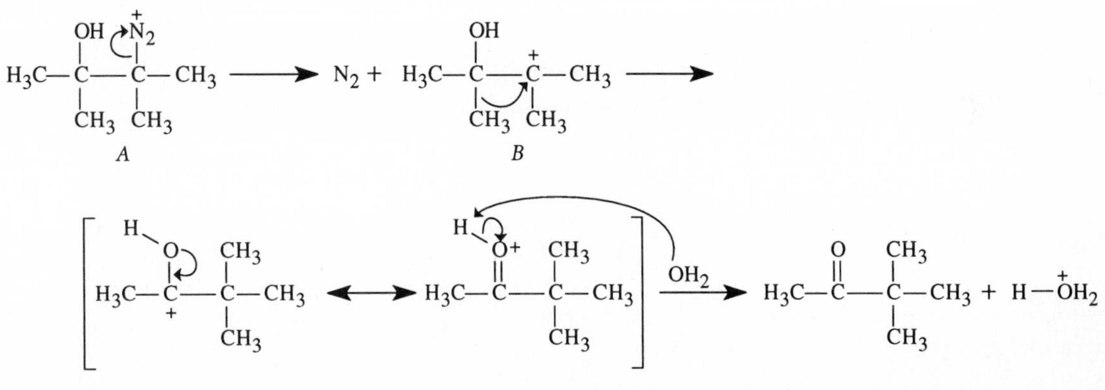

23.71 The conjugate-acid $pK_a$ of amine *C* is a normal conjugate-acid $pK_a$ for a tertiary amine. (See, for example, the $pK_a$ of the conjugate acid of triethylamine in Table 23.1 on text p. 1121.) This observation shows that incorporation of an amine into a bicyclic structure has no significant effect on its conjugate-acid $pK_a$. The conjugate-acid $pK_a$ of amine *A* is not unusual for an *N,N*-dialkylaniline. (See, for example, the $pK_a$ of $PhN(CH_3)_2$ in Table 23.1.) Recall (text p. 1124) that the conjugate-acid $pK_a$ values of aniline derivatives are considerably lower than those of tertiary alkylamines because of resonance interaction of the unshared electron pair on the nitrogen with the benzene ring. The conjugate-acid $pK_a$ of compound *B*, then, has the unusual value that requires explanation. In this compound, the unshared pair on nitrogen *cannot* overlap with the $\pi$-electron system of the benzene ring because the $\pi$ orbitals of the benzene ring and the orbital on the nitrogen are forced by the bicyclic structure out of coplanarity.

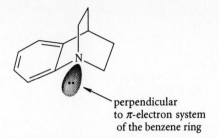

perpendicular
to $\pi$-electron system
of the benzene ring

Consequently, the basicity-lowering resonance effect is absent. For this reason, the conjugate-acid $pK_a$ of compound *B* is higher than that of compound *A*. However, the basicity-lowering polar effect of the benzene ring remains; for this reason, the conjugate-acid $pK_a$ of compound *B* is lower than that of compound *C*.

The $pK_a$ data in the problem provide good estimates of the polar and resonance effects of a benzene ring on amine basicity. These data show that polar and resonance effects are about equally important.

# Chemistry of Naphthalene and the Aromatic Heterocycles

## Terms

## Concepts

## I. Introduction to Polycyclic Aromatic Hydrocarbons

### A. GENERAL

1. Polycyclic aromatic hydrocarbons are aromatic hydrocarbons that contain two or more fused rings.
   a. The aromaticity of such compounds is not covered by the Hückel $4n + 2$ rule.
   b. The Lewis structures of these compounds consist of alternating single and double bonds.
   c. Naphthalene is the simplest example; others are shown in Figure 24.1, text page 1170.
2. Naphthalene can be represented by three resonance structures; two are equivalent and one is unique.

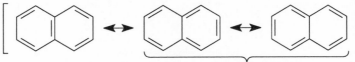

equivalent structures

3. Graphite, a carbon polymer that consists of layers of fused benzene rings, is an excellent electrical conductor because of the ease with which $\pi$ electrons can be delocalized across its structure.
4. Polycyclic aromatic compounds having ball-shaped structures of various sizes have been nicknamed fullerenes after buckminsterfullerene (the name of the first isolated compound of this type).

B. NOMENCLATURE OF NAPHTHALENE DERIVATIVES

1. In substitutive nomenclature, carbon-1 of naphthalene is the carbon adjacent to the bridgehead carbon (a vertex at which the rings are fused); substituents are given the lowest numbers consistent with this scheme and their relative priorities.
2. Naphthalene also has a common nomenclature that uses Greek letters: the 1-position is designated as $\alpha$ and the 2-position as $\beta$.
3. Common and substitutive nomenclature should never be mixed.
4. As a substituent, the naphthalene ring is called the naphthyl group.

2-chloronaphthalene
($\beta$-chloronaphthalene)         1-bromo-8-nitronaphthalene         2-naphthyl benzoate

## II.  *Introduction to the Aromatic Heterocycles*

A. GENERAL

1. Heterocyclic compounds are compounds with rings that contain more than one type of atom; the heterocyclic compounds of greatest interest to organic chemists have carbon rings containing one or two heteroatoms (atoms other than carbon).
2. The chemistry of many saturated heterocyclic compounds is analogous to that of their noncyclic counterparts.
3. A significant number of unsaturated heterocyclic compounds exhibit aromatic behavior.

B. NOMENCLATURE OF THE HETEROCYCLES

1. The names and structures of some common aromatic heterocyclic compounds are given in Fig. 24.4, text page 1178.
2. The same rules used in numbering and naming saturated heterocyclic compounds are used for numbering and naming aromatic heterocyclic compounds.
   a. In all but a few cases, a heteroatom is given the number 1 (isoquinoline is an exception).
   b. Oxygen and sulfur are given a lower number than nitrogen when a choice exists.
   c. Substituent groups are given the lowest numbers consistent with this scheme.

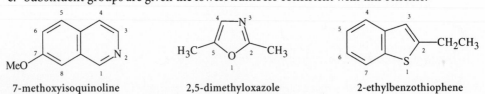

7-methoxyisoquinoline         2,5-dimethyloxazole         2-ethylbenzothiophene

C. STRUCTURE AND AROMATICITY OF THE HETEROCYCLES

1. The aromatic heterocyclic compounds furan, thiophene, and pyrrole can be written as resonance hybrids.

resonance structures of pyrrole

   a. The importance of the charge-separated structures is evident in a comparison of the dipole moments of furan and tetrahydrofuran.
      i. Electrons in the $\sigma$ bonds are pulled toward the oxygen because of its electronegativity.
      ii. The resonance delocalization of the oxygen unshared electrons into the ring tends to push electrons away from the oxygen into the $\pi$-electron system of the ring.
   b. These two effects nearly cancel in furan; thus furan has a very small dipole moment.
   c. That tetrahydrofuran has a considerably higher boiling point than furan reflects the greater dipole moment of tetrahydrofuran.
2. Pyridine can be represented by two equivalent neutral resonance structures; three additional structures of less importance reflect the relative electronegativity of nitrogen.

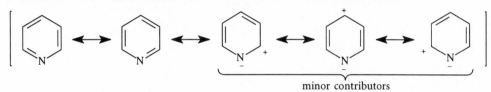

minor contributors

3. Heteroatoms involved in formal double bonds of the Lewis structure (such as the nitrogen of pyridine) contribute one $\pi$ electron to the six $\pi$-electron aromatic system; the orbital containing the unshared electron pair of the pyridine nitrogen is perpendicular to the $p$ orbitals of the ring and is therefore not involved in $\pi$ bonding.

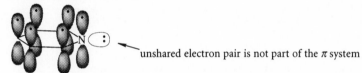

4. An unshared electron pair on a heteroatom in an allylic position (such as the unshared pair on the nitrogen of pyrrole) is part of the aromatic $\pi$ system; the hydrogen of pyrrole lies in the plane of the ring.

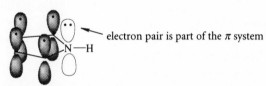

5. The oxygen of furan contributes one unshared electron pair to the aromatic $\pi$-electron system; the other unshared electron pair occupies a position analogous to the hydrogen of pyrrole (in the ring plane, perpendicular to the $p$ orbitals of the ring).

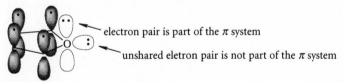

## III. *Basicity and Acidity of the Nitrogen Heterocycles*

### A. BASICITY

1. Pyridine and quinoline act as ordinary aromatic amine bases; they are less basic than aliphatic tertiary amines because of the $sp^2$ hybridization of their nitrogen unshared electron pairs.
   a. Protonation of the pyridine unshared electron pair occurs because this electron pair is not part of the $\pi$-electron system.
   b. Protonation of this electron pair does not destroy aromaticity.
2. Pyrrole and indole are not very basic.

a. These compounds are protonated only in strong acid, and protonation occurs on carbon, not nitrogen. Although protonation of the carbon of pyrrole disrupts the aromatic π-electron system, the resulting cation is resonance-stabilized.
b. Protonation of the pyrrole nitrogen would disrupt the aromatic six π-electron system by taking the nitrogen's unshared pair "out of circulation" and would give a conjugate acid that is not resonance-stabilized.

## B. Acidity

1. Pyrrole and indole are weak acids.
   a. The N—H protons of pyrrole and indole are about as acidic as alcohol O—H protons.
   b. Pyrrole and indole are acidic enough to behave as acids toward basic organometallic compounds.

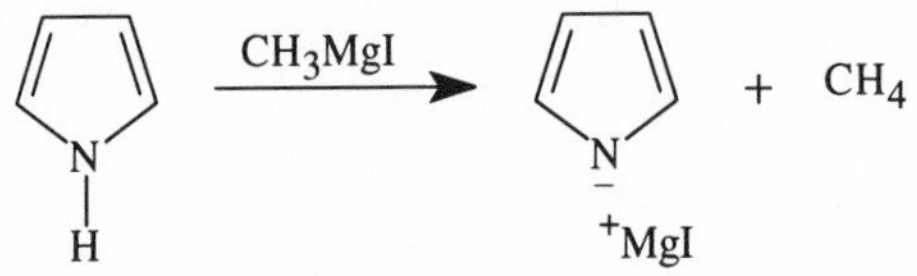

2. The greater acidity of these compounds, relative to the acidity of amines, is a consequence of the resonance stabilization and aromaticity of the conjugate-base anions.

## IV. Sources of Aromatic Nitrogen Heterocyclic Compounds

### A. Indoles

1. The Fischer indole synthesis is an important method for the preparation of indoles that involves the reaction of a phenylhydrazine derivative with an aldehyde or ketone that has α-hydrogens.
   a. The mechanism of the Fischer indole synthesis begins with the conversion of the carbonyl compound into a phenylhydrazone, a type of imine.
      i. The phenylhydrazone, which is protonated under the reaction conditions, is in equilibrium with a small amount of a protonated enamine tautomer.
      ii. The latter species undergoes a pericyclic reaction involving three electron pairs (six electrons) to give a new intermediate in which the N—H bond of the phenylhydrazone has been broken (the intermediate formed is an imine).
      iii. The imine, after protonation on the imine nitrogen, undergoes nucleophilic addition with the amine group in the same molecule with loss of ammonia.
      iv. The resulting "enamine" derivative is the product indole.

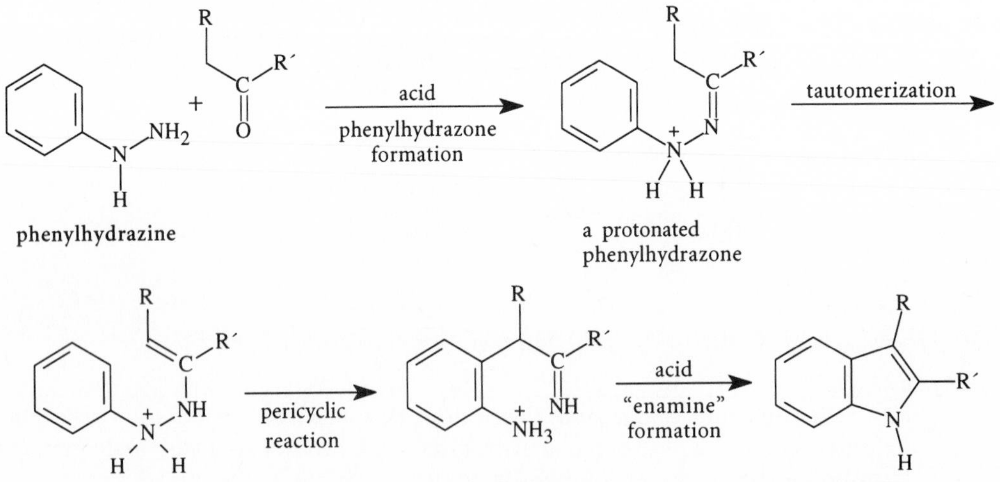

   b. The Fisher synthesis is most often used with phenylhydrazine itself to obtain indoles that are substituted at the 2- or 3-position.
2. The Reissert synthesis occurs under basic conditions; the key starting materials for this synthesis are diethyl oxalate and *o*-nitrotoluene or a substituted derivative.

a. The mechanism of the Reissert synthesis begins with the condensation of the *o*-nitrotoluene with diethyl oxalate in a variation of the Claisen condensation. (The *o*-nitro group is an essential element in the success of this reaction because its presence makes the benzylic methyl hydrogens acidic enough to be removed by ethoxide.)

   *i.* The nitro group is converted into an amino group in a separate reduction step.

   *ii.* The amino group thus formed reacts with neighboring ketone to yield, after acid-base equilibria, an "enamine," which is the product indole.

b. 2-Indolecarboxylic acid can be decarboxylated to prepare indole itself.

3. The Reissert synthesis is complementary to the Fischer indole synthesis.

   a. If substituted *o*-nitrotoluene derivatives are used in the Reissert reaction, this reaction, in conjunction with the final decarboxylation step, can be used to prepare indoles that are substituted in the benzene ring and unsubstituted at the 2- or 3-positions.

   b. Although many substituted phenylhydrazines work in the Fischer synthesis, some are difficult to prepare; thus, the Fischer synthesis is most often used to prepare indoles that are substituted at the 2- or 3-positions rather than in the phenyl ring.

## B. PYRIDINES

1. The Chichibabin reaction can be used to prepare 2-aminopyridines.

   a. In the first step of the mechanism, the amide ion (nucleophile) attacks the 2-position of the ring to form a tetrahedral addition intermediate.

      *i.* The C=N linkage of the pyridine ring is somewhat analogous to the C=O of a carbonyl group.

      *ii.* The carbon at the 2-position of pyridine has some of the character of a carbonyl carbon, and can thus be attacked by nucleophiles.

      *iii.* However, the C=N group of pyridine is much less reactive than a carbonyl group because it is part of an aromatic system.

   b. In the second step of the mechanism, the leaving group, a hydride ion, is lost; hydride ion is a very poor leaving group because it is very basic.

      *i.* The aromatic pyridine ring is re-formed; aromaticity lost in the formation of the tetrahedral addition intermediate is regained when the leaving group departs.

      *ii.* The basic sodium hydride produced in the reaction reacts with the —NH$_2$ group irreversibly to form hydrogen gas and the resonance-stabilized conjugate-base anion of 2-aminopyridine.

   c. The neutral 2-aminopyridine is formed when water is added in a separate step.

2. The 2-aminopyridines formed in the Chichibabin reaction serve as starting materials for a variety of other 2-substituted pyridines.

3. Pyridinium salts are activated toward nucleophilic displacement of groups at the 2- and 4-positions of the ring much more than are pyridines themselves, because the positively charged nitrogen is much more electronegative than the neutral nitrogen of a pyridine.

   a. When the nucleophiles in such displacement reactions are anions, charge is neutralized.

b. Pyridine-*N*-oxides are in one sense pyridinium ions, and they react with nucleophiles in much the same way as quaternary pyridinium salts.

## C. QUINOLINES

1. A number of reasonably versatile syntheses of quinolines from acyclic compounds are known.
2. One of the best known syntheses is the Skraup synthesis:
   a. In the synthesis of quinoline itself, glycerol undergoes an acid-catalyzed dehydration to provide a small but continuously replenished amount of acrolein, an $\alpha,\beta$-unsaturated aldehyde. (If acrolein itself were used as a reactant at high concentrations, it would polymerize.)
   b. Aniline undergoes a conjugate addition with the acrolein.
   c. The resulting aldehyde, after protonation, acts as an electrophile in an intramolecular electrophilic aromatic substitution reaction.
   d. Dehydration of the resulting alcohol yields a 1,2-dihydroquinoline.
   e. The 1,2-dihydroquinoline product differs from quinoline by only one degree of unsaturation, and is readily oxidized to the aromatic quinoline by mild oxidants.

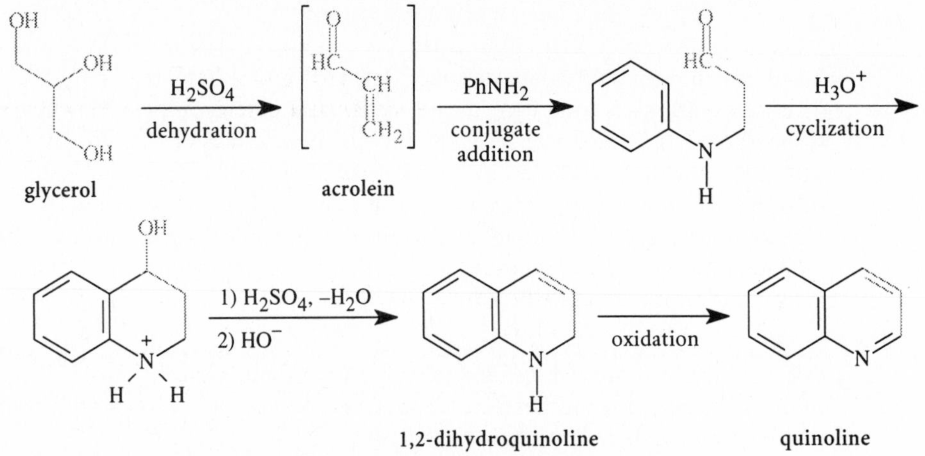

## D. NATURAL OCCURRENCE OF HETEROCYCLIC COMPOUNDS

1. A number of monosubstituted pyridines are available from natural sources.
   a. Methylpyridines (or picolines) are obtained from coal tar.
   b. Nicotinic acid (pyridine-3-carboxylic acid) can be prepared by the side-chain oxidation of nicotine, an alkaloid present in tobacco.
2. Nitrogen heterocycles occur widely in nature, for example, the alkaloids (many of which contain heterocyclic ring systems).
3. The naturally occurring amino acids proline, histidine, and tryptophan contain respectively a pyrrolidine, imidazole, and indole ring.
4. A number of vitamins are heterocyclic compounds.
5. The nucleic acids contain purine and pyrimidine rings in combined form.
6. The color of blood is due to an iron complex of heme, a heterocycle (called a porphyrin) composed of pyrrole units.
7. The green color of plants is caused by chlorophyll, a class of compounds closely related to the porphyrins.

**Reactions**

## I. Electrophilic Aromatic Substitution Reactions of Naphthalene and Its Derivatives

### A. ELECTROPHILIC AROMATIC SUBSTITUTION REACTIONS OF NAPHTHALENE

1. Naphthalene is an aromatic compound that undergoes electrophilic aromatic substitutions much like those of benzene, generally at the 1-position.

a. Seven resonance structures can be drawn for the carbocation intermediate in electrophilic aromatic substitution at the 1-position, and four of these contain intact benzene rings (black).

*i.* Structures in which benzene rings are left intact are more important than those in which the formal double bonds are moved out of the ring.

*ii.* Structures lacking the intact benzene rings are not aromatic, and are thus less stable.

b. Six resonance structures can be drawn for the carbocation intermediate in electrophilic aromatic substitution at the 2-position, and only two contain intact benzene rings (black).

c. Substitution at the 1-position gives the more stable carbocation intermediate and, by Hammond's postulate, occurs more rapidly than substitution at the 2-position.

*i.* There is nothing wrong with the 2-substitution.

*ii.* 1-Substitution is simply more favorable.

2. As with benzene, sulfonation of naphthalene is a reversible reaction.

a. Sulfonation of naphthalene under mild conditions gives mostly 1-naphthalenesulfonic acid; under more vigorous conditions, sulfonation yields mostly 2-naphthalenesulfonic acid.

  b. This is a case of kinetic *vs.* thermodynamic control of a reaction.
    *i.* At low temperature, substitution at the 1-position is observed because it is faster.
    *ii.* At higher temperature, formation of 1-naphthalenesulfonic acid is reversible, and the more stable but more slowly formed 2-naphthalenesulfonic acid is observed.
  c. In the 1-position, van der Waals repulsions occur between the large sulfonic acid group and the adjacent hydrogen in the 8-position.

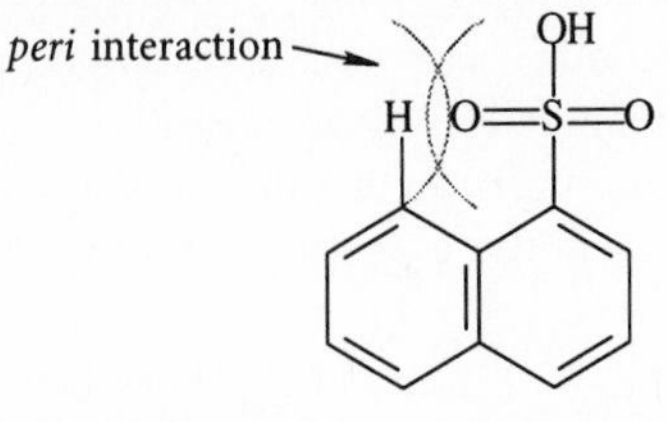

    *i.* This interaction, called a *peri* interaction, destabilizes the 1-isomer.
    *ii.* A *peri* interaction is much more severe than the interaction of the same two groups in *ortho* positions.
    *iii.* 1-Naphthalenesulfonic acid, if allowed to equilibrate, is converted into the 2-isomer to avoid this unfavorable steric interaction.
  3. Naphthalene is considerably more reactive than benzene in electrophilic aromatic substitution.
    a. Naphthalene is readily brominated in $CCl_4$ without a catalyst.
    b. The greater reactivity of naphthalene in electrophilic aromatic substitution reactions reflects the considerable resonance stabilization of the carbocation intermediate.

## B.  Electrophilic Aromatic Substitution Reactions of Substituted Naphthalenes

  1. The position of a second substitution on a monosubstituted naphthalene depends on the substituent present.
  2. The following trends are observed in most cases:
    a. When one ring of naphthalene is substituted with deactivating groups, further substitution occurs on the unsubstituted ring at an open $\alpha$-position (if available).

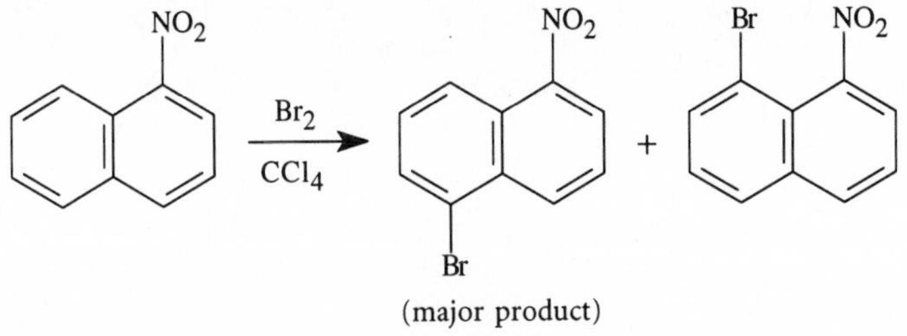

    b. When one ring of naphthalene is substituted with activating groups, further substitution occurs in the substituted ring at the *ortho* or *para* positions.

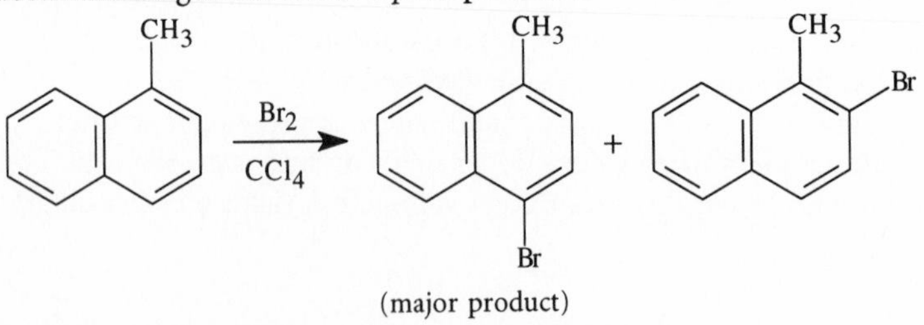

## II.  *Electrophilic Aromatic Substitution Reactions of Aromatic Heterocyclic Compounds*

### A.  Furan, Pyrrole, and Thiophene

  1. Furan, thiophene, and pyrrole undergo electrophilic substitution predominantly at the 2-position of the ring.

a. The carbocation resulting from substitution at carbon-2 has more important resonance structures, and is therefore more stable, than the carbocation resulting from attack at carbon-3.

b. There is nothing wrong with reaction at carbon-3; reaction at carbon-2 is simply more favorable.
c. Some carbon-3 substitution product accompanies the major carbon-2 substitution product in many cases.

2. Furan, pyrrole, and thiophene are all much more reactive than benzene in electrophilic aromatic substitution reactions.
   a. Precise reactivity ratios depend on the particular reaction.
   b. Milder reaction conditions must be used with more reactive compounds.
   c. The reactivity order of the heterocycles is a consequence of the relative abilities of the heteroatoms to stabilize positive charge in the intermediate carbocations.

   pyrrole > furan > thiophene >> benzene

   i. Both pyrrole and furan have heteroatoms from the first row of the periodic table.
   ii. Because nitrogen is better than oxygen at delocalizing positive charge (it is less electronegative), pyrrole is more reactive than furan.
   iii. The sulfur of thiophene is a second row element and, although it is less electronegative than oxygen, its $3p$ orbitals overlap less efficiently with the $2p$ orbitals of the aromatic $\pi$-electron system.

3. The reactivity order of the heterocycles in aromatic substitution parallels the reactivity order of the corresponding substituted benzene derivatives:

   $(CH_3)_2N$—Ph > $CH_3O$—Ph > $CH_3S$—Ph

4. The usual activating and directing effects of substituents in aromatic substitution apply; superimposed on these effects is the normal effect of the heterocyclic atom in directing substitution to the 2-position.
   a. Count around the carbon framework of the heteroatom compound, not through the heteroatom, when using the *ortho, meta, para* analogy.

   *para* to fluorine

   b. When the directing effects of substituents and the ring compete, it is not unusual to observe mixtures of products.
   c. If both 2-positions are occupied, 3-substitution takes place.

## B. PYRIDINE

1. In general, pyridine has very low reactivity in electrophilic aromatic substitutions; it is much less reactive than benzene.
   a. An important reason for this low reactivity is that pyridine is protonated under the very acidic conditions of most electrophilic aromatic substitution reactions.

   b. The resulting positive charge on nitrogen makes it difficult to form a carbocation intermediate, which would place a second positive charge within the same ring.

2. Pyridine rings substituted with activating groups such as methyl groups do undergo electrophilic aromatic substitution reactions.
   a. When substitution in pyridine does occur, it generally takes place at the 3-position.
   b. The preference for 3-substitution in pyridines can be understood by considering the resonance structures for the possible carbocation intermediates.
     *i.* Substitution in the 3-position gives a carbocation with three resonance structures.

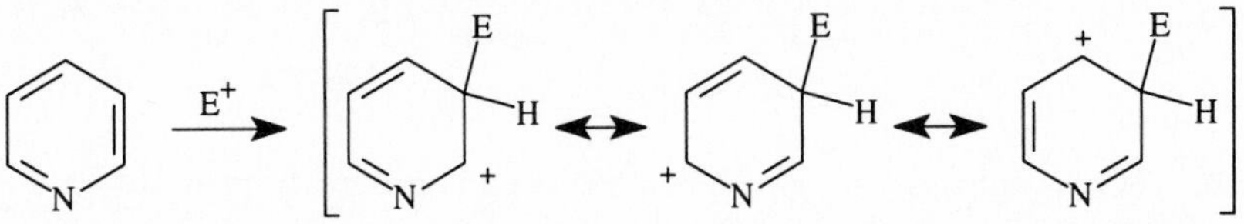

     *ii.* Substitution at the 4-position also involves an intermediate with three resonance structures, but the one shown in gray is particularly unfavorable because the nitrogen, an electronegative atom, is electron deficient.

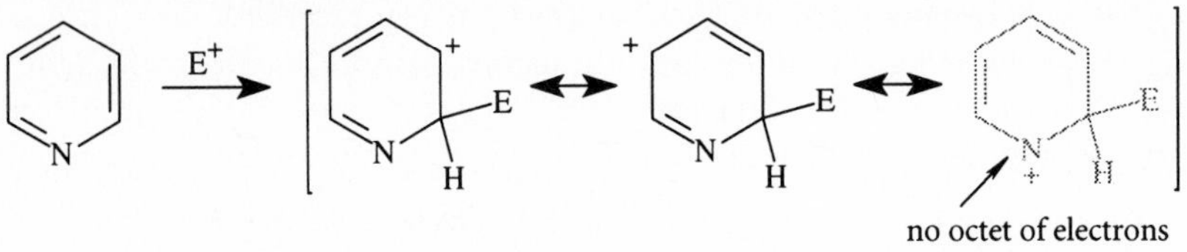

no octet of electrons

3. Pyridine-*N*-oxide is much more reactive than pyridine and undergoes useful aromatic substitution reactions, and substitution occurs in the 4-position.
   a. Pyridine-*N*-oxide is formed by oxidation of pyridine with 30% hydrogen peroxide.

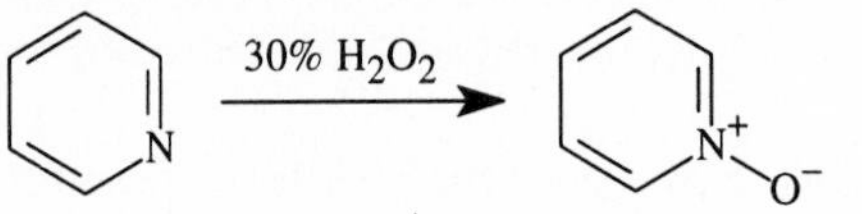

   b. The *N*-oxide function can be removed by catalytic hydrogenation, which will also reduce any nitro groups present.
   c. Reaction with trivalent phosphorous compounds, such as $PCl_3$, removes the *N*-oxide function without reducing nitro groups.

4. Quinoline-*N*-oxide undergoes reactions similar to pyridine-*N*-oxide.

## III. Nucleophilic Aromatic Substitution Reactions of Aromatic Heterocyclic Compounds

### A. PYRIDINE

1. Treatment of a pyridine derivative with the strong base sodium amide ($Na^+ {}^-NH_2$) brings about the direct substitution of an amino group for a ring hydrogen; this reaction is called the Chichibabin reaction.

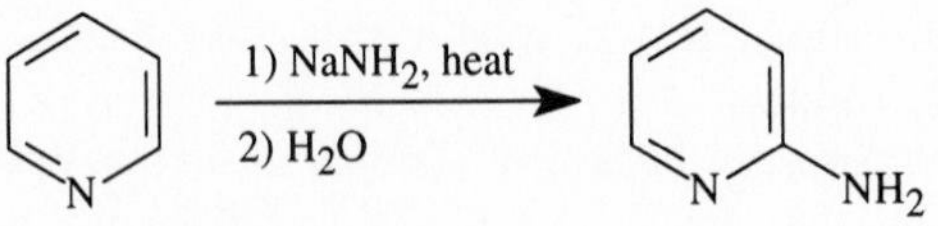

2. A reaction similar to the Chichibabin reaction occurs with organolithium reagents.

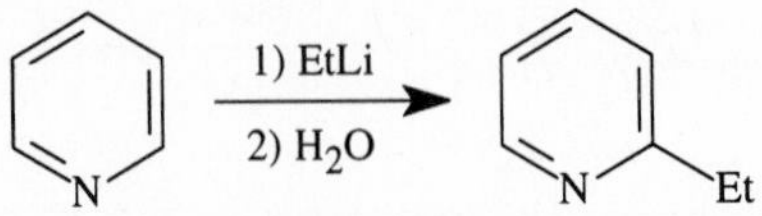

3. When pyridine is substituted with a better leaving group than hydride at the 2-position, it reacts more rapidly with nucleophiles; thus, the 2-halopyridines readily undergo substitution of the halogen by other nucleophiles under conditions milder than those used in the Chichibabin reaction.

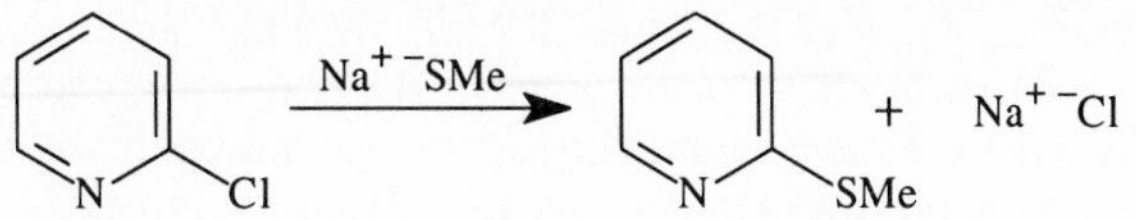

4. Nucleophilic substitution reactions on pyridine rings can be classified as nucleophilic aromatic substitution reactions.
   a. The "electron-withdrawing group" in the reaction of pyridines is the pyridine nitrogen itself; consider the ring nitrogen of pyridine as if it were a nitro group attached to a benzene ring.
   b. The tetrahedral addition intermediate is analogous to the Meisenheimer complex of nucleophilic aromatic substitution.

5. 3-Substituted pyridines are not reactive in nucleophilic substitution because negative charge in the addition intermediate cannot be delocalized onto the electronegative nitrogen.

6. Although 2-pyridone exists largely in the carbonyl form, it nevertheless undergoes some reactions reminiscent of hydroxy compounds.

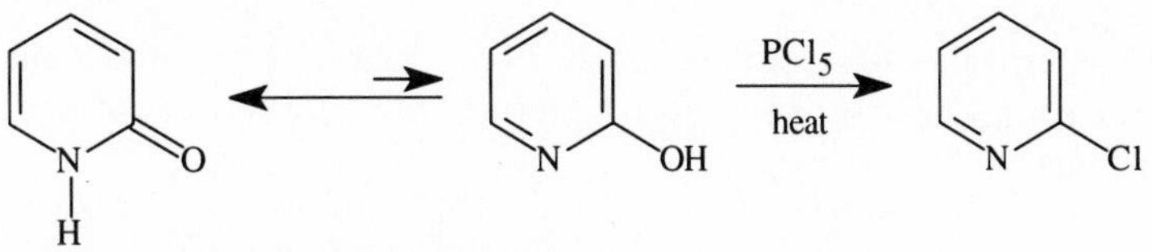

## B. PYRIDINIUM SALTS AND THEIR REACTIONS

1. Pyridine is a nucleophile and reacts in $S_N2$ reactions with alkyl halides or sulfonate esters to form quaternary ammonium salts, called pyridinium salts.

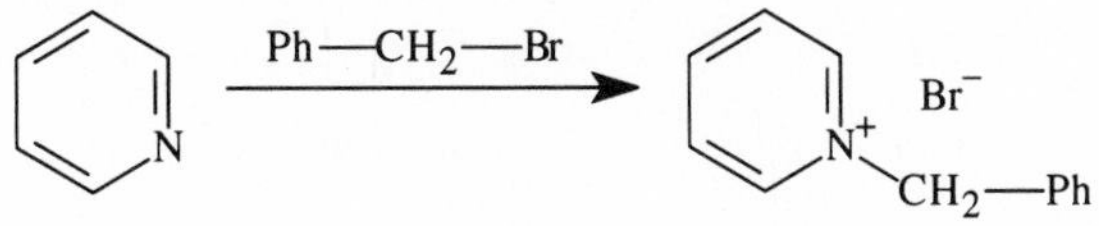

## IV. Other Reactions of Heterocyclic Aromatic Compounds

### A. ADDITION REACTIONS TO FURAN

1. Furan, pyrrole, or thiophene can be viewed as a 1,3-butadiene with its terminal carbons "tied down" by a heteroatom bridge.

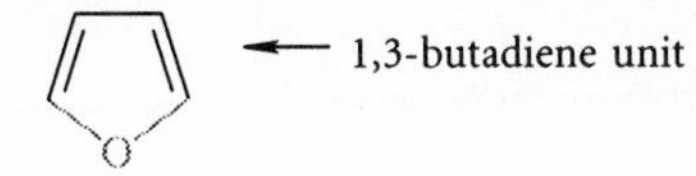

2. Of the three heterocyclic compounds furan, pyrrole, and thiophene, furan has the least resonance energy.
   a. Furan has the greatest tendency to behave like a conjugated diene.
   b. Furan undergoes some conjugate-addition reactions.
   c. Furan undergoes Diels-Alder reactions with reactive dienophiles.

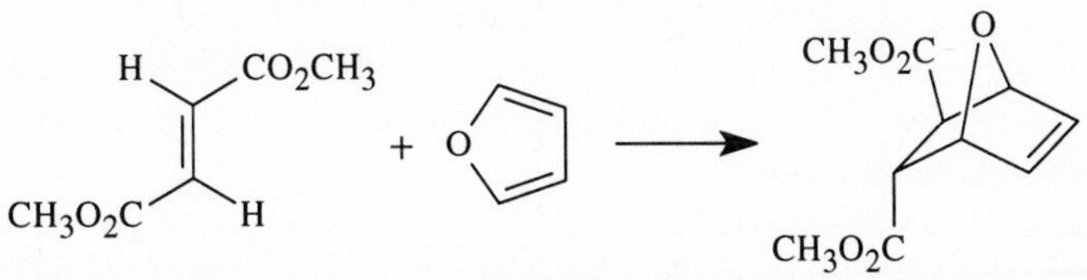

### B. SIDE-CHAIN REACTIONS OF FURAN, PYRROLE, AND THIOPHENE

1. Many reactions occur at the side chains of heterocyclic compounds without affecting the rings.
2. Decarboxylation of carboxylic acid groups directly attached to a heterocyclic ring is important in the synthesis of some unsubstituted heterocyclic compounds.

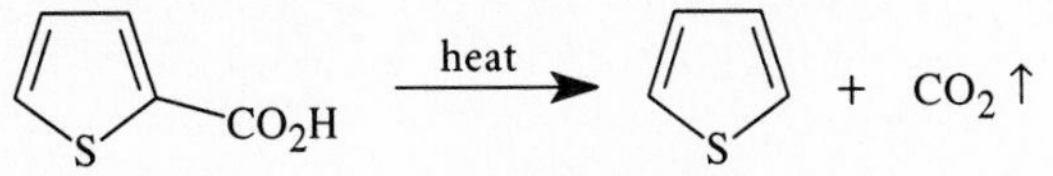

## C.  SIDE-CHAIN REACTIONS OF PYRIDINE DERIVATIVES

1.  The "benzylic" hydrogens of an alkyl group at the 2- or 4-position of a pyridine ring are more acidic than ordinary benzylic hydrogens because the electron pair (and charge) in the conjugate-base anion is delocalized onto the electronegative pyridine nitrogen.

   a.  Strongly basic reagents such as organolithium reagents or $NaNH_2$ abstract the benzylic proton from 2- or 4-alkylpyridines.

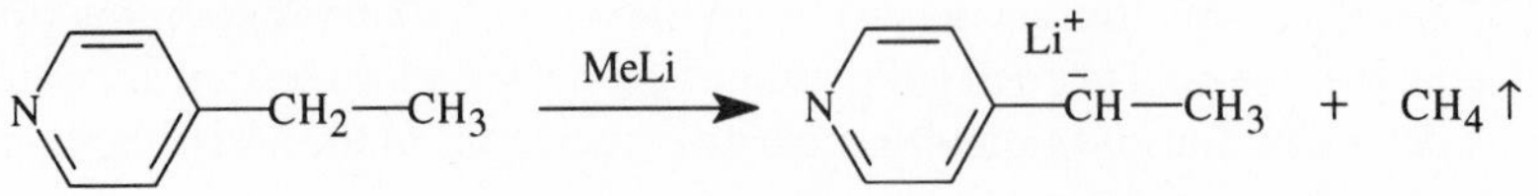

   b.  The anion formed in this way has a reactivity much like that of other organolithium reagents.

   c.  These anions are somewhat analogous to enolate ions, and thus they undergo some of the reactions of enolate anions, such as aldol condensations.

2.  The "benzylic" hydrogens of 2- or 4-alkylpyridinium salts are much more acidic than those of the analogous pyridines.

   a.  One resonance form of the conjugate-base "anion" is a neutral compound.

   b.  The conjugate base "anion" can be formed in useful concentrations by aqueous NaOH or amines.

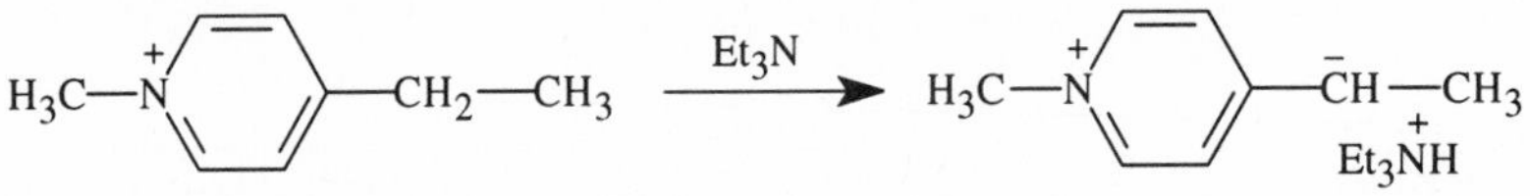

3.  Many side-chain reactions of pyridines, for example, side-chain oxidation, are analogous to those of the corresponding benzene derivatives.

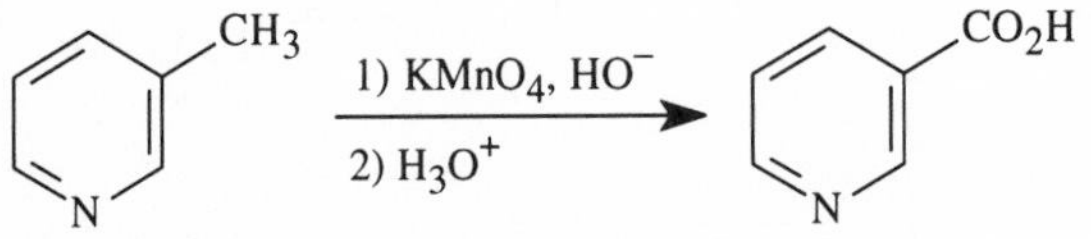

## V.  *Synthesis of Indole and Quinoline from Acyclic Starting Materials*

### A.  SYNTHESIS OF INDOLES—THE FISCHER AND REISSERT SYNTHESES

1.  In the Fischer indole synthesis, an aldehyde or ketone with at least two $\alpha$-hydrogens is reacted with a phenylhydrazine derivative in the presence of an acid catalyst and/or heat.  (See Concepts, Sec. IV.A.1., for a summary of the mechanism.)

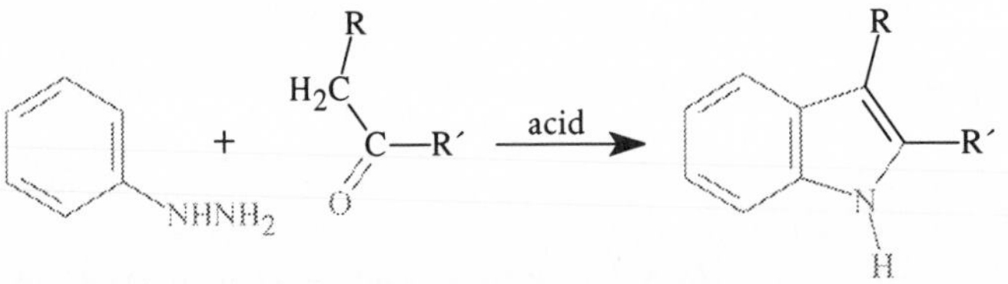

   a.  A variety of Brønsted or Lewis acid catalysts can be used.

   b.  The reaction works with many different substituted phenylhydrazines and carbonyl compounds.

   c.  Acetaldehyde, however, does not work in this reaction, probably because it polymerizes under the reaction conditions.

2.  The Reissert synthesis gives substituted 2-indolecarboxylic acids from diethyl oxalate and substituted *o*-nitrotoluenes; decarboxylation yields a substituted indole.  (See Concepts, Sec. IV.A.2., for a summary of the mechanism.)

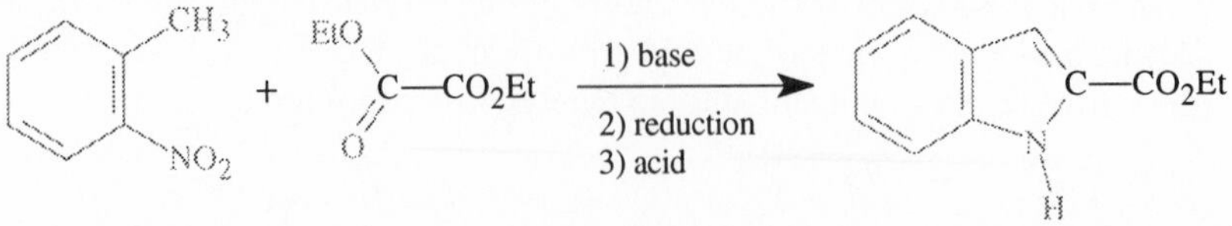

B. SYNTHESIS OF QUINOLINES

1. The Skraup synthesis of quinolines occurs under acid catalysis; the key starting materials for this reaction are glycerol and aniline or an aniline derivative. (See Concepts, Sec. IV.C.2., for a summary of the mechanism.)

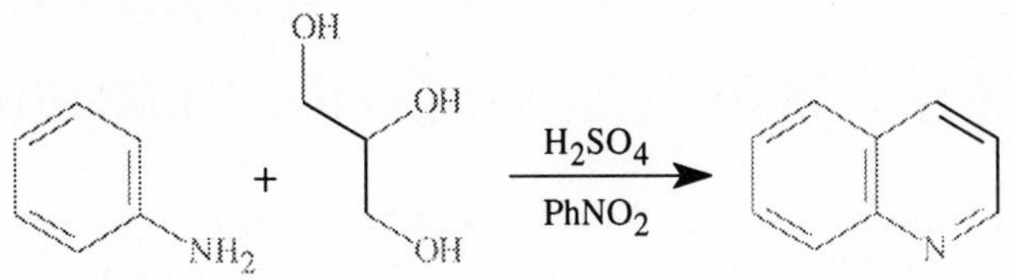

2. $\alpha,\beta$-Unsaturated aldehydes and ketones that are less prone to polymerize than acrolein can be used instead of glycerol in the Skraup synthesis to give substituted quinolines.

## Study Guide Links

### 24.1 Relative Acidities of 1,3-Cyclopentadiene and Pyrrole

It is interesting that pyrrole, a nitrogen acid, is somewhat *less* acidic than 1,3-cyclopentadiene, a carbon acid.

1,3-cyclopentadiene
$pK_a \approx 15$

pyrrole
$pK_a \approx 17$

This contrasts with the general expectation from the *element effect* (text p. 107) that hydrogens on more electronegative atoms are more acidic than those on less electronegative atoms. (For example, the $pK_a$ of $NH_3$ is about 35; the $pK_a$ of $CH_4$ is about 55; the N—H $pK_a$ of amides is about 16; the $\alpha$-hydrogen $pK_a$ of a ketone is about 19.) The greater acidity of 1,3-cyclopentadiene shows the importance of aromaticity. 1,3-Cyclopentadiene itself is not aromatic, but its conjugate-base anion is. Consequently, 1,3-cyclopentadiene "gains aromaticity" when it ionizes. This effect lowers the $pK_a$ of 1,3-cyclopentadiene significantly from what it otherwise would be. (The effect of aromaticity might be worth as much as 20 $pK_a$ units.) In contrast, both pyrrole and its conjugate-base anion are aromatic; ionization of pyrrole does not gain any aromatic stability for the molecule. In summary, then, the increase in stability resulting from the formation of an aromatic species causes the $pK_a$ of the carbon acid to be lower than that of the nitrogen acid.

### ✓24.2 Fischer Indole Synthesis

Although the reaction in Eq. 24.38c on text p. 1191 is a concerted reaction, thinking of it as a stepwise process helps to understand why it takes place.

The weakest bond in the protonated enamine is the bond between the two nitrogens; bonds between two electronegative atoms are typically rather weak. Moreover, the protonated nitrogen is especially electronegative, and electrons from the N—N bond are drawn to that nitrogen:

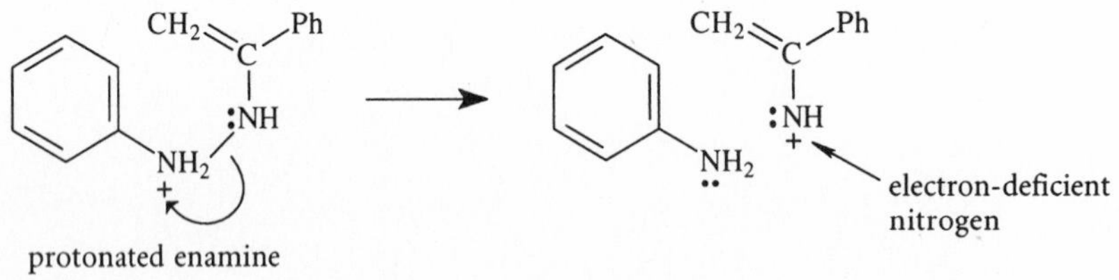

protonated enamine

electron-deficient nitrogen

The species with the electron-deficient nitrogen is really a carbocation:

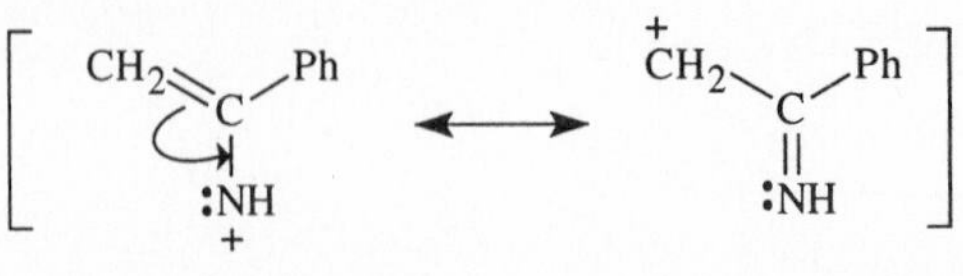

The carbocation carries out an electrophilic substitution on the neighboring aniline, which is very reactive in such reactions (Sec. 23.9 on text p. 1137):

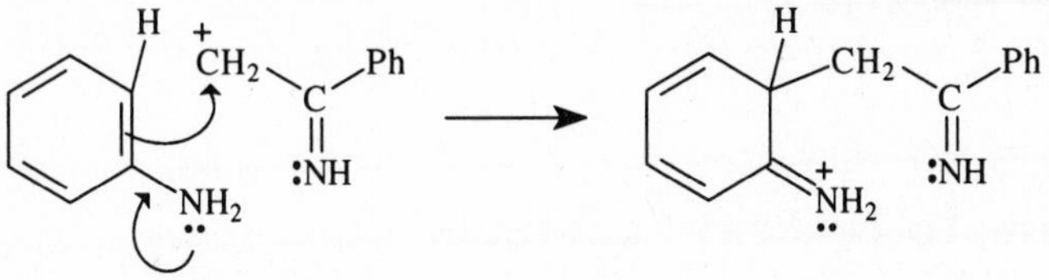

This is the product of the pericyclic process, which goes on to form the imine intermediate as shown in the last part of text Eq. 24.38c.

## 24.3  Dehydration of Glycerol

The dehydration of glycerol that is part of the Skraup quinoline synthesis is actually an example of what is called the *pinacol rearrangement*. This rearrangement was introduced in Study Problem 19.2 on text p. 882. The steps in this rearrangement can be outlined as follows; you draw the curved-arrow formalism.

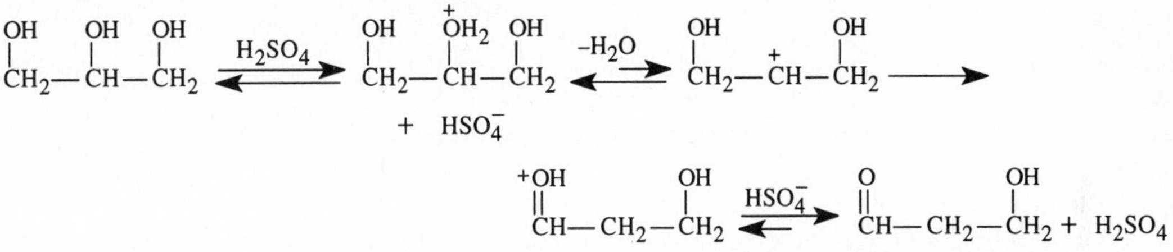

3-hydroxypropanal

Dehydration of 3-hydroxypropanal to acrolein probably proceeds by way of the enol; this type of reaction is discussed in Study Guide Link 22.3 on p. 581 of this manual.

## Solutions

### Solutions to In-Text Problems

**24.1** (a) Naphthalene has four nonequivalent sets of carbon-carbon bonds, shown as *a–d* in the structure below. Focus on *one* bond of each set and determine the number of resonance structures in which it appears as a double bond. Only bond *b* is a double bond in two of the three resonance structures (Eq. 24.1 on text p. 1171). Because no other bond has so much double-bond character, and because double bonds are shorter than single bonds, bond *b* is the shortest bond. (This hypothesis is confirmed by the data: the bond lengths are *a* and *d*, 1.42 Å; *b*, 1.36 Å, and *c*, 1.41 Å.)

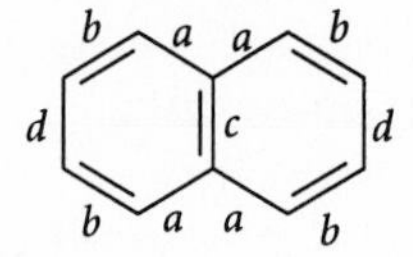

**24.2** (a) 5,8-Dinitro-2-naphthalenecarboxylic acid
(c) 2-Naphthyl ethanoate (common: β-naphthyl acetate)

**24.3** Because the substituted ring is deactivated, and because sulfonation is reversible, the sulfonic acid group in (a) is introduced into the β positions of the unsubstituted ring. Because the substituted ring in (c) is activated, and because the hydroxy group is an *ortho, para*-directing substituent, diazo coupling occurs at the 4-position.

(a)

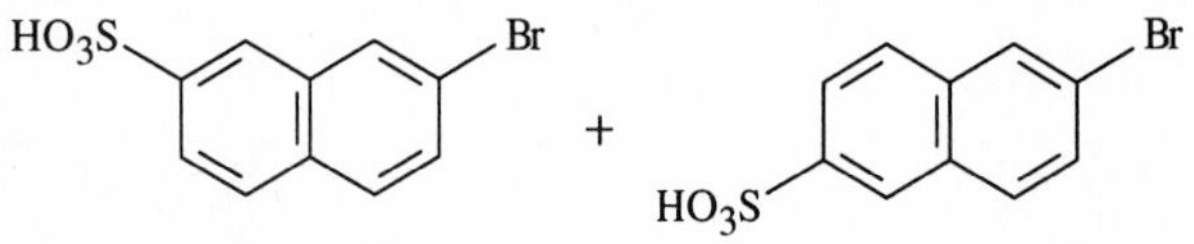

(c)

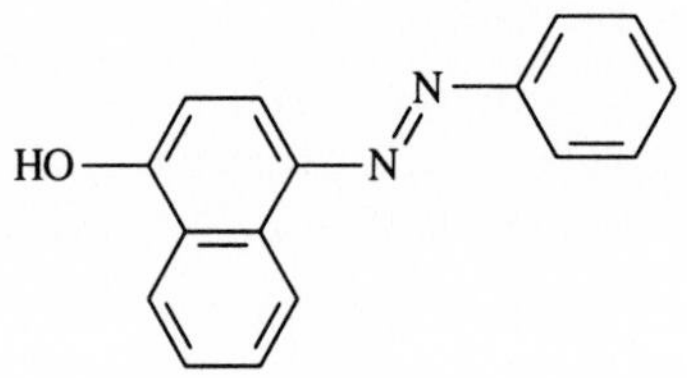

**24.4** Use a Grignard reaction to form the carbon-carbon bond.

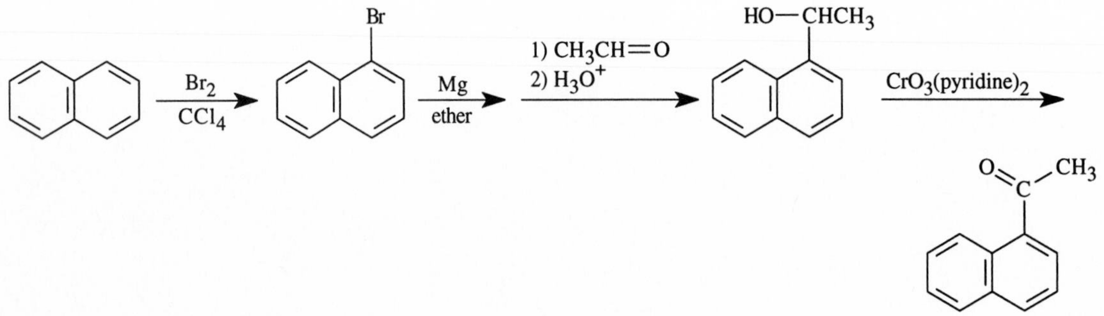

**24.5** (a)

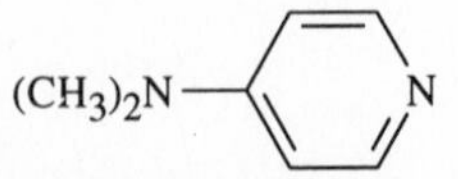

4-(dimethylamino)pyridine

**24.6** (a) 5-Bromo-2-methylthiazole
(c) 8-Methoxyquinoline

**24.7** The resonance structures of pyrrole:

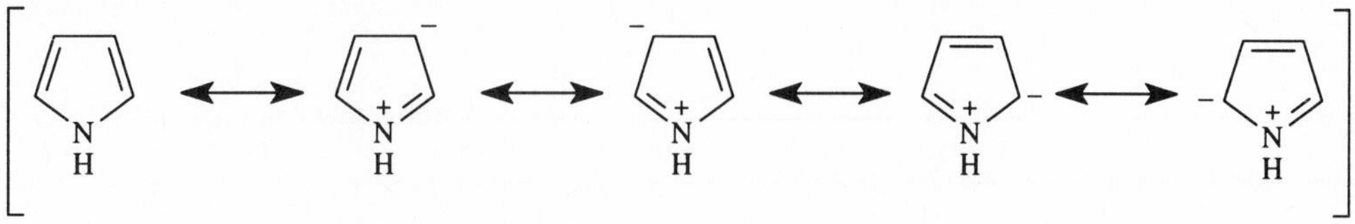

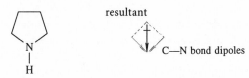

**24.8** (a) Because nitrogen is an electronegative atom, the C—N bond dipoles in pyrrolidine are directed towards the nitrogen, and their resultant is also directed towards the nitrogen:

It is given that the dipole moment of pyrrole is in the opposite direction, that is, directed away from the nitrogen. Although there is a dipole component directed towards the nitrogen as shown for pyrrolidine, the major component is due to the charge separation shown in the resonance structures given in the solution to Problem 24.7. In all resonance structures but the first, positive charge resides on the nitrogen and negative charge is on one of the carbons. Evidently, these contributions significantly outweigh those from the carbon-nitrogen bond dipoles.

(b) Oxygen is more electronegative than nitrogen. Consequently, the contribution of the carbon-oxygen bond dipoles in furan is greater than the contribution of the carbon-nitrogen bond dipoles in pyrrole, and the contribution of the charge-separated resonance structures for furan is smaller than the contribution of the analogous structures for pyrrole. Notice that the resonance energy of furan is less than that of pyrrole (Table 24.1 on text p. 1181); this is consistent with the idea that the resonance structures of furan are less important than those of pyrrole.

In summary: In furan, the sum of the C—O bond dipoles outweighs the dipole-moment contribution of $\pi$-electron delocalization (see Eq. 24.13 on text p. 1179). In pyrrole, the dipole-moment contribution of $\pi$-electron delocalization outweighs the sum of its C—N bond dipoles.

**24.9** Both pyrrole and pyridine are aromatic, and therefore both should have ring current-induced downfield chemical shifts. Consequently, the two larger shifts are associated with the two aromatic compounds, and the $\delta$ 2.82 resonance is that of pyrrolidine. This chemical shift is not far from the value predicted for the $\alpha$-proton resonance of an alkylamine. (See Table 13.2 on text p. 590.) The resonance structures for pyrrole (see the solution to Problem 24.7, above) as well as its dipole moment (see the solution to Problem 24.8(a), above) show that pyrrole has partial negative charge on carbon-2; in contrast, the resonance structures of pyridine (Eq. 24.14, text p. 1180) show that carbon-2 has partial positive charge. Since electron deficiency is associated with downfield chemical shifts, it follows that the $\delta$ 8.51 resonance is that of pyridine and the $\delta$ 6.41 resonance is that of pyrrole.

**24.10** (a) The pyridine nitrogen is basic. Since bases can accept hydrogen bonds, pyridine can accept hydrogen bonds from water. This hydrogen-bonding capability increases the solubility of pyridine in water. Pyrrole, in contrast, is not basic, and the N—H hydrogen is not particularly acidic. Hence, pyrrole does not form strong hydrogen bonds with water, and it is therefore not very soluble.

**24.11** (a) Protonation on the pyridine nitrogen occurs because the conjugate acid is resonance-stabilized. Protonation on the dimethylamino nitrogen, in contrast, gives a conjugate acid that is not resonance-stabilized.

*(solution continues)*

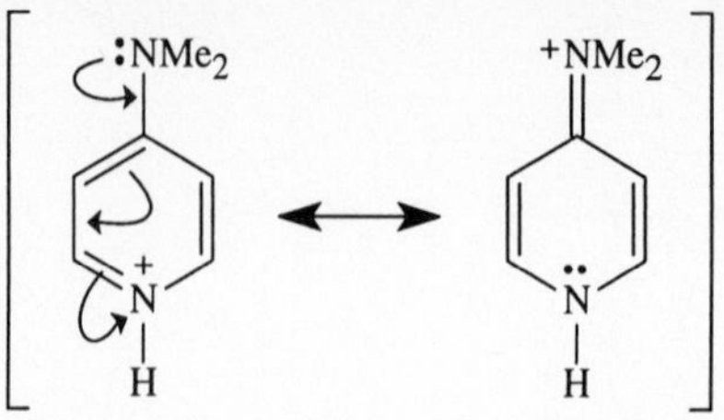

conjugate acid of
4-(dimethylamino)pyridine

4-(Dimethylamino)pyridine is more basic than pyridine because of the electron-donating resonance interaction of the dimethylamino substituent in the conjugate-base cation, shown above. The conjugate-base cation of pyridine itself, of course, lacks this stabilizing interaction.

24.12 The unshared electron pair on the nitrogen of aniline is conjugated with the $\pi$-electron system of the ring. When aniline is protonated, this electron pair is no longer available for conjugation with the ring. Consequently, the reduction in conjugation causes a change in the UV spectrum. (See the solution to Problem 23.64 on pp. 648–649 of this manual.) The pyridine unshared electron pair, in contrast, is not conjugated with the $\pi$-electron system of the ring because its orbital is in the plane of the ring. Consequently, protonation does not affect the conjugated system and therefore does not affect the UV spectrum.

24.13 (a)                              (c)

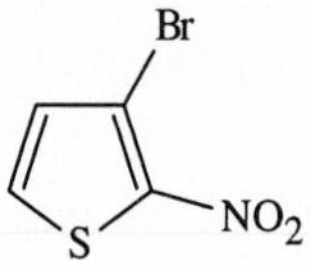

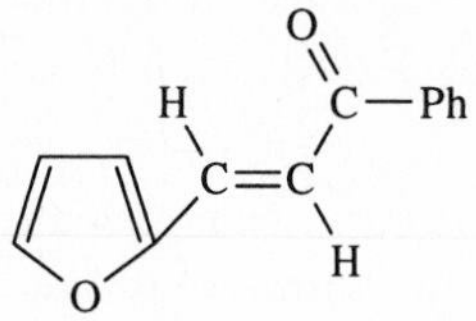

In part (a), the bromine directs the incoming nitro group to the *"ortho"* position, and the thiophene ring is more activated at the same position. Part (c) is an aldol condensation.

24.14 This is an electrophilic aromatic substitution reaction in which the electrophile is the protonated aldehyde. (The carbonyl-protonated aldehyde shown in the mechanism below is undoubtedly in equilibrium with its nitrogen-protonated isomer, but the latter is not shown since it is not involved directly in the mechanism.)

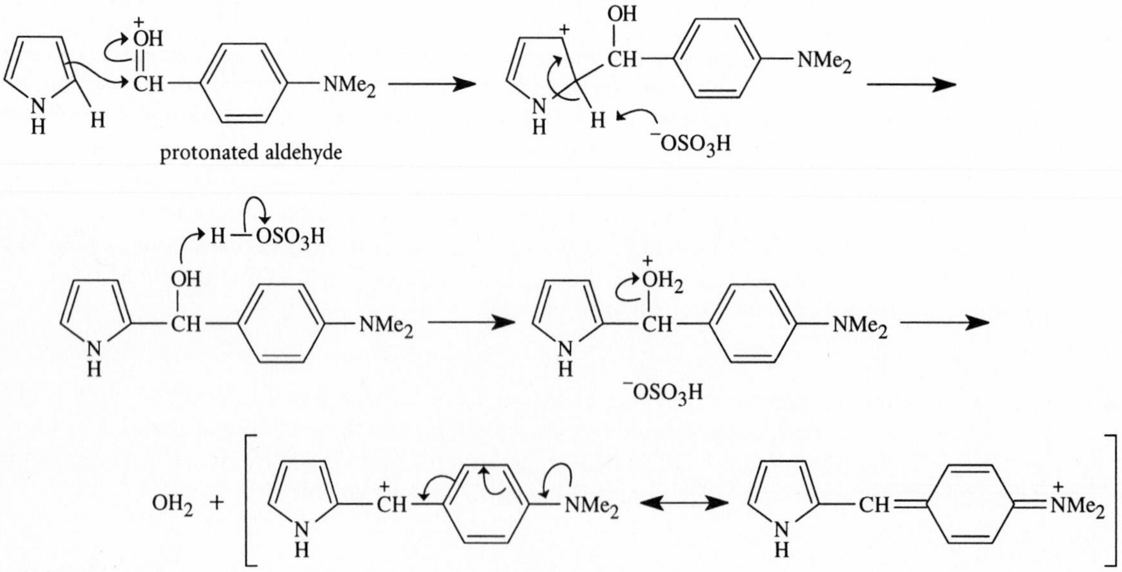

protonated aldehyde

**24.15** (a)

(c)

PhNHNH$_2$ + HCCH$_2$CH(CH$_3$)$_2$          PhNHNH$_2$ +

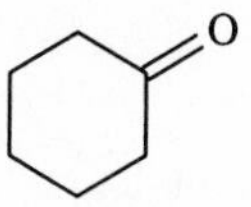

In addition to these starting materials an acid catalyst is required in each case.

**24.16** In the mechanistic step of Eq. 24.38c, a protonated enamine is involved. In the case of 2-butanone, there are two possible enamines and, hence, two possible products.

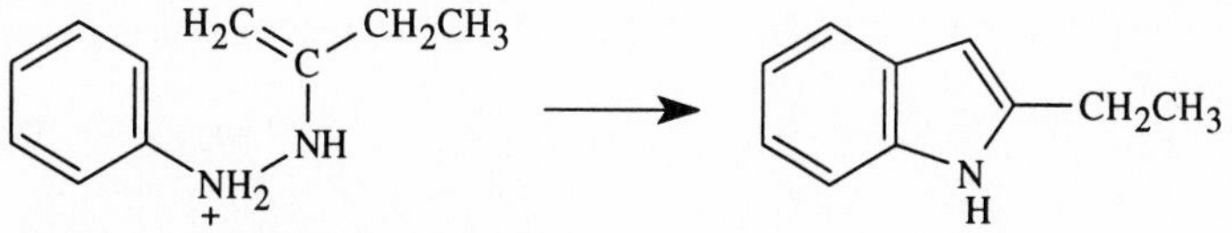

one protonated enamine intermediate

one product of the Fischer indole synthesis with 2-butanone

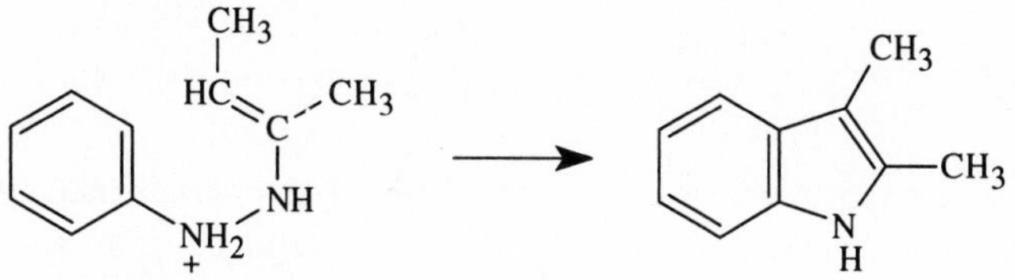

the other protonated enamine intermediate

the other product of the Fischer indole synthesis with 2-butanone

**24.17** (a)

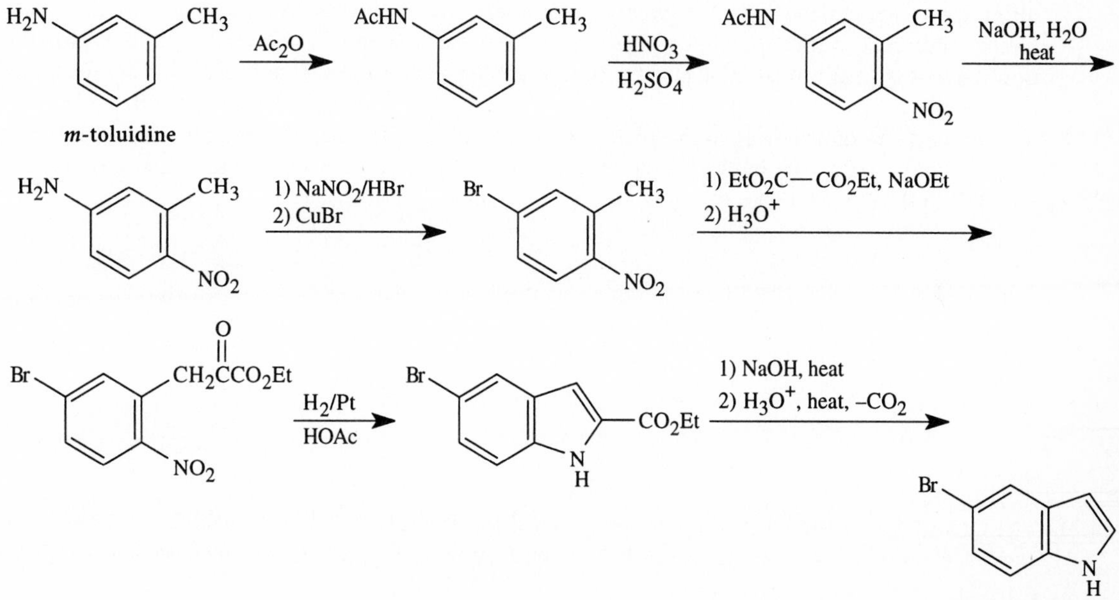

*m*-toluidine

5-bromoindole

**24.18** The structures of the picolines are given on text p. 1194. $\alpha$-Picoline is more reactive because both the methyl substituent group and the ring direct substitution to the 3-position. The carbocation intermediate for substitution at the 3-position has a resonance structure in which the positive charge is on the carbon bearing the methyl group:

*(solution continues)*

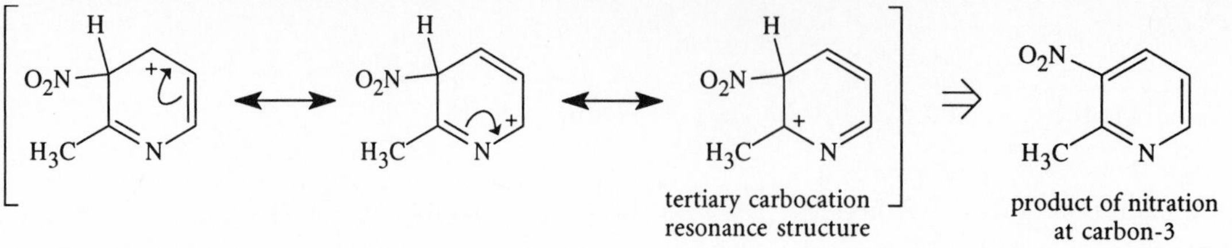

In contrast, nitration of β-picoline gives a carbocation intermediate for which all the resonance structures are secondary carbocations.

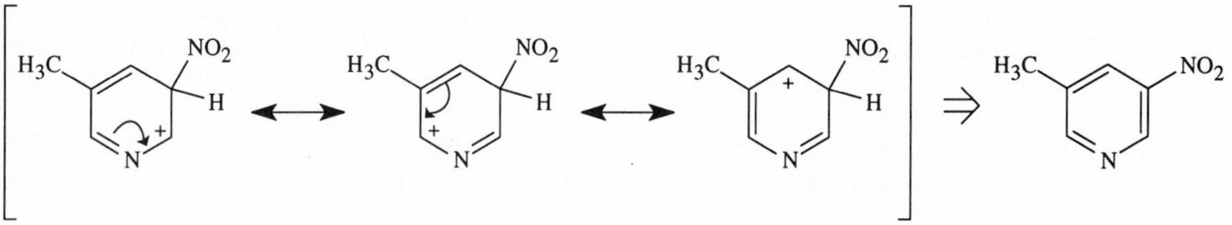

(Nitration at the 2- or 4-position gives a carbocation that has a tertiary resonance structure; however, another resonance structure is unimportant because it has an electron-deficient nitrogen.)

24.20 Electrophilic aromatic substitution on a substituted naphthalene derivative occurs on the more activated ring. Because a pyridine ring is deactivated toward electrophilic aromatic substitution, the pyridine ring of quinoline is less reactive than the phenyl ring; hence, electrophilic substitution in quinoline occurs in the phenyl ring. In naphthalene, the α-positions are more activated towards electrophilic aromatic substitution than the β-positions. Thus, electrophilic aromatic substitution of quinoline occurs on the two α-positions of the phenyl ring. Consequently, nitration of quinoline yields a mixture of nitroquinolines in which the two possible α-positions of the phenyl ring are nitrated, that is, 5-nitroquinoline and 8-nitroquinoline.

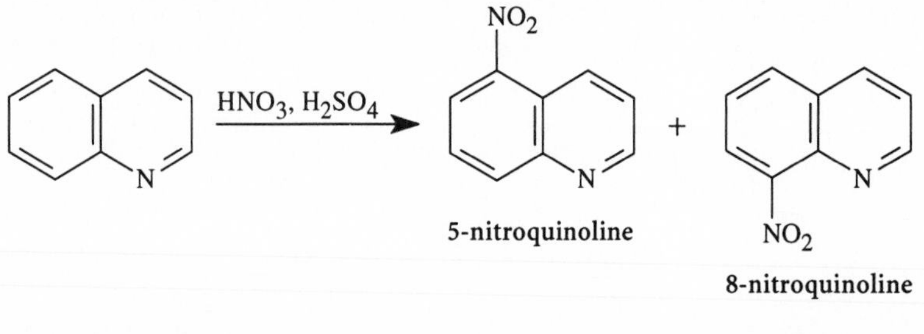

24.21 The curved-arrow mechanism for the nucleophilic aromatic substitution reaction of 4-chloropyridine with methoxide ion is as follows. Notice that negative charge in the anionic intermediate is delocalized onto the nitrogen, an electronegative atom.

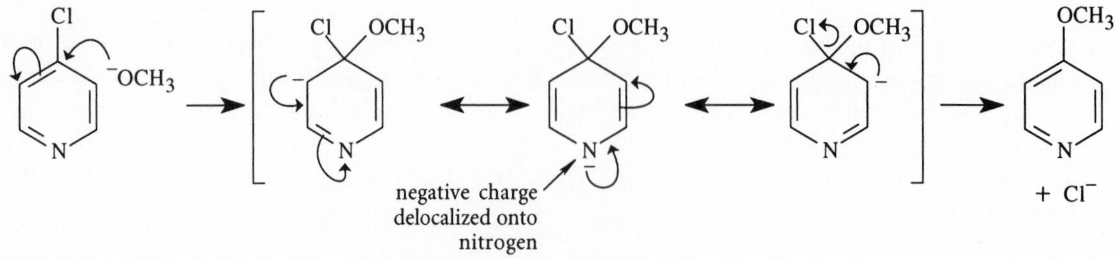

**24.23** (a)

(c)

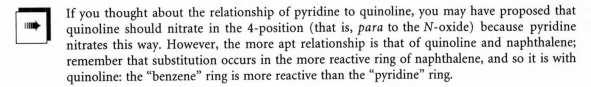

> If you thought about the relationship of pyridine to quinoline, you may have proposed that quinoline should nitrate in the 4-position (that is, *para* to the *N*-oxide) because pyridine nitrates this way. However, the more apt relationship is that of quinoline and naphthalene; remember that substitution occurs in the more reactive ring of naphthalene, and so it is with quinoline: the "benzene" ring is more reactive than the "pyridine" ring.

**24.24** (a) Because nitration of a pyridine ring generally occurs in a 3- or 5-position, the pyridine must first be oxidized to a pyridine-*N*-oxide so that nitration is directed to the 4-position. Once nitrated, the *N*-oxide is reduced back to a substituted pyridine. (See Eq. 24.47 on text p. 1197.)

3-picoline
(β-picoline)

$30\%\ H_2O_2$

$HNO_3$
$H_2SO_4$

$PCl_3$

3-methyl-4-nitropyridine

(c) The "anion" formed by abstraction of a methyl proton by butyllithium can be carbonated with $CO_2$ much as any other organolithium reagent or Grignard reagent can be carbonated.

2-picoline
(α-picoline)

$LiCH_2CH_2CH_2CH_3$

$+\ CH_3CH_2CH_2CH_3$

1) $CO_2$
2) $H_3O^+$

(d) A Hofmann or Curtius rearrangement can be used to introduce an amino group with loss of a carbon atom.

3-picoline
(β-picoline)

$HNO_3$, heat

1) $SOCl_2$
2) $NH_3$

$Br_2$, NaOH

3-aminopyridine

**24.25** (a) The formula of the product shows that one methyl has been introduced. The conjugate-base anion of 3,4-dimethylpyridine is formed by proton removal from the 4-methyl group, and this anion is alkylated by methyl iodide. A 4-methyl hydrogen is considerably more acidic than a 3-methyl hydrogen because the anion *A* that results from removal of a 4-methyl hydrogen is more stable. It is more stable because the negative charge can be delocalized to the pyridine nitrogen. (Draw the appropriate resonance structures.) The negative charge in the anion derived from ionization of the 3-methyl group cannot be delocalized to the nitrogen.

*(solution continues)*

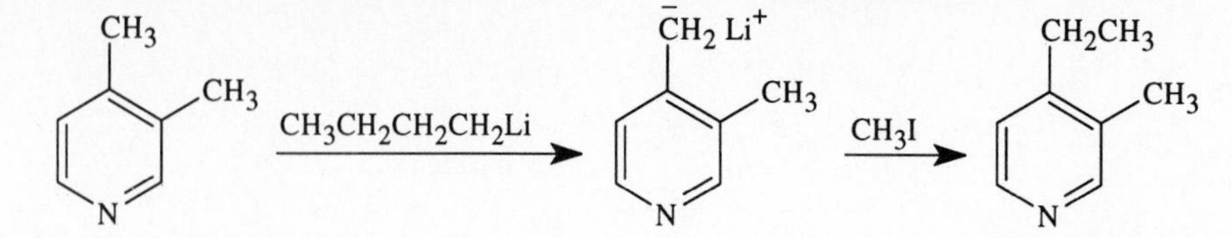

CH₃CH₂CH₂CH₂Li

3,4-dimethylpyridine    →    A    →(CH₃I)    4-ethyl-3-methylpyridine

In early printings of the text, there is an error in Eq. 24.68b (p. 1205). The structure following step *(c)* should be as follows:

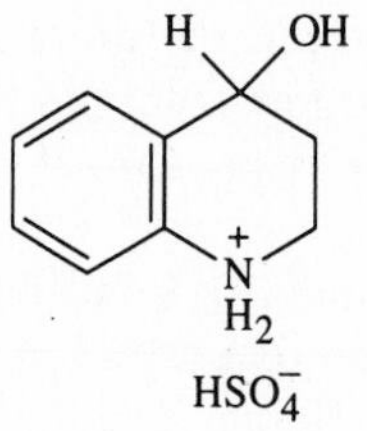

$HSO_4^-$

In addition, the text over the final arrow of the equation should be 1) $H_2SO_4$, $-H_2O$; the $H_2O$ is missing.

**24.26**  (a)

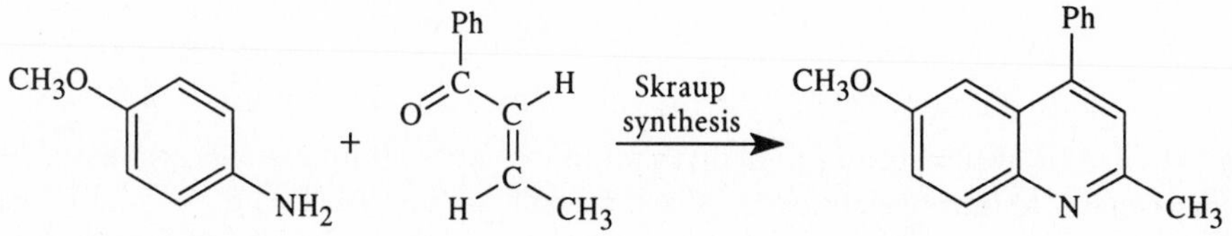

*p*-methoxyaniline
(*p*-anisidine)    1-phenyl-2-buten-1-one    →(Skraup synthesis)

**24.27**  The reactants required for a Skraup synthesis of 6-chloro-3,4-dimethylquinoline are *p*-chloroaniline and 3-methyl-3-buten-2-one.

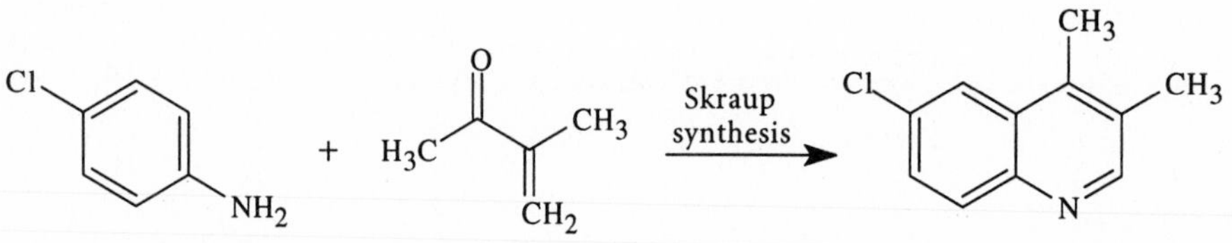

*p*-chloroaniline    +    3-methyl-3-buten-2-one    →(Skraup synthesis)    6-chloro-3,4-dimethylquinoline

## Solutions to Additional Problems

**24.29**  (a)                    (c)

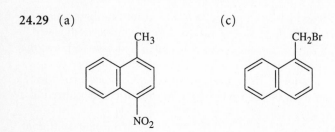

**24.30** (a)                    (c) 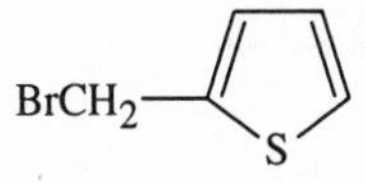                    (d) no reaction    (f)

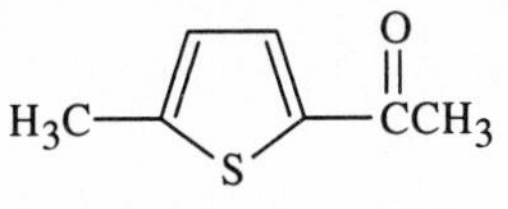

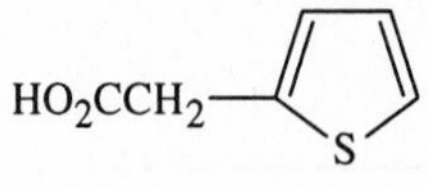

**24.31** (a) 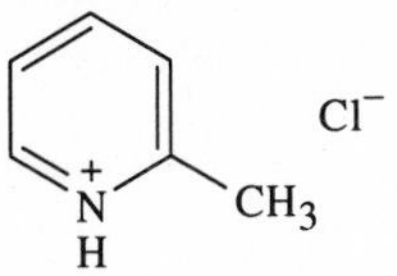                    (c)                    (e)

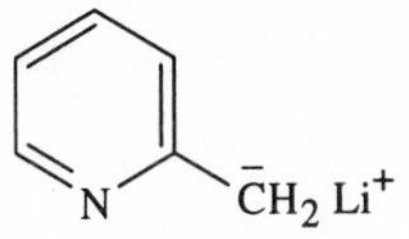

+ CH₃CH₂CH₂CH₃

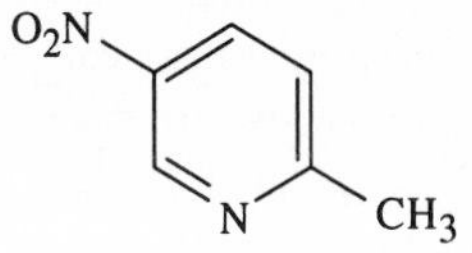

(f) 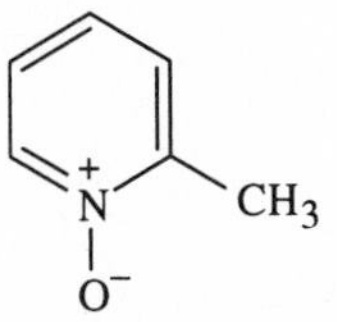                    (h) 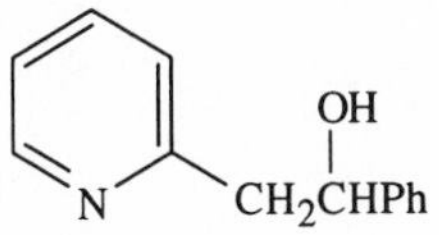                    (j)

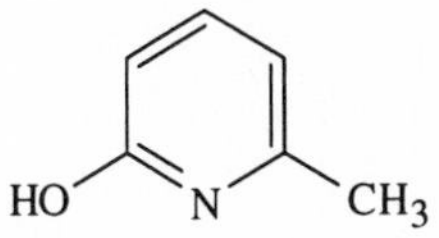

**24.32** (a)  The order of increasing reactivity is

pyridine < quinoline < naphthalene

The reason for the low reactivity of pyridine is discussed on text p. 1195, and the reason for the high reactivity of naphthalene is discussed on text p. 1175. Quinoline is more reactive than pyridine for the same reasons that naphthalene is more reactive than benzene. However, quinoline is also less reactive than naphthalene because the carbocation intermediates in electrophilic aromatic substitution reactions are destabilized by the polar effect of nitrogen, an electronegative atom.

**24.33**  The order of increasing $S_N1$ reactivity is as follows:

$B < A < D < C$

These reactivities parallel the corresponding reactivities of the respective ring systems themselves in electrophilic aromatic substitution reactions. For example, furan is more reactive than benzene in electrophilic aromatic substitution, and thus compound *C* is more reactive than compound *A* in $S_N1$ reactions, because of the involvement of an oxygen unshared electron pair in the resonance stabilization of the carbocation intermediate:

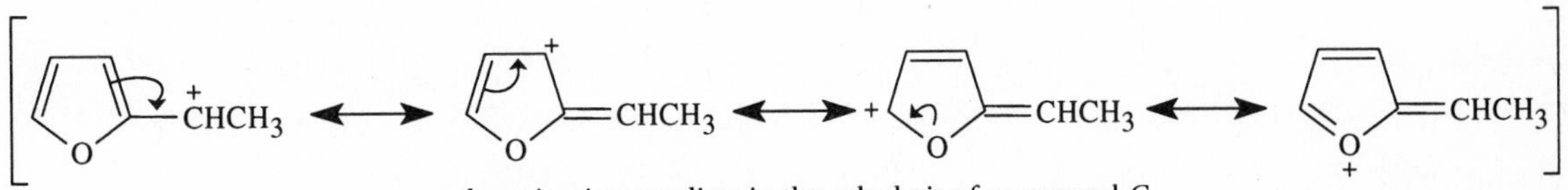

carbocation intermediate in the solvolysis of compound *C*

Just as a *para*-methoxy group accelerates solvolysis reactions (see Eq. 17.6 on text p. 792), atoms that donate electrons by resonance within the ring also accelerate solvolysis reactions. In contrast, a pyridine ring retards solvolysis reactions because the nitrogen unshared electron pair *cannot* be used to stabilize the carbocation by resonance; this electron pair is in the plane of the ring and cannot overlap with the rest of the π-electron system. Furthermore, one resonance structure of the carbocation intermediate places positive charge and electron deficiency on the nitrogen, an electronegative atom. Such a structure is not important. Thus, the nitrogen exerts only its rate-retarding polar effect.

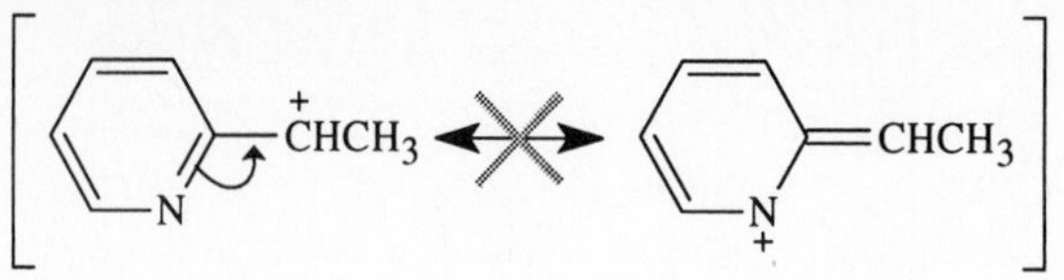

Finally, solvolysis of the thiophene derivative *D* is faster than that of benzene because of the activating effect of the sulfur, which is similar to the effect of the oxygen in the furan derivative; but because resonance involving sulfur orbitals is less effective than resonance involving oxygen orbitals, the thiophene derivative is less reactive than the furan derivative.

**24.34** (a)  The carbocation intermediate involved in the Friedel-Crafts acylation of benzofuran at carbon-2:

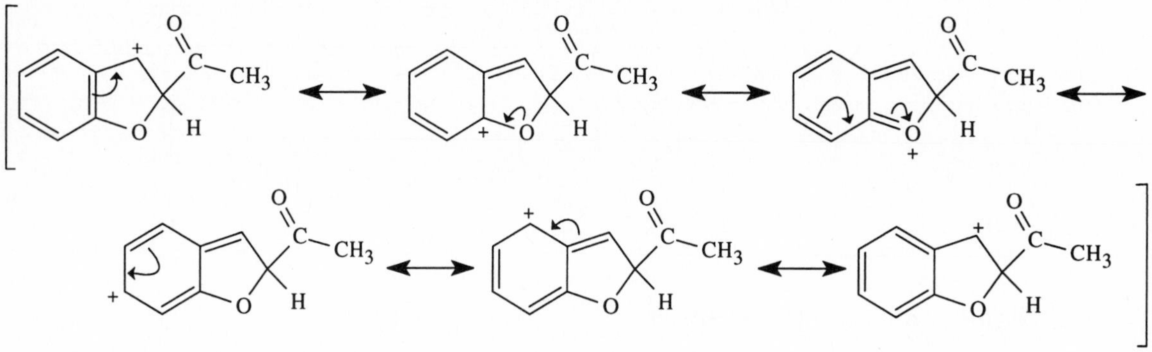

**24.35**  The hydroxy tautomer in each case is aromatic, and the carbonyl tautomer is not. The principle to apply is that the compound with the greater resonance energy (Table 24.1 on text p. 1181) has the greater amount of hydroxy tautomer.

(a)  Because furan has less resonance energy (that is, less stabilization due to aromaticity) than pyrrole, 2-hydroxypyrrole contains a greater percentage of hydroxy tautomer than 2-hydroxyfuran.

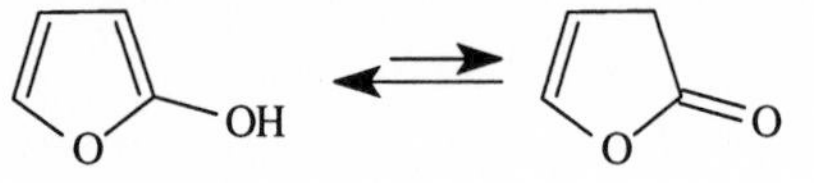

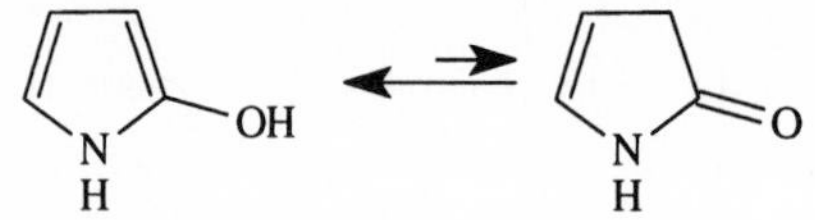

**24.36**  The following three isomers are formed in the bromination of 1,6-dimethylnaphthalene.

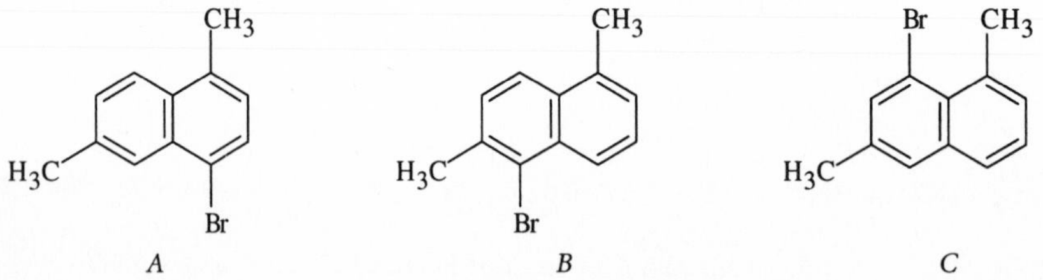

All products result from bromination at an $\alpha$-position; recall that $\alpha$-positions of naphthalene rings are more activated than are $\beta$-positions towards electrophilic aromatic substitution (text p. 1173–1174). Compounds *A* and *B* result from bromination at an $\alpha$-position that is also activated by the methyl substituent. The substituent effect of the methyl group and the tendency of the ring to brominate in an $\alpha$-position are in opposition in product *C*; furthermore, there is a severe methyl-bromine *peri* interaction in compound *C*. Thus, compound *C* is formed in least amount. Compound *A* is probably formed in greatest amount because there is no methyl-bromine *ortho* van der Waals repulsion as there is in compound *B*.

**24.38** (a)  The order of increasing basicity (increasing conjugate-acid p$K_a$) is as follows:

5-methoxyindole $\ll$ 3-methoxypyridine $<$ pyridine $<$ 4-methoxypyridine

All pyridines are much more basic than the indole, which is not basic at nitrogen. 4-Methoxypyridine is more basic than either of the other pyridines, because an unshared electron pair of the methoxy substituent is available to stabilize the conjugate acid by resonance:

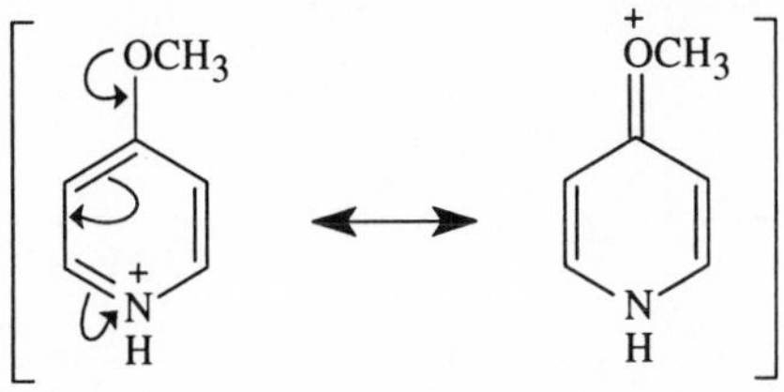

conjugate acid of
4-methoxypyridine

Such resonance is not possible when the methoxy group is in the 3-position; hence, the effect of a 3-methoxy substituent is limited to its acid-strengthening (base weakening) polar effect. Pyridine itself has neither the base-strengthening resonance effect nor the base-weakening polar effect of a methoxy substituent. Hence, its conjugate-acid $pK_a$ is between those of the methoxypyridines.

(c) The first "anion" in the problem is actually a neutral compound.

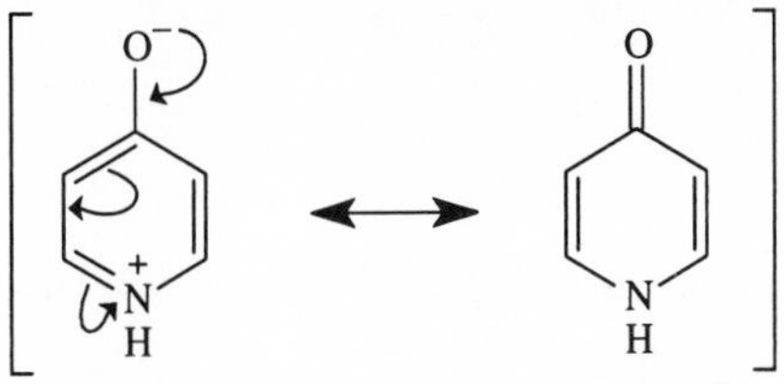

That is, it is a carbonyl compound. It has much of the characteristics of an amide, because, as structures $A$ and $B$ above show, a nitrogen unshared electron pair is delocalized into the carbonyl group, as in an amide. Thus, the $pK_a$ of its conjugate acid is near $-1$. The other anion in the problem is a phenolate ion; its conjugate acid is a phenol, with a $pK_a$ of about 10. Consequently, the phenolate ion is considerably more basic.

(d) The resonance structures of protonated imidazole are shown in Eq. 24.19 on text p. 1183. Compare the structure of protonated imidazole with the structure of protonated oxazole:

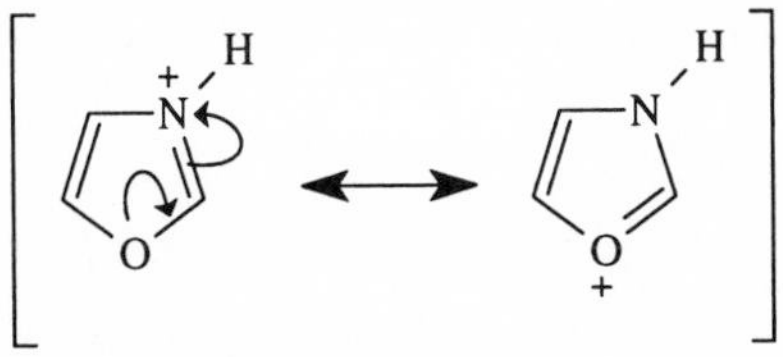

conjugate acid
of oxazole

In protonated oxazole, the positive charge is shared between an oxygen and a nitrogen; in protonated imidazole, the positive charge is shared between two nitrogens. Because oxygen is more electronegative than nitrogen, it supports positive charge less effectively; consequently, oxazole is less basic, and imidazole is more basic.

**24.39** The conjugate acid is formed by protonation on the nitrogen of the double bond. Protonation on either of the other nitrogens does not yield a resonance-stabilized cation.

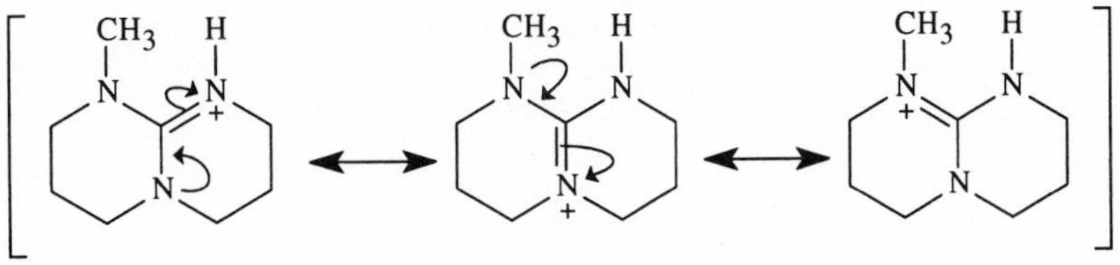

**24.40**  (a)  Because alkylamines are more basic than pyridines (or quinolines), the conjugate acid of quinine is protonated on the nitrogen of the bridged bicyclic ring:

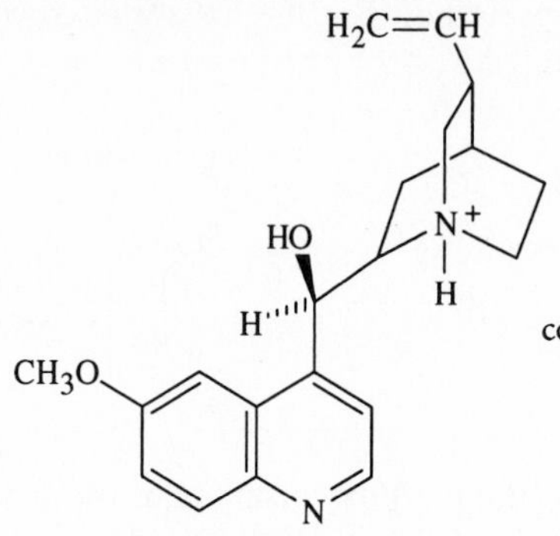

conjugate acid
of quinine

(c)  Indoles are not appreciably basic; tryptamine protonates on the nitrogen of the primary amino group.

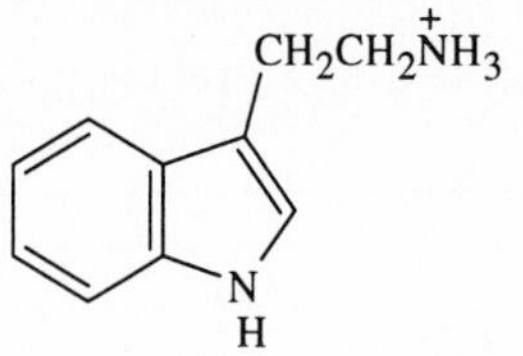

conjugate acid
of tryptamine

(d)  Although it might seem that the nitrogen of an amino group might be more basic than the nitrogen of a pyridine ring, the pyridine nitrogen protonates because the conjugate acid is resonance-stabilized. (See the solution to Problem 24.11(a) on page 669 of this manual for a similar situation.)

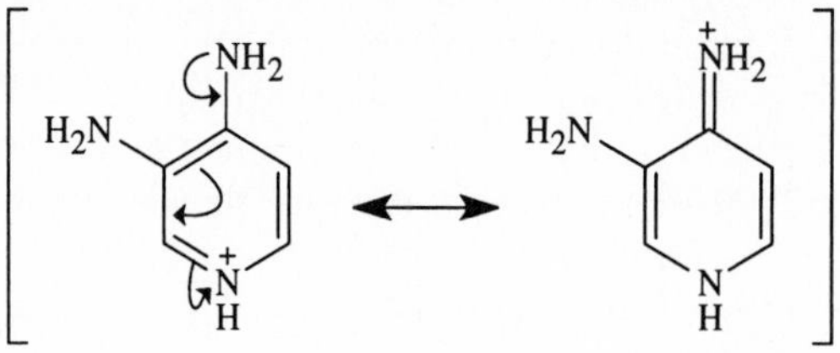

conjugate acid of
3,4-diaminopyridine

(f)  Think of a 1,2,3-triazole as an imidazole with an extra nitrogen. Protonation occurs on the terminal nitrogen of the double bond as it does in imidazole because a resonance-stabilized cation is formed:

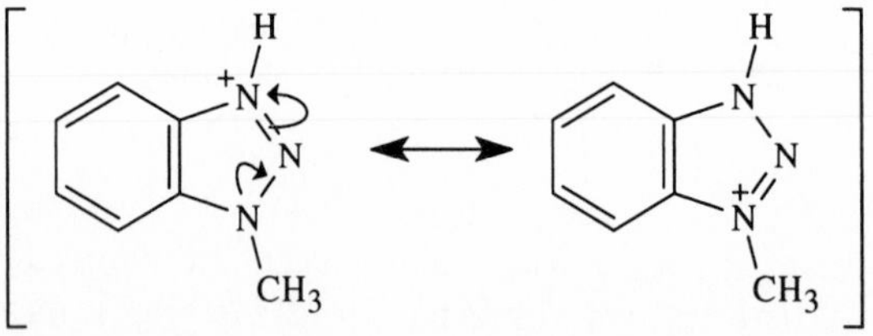

conjugate acid of
1-methyl-1,2,3-benzotriazole

Protonation on the central nitrogen, in contrast, gives a cation that has an electron-deficient nitrogen, and protonation on the nitrogen bearing the methyl group gives a cation that is not resonance-stabilized.

**24.41**  (a)  The text has stressed that many reactions of pyridines can be rationalized by thinking of the N=C bond of the pyridine as if it were a carbonyl group. Thus, the hydrogens of the methyl group, like the $\alpha$-hydrogens of a methyl ketone, are acidic enough to exchange for deuterium in basic $D_2O$.

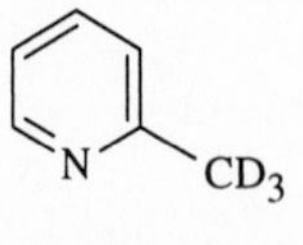

(c)  Naphthalene rings are more activated towards electrophilic aromatic substitution than are phenyl rings; furthermore, a phenyl group is an activating, *ortho, para*-directing substituent. (The rationale for this conclusion is explored in detail in the solution to Problem 16.23 on p. 375, Vol. 1, of this manual.)

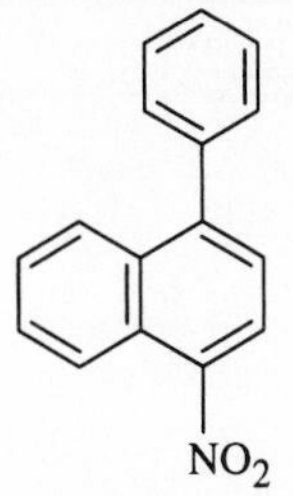

(e)  The fact that the amino group is part of a five-membered ring does not alter the fact that it is an activating, *ortho, para*-directing substituent. Under the conditions shown, the aromatic ring brominates once. (See Eq. 18.40, text p. 846, for bromination of a phenol under similar conditions.)

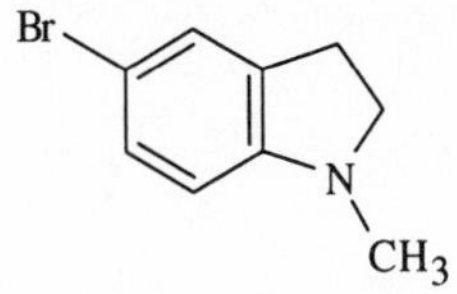

(g)  The ethoxycarbonylamino substituent and the pyridine ring itself direct aromatic nitration to carbon-5. The stringent conditions are necessary because the pyridine ring is protonated, and thus is highly deactivated, under the acidic conditions of the reaction.

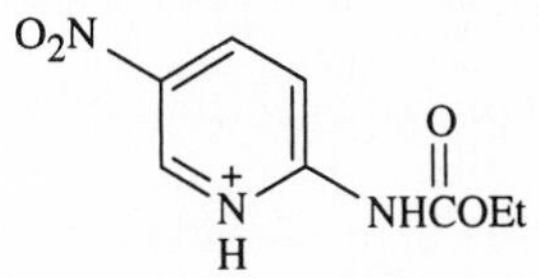

(h)  Because furan is more activated than thiophene towards electrophilic aromatic substitution (Eq. 24.26 on text p. 1185), substitution occurs predominantly in the furan ring. Notice that the directing effect of the electron-releasing ring oxygen overrides the directing effect of the carbonyl group, which, by itself, would result in substitution at the 4-position of the furan ring. (See the discussion of the relative importance of directing effects in the last paragraph of text p. 772.)

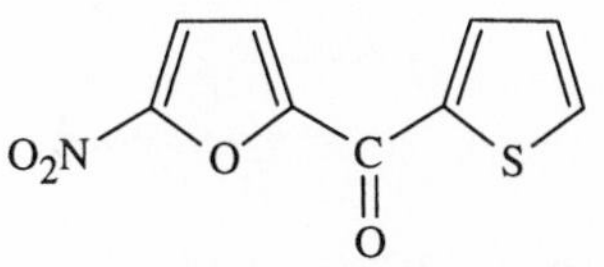

(j)  Wolff-Kishner reductions occur on acylated pyridines just as they do on acylated benzenes.

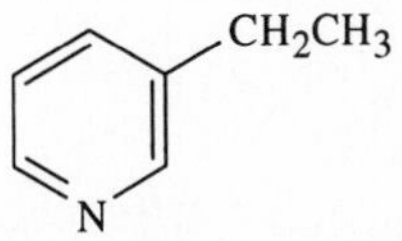

(k)  This is a Fischer indole synthesis in which the product is an *N*-methyl indole:

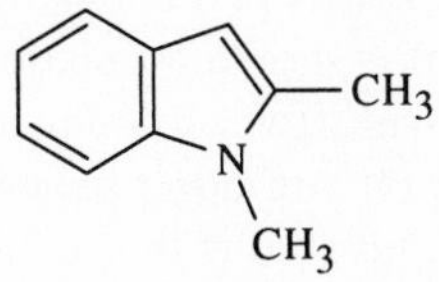

1,2-dimethylindole

**24.42** The *p*-nitrobenzenediazonium ion, which is formed in the diazotization reaction, undergoes an electrophilic aromatic substitution at carbon-3 of indole:

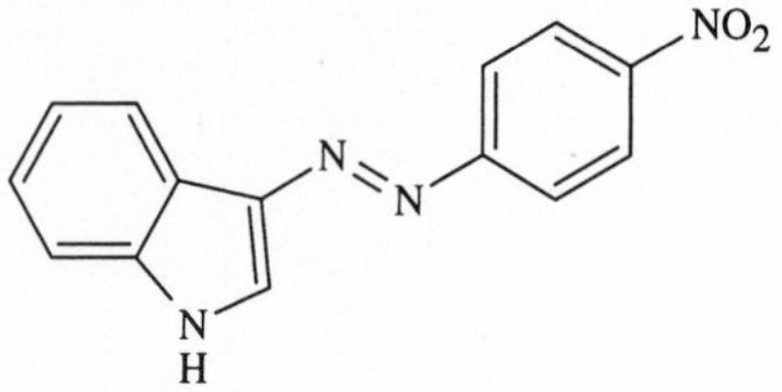

**24.43** (a) Given that anthracene adds benzyne across carbons 9 and 10, the structure of triptycene can be deduced by the curved-arrow formalism:

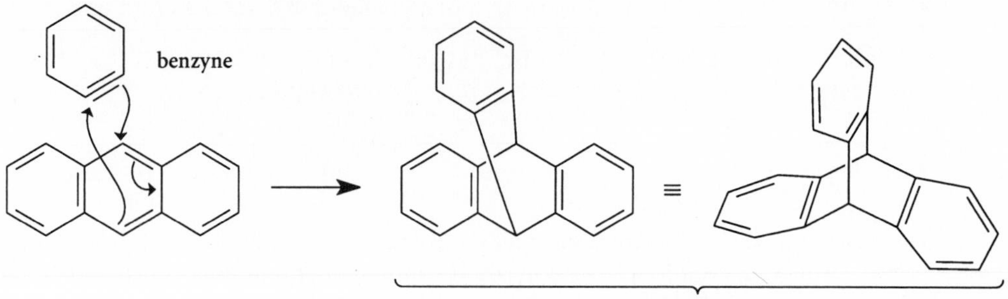

triptycene

**24.44** Use the principle that the carbocation intermediate with the greatest number of resonance structures is most stable, and that the reaction which involves the most stable carbocation intermediate is fastest. First, consider the carbocation intermediate in bromination at an $\alpha$-position of a terminal ring:

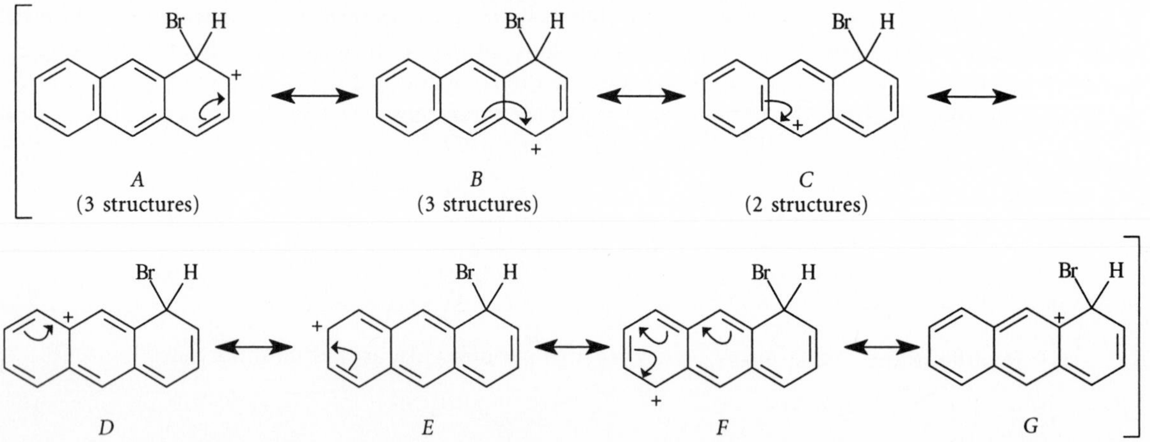

For each structure in which there is an intact naphthalene ring there are two additional structures (as in Eq. 24.1 on text p. 1171). Thus, structures *A* and *B* actually represent three resonance structures each. For the structure *C*, in which there is an intact benzene ring, there is one additional structure (as in Eq. 15.39 on text p. 717). Thus, structure *C* actually represents two resonance structures. Hence, the carbocation shown above has twelve resonance structures. It is shown on text p. 1174 that substitution at a $\beta$-position of a naphthalene ring results in a carbocation with fewer resonance structures; the same is true for substitution at the $\beta$-position of an anthracene ring. (If you don't believe it, draw them!) The one remaining possibility is bromination at a central carbon. This results in the following carbocation intermediate:

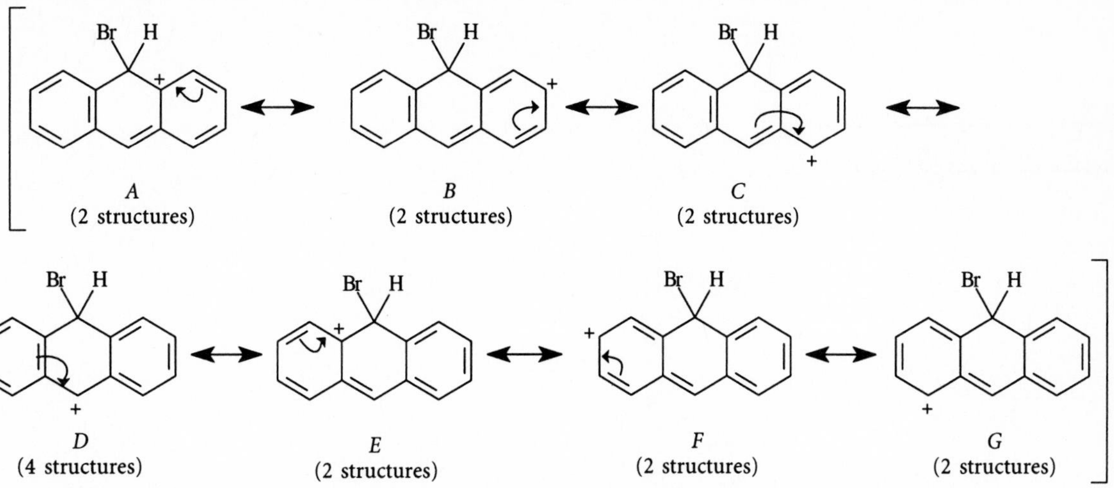

*A*
(2 structures)

*B*
(2 structures)

*C*
(2 structures)

*D*
(4 structures)

*E*
(2 structures)

*F*
(2 structures)

*G*
(2 structures)

Again, each intact benzene ring represents two structures. This carbocation has a total of sixteen resonance structures! Because this carbocation is the more stable intermediate, bromination therefore occurs at carbon-9 of anthracene.

**24.45** (a)

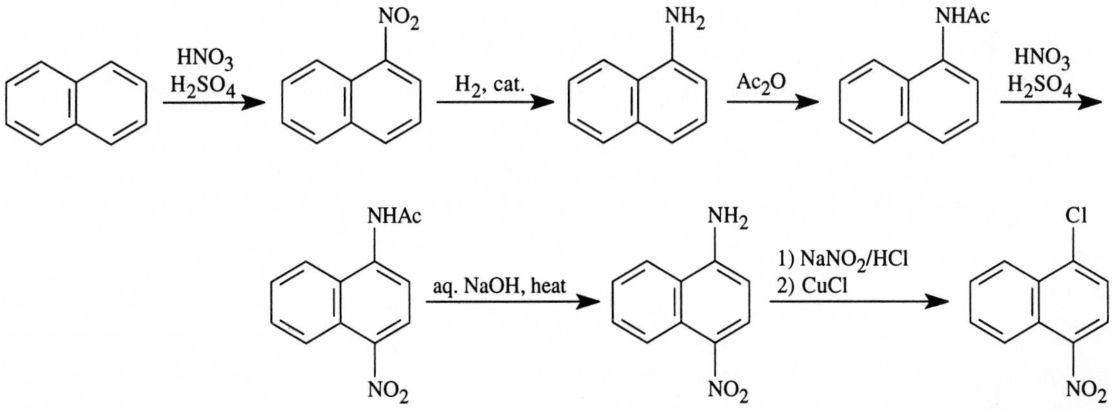

1-chloro-4-nitronaph-
thalene

Note that chlorination of naphthalene followed by nitration would not work. (Why?)

(c)

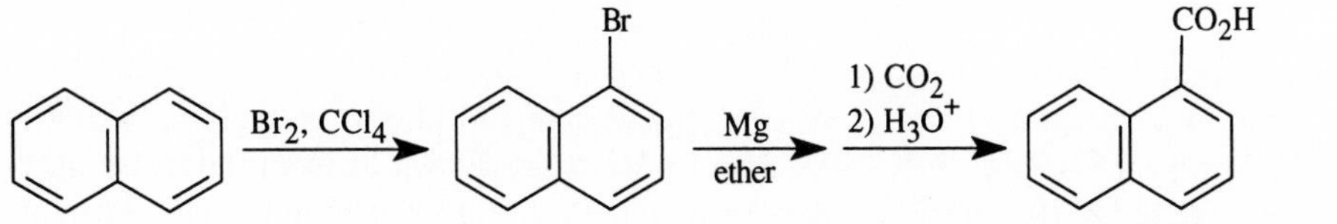

1-naphthalenecarboxylic
acid

(e)

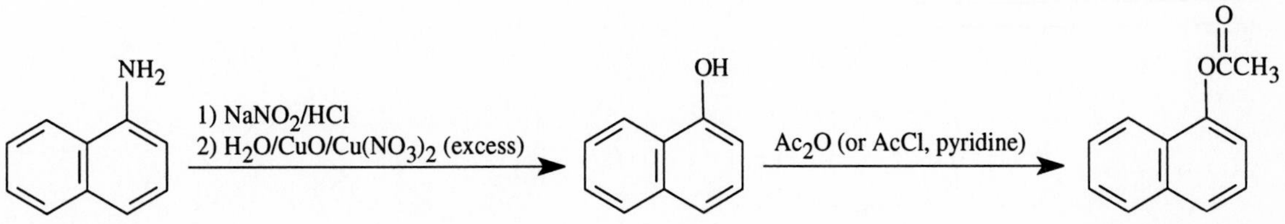

prepared in part (a)

1-naphthyl acetate

(g)

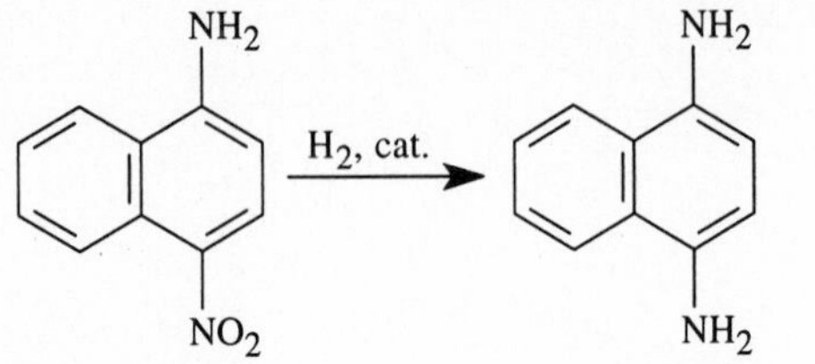

prepared in part (a)     1,4-naphthalenediamine

(h)

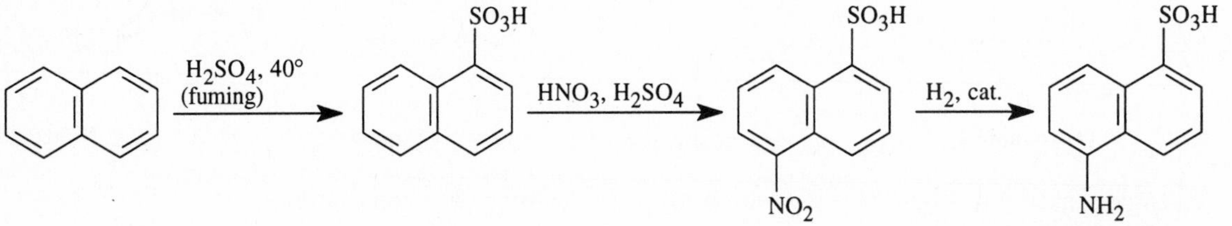

5-amino-1-naph-
thalenesulfonic
acid

**24.46** (a) Doreen is attempting to apply a reaction of pyridines and quinolines to indole. Because attack on a double bond of indole by the amide ion cannot result in charge delocalization to the nitrogen, the reaction does not take place. The reaction that occurs instead is removal of the N—H proton of indole.

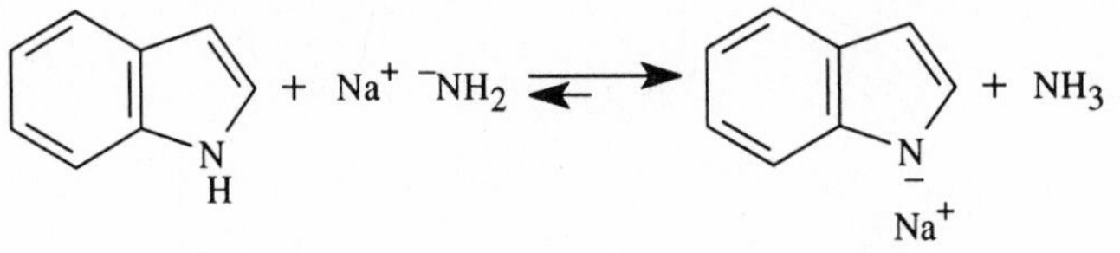

(b) Doreen is attempting the Chichibabin reaction on a pyridine derivative, but this derivative contains a chlorine in the 2-position, which, as chloride ion, is a vastly superior leaving group to hydride. The chlorine is lost rather than hydride to give 2-aminopyridine.

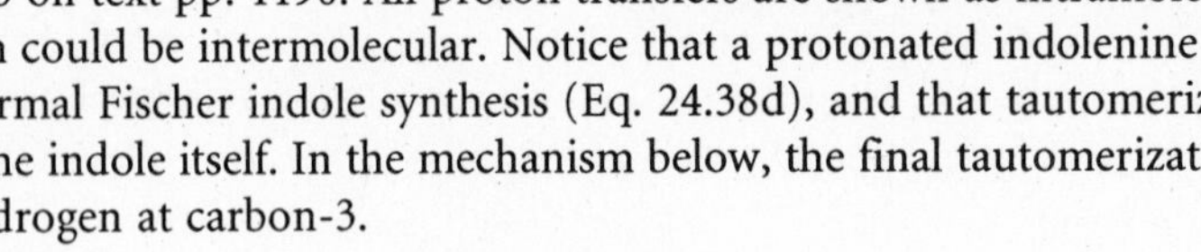

2-aminopyridine

**24.48** This is a variation of the Fischer indole synthesis. The mechanism that follows begins with the protonated enamine, formed in an acid-catalyzed reaction from the ketone and phenylhydrazine starting materials as shown in Eqs. 24.38a–b on text pp. 1190. All proton transfers are shown as intramolecular processes, although some of them could be intermolecular. Notice that a protonated indolenine is formed as an intermediate in the normal Fischer indole synthesis (Eq. 24.38d), and that tautomerization of this intermediate leads to the indole itself. In the mechanism below, the final tautomerization is not possible because there is no hydrogen at carbon-3.

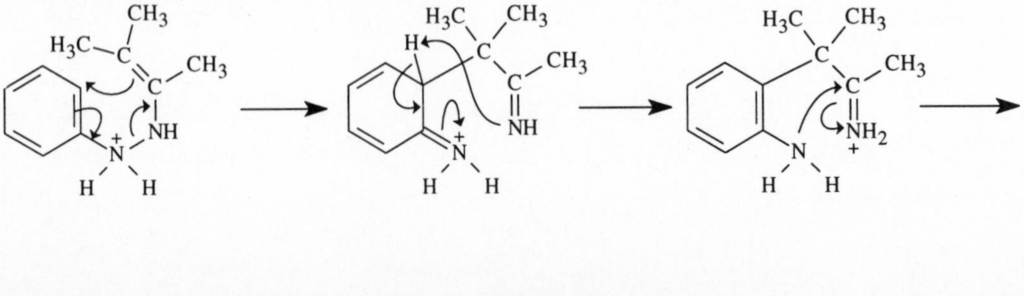

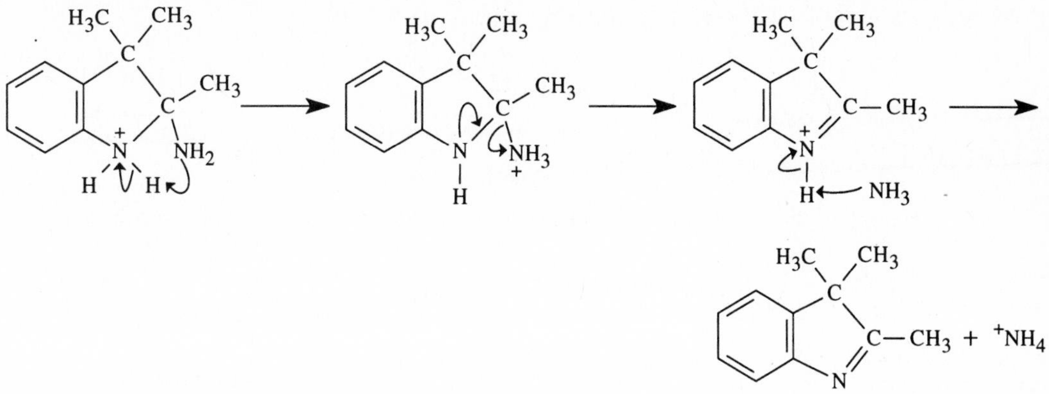

**24.49** (a) Protonation of one pyrrole molecule at carbon-2 gives a carbocation that reacts with a second indole. The resulting product *X* undergoes an acid-catalyzed double-bond shift to give *A*. This rearrangement is favorable because, in compound *A*, the double bond is conjugated with a nitrogen unshared electron pair whereas, in compound *X*, such conjugation is absent.

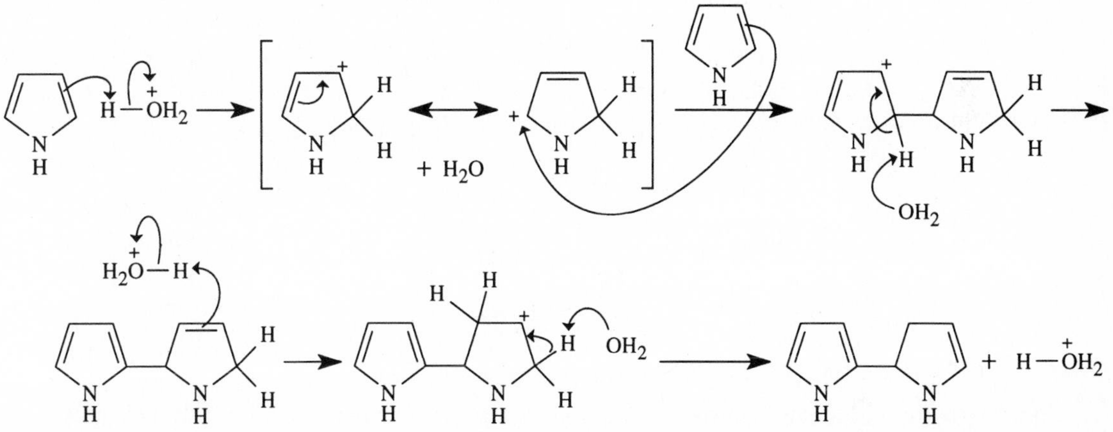

If you solved this problem by protonating carbon-3 of pyrrole in the initial step, thus avoiding the necessity for the final tautomerization, you have also written a reasonable and more direct mechanism. However, because pyrrole adds electrophiles (including the proton) predominantly at carbon-2, the mechanism above is probably preferable.

**24.50** (a) First prepare 4-bromo-2-nitrotoluene and then use it in a Reissert synthesis.

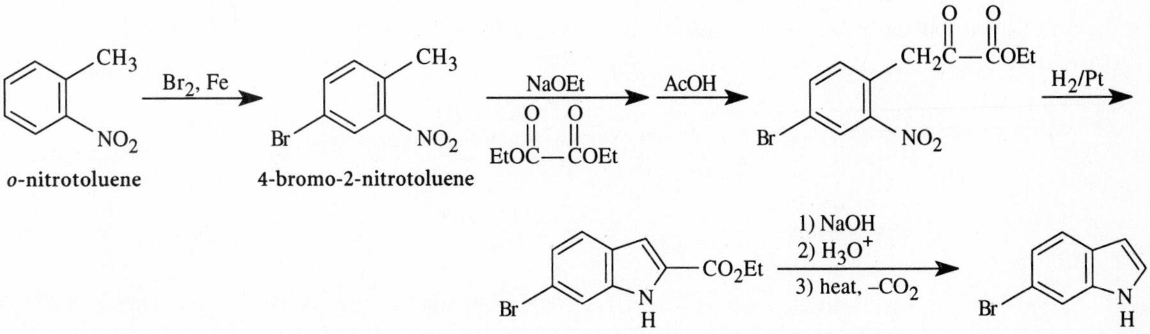

(c) Prepare pyridine-2-carbonitrile and reduce it to the amine.

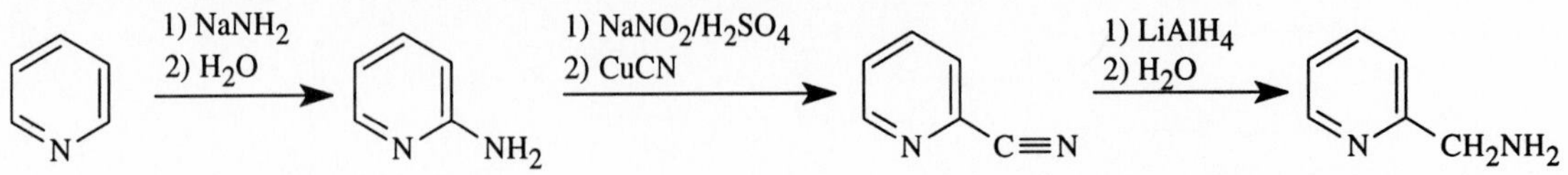

(e) Prepare 1-(2-furyl)-1-ethanone ("2-acetylfuran") and carry out an aldol condensation with furfural.

furfural

"2-acetylfuran"

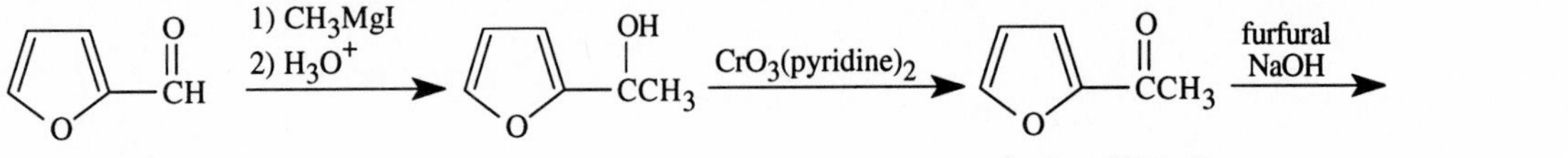

(g) Oxidize the methyl group to carboxylic acid, convert it into an acyl azide, and carry out a Curtius rearrangement to the isocyanate. Divide the isocyanate in half; convert one part into the amine, and then let the amine react with the remaining isocyanate to give the urea.

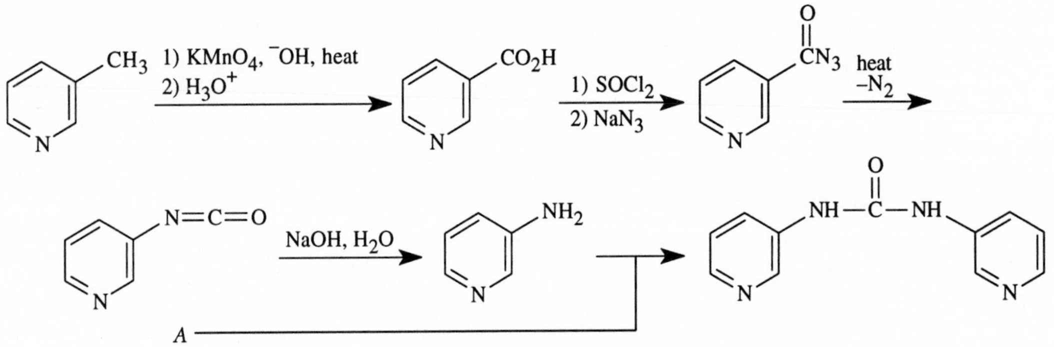

(h) Prepare 1-aminonaphthalene and use it as the "aniline" derivative in the Skraup synthesis.

1-aminonaphthalene

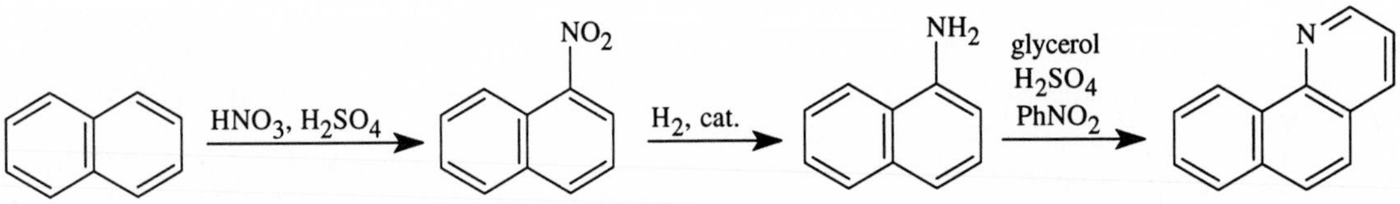

(i) Use the acidity of *N*-methylpyridinium salts to form a nucleophilic anion at the benzylic carbon of the ethyl group, which is then cyanoethylated twice. (Compare to the reactions in Eqs. 24.65 and 24.66 on text p. 1203. Note that in early printings of the text, a hydrogen is missing at the end of the "dangling bond" in the first structure of Eq. 24.65.) Cyanoethylation is discussed at the top of text p. 1087.

4-ethylpyridine

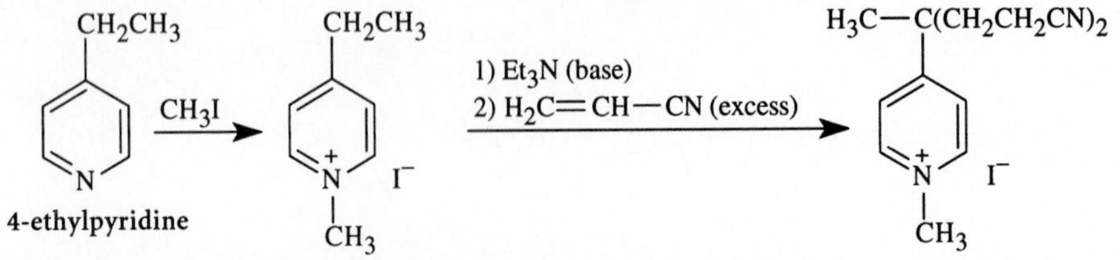

Note that if a $\beta$-halo nitrile such as $ICH_2CH_2C{\equiv}N$ were used as the alkylating agent, $H_2C{=}CHC{\equiv}N$ would be formed and would be the effective alkylating agent. (Why? See Sec. 17.3B on text p. 802.)

**24.51**  The unsaturation number of compound *A* is 4; its oxidation to nicotinic acid shows that it contains a pyridine ring with a 3-substituent. The pyridine ring accounts for all four degrees of unsaturation. Because pyridine contains five carbons, the side-chain has three carbons. Compound *A* cannot be a primary alcohol, because the oxidation product of such an alcohol would be an aldehyde, which could not have five exchangeable α-hydrogens. (An aldehyde of the form RCH₂CH═O has only two α-hydrogens.) Therefore, compound *A* is a secondary alcohol, and ketone *B* is its oxidation product.

$$\text{A} \qquad\qquad \text{B}$$

**24.53**  (a)  The reaction of thiophene with chlorosulfonic acid gives compound *A*, 2-thiophenesulfonyl chloride (see Eq. 20.27 on text p. 953). Because the chlorosulfonyl group exerts a strong *meta*-directing effect, nitration of *A* gives the 4-nitro derivative *B*. Heating *B* in water converts the sulfonyl chloride into the sulfonic acid, which then loses its sulfonyl group to give 3-nitrothiophene, *C*. (Recall that sulfonation is reversible; see Eq. 24.4, text page 1174.)

$$A \xrightarrow{\text{conc. HNO}_3} B \xrightarrow[\text{heat}]{\text{H}_2\text{O,}} + \text{HCl} \xrightarrow[\text{–H}_2\text{SO}_4]{\text{H}_3\text{O}^+,\ \text{heat}} C$$

(b)  3-Nitrothiophene cannot be made directly from thiophene because thiophene nitrates mostly in the 2-position to give 2-nitrothiophene. (See Eq. 24.25 on text page 1185.)

 Note that the sulfonyl chloride group is used as a *protecting group* in this reaction. It blocks the 2-position of the thiophene ring and directs nitration to the 4-position. It is then removed when no longer needed.

**24.55**  (a)  Diethylamine and formaldehyde react to form an imminium ion by a mechanism completely analogous to the one shown in Eq. 23.26 on text p. 1131. This ion then serves as the electrophile in an electrophilic aromatic substitution reaction to the 3-position of indole.

Note that indole, benzofuran, and benzothiophene tend to give a much greater percentage of 3-substitution than pyrrole, furan, and thiophene.

(c)  This reaction is a carbocation rearrangement known as the *indolenine rearrangement.* The driving force for this reaction is the aromaticity of the product.

*(solution continues)*

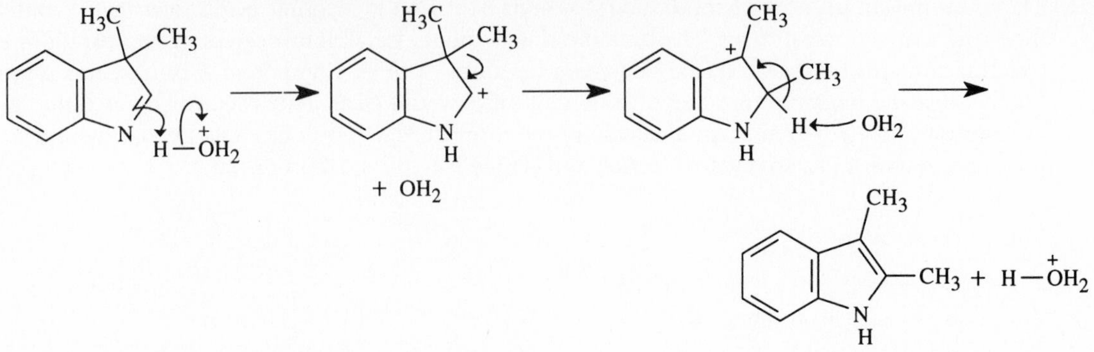

(d) Alkylation of the nitrogen activates the carbon-nitrogen double bond toward nucleophilic attack by hydroxide ion, which, in turn, leads to ring opening.

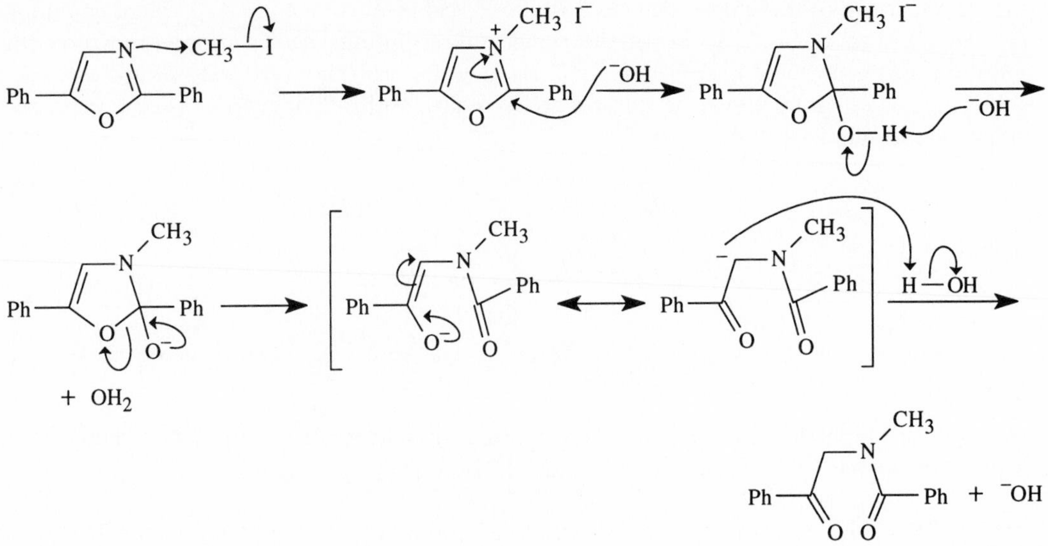

(e) Sulfonylation of the *N*-oxide by the sulfonyl chloride activates the carbon-nitrogen double bond toward attack by chloride ion, which is produced in the sulfonation reaction. Hydroxide-promoted elimination forms a double bond and liberates chloride ion, which then attacks the terminal carbon of the double bond and displaces tosylate ion in an allylic variation of an $S_N2$ reaction (sometimes called an $S_N2'$ reaction). This last step is undoubtedly driven by the aromatic stability of the product. The mechanism below begins with the sulfonate ester and chloride ion, produced in the formation of the sulfonate ester. (For the mechanism of sulfonate ester formation, see Study Guide Link 10.3 on p. 227, Vol. 1, of this manual.) The toluenesulfonate group is abbreviated —OTs.

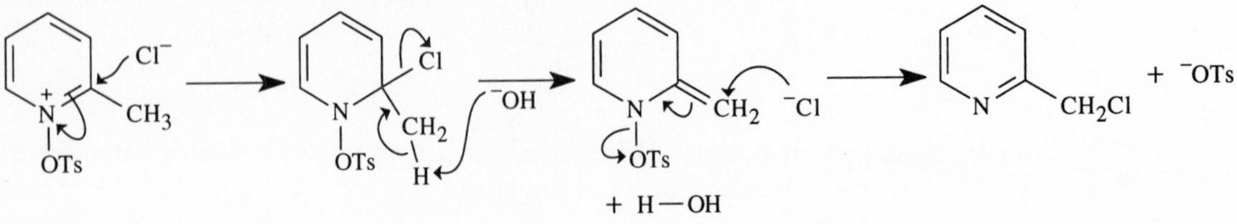

(g) This reaction is a nucleophilic aromatic substitution reaction by the benzenethiolate anion (PhS⁻), which is formed by the reaction of benzenethiol with triethylamine.

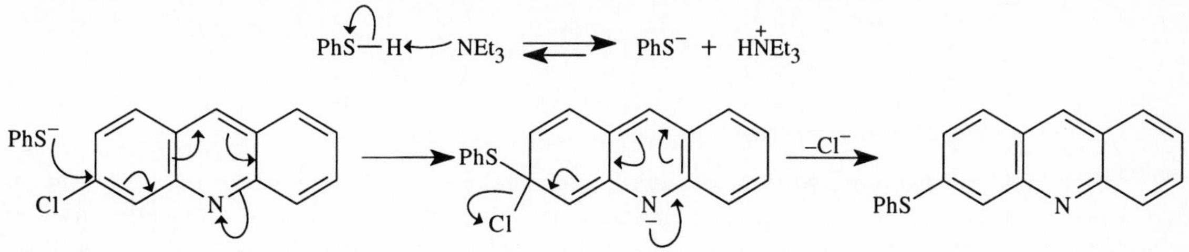

$$\text{PhS} \overset{\frown}{-} \text{H} \longleftarrow \text{NEt}_3 \;\rightleftharpoons\; \text{PhS}^- + \text{HNEt}_3^+$$

**24.56** (a) Exchange occurs when the conjugate-base anion is formed by abstraction of one of the colored protons, and this anion removes a deuteron (a deuterium nucleus) from the solvent. The anionic intermediate involved in this exchange is resonance-stabilized in such a way that negative charge is delocalized onto two electronegative atoms—one of the nitrogens and the carbonyl oxygen. In contrast, removal of a proton from the other methyl group gives an anion in which the negative charge is not delocalized onto electronegative atoms; consequently, this anion is less stable. Thus, exchange of a colored hydrogen occurs because this exchange involves the more stable anionic intermediate.

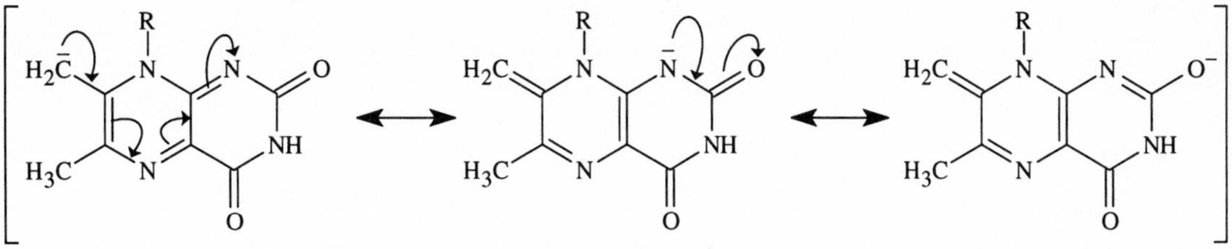

anionic intermediate involved in the exchange of the colored protons

The imide hydrogen is also exchanged by a base-catalyzed mechanism that involves the conjugate-base anion of the imide. (Imides, like other $\beta$-dicarbonyl compounds, are acidic; see Problem 22.2 on text p. 1042.)

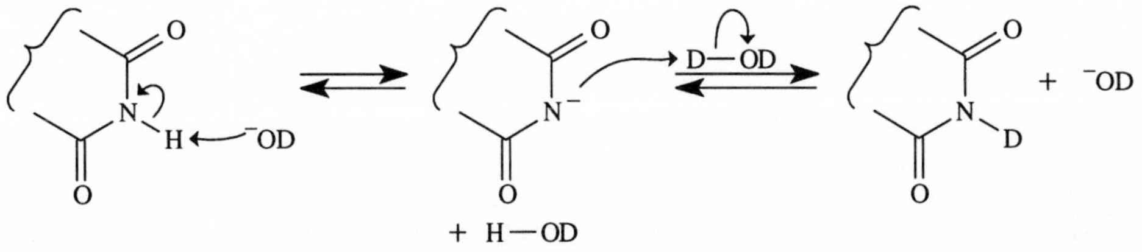

(c) The positive charge on the nitrogen activates the carbon-nitrogen double bond to nucleophilic attack by hydroxide ion. This forms a tetrahedral addition intermediate that breaks down to a substituted formamide $A$, which then undergoes amide hydrolysis to the products shown in the problem.

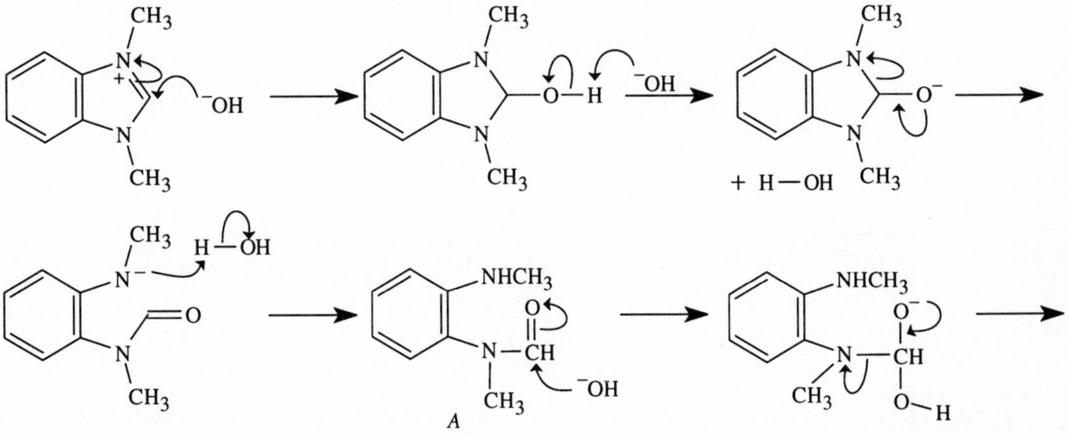

*(equation continues)*

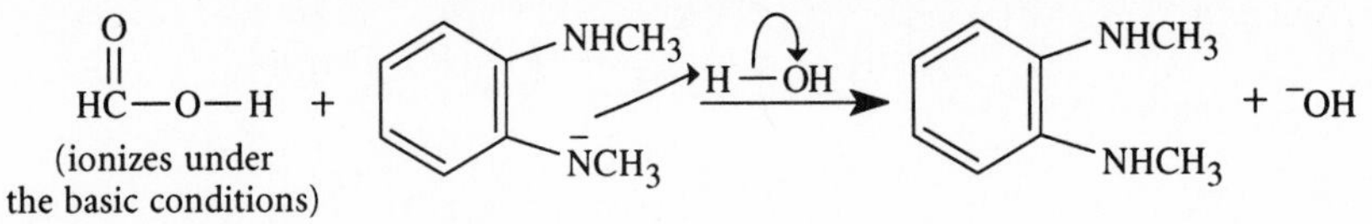

(d) 2-Pyridone does not hydrolyze because it undergoes a different reaction: ionization to its conjugate-base anion. This ionization occurs because the resulting anion is aromatic and is therefore very stable. Ionization of γ-butyrolactam is much less favorable; consequently, ionization does not compete with the hydrolysis reaction.

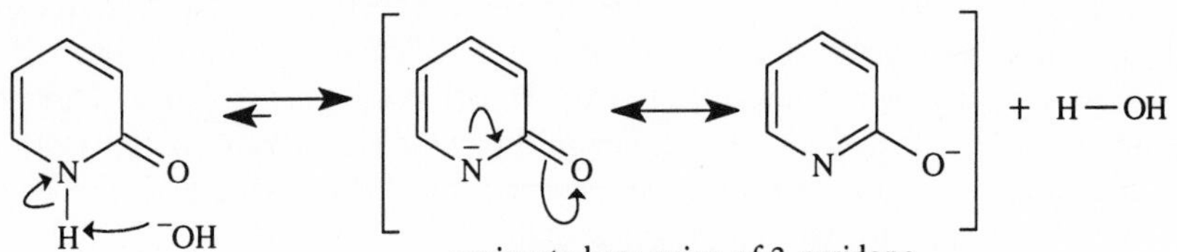

conjugate-base anion of 2-pyridone,
an aromatic species

In early printings of the text, there is an error in part (e). The second line should read, "... treatment of 3-chloropyridine under the same conditions ... "

(f) The observation of *cine* substitution suggests that a pyridine analog of benzyne (3,4-pyridyne) is involved as an intermediate. (You should supply the curved-arrow formalism.)

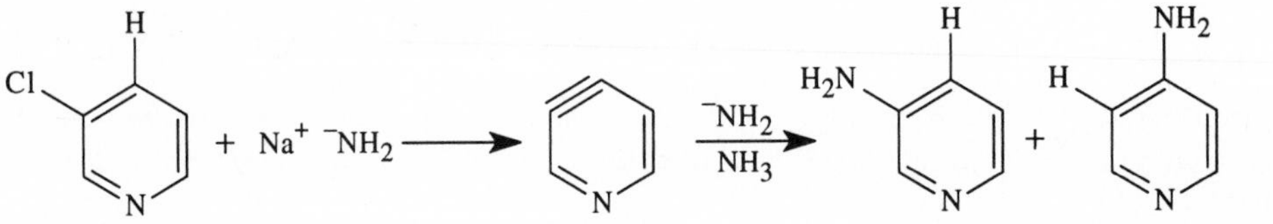

3-chloropyridine       3,4-pyridyne       3-aminopyridine   4-aminopyridine

24.57 (a) Bromine adds to the double bond, and KOH saponifies the resulting dibromo lactone. An internal nucleophilic substitution reaction then closes the five-membered ring with loss of one bromine as bromide ion. (Recall that α-halo carbonyl compounds are particularly reactive in nucleophilic substitution reactions; see Sec. 22.3D on text p. 1053.) An E2 reaction forms the double bond with loss of the second bromine as bromide ion. (This is a very favorable E2 reaction; why? See Sec. 17.3B on text p. 802.) The mechanism of bromine addition, not shown below, is exactly analogous to the mechanism shown in Eqs. 5.3–5.4 on text p. 176; and the mechanism of saponification, also not shown below, is exactly analogous to the mechanism shown in Eqs 21.9a–b on text p. 990.)

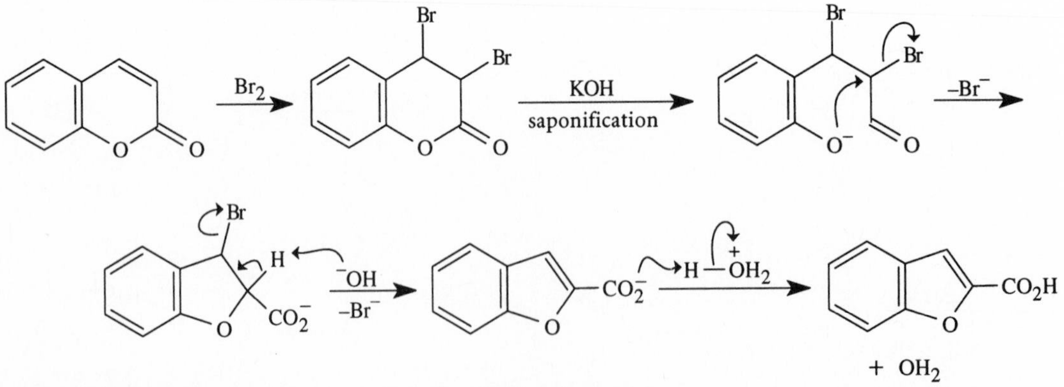

(c) The mechanism shown below starts with the enamine formed between the amino group and the carbonyl group of 2-butanone. (The mechanism of enamine formation is discussed on text p. 908.) Note that, although primary amines form imines with ketones, imines are in equilibrium with enamines just as ketones are in equilibrium with enols. Water is produced as a by-product of enamine formation. The conjugate acid of water, $H_3O^+$, is shown as the catalyzing acid, but any of the other acids present could also catalyze the reaction.

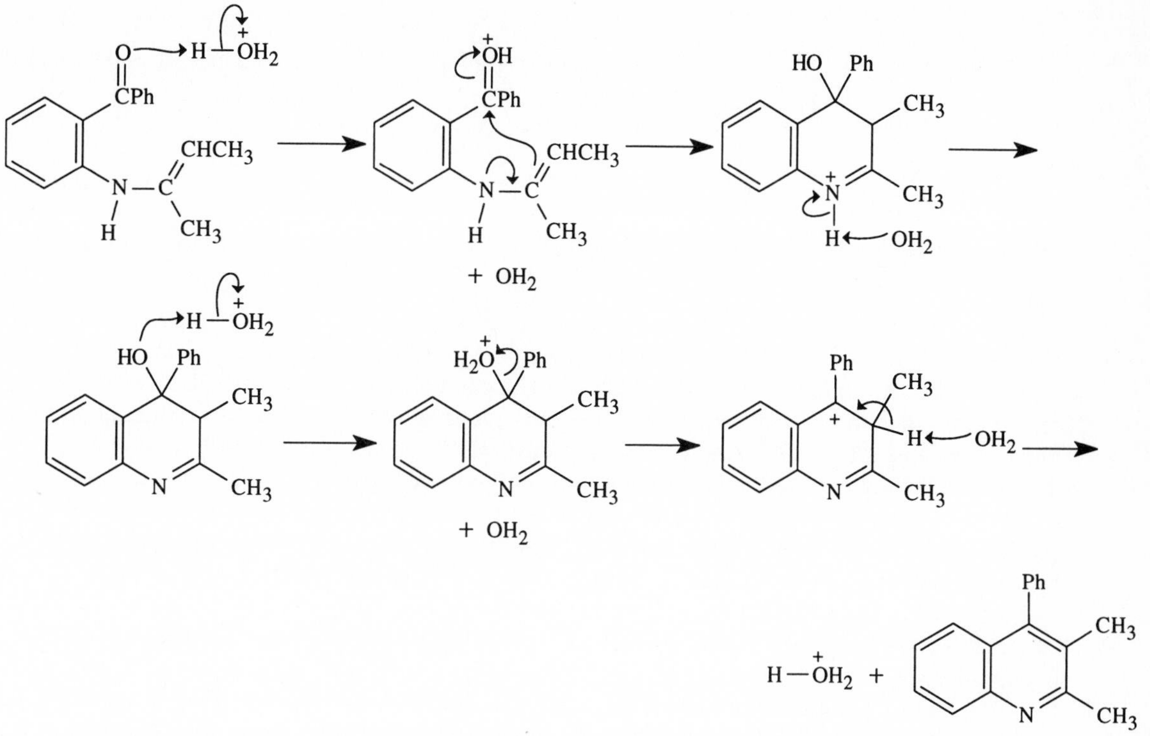

$+ OH_2$

$+ OH_2$

(e) Diethylamine serves as a base catalyst; its first function is to form the conjugate-base enolate ion *A* of ethyl acetoacetate:

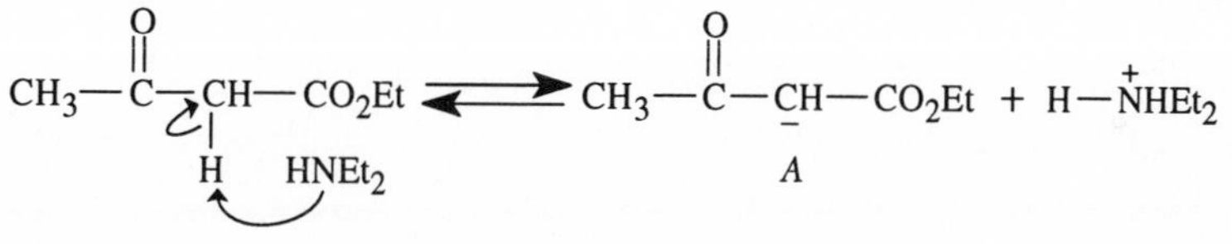

The first stage of the mechanism is a crossed-aldol condensation in which enolate *A* condenses with formaldehyde. (The mechanism of the aldol condensation is given in Eqs. 22.38b and 22.42 on text pp. 1055 and 1057, respectively.) The product of this aldol condensation then undergoes a conjugate-addition reaction (that is, a Michael addition) with a second enolate ion *A*. Ammonia reacts with one of the ketone functional groups to form an enamine, and the ring is closed when the amino group of that enamine forms an intramolecular enamine with the second ketone group. (The mechanism of enamine formation is discussed on text p. 908; see the note about the equilibration of imines and enamines in the solution to part (c) of this problem.)

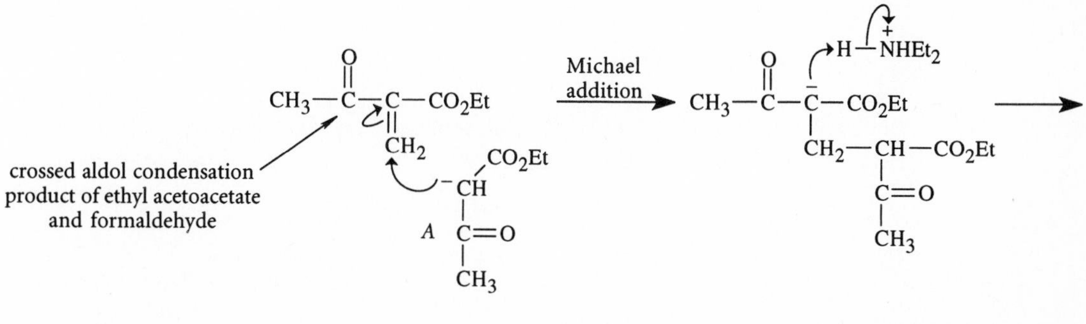

crossed aldol condensation
product of ethyl acetoacetate
and formaldehyde

*(equation continues)*

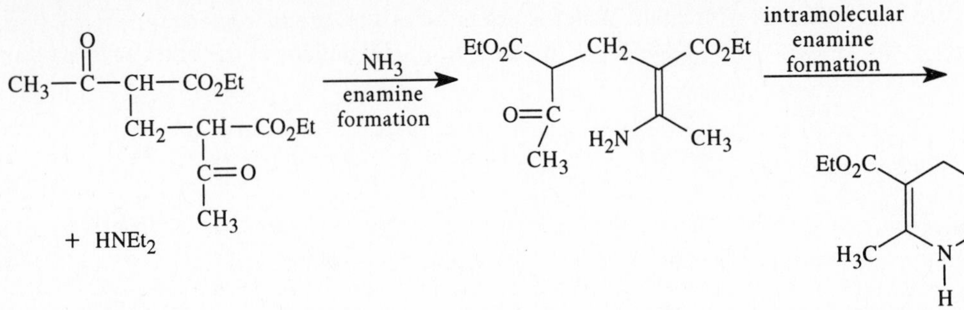

Why are the ketone carbonyl groups rather than the ester carbonyl groups the objects of nucleophilic attack by ammonia? See Sec. 21.9E on text p. 1014.

24.58   Step (1) is a nitration reaction. Because 2-pyridone derivatives such as this starting material contain significant amounts of their aromatic hydroxypyridine tautomers (Eq. 24.54 on text p. 1200), the nitro group can be introduced by aromatic nitration. The conditions of step (1) are $HNO_3$ and $H_2SO_4$. Step (2) can be effected with $PCl_5$ and heat. (See Eq. 24.55 on text p. 1200.) Step (3) is a reduction of both the nitro group and the nitrile to amine groups; catalytic hydrogenation is the method of choice. Evidently, the chlorine is also removed under the conditions of this hydrogenation. To see why this is reasonable, recall that 2-chloropyridines have some of the reactivity characteristics of acid chlorides, and that acid chlorides are reduced to aldehydes (that is, the chlorine is replaced by a hydrogen) in a type of catalytic hydrogenation (Rosenmund reduction; text p. 1012). Step (4) is an ether cleavage, which can be effected with aqueous $H_2SO_4$; water is the nucleophile in this reaction. This ether is activated toward cleavage by either an $S_N1$ or an $S_N2$ mechanism because it is benzylic. Neutralization with aqueous hydroxide gives the free amine shown. Finally, step (5) can be realized by diazotization with $NaNO_2$ in aqueous $H_2SO_4$, which converts the benzylic amine into a benzylic diazonium ion, and the arylamine into an aryldiazonium ion; nitrogen is rapidly displaced from the benzylic diazonium ion by water. Elimination, usually a competing reaction in the decomposition of alkyldiazonium ions, is not observed here because there are no $\beta$-hydrogens. The aryldiazonium ion is also converted into the phenolic hydroxy group by heating the aqueous solution. (See Eq. 23.50 on text p. 1141.)

# Pericyclic Reactions

## Terms

## Concepts

### I. Molecular Orbitals of Conjugated π-Electron Systems

#### A. MOLECULAR ORBITALS OF CONJUGATED ALKENES

1. Atomic $p$ orbitals on adjacent atoms within molecules overlap to give $\pi$ molecular orbitals.
2. The $\pi$ molecular orbitals for ethylene and conjugated alkenes can be constructed according to the following generalizations:
   a. When a number (say $m$) of $p$ orbitals interact, the resulting $\pi$-electron system contains the same number $m$ of molecular orbitals (MOs), all with different energies.
   b. Ethylene and conjugated alkenes have two types of $\pi$ molecular orbitals:
      i. Bonding $\pi$ molecular orbitals (half of the $\pi$ molecular orbitals) have lower energy than the isolated $p$ orbitals.
      ii. Antibonding $\pi$ molecular orbitals (the other half of the $\pi$ molecular orbitals) have higher energy than the isolated $p$ orbitals. (Antibonding MOs are indicated with asterisks.)

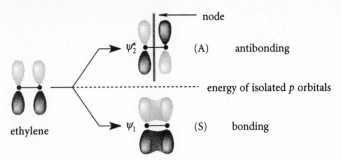

isolated *p* obritals        molecular orbital        (See Figs. 25.1 and 25.2, text pp. 1223–1224)

   c. The bonding molecular orbital of lowest energy, $\psi_1$, has no nodes between the atoms; each molecular orbital of increasingly higher energy has one additional node.
     *i.* A node is a plane at which any wave is zero; that is, when an electron is in a given MO there is zero probability of finding the electron, and thus zero electron density, at the node.
     *ii.* The electron wave has a peak on one side of the node and a trough on the other side.
     *iii.* The orbital changes phase at the node.
   d. The nodes occur between atoms and are arranged symmetrically with respect to the center of the $\pi$-electron system.
   e. With respect to an imaginary reference plane at the center of the $\pi$-electron system and perpendicular to the plane of the molecule:
     *i.* odd-numbered MOs are symmetric (S) (molecular orbitals in which peaks reflect across the reference plane into peaks, and troughs into troughs)
     *ii.* even-numbered MOs are antisymmetric (A) (molecular orbitals in which peaks reflect across the reference plane into troughs, and troughs into peaks)
     *iii.* Within any symmetric MO, the phase at the two terminal carbons is the same; within any antisymmetric MO the phase at the two terminal carbons is different.
   f. Electrons are placed pairwise into each molecular orbital, beginning with the orbital of lowest energy (Aufbau principle).
3. The presence of unconjugated substituents, to a useful approximation, does not alter the $\pi$ molecular orbital structure of a conjugated alkene.
4. The $\pi$-electron contribution to the energy of a molecule is determined by the energies of its occupied MOs. Bonding MOs have lower energy than isolated *p* orbitals; hence, there is an energetic advantage to $\pi$-molecular orbital formation.
5. Two MOs are of particular importance in understanding pericyclic reactions:
   a. The highest occupied $\pi$ molecular orbital (HOMO).
   b. The lowest unoccupied $\pi$ molecular orbital (LUMO).
   c. The HOMO and LUMO of a conjugated alkene have opposite symmetry.
   d. The HOMO and LUMO are sometimes collectively termed frontier orbitals because they are the molecular orbitals at the high- and low-energy extremes, respectively, of the occupied and unoccupied molecular orbitals. The HOMO has a lower energy than the LUMO.

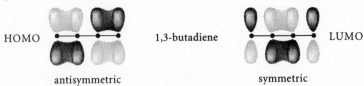

HOMO        1,3-butadiene        LUMO

antisymmetric        symmetric

## B. Molecular Orbitals of Conjugated Ions and Radicals
1. Conjugated unbranched ions and radicals have an odd number of carbon atoms.
2. The MOs of such species follow many of the same patterns as those of conjugated alkenes, with two important differences:
   a. One MO is neither bonding nor antibonding but has the same energy as isolated *p* orbitals.
     *i.* This MO is called a nonbonding molecular orbital.

      *ii.* The remaining orbitals are either bonding or antibonding, and there is an equal number of each type.

    b. In some of the MOs, nodes pass through carbon atoms.

3. The charge in a conjugated carbanion can be associated with the electrons in its HOMO.

4. Cations, radicals, and anions involving the same $\pi$ system have the same MOs; they differ only in the number of $\pi$ electrons.

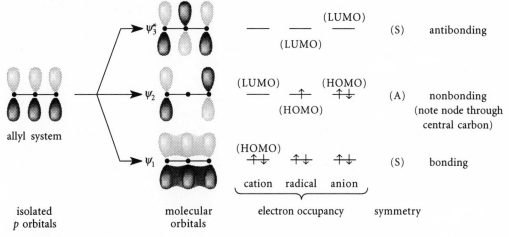

## C. EXCITED STATES

1. The normal electronic configuration of any molecule is called the ground state.

2. Energy from absorbed light is used to promote an electron from the HOMO of the ground state into the LUMO.

    a. The species with the promoted electron is called an excited state.

    b. The HOMOs of the ground state and the excited state have opposite symmetries.

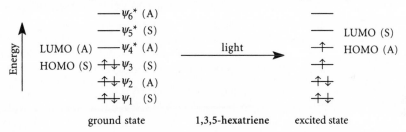

## II. *Pericyclic Reactions*

### A. INTRODUCTION

1. Pericyclic reactions occur by a concerted cyclic shift of electrons; that is, reactant bonds are broken and product bonds are formed at the same time, without intermediates.

2. There are three major types of pericyclic reactions:

    a. Electrocyclic reaction: an intramolecular reaction of an acyclic $\pi$-electron system in which a ring is formed with a new $\sigma$ bond, and the product has one fewer $\pi$ bond than the starting material.

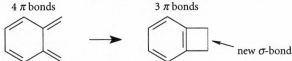

    b. Cycloaddition reaction: a reaction of two separate $\pi$-electron systems in which a ring is formed with two new $\sigma$ bonds, and the product has two fewer $\pi$ bonds than the reactants.

3 π bonds   1 π bond

c. Sigmatropic reaction: a reaction in which a σ bond at one end of a π-electron system appears to migrate to the other end of the π-electron system; the π bonds change position in the process, and their total number is unchanged.

migrating σ-bond

3. Three features of any type of pericyclic reaction are intimately related; specifying any two of the features specifies the third:
   a. The way the reaction is activated (heat or light)
      i. Many pericyclic reactions require no catalysts or reagents other than the reacting partners.
      ii. Such reactions take place either on heating or on irradiation with ultraviolet light.
      iii. Many reactions activated by heat are not activated by light, and vice-versa.
   b. The number of electrons involved in the reaction
      i. The number of electrons involved in a pericyclic reaction is twice the number of curved arrows required to write the reaction mechanism in the curved-arrow notation.
      ii. The direction of "electron flow" in pericyclic reactions indicated by the curved arrows is arbitrary.
   c. The stereochemistry of the reaction

B. EXCITED-STATE PERICYCLIC REACTIONS
   1. When a molecule absorbs light, it reacts through its excited state.
   2. The mode of ring closure in photochemical electrocyclic and cycloaddition reactions differs from that of thermal electrocyclic reactions; the HOMO of the excited state is different from the HOMO of the ground state and has different symmetry.
      a. Photochemical reactions are reactions that occur through electronically excited states.
      b. Thermal reactions occur through the ground state.

C. CLASSIFICATION OF SIGMATROPIC REACTIONS
   1. Sigmatropic reactions are classified by using bracketed numbers to indicate the number of atoms over which a σ bond appears to migrate. (Count the point of original attachment as atom #1.)
      a. In some reactions, each end of a σ bond migrates.
      b. In other reactions, one end of a σ bond remains fixed to the same group and the other end migrates.

D. FLUXIONAL MOLECULES
   1. A number of compounds continually undergo rapid sigmatropic rearrangements at room temperature.
   2. Molecules that undergo rapid bond shifts are called fluxional molecules; their atoms are in a continual state of motion associated with the rapid changes in bonding.

E. PERICYCLIC SELECTION RULES; SUMMARY
   1. It is important to understand that the selection rules refer to the rates of pericyclic reactions, but have nothing to say about the positions of the equilibria involved.
      a. It is common for a photochemical reaction to favor the less stable isomer of an equilibrium because the energy of light is harnessed to drive the equilibrium energetically "uphill."
      b. The selection rules do not indicate which component of an equilibrium will be favored, only whether the equilibrium will be established at a reasonable rate.
      c. According to the principle of microscopic reversibility, the selection rules apply equally well to the forward and reverse of any pericyclic reaction, because the reaction in both directions must proceed through the same transition state.

d. "Allowed" reactions are sometimes prevented from occurring for reasons having nothing to do with the selection rules.

e. The allowed stereochemistry of a pericyclic reaction follows from the phase relationships within the molecular orbitals involved.

2. The selection rules for the three types of pericyclic reactions are summarized in:

a. Table 25.1 (text page 1232) for electrocyclic reactions

b. Table 25.2 (text page 1238) for cycloaddition reactions

c. Table 25.3 (text page 1249) for sigmatropic reactions

3. A convenient way to remember the selection rules is to assign either +1 or −1 to each of the following aspects of the reaction:

a. For the mode of activation assign

i. +1 for a thermal reaction

ii. −1 for a photochemical reaction

b. For the number of reacting electrons (twice the number of arrows needed to describe the reaction in conventional curved-arrow notation) assign

i. +1 for $4n + 2$ electrons

ii. −1 for $4n$ electrons

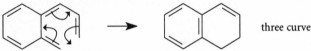

three curved arrows ⇒ six electrons

c. For the stereochemistry of each component assign

i. +1 for a suprafacial or disrotatory reaction or for retention of stereochemistry

ii. −1 for an antarafacial or conrotatory reaction or for inversion of stereochemistry

d. Multiply together the resulting numbers:

i. if the product is +1, the reaction is allowed.

ii. if the product is −1, the reaction is forbidden.

## F. SUMMARY OF SIGMATROPIC REACTION SELECTION RULES

1. The stereochemistry of sigmatropic reactions is a function of the number of electrons involved which, in turn, is determined from the curved-arrow formalism by counting the curved arrows and multiplying by two. (See Table 25.4, text page 1250.)

a. All-suprafacial sigmatropic reactions occur when $4n + 2$ electrons are involved in the reaction (an odd number of electron pairs, or curved arrows).

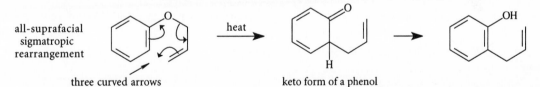

all-suprafacial sigmatropic rearrangement

three curved arrows

keto form of a phenol

b. A sigmatropic reaction must be antarafacial on one component and suprafacial on the other when $4n$ electrons are involved (an even number of electron pairs, or curved arrows).

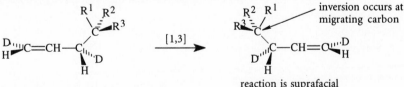

inversion occurs at migrating carbon

reaction is suprafacial with respect to the alkene

c. For a single migrating carbon, the term "suprafacial" is taken to mean "retention of configuration," and the term "antarafacial" is taken to mean "inversion of configuration."

## R  Reactions

............................................................................

### I.  Electrocyclic Reactions

1. When an electrocyclic reaction takes place, the carbons at each end of the conjugated $\pi$ system turn in a concerted fashion so that the *p* orbitals can overlap (and rehybridize) to form the $\sigma$ bond that closes the ring.

2. This turning can occur in two stereochemically distinct ways.

   a. In a conrotatory closure the two carbon atoms turn in the same direction, either both clockwise or both counterclockwise.

   conrotatory closure                  or

   b. In a disrotatory closure the two carbon atoms turn in opposite directions so that either the upper lobes of the *p* orbitals overlap or the lower lobes of the *p* orbitals overlap.

   disrotatory closure          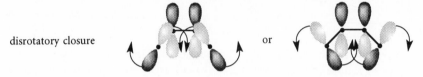        or

3. The HOMO of the conjugated alkene contains the $\pi$ electrons of highest energy and governs the course of pericyclic reactions.

   a. When the ring closure takes place, the two *p* orbitals on the ends of the $\pi$ system must overlap in phase.

   b. The wave peak on one carbon must overlap with the wave peak on the other, or a wave trough must overlap with a wave trough.

4. The relative orbital phase at the terminal carbon atoms of the HOMO (the orbital symmetry) determines whether the reaction is conrotatory or disrotatory.

   a. Conjugated alkenes with $4n$ $\pi$ electrons ($n$ = any integer) have antisymmetric HOMOs and undergo conrotatory ring closure.  Conrotatory ring closure is allowed for systems with $4n$ $\pi$ electrons; it is forbidden for systems with $4n + 2$ $\pi$ electrons.

   b. Conjugated alkenes with $4n + 2$ $\pi$ electrons have symmetric HOMOs and undergo disrotatory ring closure.  Disrotatory ring closure is allowed for systems with $4n + 2$ $\pi$ electrons; it is forbidden for systems with $4n$ $\pi$ electrons.

### II.  Cycloaddition Reactions

1. Cycloadditions are classified, first, by the number of electrons involved in the reaction with respect to each component.

   a. The number of electrons involved is determined by writing the reaction mechanism in the curved-arrow notation.

   b. The number of electrons contributed by each given reactant is equal to twice the number of curved arrows originating from that component (two electrons per arrow).

   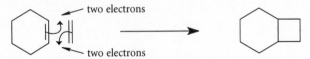

2. Cycloaddition reactions are also classified by their stereochemistry with respect to the plane of each reacting molecule and may in principle occur either across the same face, or across opposite faces, of the planes in each reacting component.

a. If the reaction occurs across the same face of a $\pi$ system, the reaction is said to be suprafacial with respect to that $\pi$ system; that is, a *syn* addition that occurs in a single mechanistic step.

b. If the reaction bridges opposite faces of a $\pi$ system, it is said to be antarafacial; that is, it is an *anti* addition that occurs in one mechanistic step.

3. In order for a cycloaddition to occur, bonding overlap must take place between the $p$ orbitals at the terminal carbons of each $\pi$ electron system.
   a. The bonding overlap begins when the HOMO of one component interacts with the LUMO of the other.
   b. The LUMO of the other component is the empty orbital of lowest energy into which the electrons from the HOMO must flow.
   c. These two frontier MOs involved in the interaction must have matching phases if bonding overlap is to be achieved.
   d. It does not matter which component provides the HOMO and which provides the LUMO.
4. All-suprafacial cycloadditions are allowed thermally for systems in which the total number of reacting electrons is $4n + 2$, and they are allowed photochemically for systems in which the number is $4n$.

## III. *Sigmatropic Reactions*

### A. Stereochemistry of Sigmatropic Reactions

1. Sigmatropic reactions can be classified by their stereochemistry according to whether the migrating bond moves over the same face, or between opposite faces, of the $\pi$-electron system.
   a. If the migrating bond moves across one face of the $\pi$ system the reaction is said to be suprafacial.
   b. If the migrating bond moves from one face of the $\pi$ system to the other the reaction is said to be antarafacial.
   c. When both ends of a $\sigma$ bond migrate, the reaction can be suprafacial or antarafacial with respect to each $\pi$ system.
2. The stereochemistry of a sigmatropic reaction is revealed experimentally only if the molecules involved have stereocenters at the appropriate carbons.
3. Molecular orbital theory provides the connection between the type of sigmatropic reaction and its stereochemistry.

### B. [1,3] and [1,5] Sigmatropic Rearrangements

1. The interaction of the LUMO of the migrating group with the HOMO of the $\pi$ system, or vice-versa, controls the stereochemistry of the reaction.
   a. In the migration of a hydrogen, the orbital involved is a $1s$ orbital, which has no nodes.
   b. In the migration of a carbon, the orbital involved is a $2p$ orbital, which has one node.
   c. The shift of a carbon can occur in two stereochemically distinct ways:
      i. Migration with retention of configuration.
      ii. Migration with inversion of configuration.
   d. In the allyl anion, the HOMO is antisymmetric:

allyl anion

      i. The suprafacial [1,3] shift of a hydrogen is forbidden by orbital symmetry.
      ii. The antarafacial [1,3] migration of hydrogen is allowed by orbital symmetry but is virtually nonexistent in organic chemistry because unreasonably long bonds would be required in the transition state.

    *iii.* The suprafacial [1,3] shift of a carbon with inversion of configuration is allowed by orbital symmetry.

    *iv.* The antarafacial [1,3] shift of a carbon with retention of configuration is allowed by orbital symmetry but is rarely observed.

  e.  In the 2,4-pentadienyl anion, the HOMO is symmetric:

2,4-pentadienyl anion

    *i.* The suprafacial [1,5] migration of a hydrogen is allowed by orbital symmetry.

    *ii.* The antarafacial [1,5] migration of a hydrogen is forbidden by orbital symmetry.

    *iii.* The suprafacial [1,5] migration of a carbon with retention of configuration is allowed by orbital symmetry.

    *iv.* The antarafacial [1,5] migration of a carbon with inversion of configuration is allowed by orbital symmetry.

## C. SIGMATROPIC REARRANGEMENTS; COPE AND CLAISEN REARRANGEMENTS

1.  A [3,3] sigmatropic rearrangement is a reaction in which both ends of a σ bond change positions (migrate).

  a.  The transition state of a [3,3] sigmatropic rearrangement can be visualized as the interaction of two allylic systems, one a cation and one an anion.

  b.  The two MOs involved achieve bonding overlap when the [3,3] sigmatropic rearrangement occurs suprafacially on both components.

2.  The Cope rearrangement is a [3,3] sigmatropic rearrangement in which a 1,5-diene isomerizes.

$$165\text{--}185°$$

3.  The oxyCope reaction involves the initial formation of an enol; tautomerization of the enol into the corresponding carbonyl compound is a very favorable equilibrium that drives the reaction to completion.

4.  In the Claisen rearrangement, an ether that is both allylic and vinylic (or an allylic aryl ether) undergoes a [3,3] sigmatropic rearrangement. If both *ortho* positions of an aryl allylic ether are blocked by substituent groups, the *para*-substituted derivative is obtained by a sequence of two Claisen rearrangements followed by tautomerization of the product to the phenol.

slow      tautomerization

# Study Guide Links

## ✓25.1   Thermal Activation of Pericyclic Reactions

It is important to realize that the amount of heat required to activate a pericyclic reaction can vary significantly from reaction to reaction. Thus, some "thermal" reactions are so fast that they readily occur at room temperature or below. Others require strong heating. A reaction in which the reactants are very unstable may be very fast; one in which the products and, by implication, the transition state, have a large amount of strain may be very slow.

## ✓25.2   Frontier Orbitals

You've seen that the analysis of pericyclic reactions in the text has focused on frontier orbitals. The discussion of electrocyclic reactions focused on one of the frontier orbitals, the HOMO, and the analysis of cycloadditions focused on the HOMO of one component and the LUMO of the other. Yet each of the frontier orbitals houses only two of the many reacting $\pi$ electrons. Because *all* $\pi$ electrons in the reacting molecules are involved in pericyclic reactions, it can perhaps be appreciated that all of the $\pi$ molecular orbitals are involved as well. Why focus, then, on frontier orbitals? This focus is justified because it has been shown that frontier-orbital interactions play the *major* role in determining the energies of transition states for concerted pericyclic reactions. However, these are not the only orbital interactions that occur; they are simply the *most important* interactions.

Some students try to associate frontier molecular orbital interactions with formation of *a particular bond* in the reactant or product. This temptation is understandable because of the association of a $\pi$ molecular orbital with the double bond in a simple alkene, or the association of a $\sigma$ molecular orbital with a particular single bond in saturated molecules. These early encounters with molecular orbitals fail to take account of the fact that all molecular orbitals within a molecule must have certain symmetry properties, namely, they must be symmetric or antisymmetric with respect to the symmetry elements of the molecule itself. (For example, if a molecule contains an internal mirror plane, molecular orbitals of the molecule must be either antisymmetric or symmetric with respect to that plane.) For this reason, a one-to-one correspondence in general does not exist between molecular orbitals and the localized bonds used to write Lewis structures. Molecular orbitals are orbitals of *molecules,* and changes in a particular molecular orbital may involve more than one bond, and vice-versa.

## ✓25.3   Orbital Analysis of Sigmatropic Reactions

Sigmatropic reactions are concerted and do not involve free ions, but the orbital relationships are nevertheless understood by thinking of the two components of the transition state as ions of opposite charge. However, we could just as well think of the reaction as the migration of a hydrogen atom across a 2,4-pentadien-1-yl radical, or as the migration of a hydrogen anion (hydride ion) across a 2,4-pentadien-1-yl cation; these different approaches give the same predictions. Let's verify this point by viewing the transition state of the [1,5] hydrogen migration as the migration of a hydrogen atom across a 2,4-pentadien-1-yl radical:

$$\left[ \; \cdot CH_2 \overset{\displaystyle\frown}{\phantom{xx}} CH_2 \quad \underset{H\,\cdot}{} \longleftrightarrow \quad CH_2 \overset{\displaystyle\frown}{\phantom{xx}} \cdot CH_2 \quad \underset{\cdot H}{} \; \right]^{\ddagger}$$

In this case, analyze the symmetries of the orbitals containing the unpaired electrons, that is, the HOMO of each radical. For a hydrogen atom, this orbital is just the $1s$ orbital. For the 2,4-pentadien-1-yl radical, this is $\psi_3$ (see Fig. 25.5 on text p. 1228). Because this molecular orbital is symmetric, the migration is predicted to occur suprafacially.

Now let's analyze the reaction as if it were the migration of a hydrogen anion (hydride ion, H:$^-$) across the termini of a 2,4-pentadien-1-yl cation and show that such an analysis makes the same prediction. In this case, the $1s$ orbital of the migrating hydride contains two electrons; thus, this orbital is classified as the HOMO of the hydride ion. It interacts with the LUMO of the 2,4-pentadien-1-yl cation, which, from Fig. 25.5 on text p. 1228, is $\psi_3$. Because this molecular orbital is symmetric, the migration hydride is predicted to occur suprafacially.

Notice that whether we deal with with the migration of a proton across an anion, a radical across another radical, or an anion across a cation, the relevant orbitals of the species involved are the same: the $1s$ orbital of the hydrogen and $\psi_3$ of the pentadienyl system. Hence, the predictions of orbital symmetry must also be the same.

## Solutions

### Solutions to In-Text Problems

**25.1**   (a)  This is an electrocyclic reaction.

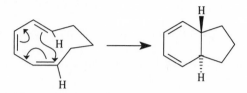

(c)  This is an intramolecular cycloaddition reaction.

(e)  This is an electrocyclic reaction.

**25.2**   (a)  1,3,5-Hexatriene has six $\pi$ molecular orbitals.

(b)  Questions about the symmetry of molecular orbitals refer to the symmetry about a reference plane like the one shown in Fig. 25.3 on text p. 1225. The symmetry of the 1,3,5-hexatriene MOs alternates; that is, $\psi_1$, $\psi_3$, and $\psi_5$ are symmetric, and $\psi_2$, $\psi_4$, and $\psi_6$ are antisymmetric.

(c)  $\psi_1$, $\psi_2$, and $\psi_3$ are bonding and $\psi_4$, $\psi_5$, and $\psi_6$ are antibonding.

(d)  Because 1,3,5-hexatriene has six $\pi$ electrons, the highest occupied molecular orbital (HOMO) is $\psi_3$; the lowest unoccupied molecular orbital (LUMO) is $\psi_4$.

(e)  As indicated in the solution to part (b), the HOMO ($\psi_3$) is symmetric; therefore, the phase of the HOMO at the terminal carbons is the same.

(f)  The phase of the LUMO ($\psi_4$) at the terminal carbons is different.

**25.4**   (a)  The MO $\psi_4$ is nonbonding. (If there is an odd number $n$ of molecular orbitals in an unbranched acyclic system, the nonbonding MO is always the one with the number $(n + 1)/2$.)

(b)  The symmetry of the MOs alternates; that is, $\psi_1$, $\psi_3$, $\psi_5$, and $\psi_7$ are symmetric, and $\psi_2$, $\psi_4$, and $\psi_6$ are antisymmetric.

(c)  According to its resonance structures, this cation has sites of positive charge at alternating carbons:

$$\left[ H_2C{=}CH{-}CH{=}CH{-}CH{=}CH{-}\overset{+}{C}H_2 \longleftrightarrow H_2C{=}CH{-}CH{=}CH{-}\overset{+}{C}H{-}CH{=}CH_2 \longleftrightarrow \right.$$

$$\left. H_2C{=}CH{-}\overset{+}{C}H{-}CH{=}CH{-}CH{=}CH_2 \longleftrightarrow H_2\overset{+}{C}{-}CH{=}CH{-}CH{=}CH{-}CH{=}CH_2 \right]$$

In molecular orbital terms, the positive charge in the cation results from the absence of an electron in the LUMO; the LUMO *is* the nonbonding MO. Positive charge is shared at alternating carbons because the LUMO has nodes at the other carbons. That is, positive charge can only exist on carbons at which there is no node in the LUMO. If this is so, the LUMO must have three nodes: one at each carbon that does not share the positive charge. As shown in the solution to part (a), $\psi_4$ is the nonbonding MO and is therefore the LUMO, and it has three nodes, one at each carbon that does not share positive charge in the above structures.

25.6 (a) This is a disrotatory reaction involving $4n + 2$ electrons, and is allowed. Therefore it can occur readily by a concerted mechanism. The reaction is disrotatory because the hydrogens shown in the product are both on the outside of the starting material. (Draw these hydrogens in the starting material, if necessary.) In order for these hydrogens to end up on the same face of the product, the ends of the $\pi$-electron system must turn in opposite directions.

(b) This is a conrotatory reaction involving $4n + 2$ electrons, and is not allowed; it does not occur readily.

25.8 The thermal ring opening of *trans*-3,4-dimethylcyclobutene must be a conrotatory process. The two possible conrotatory processes are as follows:

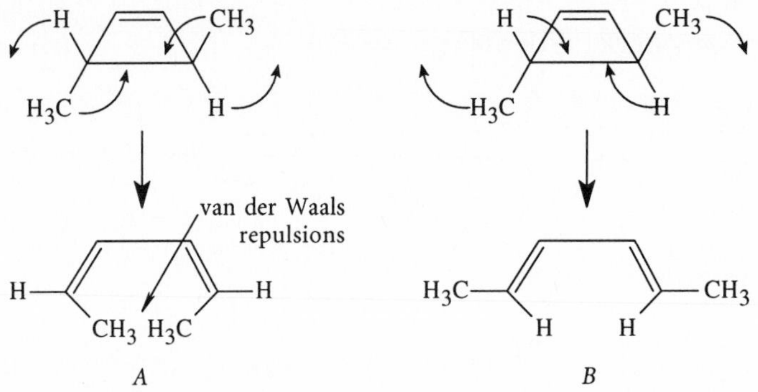

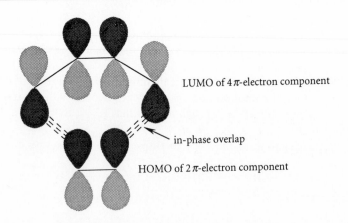

Product A has severe van der Waals repulsions between the methyl groups; product B does not. The repulsions raise the energy of product A and the transition state for its formation; consequently, product B is observed.

25.10 The HOMO of the $2\pi$-electron component is symmetric, and the LUMO of the $4\pi$-electron component is symmetric. As the diagram in Fig. SG25.1 shows, the ends of these two MOs have matching phases in a $[4s + 2s]$ cycloaddition.

LUMO of $4\pi$-electron component

in-phase overlap

HOMO of $2\pi$-electron component

**Figure SG25.1** *Orbital diagram to accompany the solution to Problem 25.10. Interaction of the LUMO of the $4\pi$-electron component and the HOMO of the $2\pi$-electron component in a $[4s + 2s]$ cycloaddition gives in-phase overlap at both ends of the two $\pi$-electron systems.*

**25.12** The [8s + 2s] cycloaddition and its product:

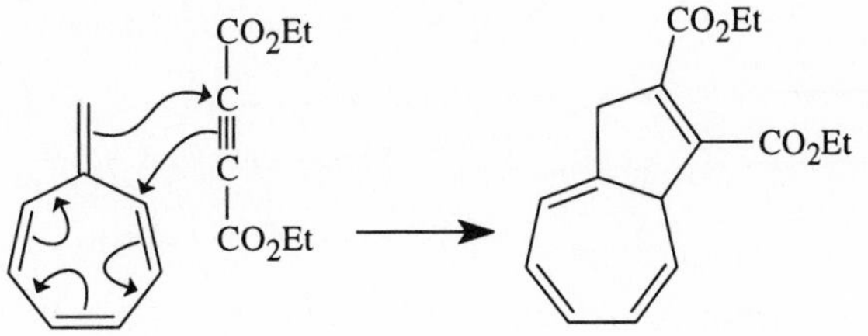

**25.13** The four products are the diastereomeric 1,2,3,4-tetramethylcyclobutanes. Each of the pure alkene stereoisomers can undergo an allowed photochemical [2s + 2s] cycloaddition in two distinguishable ways. *Cis*-2-butene reacts to give compounds A and B; *trans*-2-butene reacts to give compounds B and C; and the mixture of *cis*- and *trans*-2-butene can give these three compounds plus a fourth, compound D, that results from the [2s + 2s] cycloaddition of *cis*-2-butene to *trans*-2-butene.

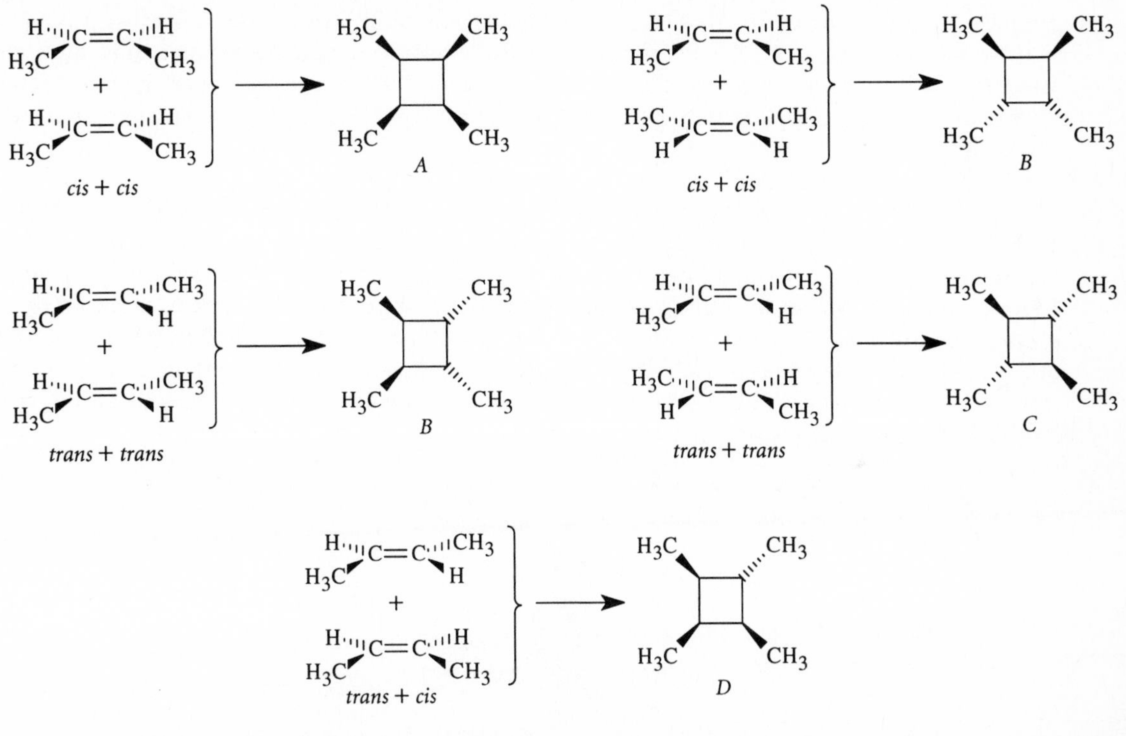

**25.14** (a) Since exchanging the positions of two groups at a stereocenter changes the configuration of the stereocenter, exchange the positions of the T and the D in the starting material; this will lead to product with the opposite, that is, the R, configuration at the indicated carbon.

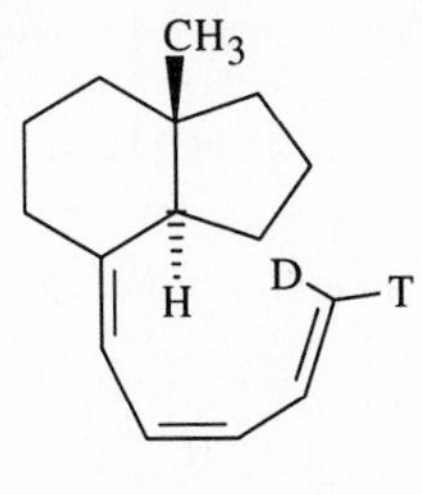

**25.15** (a) This is a [1,4] sigmatropic rearrangement.

migration of this end of the
bond does not occur, that is, it
"migrates" from carbon-1* to carbon-1*.

migration of this end of
the bond occurs from
carbon-1 to carbon-4.

(c) This is a [5,5] sigmatropic rearrangement.

This bond migrates
from carbons 1 and 1*
to carbons 5 and 5*,
respectively.

**25.16** Reaction (1) occurs readily because it is a [1,5] sigmatropic rearrangement, which is allowed when the hydrogen migrates suprafacially over the $\pi$-electron system. This reaction pathway does not introduce significant strain or twist into the $\pi$-electron system. Reaction (2) does not occur because it is a [1,3] sigmatropic rearrangement, which is allowed only if the hydrogen migrates antarafacially. Such an antarafacial migration requires that the hydrogen simultaneously bridge the upper and lower faces of the$\pi$-electron system. This requirement cannot be met while at the same time maintaining normal bond lengths and angles.

**25.18** (a) Think of the five-carbon $\pi$-electron system as a 2,4-pentadienyl cation, whose MOs are shown in Fig. 25.5 on text p. 1228, and think of the migrating group as a carbon anion in which the unshared electron pair resides in a $p$ orbital. The orbitals involved in the rearrangement are the LUMO of the pentadienyl system and the HOMO of the carbon anion. The LUMO of the pentadienyl system, as shown by Fig. 25.5, is symmetric; that is, at each end of the system, the LUMO has the same phase on a given face. The HOMO of the carbon anion is simply the filled $p$ orbital. The migration is allowed only if the carbon anion migrates suprafacially such that each end of the pentadienyl LUMO interacts with the same lobe of the carbon anion $p$ orbital. (Fig. SG25.2)

HOMO of migrating carbon anion

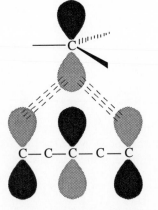

Migration occurs suprafacially on
both systems (and thus with retention
of configuration at the migrating
carbon).

LUMO of 2,4-pentadienyl cation

**Figure SG25.2** *Orbital diagram to accompany the solution to Problem 25.18(a). In order for continuous orbital overlap to occur, migration of the carbon must occur suprafacially on the pentadienyl system and at the same lobe of the p orbital of the migrating carbon (that is, suprafacially and with retention of configuration).*

In the equation of part (b), early printings of the text have a transition-state symbol on the bracketed intermediate; this should be deleted. This is an unstable intermediate, not a transition state.

**25.19** This transformation involves two successive [1,5] sigmatropic hydrogen migrations—sort of a "hydrogen walk," as shown by the following curved arrows. These are allowed suprafacial processes.

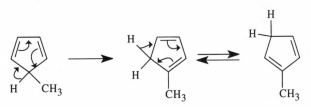

Why does the equilibrium favor the two products over the starting material? See Sec. 4.5B on text p. 141.

**25.20** (a) Deduce the product by using the curved-arrow formalism:

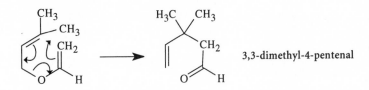

3,3-dimethyl-4-pentenal

**25.21** See Fig. 25.4 on text p. 1227 for the relevant orbital diagram. The HOMOs of the two allylic radicals ($\psi_2$) interact because it is these molecular orbitals that contain the unpaired electrons. Since the two orbitals are the same, they have the same symmetry, and therefore interact suprafacially on both components with positive overlap.

**25.23** (a) Thermal reaction: (+1); stereochemistry: (+1)(−1); number of electrons (8 = 4n): (−1). Result = +1; the reaction is allowed.

(c) Photochemical (excited state) reaction: (−1); stereochemistry: (+1); number of electrons (8 = 4n): (−1). Result = +1; the reaction is allowed.

**25.24** (a) The $\sigma$ bond that "moves" is indicated with an asterisk (*). It moves suprafacially on both $\pi$ systems; [3,3] sigmatropic rearrangements are thermally allowed processes.

**25.25** Note that previtamin $D_2$ is the same as previtamin $D_3$ except for the R-group. (See the last sentence on text p. 1253.) Comparing the structures of previtamin $D_3$ (or $D_2$) in Eq. 25.38 on text p. 1252, ergosterol at the bottom of text p. 1253, and lumisterol in the problem reveals that the latter two compounds result from conrotatory electrocyclic ring closures, ergosterol being formed by a clockwise conrotatory motion (as viewed edge-on by the "eye" below) and the lumisterol from a counterclockwise conrotatory motion. Photochemical conrotatory electrocyclic reactions involving $4n + 2$ electrons are allowed.

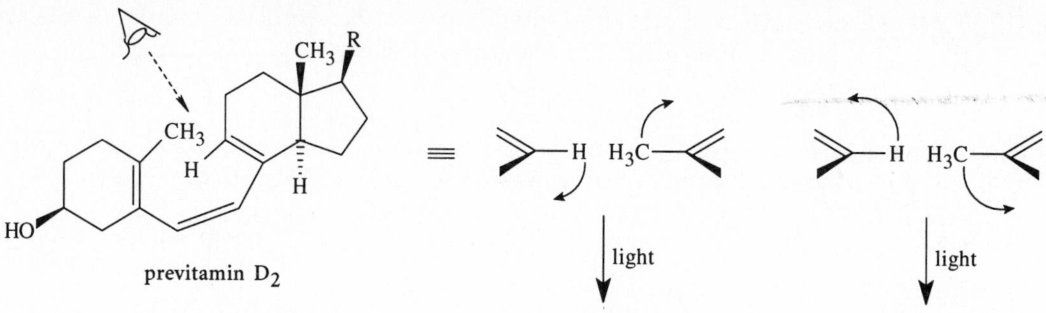

previtamin D$_2$

ergosterol          lumisterol

25.27  First determine why irradiation of *A* or *B* does not give back previtamin D$_2$. One reason is mechanistic. Since *A* and *B* originate from *heating* previtamin D$_2$, then previtamin D$_2$ can only result from the exact reverse of the same reaction, which must occur as a disrotatory process. However, ring opening by irradiation, if it were to occur, would have to take place with opposite stereochemistry, that is, as a conrotatory process. Either of the two possible conrotatory processes would result in a product containing a *trans* double bond within a six-membered ring. Even irradiation does not provide enough energy for this to occur.

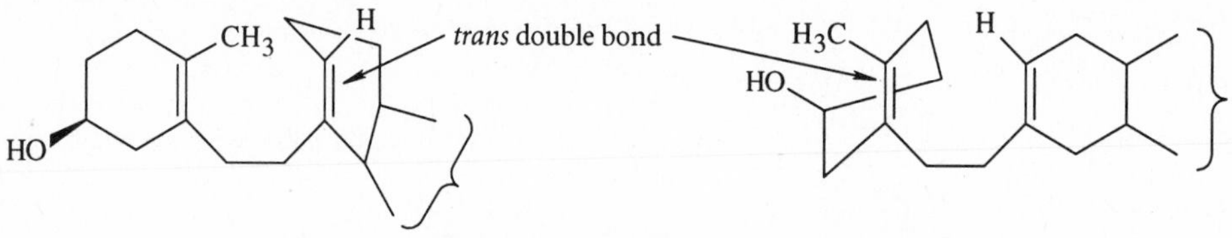

the two possible products of conrotatory ring opening

The only alternative to this process is a four-electron disrotatory process that gives a cyclobutene:

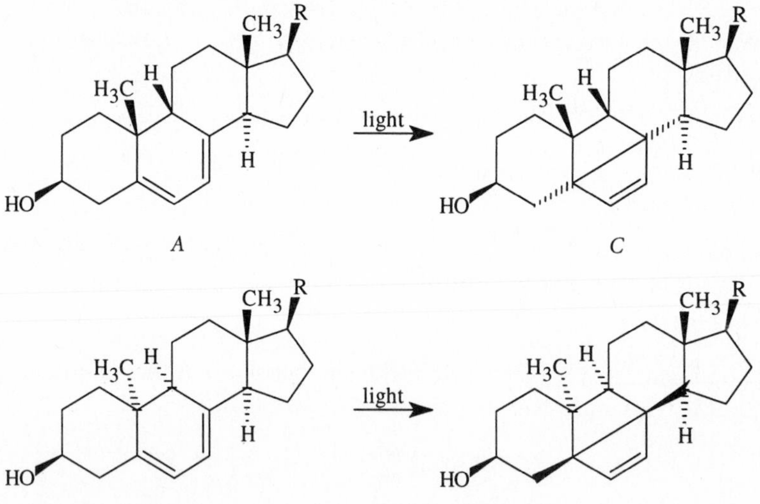

*A*          *C*

*B*          *D*

## Solutions to Additional Problems

**25.28** (a) Follow the procedure used in Study Problem 25.3 on text p. 1251. This reaction is a thermal disrotatory process that involves $4n$ electrons. Result: $(+1)(+1)(-1) = (-1)$; the reaction is forbidden.

(c) The highest occupied molecular orbital (HOMO) of a conjugated triene is $\psi_3$. Since orbitals of conjugated dienes alternate in symmetry, this MO is symmetric. (See the solution to Problem 25.2 on p. 701 of this manual.) The methyl groups, to a useful approximation, have no effect on the nodal properties of the molecular orbitals.

**25.29** The pericyclic selection rules say *absolutely nothing* about the position of equilibrium in each case. These rules refer to *rates* of reactions, not to equilibrium constants. Other considerations must be used to decide on the position of equilibrium. The point of the problem is for you to use what you have learned about the relative energies of molecules; the molecule of lower energy is favored in each equilibrium.

(a) The right side of the equation is favored at equilibrium because the double bonds are conjugated with the phenyl rings.

(c) The right side of the equation is favored at equilibrium because a $C=O$ double bond is formed at the expense of a $C=C$ double bond. ($C=O$ bonds are stronger than $C=C$ bonds; see text page 1089.)

(e) The left side of the equation is favored at equilibrium because the double bonds are conjugated not only with each other but also with the carbonyl group.

**25.30** (a) Because it involves $4n$ $\pi$ electrons and thermal conditions, the electrocyclic reaction that converts compound *A* into compound *B* is conrotatory. (A "thermal" process can occur at low temperature if the process is particularly rapid.) Hence, the methyl groups in compound *B* are *trans*. The thermal process that converts compound *B* into compound *C* involves six $(4n + 2)$ electrons and is therefore disrotatory. (Both compounds *B* and *C* are, of course, formed as racemates.)

**25.31** (a) This reaction can be viewed as either a [1,9] or a [1,13] sigmatropic rearrangement. In either of these classifications the reaction is an allowed process that is suprafacial on the $\pi$-electron system and occurs with retention at the migrating carbon.

**25.32** (a) The formula indicates that an addition has taken place. This is an allowed photochemical [2s + 2s] cycloaddition:

(b) This allylic vinylic ether undergoes a [3,3] sigmatropic rearrangement (that is, a Claisen rearrangement) on heating:

*(solution continues)*

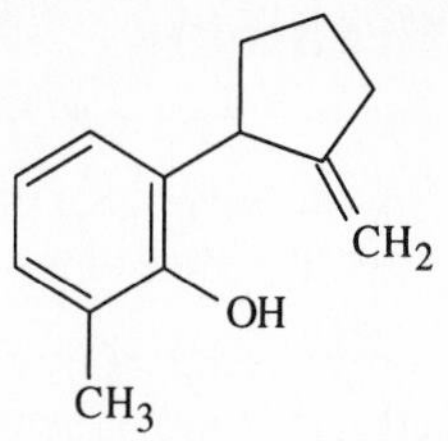

(c)  This is a [4s + 2s] cycloaddition, that is, a Diels-Alder reaction. The exocyclic double bond is not involved in the reaction, and *endo* stereochemistry is assumed:

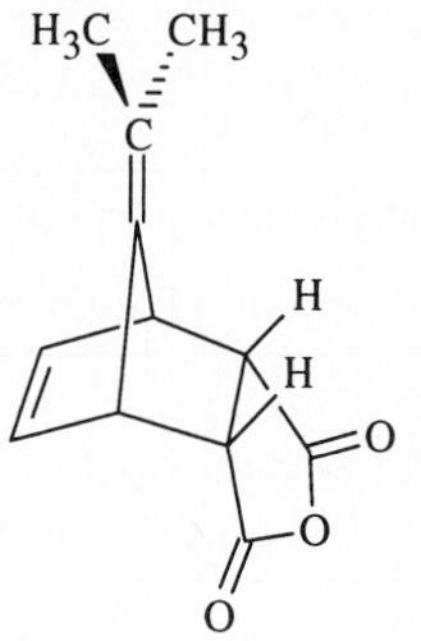

Two other reactions are allowed in principle that *do* involve the exocyclic double bond. One is a [2s + 6a] cycloaddition to give product *A*, and the other is a [2a + 6s] cycloaddition to give product *B*:

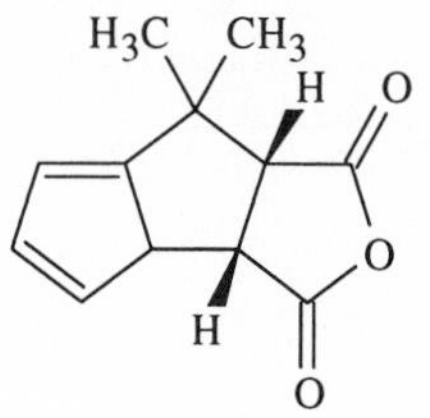

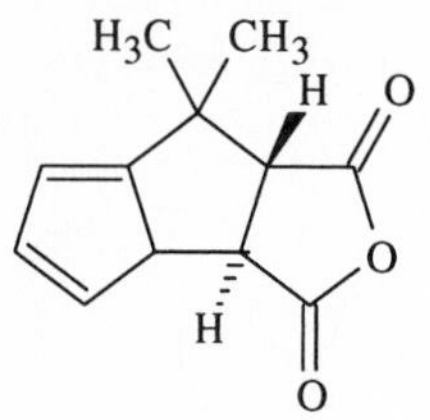

If you construct models of the starting materials, and if you bring them together so that the appropriate orbital interactions can occur, you will see that the transition states required to form these products (as well as the products themselves) are considerably strained. Hence, the reactions that form these products are much slower than the Diels-Alder process.

25.33  (a)  The structure of the ozonolysis product *C* shows that compound *B* is a cyclobutene, which must be formed in a disrotatory photochemical electrocyclic reaction. This defines the stereochemistry of *B*, which, in turn, defines the stereochemistry of *C*:

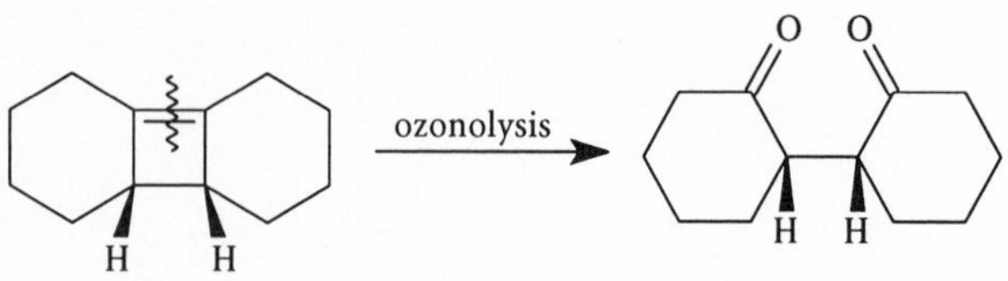

25.34  This cycloaddition reaction involves sixteen electrons; hence, it must be suprafacial on one component and antarafacial on the other. Thus, it must be a [14s + 2a] or a [14a + 2s] cycloaddition. The heptafulvene molecule is large enough that its π-electron system can twist without introducing too much strain or without losing too much π-electron overlap; hence, the cycloaddition is a [14a + 2s] process. In either case, the product has the following stereochemistry:

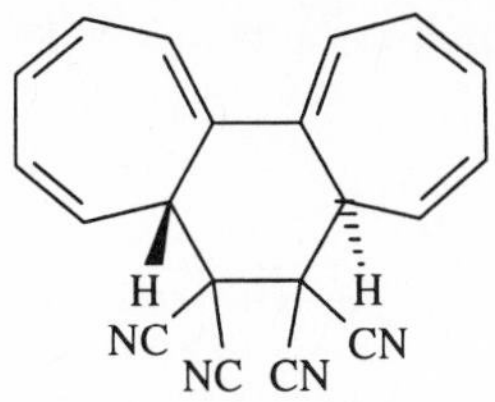

**25.35** (a) This is a [3,3] sigmatropic rearrangement, that is, a Claisen rearrangement.

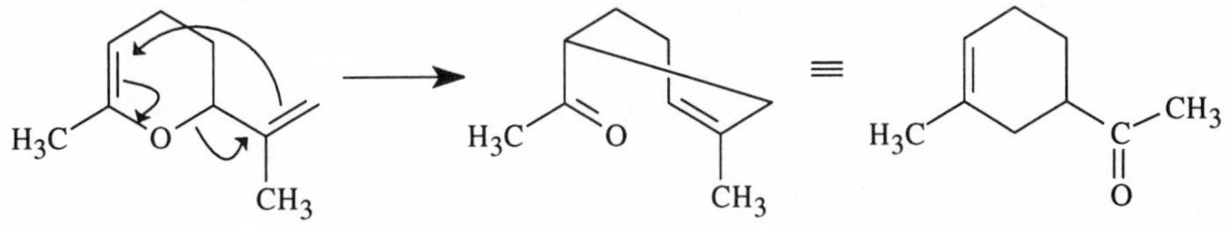

The solution to the foregoing part and the solution to part (c) that follows illustrate a useful two-step process for drawing the products of complicated rearrangements. It usually makes sense to draw the bond connections first and thus obtain a highly distorted structure of the product. Then convert the distorted structure into a more conventional structure, using models if necessary.

(c) This is a [3,3] sigmatropic rearrangement in which the product requires a little redrawing!

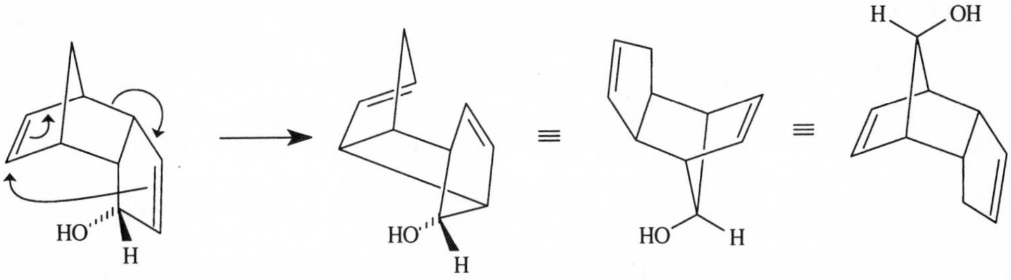

**25.36** Take a cue from the solution to Problem 25.30(a) on p. 707 of this manual, in which the last step involves a thermal reaction of 1,3,5-cyclooctatriene with two methyl groups. Replace the methyl groups with hydrogens and the result is the same. The resulting compound *B* undergoes a Diels-Alder reaction (that is, a [4s + 2s] cycloaddition) with the alkyne, and the product of that reaction, *C*, undergoes a reverse Diels-Alder to generate the final products.

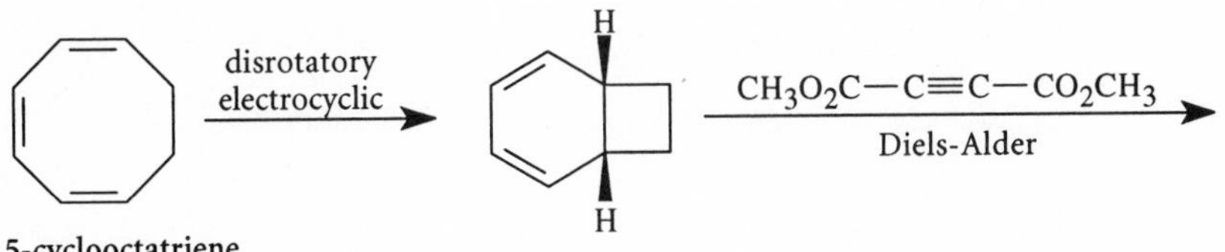

1,3,5-cyclooctatriene
(compound *A*)

*B*

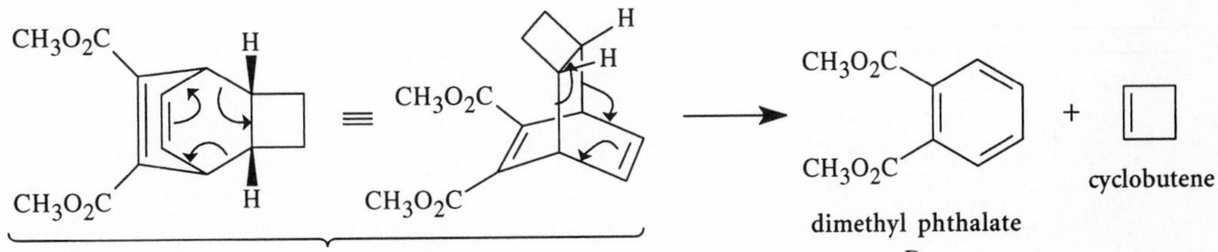

$CH_3O_2C-C\equiv C-CO_2CH_3$
Diels-Alder

*C*

dimethyl phthalate

*D*

cyclobutene

**25.38** Heating an allylic vinylic ether generally results in a [3,3] sigmatropic (Claisen) rearrangement.

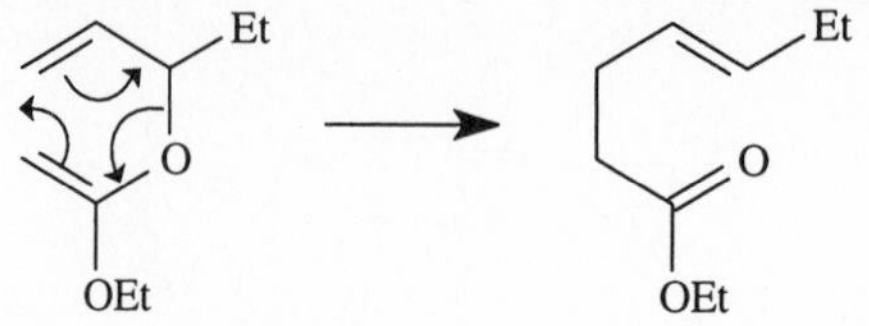

(substance *A*)

Why the product should have *E* stereochemistry is not obvious from the mechanism shown above, and an answer that gives either the *E* or the *Z* stereoisomer (or a mixture of both) is satisfactory. If you work Problem 25.44(a) you will find that the Claisen rearrangement proceeds through a chairlike conformation. If the ethyl group is placed in the more favorable equatorial position of the chairlike conformation, the product is the *E* stereoisomer of substance *A* shown above.

**25.40** The structure of the ozonolysis product *D* and the formula of elemicin show that elemicin has the structure *C*. Compound *B* and elemicin differ by one methyl group, and compound *B* ionizes in base. Consequently, a reasonable hypothesis is that one of the three oxygens of compound *B* is part of a phenol —OH group, and that this is methylated by dimethyl sulfate treatment to give elemicin. Compound *A* is *not* a phenol because it is insoluble in base, but forms a phenol *B* when heated. This series of observations suggests that compound *A* is an allyl ether that undergoes a Claisen rearrangement when heated. Which oxygen is part of the allyl ether? If one of the two outer oxygens of *A* is an allyl ether, its rearrangement would result in transfer of the allyl group to the adjacent *ortho* position. Since the allyl group in *C* is not *ortho* to one of the methoxy groups, the rearrangement must have occured from the central oxygen. Since the *ortho* positions are blocked, rearrangement occurs to the *para* position. The structures of *A* and *B* are thus defined.

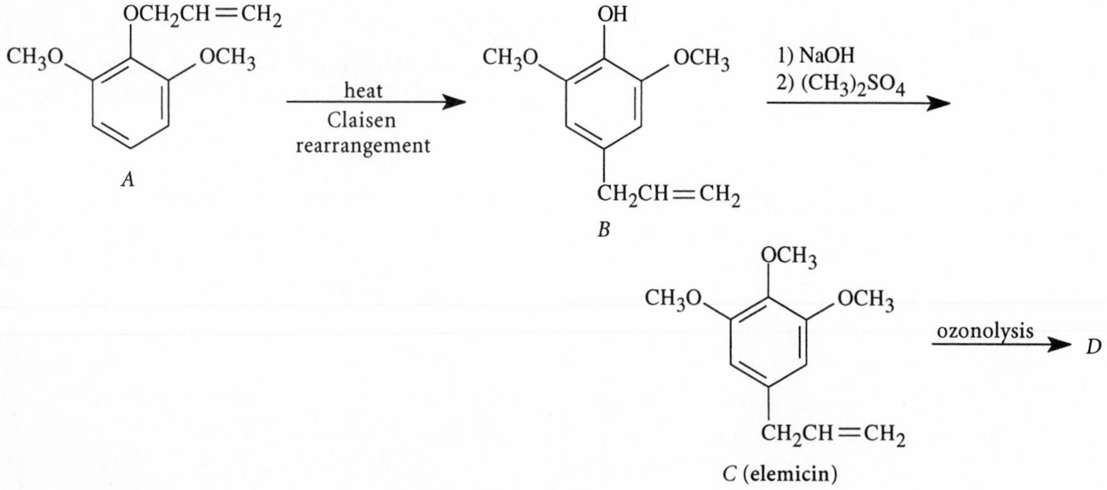

**25.41** Claisen rearrangements are key steps in both parts (a) and (b).

(a)

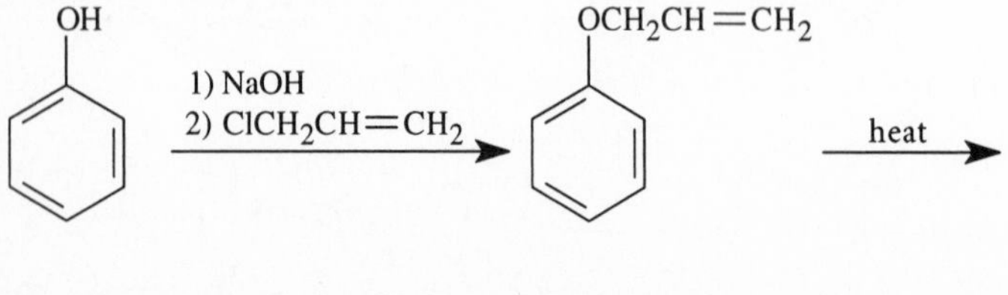

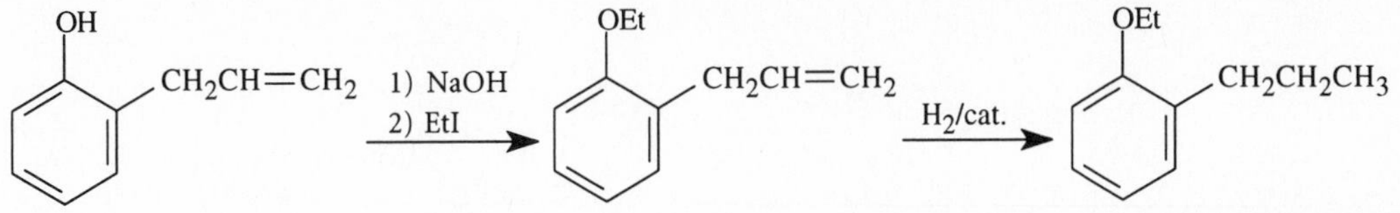

1-ethoxy-2-propylbenzene

**25.42**  (a)  The first transformation is a thermal conrotatory electrocyclic ring opening, and the second transformation is a thermal disrotatory electrocyclic ring closure.

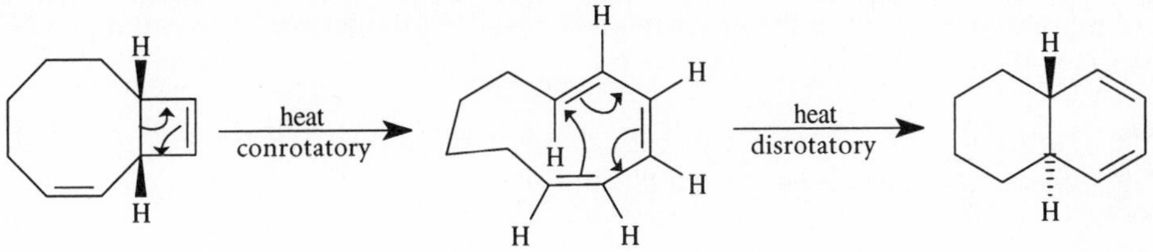

*X*

**25.43**  (a)  The approach to solving this type of problem is described in the solution to Problem 15.58(a) on p. 356, Vol. 1, of this manual. The intermediate *X* is formed from the starting material by an allowed thermal disrotatory ring closure. Because *X* is a conjugated diene, it reacts in a Diels-Alder reaction with maleic anhydride, a good dienophile, to give the product shown in the problem.

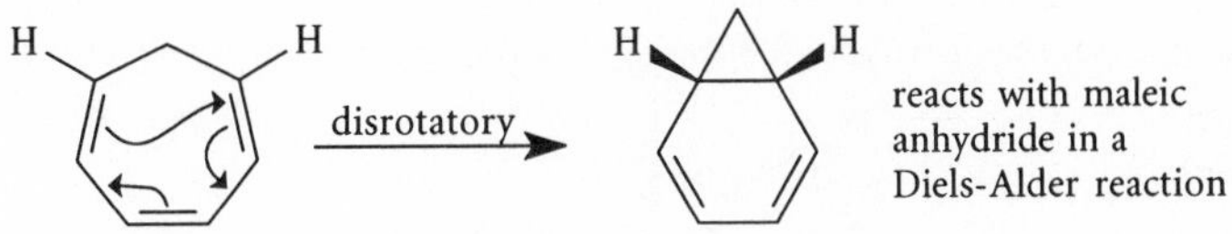

reacts with maleic anhydride in a Diels-Alder reaction

*X*

(b)  In this case maleic anhydride is used to trap the trienone intermediate *Y* formed in the first of what would, in the absence of maleic anhydride, be two successive [3,3] sigmatropic rearrangements. (See Eqs. 25.36a–b on text p. 1248.) Maleic anhydride reacts with the conjugated-diene unit of intermediate *Y*.

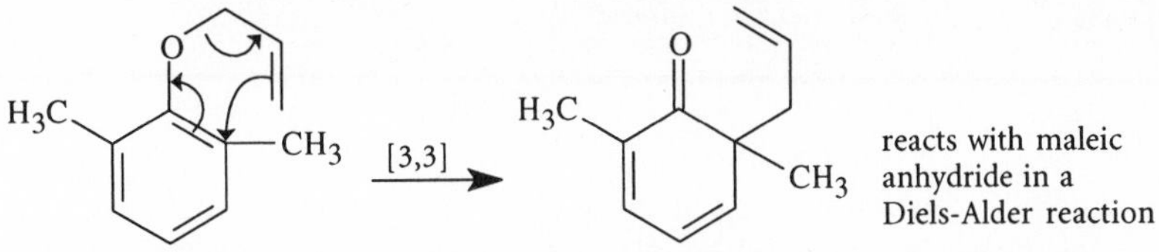

reacts with maleic anhydride in a Diels-Alder reaction

*Y*

**25.44**  (a)  Place the starting material in a chairlike conformation, carry out the [3,3] sigmatropic rearrangement, and compare the stereochemistry of the product with that obtained experimentally. (Models are useful to ensure that the asymmetric carbons have the proper configuration.)

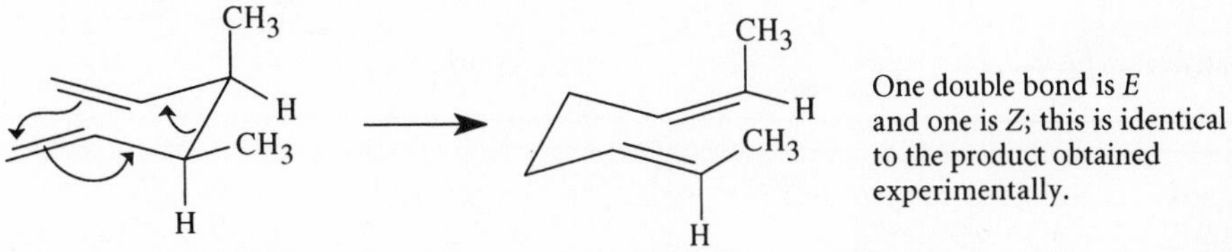

One double bond is *E* and one is *Z*; this is identical to the product obtained experimentally.

This verifies that the product could be formed from a chairlike conformation. However, the boatlike conformation must also be examined:

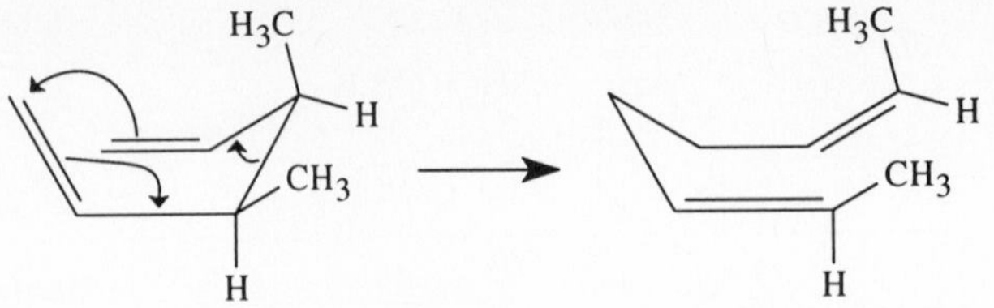

Both double bonds are
Z; this is different from
the product obtained
experimentally.

(A second boatlike conformation is possible, and it would lead to formation of the all-*E* product, which is also different from that observed experimentally.) Thus, of the two possible conformations, only a chairlike transition state fits the observations. You should use similar reasoning to verify that a chairlike conformation of the second reactant in the problem gives the all-*E* product observed and that a boatlike conformation does not.

 Notice that experiments with *both* alkene stereoisomers are required to establish that a chairlike transition state is a general phenomenon.

25.45   A thermal electrocyclic reaction of the carbocation, which is formed by protonation of the ketone starting material, is followed by loss of a proton to give an enol, which spontaneously tautomerizes to the corresponding ketone. Since four electrons are involved, the reaction is predicted to be conrotatory, and the methyl groups in the product are therefore *trans:*

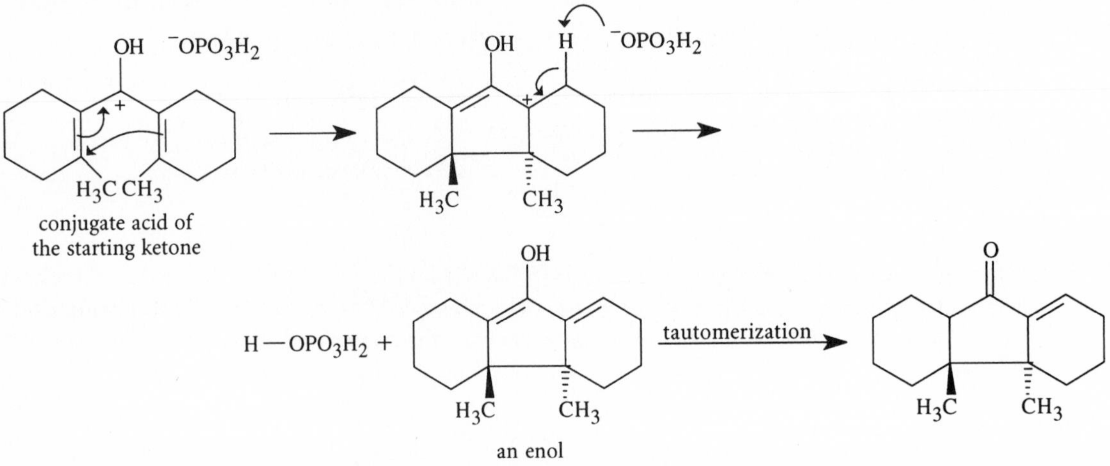

conjugate acid of
the starting ketone

$H—OPO_3H_2$ +

an enol

tautomerization

25.46   (a)  A (presumably) conrotatory ring opening gives intermediate *A*. Although this compound is no longer aromatic, its formation is driven by the relief of strain in the four-membered ring. Compound *A* undergoes an *intramolecular* Diels-Alder reaction that restores aromaticity and gives the product with precisely the desired stereochemistry.

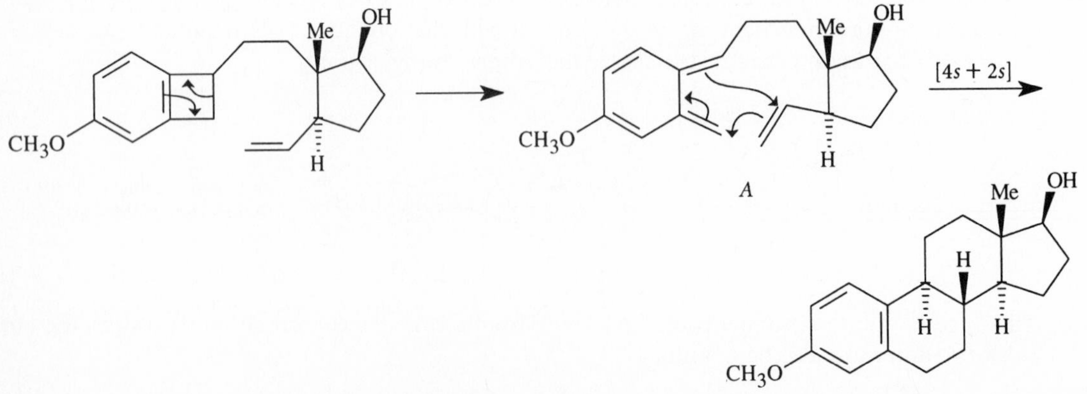

*A*

$[4s + 2s]$

(b) Oxidation of the secondary alcohol and cleavage of the ether gives estrone:

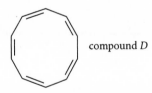

estrone

**25.48** (a) If compound *B* is aromatic, then it must be planar. If it is planar, the "inner" hydrogens (the ones shown explicitly in the problem) fall essentially on top of each other. (A model can be used to demonstrate this point.) The resulting van der Waals repulsions are so severe that compound *B* does not exist in spite of its aromaticity.

(c) There are two ways in which compound *C* could be formed. First, compound *B* could be formed as an unstable intermediate, and then compound *C* could be formed from *B* by an allowed *thermal* disrotatory electrocyclic reaction. Such a reaction would be very rapid even at low temperature because of the instability of compound *B*. The second pathway is that the all-*cis* pentaene *D* could be formed from compound *A* in an allowed photochemical conrotatory electrocyclic reaction. (Note that two different allowed conrotatory reactions of compound *A* are possible: one gives compound *B*, and one gives compound *D*.) Compound *C* could then be formed from compound *D* by a thermal disrotatory electrocyclic reaction. (One could potentially distinguish between the two pathways by preparing compound *D* by another route and determining whether it spontaneously closes to compound *C*. If it does not, then the first pathway must be operating.)

compound *D*

**25.49** (a) If the formation of benzene were concerted, it would have to be a thermal disrotatory electrocyclic ring-opening reaction involving four electrons. This is not allowed by the selection rules.

 Prismane, or Ladenburg benzene, which is discussed on text p. 716, is another very unstable constitutional isomer of benzene that is effectively trapped into existence because its concerted conversion into benzene would violate the selection rules for pericyclic reactions. Thus, it has been referred to as a molecular "caged tiger."

**25.50** (a) Bromine addition to the starting material is followed by four successive base-promoted E2 reactions to give compound *A*, which spontaneously forms compound *B* by an allowed disrotatory thermal electrocyclic reaction. Notice that compound *B* is a "tied-down" variation on hydrocarbon *B* in Problem 25.48. Because the offending inner hydrogens are absent in *B*, its continuous $\pi$-electron system can nearly achieve planarity.

*(solution continues)*

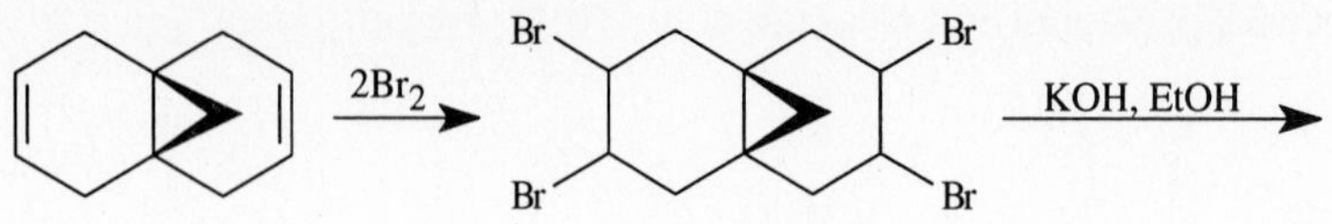

(probably a mixture
of stereoisomers)

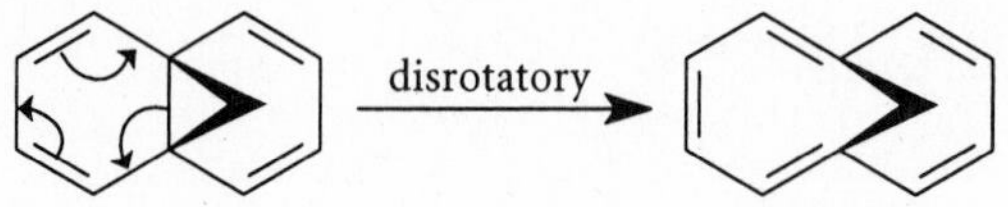

*A*                                    *B*

(b) As discussed in part (a), compound *B* contains a planar (or nearly planar) $4n + 2$ $\pi$-electron system. Consequently, it shows the typical ring-current effects expected of aromatic compounds in its NMR spectra. (See Fig. 16.2 on text p. 742.) The protons on the double bonds, like those in benzene, are deshielded and show typical aromatic chemical shifts near $\delta$ 7.1. The two methylene protons, however, occupy the region that is strongly shielded by the ring current. Thus, it is these protons that have the negative chemical shift.

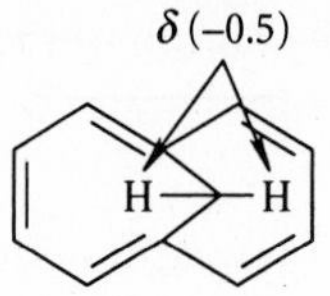

$\delta\,(-0.5)$

compound *B*

# 26

# Amino Acids, Peptides, and Proteins

## Terms

# Concepts

## I. Introduction to Amino Acids and Peptides

### A. GENERAL

1. Compounds that contain both an amino group and a carboxylic acid group are called amino acids.
   a. An amino acid with an overall charge of zero (neutral) can contain within the same molecule two groups of opposite charge.
   b. Molecules containing oppositely charged groups are known as zwitterions.
2. An $\alpha$-amino acid has an amino group on the $\alpha$-carbon, the carbon adjacent to the carboxylic acid group.

$$H_3\overset{+}{N}-\underset{\underset{R}{|}}{CH}-\overset{\overset{O}{\|}}{C}-O^-$$

an $\alpha$-amino acid in zwitterionic form

3. Peptides are polymers in which $\alpha$-amino acids are joined into chains through amide bonds (peptide bonds)

$$-NH-\underset{\underset{R}{|}}{CH}-\overset{\overset{O}{\|}}{C}-NH-\underset{\underset{R}{|}}{CH}-\overset{\overset{O}{\|}}{C}-NH-\underset{\underset{R}{|}}{CH}-\overset{\overset{O}{\|}}{C}-NH-\underset{\underset{R}{|}}{CH}-\overset{\overset{O}{\|}}{C}-$$

(R groups may be the same or different)

   a. A peptide bond is derived from the amino group of one amino acid and the carboxylic acid group of another.
   b. Proteins are large peptides, and some proteins are aggregates of more than one peptide.

### B. NOMENCLATURE OF AMINO ACIDS

1. Some amino acids are named substitutively as uncharged compounds (carboxylic acids with amino substituents).
2. Twenty $\alpha$-amino acids occur commonly as constituents of most proteins and are known by widely accepted traditional names which, with their structures, are given in Table 26.1, text pp. 1266-1267.
   a. With the exception of proline, all $\alpha$-amino acids have the same general structure, differing only in the identity of the side chain R.
      i. Proline is the only naturally occurring amino acid with a secondary amino group.
      ii. In proline the —NH— and the side chain are "tied together" in a ring.

$$H-\overset{\overset{H}{|}}{\underset{+}{N}}-\underset{}{CH}-\overset{\overset{O}{\|}}{C}-O^-$$

proline
(in zwitterionic form)

   b. The amino acids can be organized into six groups according to the nature of their side chains:
      i. hydrogen or aliphatic hydrocarbons
      ii. aromatic groups
      iii. thiol, sulfide, or alcohol groups
      iv. carboxylic acid or amide groups
      v. basic side chains
      vi. proline.
3. The $\alpha$-amino acids are often designated by either three-letter or single-letter abbreviations, which are given in Table 26.1, text pp. 1266-1267.

### C. NOMENCLATURE OF PEPTIDES

1. The peptide backbone is the repeating sequence of nitrogen, $\alpha$-carbon, and carbonyl groups.

a. The characteristic amino acid side chains are attached to the peptide backbone at the respective α-carbon atoms.

b. Each amino acid unit in a peptide is called a residue.

c. The ends of a peptide are labeled as the amino end or amino terminus ($Pep^N$) and the carboxy end or carboxy terminus ($Pep^C$).

d. A peptide can be characterized by the number of residues it contains.

  i. A prefix *di*, *tri*, *tetra*, etc., is attached to the word *peptide* to indicate the number of amino acids that are contained within the peptide.

  ii. A relatively short peptide of unspecified length containing a few amino acids is sometimes referred to as an oligopeptide.

2. A peptide is conventionally named by giving successively the names of the amino acid residues, starting at the amino end.

a. The names of all but the carboxy-terminal residue are formed by dropping the final ending of the amino acid and replacing it with *yl*.

b. This type of nomenclature is used only for the smallest peptides.

3. Peptides can be represented by connecting with hyphens the three-letter (or one-letter) abbreviations of the component amino acid residues beginning with the amino-terminal residue.

alanylserylglycylaspartylphenylalanine (abbreviated **Ala-Ser-Gly-Asp-Phe** or **A-S-G-D-F**)

4. Large peptides and proteins of biological importance are known by common names.

D. STEREOCHEMISTRY OF THE α-AMINO ACIDS

1. With the exception of glycine, all common naturally occurring α-amino acids have an asymmetric carbon atom; all have the *S* configuration at the α-carbon except for cysteine.

2. The stereochemistry of α-amino acids is often specified with an older system, the D,L system.

a. An L amino acid by definition has the amino group on the left and the hydrogen on the right when the carboxylic acid group is up and the side chain is down in a Fischer projection of the α-carbon.

b. The naturally occurring amino acids have the L configuration (the correspondence between *S* and L is not general).

c. The D or L designation for an α-amino acid refers to the configuration of the α-carbon regardless of the number of asymmetric carbons in the molecule.

d. In the D,L system, diastereomers are given different names, for example, threonine and allothreonine.

## II. *Acid-Base Properties of Amino Acids and Peptides*

A. ZWITTERIONIC STRUCTURES OF AMINO ACIDS AND PEPTIDES

1. The major neutral form of any α-amino acid is the zwitterion.

a. The high melting points of amino acids and their greater solubilities in water than in ether are characteristics of salts (charged species).

  b. Water is the best solvent for most amino acids because it solvates ionic groups.

  c. The dipole moments of the amino acids are very large—much larger than those of similar-sized molecules with only one amine or carboxylic acid group—and suggest a great deal of separated charge.

  d. The p$K_a$ values for amino acids are what one would expect for the zwitterionic forms of the neutral molecules.

 2. Peptides also exist as zwitterions; at neutral pH, the amino groups are protonated and the carboxylic acid groups are ionized.

## B. Isoelectric Points of Amino Acids and Peptides

 1. Amino acids and peptides are amphoteric substances; they contain both acidic and basic groups.

 2. An important measure of the acidity or basicity of an amino acid or peptide is its isoelectric point, or isoelectric pH; this is the pH of a dilute aqueous solution of the amino acid or peptide at which the total charge on all molecules is exactly zero. (The isoelectric points of the $\alpha$-amino acids are given in Table 26.1, text p. 1266–1267.)

 3. At the isoelectric point, two conditions are met:

  a. The concentration of negatively charged species equals the concentration of positively charged species.

  b. The relative concentration of the neutral form is greater than at any other pH.

 4. The isoelectric point, p$I$, of an amino acid is the average of the two p$K_a$ values of the amino acid:

$$\text{isoelectric point} = \text{p}I = \frac{\text{p}K_{a1} + \text{p}K_{a2}}{2}$$

 5. The isoelectric point indicates not only the pH at which a solution of the amino acid or peptide contains the greatest amount of neutral form but also the sign of the net charge on the amino acid or peptide at any pH.

  a. At pH values less than the isoelectric point, more molecules of the amino acid or peptide are positively charged than are negatively charged.

  b. At pH values greater than the isoelectric point, more molecules of an amino acid or peptide are negatively charged than are positively charged.

 6. When an amino acid or peptide has a side chain containing an acidic or basic group, the isoelectric point is markedly changed.

  a. With a basic group, the isoelectric point is the average of the two highest p$K_a$ values.

  b. With an acidic group, the isoelectric point is the average of the two lowest p$K_a$ values.

   *i.* Amino acids with high isoelectric points are classified as basic amino acids.

   *ii.* Amino acids with low isoelectric points are classified as acidic amino acids.

   *iii.* Amino acids with isoelectric points near 6 are classified as neutral amino acids.

 7. A peptide can be classified as acidic, basic, or neutral by examining the number of acidic and basic groups that it contains.

  a. A peptide with more amino and guanidino groups than carboxy groups will have a high isoelectric point.

  b. A peptide with more carboxy groups than amino and guanidino groups will have a low isoelectric point.

## C. Separation of Amino Acids and Peptides Using Acid-Base Properties

 1. Their isoelectric points are often used to design separations of amino acids and peptides.

  a. Most peptides and amino acids are most soluble when they carry a net charge and are least soluble in their neutral forms.

  b. Some peptides, proteins, and amino acids precipitate from water when the pH is adjusted to their isoelectric points; these same compounds are more soluble at pH values far from their isoelectric points, because they carry a net charge at these pH values.

 2. A separation technique used a great deal in amino acid and peptide chemistry is ion-exchange chromatography, which depends on the isoelectric points of amino acids and peptides.

  a. In this technique, a hollow tube or column is filled with a buffer solution in which is suspended a finely powdered, insoluble polymer called an ion-exchange resin; the resin bears charged groups.

    *i.* Resins that bear negatively charged pendant groups absorb cations, and are called cation-exchange resins.

    *ii.* Resins that bear positively charged pendant groups absorb anions, and are called anion-exchange resins.

  b. Whether an amino acid or peptide is absorbed by the column depends on its charge, which, in turn, depends on the relationship of its isoelectric point to the pH of the buffer.

## III. Structures of Peptides and Proteins

### A. PRIMARY STRUCTURE

1. One description of a peptide or protein structure is its covalent structure or primary structure.
2. The most important aspect of any primary structure is the amino acid sequence.
3. Disulfide bonds link the cysteine residues in different parts of a sequence and serve as cross-links between different parts of a peptide chain.

$$\text{—Val—Ala—Cys—Arg—}$$
$$|$$
$$S$$
$$|\quad\longleftarrow\quad\text{disulfide linkage}$$
$$S$$
$$|$$
$$\text{—Asn—Ser—Cys—His—Lys—}$$

  a. The disulfide bonds of a protein are readily reduced to free cysteine thiols by other thiols.

  b. Two commonly used thiol reagents are 2-mercaptoethanol ($HSCH_2CH_2OH$), and dithiothreitol (DTT, or Cleland's reagent), $HSCH_2CH(OH)CH(OH)CH_2SH$.

### B. SECONDARY STRUCTURE

1. The description of a peptide or protein in terms of the conformations of its peptide chains is called secondary structure.
2. Three conformations occur commonly:

  a. In a right-handed $\alpha$-helix, the side-chain groups are positioned on the outside of the helix, and the helix is stabilized by hydrogen bonds between the amide N—H of one residue and the carbonyl oxygen four residues further along the helix.

helix axis →

right handed $\alpha$-helix

  b. In a $\beta$-structure (pleated sheet), the peptide chain adopts an open, zigzag conformation, and it is engaged in hydrogen bonding with another peptide chain (or a different part of the same chain) in a similar conformation.

    *i.* Successive hydrogen-bonded chains can run (in the amino terminal to carboxy terminal sense) in the same direction (parallel pleated sheet) or in opposite directions (antiparallel pleated sheet).

    *ii.* The positions of the side-chain R-groups alternate above and below the sheet.

c. Peptides that adopt a random-coil conformation show no discernible pattern in their conformation.

3. Some peptides and proteins exist entirely as $\alpha$-helix or pleated sheet; however, most proteins contain different types of secondary structure in different parts of their peptide chains.

## C. TERTIARY AND QUATERNARY STRUCTURES

1. The complete three-dimensional description of protein structure at the atomic level is called tertiary structure and is determined by X-ray crystallography.
2. The tertiary structure of any given protein is an aggregate of $\alpha$-helix, $\beta$-sheet, random-coil, and other structural elements.
   a. In many proteins certain higher-order structural motifs are common.
      i. A common motif is a bundle of four helices (called a four-helix bundle), each running approximately antiparallel to the next and separated by short turns in the peptide chain.
      ii. Another common structural motif is the beta barrel, a "bag" consisting of $\beta$-sheets connected by short turns.
   b. Several motifs can occur within a given protein, so that a protein might consist of several smaller, relatively ordered structures connected by short random loops; these ordered sub-structures are sometimes called domains.
3. In general, the tertiary structures of proteins are determined by the noncovalent interactions between groups within protein molecules, and between groups of the protein with the surrounding solvent:
   a. Hydrogen bonds
      i. Hydrogen bonds stabilize both $\alpha$-helices and $\beta$-sheets as well as other conformations.
      ii. Protein conformations are also stabilized in part by hydrogen bonding of certain groups to solvent water.
   b. Van der Waals interactions
      i. Van der Waals interactions, or dispersion forces, are the same interactions that provide the cohesive force in a liquid hydrocarbon, and can be regarded as examples of the "like-dissolves-like" phenomenon.
      ii. The van der Waals interactions between hydrocarbonlike residues are sometimes called hydrophobic bonds, because the hydrocarbon groups energetically prefer each other to water.
      iii Residues such as the side chain of phenylalanine or isolucine are sometimes called hydrophobic residues. (In contrast, polar residues are often found to interact with other polar residues or with the aqueous solvent and are sometimes termed hydrophilic residues.)
   c. Electrostatic interactions are noncovalent interactions between charged groups governed by the electrostatic law.
4. A protein adopts a tertiary structure in which favorable interactions are maximized and unfavorable interactions are minimized.
   a. Although there are exceptions, most soluble proteins are globular and compact rather than extended, because a near-spherical shape minimizes the amount of protein surface exposed to water.

   b. The reason for minimizing the exposed surface is that the majority of the residues in most proteins are hydrophobic, and the interactions of hydrophobic side chains with solvent water is unfavorable.

      *i.* The hydrocarbonlike amino acid residues tend to be found on the interior of a protein, away from solvent water.

      *ii.* The polar residues tend to be on the exterior of a protein where they can form hydrogen bonds with water.

5. When a protein is denatured, it is converted entirely into a random-coil structure; chemical denaturation of a protein is typically brought about by breaking its disulfide bonds with thiols, such as DTT or 2-mercaptoethanol, and treating it with 8 *M* urea, detergents, or heat.

6. The amino acid sequence of a protein specifies its conformation; that is, the native structure is the most stable structure. In many cases it appears that primary structure dictates tertiary structure.

7. Some proteins are aggregates of other proteins, called subunits; the description of subunit arrangement in a protein is called quaternary structure.

## D.   ENZYMES—BIOLOGICAL CATALYSTS

1. Many of the proteins that occur naturally are enzymes, which are catalysts for biological reactions.

   a. Enzymes are true catalysts; their concentrations are typically much lower than the concentrations of the compounds in the reactions they catalyze.

   b. They do not affect the equilibrium constants of the reactions they catalyze.

   c. They catalyze equally both the forward and reverse reactions of an equilibrium.

2. Two characteristics of all enzymes are

   a. catalytic efficiency

   b. specificity.

3. When an enzyme catalyzes a reaction of a certain compound, called a substrate of the enzyme, it acts on the substrate in at least three stages:

   a. The enzyme binds the substrate in a noncovalent enzyme-substrate complex.

      *i.* The binding occurs at a part of the enzyme called the active site.

      *ii.* Within the active site are groups that attract the substrate by interacting favorably with it; the interactions involved are electrostatic interactions, hydrogen bonding, and van der Waals attractions (hydrophobic bonds).

   b. The enzyme promotes the appropriate chemical reaction(s) on the bound substrate to give an enzyme-product complex.

      *i.* The necessary chemical transformations are brought about by groups in the active site of the enzyme.

      *ii.* In some cases, other molecules, called coenzymes, are also required.

   c. The product(s) depart from the active site, leaving the enzyme ready to repeat the process on a new substrate molecule.

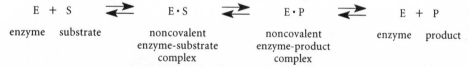

4. Enzymes interact differently with enantiomers because enzymes are chiral reagents; the enzyme-substrate complex derived from the enantiomer of the substrate is a diastereomer of the complex derived from the substrate itself.

   a. Because the two complexes are diastereomers, they have different energies, and one is more stable than the other.

   b. The rates at which these complexes are converted into products also differ.

   c. In most cases enzymes catalyze the reaction of only one enantiomer of a pair.

# Reactions

......................

## I. *Synthesis and Optical Resolution of α-Amino Acids*

### A. ALKYLATION OF AMMONIA

1. Some α-amino acids can be prepared by alkylation of ammonia with α-bromo carboxylic acids.

$$\underset{\underset{\displaystyle \overset{|}{\text{C}}}{\text{Br}}}{\overset{|}{\text{C}}}\text{—CO}_2\text{H} \quad \xrightarrow{\text{NH}_3 \text{ (large excess)}} \quad \overset{\overset{+}{\text{NH}_3}}{\underset{|}{\overset{|}{\text{C}}}}\text{—CO}_2^- \; + \; \overset{+}{\text{NH}_4} \; + \; \text{Br}^-$$

   a. This is an $S_N2$ reaction in which ammonia acts as the nucleophile.
   b. The use of a large excess of ammonia in the synthesis favors monoalkylation.
2. Amino acids are less reactive toward alkylating agents than simple alkylamines because the amino groups of amino acids are less basic (nucleophilic) than ammonia and simple alkylamines, and because branching in amino acids retards further alkylation.

### B. ALKYLATION OF AMINOMALONATE DERIVATIVES

1. Another method for preparing α-amino acids is a variation of the malonic ester synthesis.
2. The malonic ester derivative used is one in which a protected amino group is already in place: diethyl α-acetamidomalonate.
   a. Treatment of diethyl α-acetamidomalonate with sodium ethoxide in ethanol forms the conjugate base enolate ion, which is then alkylated with an alkyl halide.
   b. The resulting compound is then treated with hot aqueous HCl or HBr which accomplishes three things:
      *i.* The ester groups are hydrolyzed to give a substituted malonic acid.
      *ii.* The malonic acid derivative is decarboxylated under the reaction conditions.
      *iii.* The acetamido group, an amide, is also hydrolyzed.
   c. Neutralization affords the α-amino acid.

$$\text{AcNH}-\overset{\overset{\displaystyle \text{CO}_2\text{Et}}{|}}{\underset{\underset{\displaystyle \text{H}}{|}}{\text{C}}}-\text{CO}_2\text{Et} \quad \xrightarrow[\text{2) R—X}]{\text{1) NaOEt/HOEt}} \quad \text{AcNH}-\overset{\overset{\displaystyle \text{CO}_2\text{Et}}{|}}{\underset{\underset{\displaystyle \text{R}}{|}}{\text{C}}}-\text{CO}_2\text{Et} \quad \xrightarrow[\text{heat}]{\text{H}_3\text{O}^+} \quad \text{AcNH}-\overset{}{\underset{\underset{\displaystyle \text{R}}{|}}{\text{CH}}}-\text{CO}_2^- \; + \; 2\,\text{EtOH} \; + \; \text{CO}_2{\uparrow} \; + \; \text{HOAc}$$

### C. STRECKER SYNTHESIS

1. Hydrolysis of α-amino nitriles to give α-amino acids is called the Strecker synthesis.

$$\text{H}_2\text{N}-\overset{}{\underset{\underset{\displaystyle \text{R}}{|}}{\text{CH}}}-\text{C}{\equiv}\text{N} \quad \xrightarrow[\text{heat}]{\text{H}_3\text{O}^+} \quad \overset{+}{\text{H}_3\text{N}}-\overset{}{\underset{\underset{\displaystyle \text{R}}{|}}{\text{CH}}}-\text{CO}_2^- \; + \; \overset{+}{\text{NH}_4}$$

2. α-Amino nitriles are prepared by treatment of aldehydes with ammonia in the presence of cyanide ion, a reaction that probably involves an imine intermediate.
   a. The conjugate acid of the imine reacts with cyanide under the conditions of the reaction to give the α-amino nitrile.
   b. The addition of cyanide to an imine is analogous to the formation of a cyanohydrin from an aldehyde or ketone.

$$\text{O}{=}\overset{}{\underset{\underset{\displaystyle \text{R}}{|}}{\text{CH}}} \quad \xrightarrow[\text{NaCN}]{\text{NH}_4\text{Cl}} \quad \text{HN}{=}\overset{}{\underset{\underset{\displaystyle \text{R}}{|}}{\text{CH}}} \quad \longrightarrow \quad \text{H}_2\text{N}-\overset{}{\underset{\underset{\displaystyle \text{R}}{|}}{\text{CH}}}-\text{C}{\equiv}\text{N}$$

### D. ENANTIOMERIC RESOLUTION OF α-AMINO ACIDS

1. α-Amino acids synthesized by common laboratory methods are racemic; since many applications require the pure enantiomers, the racemic compounds must be resolved.

2. An alternative approach to the preparation of enantiomerically pure amino acids is the synthesis of amino acids by microbiological fermentation.
3. Certain enzymes can be used to resolve racemic amino acid derivatives by selectively catalyzing a reaction of one enantiomer.

## II. Reactions of Amino Acids and Synthesis of Peptides

### A. ACYLATION AND ESTERIFICATION

1. Amino acids undergo many of the characteristic reactions of both amines and carboxylic acids.
   a. They can be acylated by acid chlorides or anhydrides.
   b. They are easily esterified by heating with an alcohol and a strong acid catalyst.

### B. SOLID-PHASE PEPTIDE SYNTHESIS

1. In a method called solid-phase peptide synthesis, the carboxy-terminal amino acid is covalently anchored to an insoluble polymer, and the peptide is "grown" by adding one residue at a time to this polymer.
   a. Solutions containing the appropriate reagents are allowed to contact the polymer with shaking.
   b. At the conclusion of each step, the polymer containing the peptide is simply filtered away from the solution, which contains soluble by-products and impurities.
   c. The completed peptide is removed from the polymer by a reaction that cleaves its bond to the resin.
2. In a solid-phase peptide synthesis, the amino group of the amino acids is protected with a special acyl group, the *tert*-butyloxycarbonyl (Boc) group.
   a. The Boc group is introduced by allowing an amino acid to react with the anhydride di-*tert*-butyl dicarbonate.

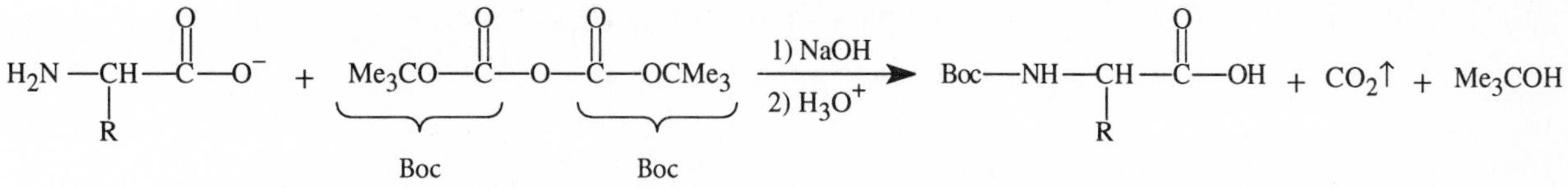

   b. The amino group of the amino acid rather than the carboxylate group reacts with the anhydride because the amino group is the more basic, and therefore the more nucleophilic, group.
3. At the start of the solid-phase peptide synthesis, a Boc-protected amino acid is anchored to the insoluble solid support (called a Merrifield resin) using the reactivity of its free carboxylic acid group.
   a. An $S_N2$ reaction between the cesium salt of the Boc-amino acid and the resin results in the formation of an ester linkage to the resin.
   b. Once the Boc-amino acid is anchored to the resin, the Boc protecting group is removed with anhydrous trifluoroacetic acid. This deprotection step, followed by neutralization of the resulting ammonium salt, exposes the free amino group of the resin-bound amino acid, which is used as a nucleophile in the next reaction.

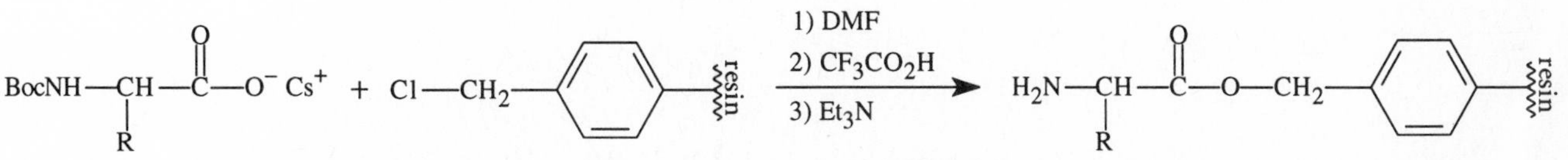

4. Coupling of another Boc-protected amino acid to the free amino group of the resin-bound amino acid is effected by the reagent *N,N′*-dicyclohexylcarbodiimide (DCC). (For details, see Eqs. 26.29a–c and Eqs. 26.30ab on text pp. 1295–1296.)

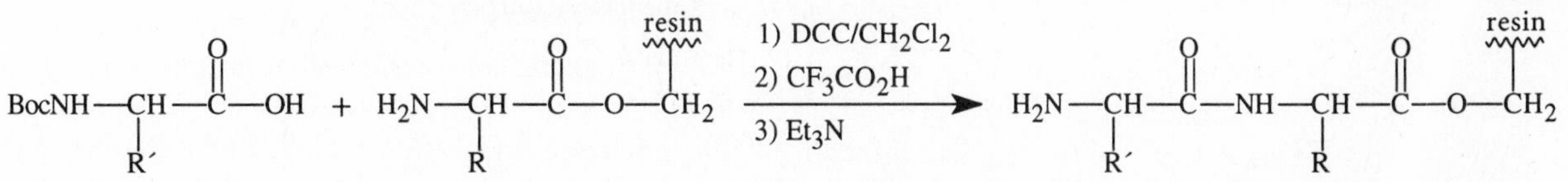

5. Completion of the peptide synthesis requires deprotection of the resin-bound peptide in the usual way, a final coupling step with the amino-terminal Boc-amino acid and DCC, and removal of the peptide from the resin.
   a. The ester linkage that connects the peptide to the resin, like most esters, is more easily cleaved than the peptide (amide) bonds, and is typically broken by liquid HF.
   b. This acidic reagent also removes the Boc group from the product peptide.

$$\text{BocNH—CH—}\overset{\displaystyle O}{\overset{\|}{C}}\text{—OH} \ + \ \text{H}_2\text{N—Pep}^C \ \xrightarrow[\text{3) HF}]{\substack{\text{1) DCC/CH}_2\text{Cl}_2 \\ \text{2) CF}_3\text{CO}_2\text{H}}} \ \text{H}_3\overset{+}{\text{N}}\text{—CH—}\overset{\displaystyle O}{\overset{\|}{C}}\text{—NH—Pep}^C \ + \ \text{resin}$$

(benzylic ester linkage → resin; R groups on CH)

6. Another important method involves a conceptually similar stepwise approach but employs a different protecting group, the (9-fluorenylmethyloxy)carbonyl (Fmoc) group, that can be removed by a mild base (pyridine in DMF). (See Eq. 26.33 on text p. 1297.)
7. The advantage of the solid-phase method is the ease with which dissolved impurities and by-products are removed from the resin-bound peptide by simple filtration.
   a. The same reagents used in solid-phase peptide synthesis can also be used for peptide synthesis in solution, but removal of by-products from the product peptide is sometimes difficult.
   b. In order to avoid impurities, each step in the solid-phase synthesis must occur with virtually 100% yield, an ideal that is often approached in practice.

## III. *Analytically Important Reactions of Peptides*

### A. Hydrolysis of Peptides; Amino Acid Analysis

1. An important reaction used to determine the structures of unknown peptides is hydrolysis of the peptide (amide) bonds of a peptide to give its constituent amino acids.
2. When a peptide or protein is hydrolyzed, the product amino acids can be separated, identified, and quantitated by a technique called amino acid analysis.
   a. The amino acids in the hydrolyzed mixture are separated by passing them through a cation-exchange column under very carefully defined conditions; the time at which each amino acid emerges from the column is accurately known.
   b. As each amino acid emerges, it is mixed with ninhydrin to give an intense blue-violet dye called Ruhemann's purple; the intensity of the resulting color is proportional to the amount of the amino acid present.

ninhydrin      $\xrightleftharpoons[\text{H}_2\text{O}]{-\text{H}_2\text{O}}$

$$2 \ (\text{ninhydrin, ketone form}) \ + \ \text{H}_3\overset{+}{\text{N}}\text{—CH—CO}_2^- \ \longrightarrow \ (\text{indanedione—N=indanedione}) \ + \ \overset{\displaystyle O}{\overset{\|}{\text{CH}}}_{\text{R}} \ + \ \text{CO}_2\uparrow \ + \ \text{H}_3\text{O}^+$$

Ruhemann's purple

   c. The color intensity is recorded as a function of time; the area of the peak is proportional to the amount of the amino acid.
3. Amino acid analysis can determine the identity and relative amounts of amino acid residues, but not their relative orders within a peptide.

## B. Sequential Degradation of Peptides

1. The actual arrangement, or sequential order, of amino acid residues in a peptide is called the amino acid sequence or primary sequence of the peptide.

2. It is possible to remove one residue at a time from the amino end of the peptide, identify it, and then repeat the process sequentially on the remaining peptide; the standard method for implementing this strategy is called the Edman degradation.

   a. The peptide is treated with phenyl isothiocyanate (Edman reagent), with which it reacts at its amino groups to give a thiourea derivative. (Only the reaction at the terminal amino group is relevant to the degradation.)

   b. Any remaining phenyl isothiocyanate is removed, and the modified peptide is then treated with anhydrous trifluoroacetic acid.

   c. The sulfur of the thiourea, which is nucleophilic, displaces the amino group of the adjacent residue to yield a five-membered heterocycle called a thiazolinone; the other product of the reaction is a peptide that is one residue shorter.

   d. When treated subsequently with aqueous acid, the thiazolinone derivative forms an isomer called a phenylthiohydantoin (PTH).

   e. Because the PTH derivative carries the characteristic side chain of the amino-terminal residue, the structure of the PTH derivative identifies the amino acid residue that was removed.

3. In practice, this type of analysis is limited to 20–60 consecutive residues because the yields at each step are not perfectly quantitative; hence, an increasingly complex mixture of peptides is formed with each successive step in the cleavage, and after a number of such steps the results become ambiguous.

## C. Specific Cleavage of Peptides

1. The amino acid sequence of most large proteins is determined by breaking the protein into smaller peptides and sequencing these peptides individually. (The sequence of the protein is reconstructed from the sequences of the peptides.)

2. When breaking a larger protein into smaller peptides, it is desirable to use reactions that cleave the protein in high yield at well-defined points so that a relatively small number of peptides are obtained.

3. One method uses ordinary chemical reagents; another method involves the use of enzymes to catalyze peptide-bond hydrolysis.

## D. Peptide Cleavage at Methionine with Cyanogen Bromide

1. When a peptide reacts with cyanogen bromide ($Br—C≡N$) in aqueous HCl, a peptide bond is cleaved specifically at the carboxy side of each methionine residue.

   a. The amino-terminal fragment formed in the cleavage has a carboxy-terminal homoserine lactone residue instead of the starting methionine.

   b. Methionine is a relatively rare amino acid; hence, when a typical protein is cleaved with BrCN, relatively few cleavage peptides are obtained, and all of them are derived from cleavage at methionine residues.

2. Although cyanogen bromide has the character of an acid chloride, under acidic conditions only its reaction at methionine leads to a peptide cleavage.

a. The sulfur in the methionine side chain acts as a nucleophile, displacing bromide from cyanogen bromide to give a type of sulfonium ion.
b. The sulfonium ion, with its electron-withdrawing cyanide, is an excellent leaving group, and is displaced by the oxygen of the neighboring amide bond to form an imminium ion containing a five-membered ring; only methionine has a side chain that can form a five-membered ring by such a mechanism.
c. Hydrolysis of the imminium ion formed cleaves the peptide bond.

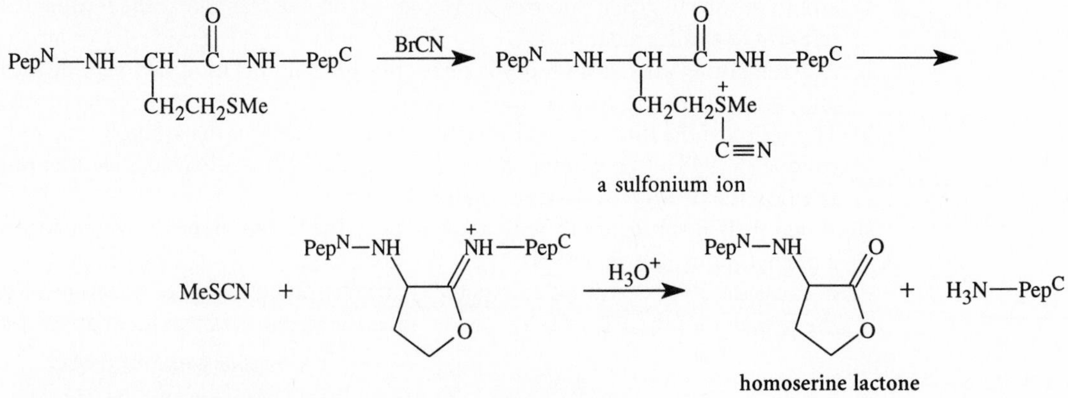

$$Pep^N-NH-CH-\overset{\overset{\displaystyle O}{\|}}{C}-NH-Pep^C \xrightarrow{BrCN} Pep^N-NH-CH-\overset{\overset{\displaystyle O}{\|}}{C}-NH-Pep^C \longrightarrow$$

with side chains $CH_2CH_2SMe$ and $CH_2CH_2\overset{+}{S}Me$ with $C\equiv N$

a sulfonium ion

$$MeSCN \quad + \quad \text{(ring with } Pep^N-NH \text{ and } \overset{+}{N}H-Pep^C) \xrightarrow{H_3O^+} Pep^N-NH \text{ (lactone ring)} \quad + \quad H_3N-Pep^C$$

homoserine lactone

## E. PEPTIDE CLEAVAGE WITH PROTEOLYTIC ENZYMES

1. A number of enzymes (called proteases, peptidases, or proteolytic enzymes) catalyze the hydrolysis of peptide bonds at specific points in an amino acid sequence.
2. One of the most widely used proteases is the enzyme trypsin, which catalyzes the hydrolysis of peptides or proteins at the carbonyl group of arginine or lysine residues provided that:
   a. these residues are not at the amino end of the protein.
   b. these residues are not followed by a proline residue.
3. Because trypsin catalyzes the hydrolysis of peptides at internal rather than terminal residues, it is called an endopeptidase; enzymes that cleave peptides only at terminal residues are termed exopeptidases.
4. Chymotrypsin, a protein related to trypsin, is used to cleave peptides at amino acid residues with aromatic side chains and, to a lesser extent, residues with large hydrocarbon side chains; thus, chymotrypsin cleaves peptides at Phe, Trp, Tyr, and in some cases Leu and Ile residues.

◆ **Study Guide Links**
SGL

### ✓26.1   Neutral Amino Acids

It is very important in this chapter to make a clear distinction between the terms *neutral* and *uncharged* when they are used with amino acids and their derivatives. A *neutral* amino acid can have charged atoms; if so, then it is a *zwitterion*, that is, it bears two equal charges of opposite sign (+ and −). An uncharged amino acid has no atoms bearing formal charge.

### ✓26.2   Names of the Amino Acids

A student in the author's organic chemistry course proposed the following mnemonics for remembering the twenty common naturally occurring amino acids. (The groups are those listed at the top of text p. 1265.)

| | |
|---|---|
| *Group 1 and proline:* | LIAGVP (*Loudon Is A Ghastly Volleyball Player*) |
| *Group 2:* | HTTP (*Hail To Terrific Purdue*) |
| *Group 3:* | CTMS (*Chemistry Takes Much Study*) |
| *Groups 4 and 5:* | AGAGLA (*All Girls And Guys Love Artichokes*) |

The substitution of other adjectives in the group 1 and group 2 mnemonics are, of course, acceptable, provided that they begin with the same letter.

### 26.3   Reaction of α-Amino Acids with Ninhydrin

To understand the reaction of ninhydrin with α-amino acids, first notice that *two* molecules of ninhydrin react per molecule of amino acid. Second, notice that ninhydrin is the hydrate of a tricarbonyl compound.

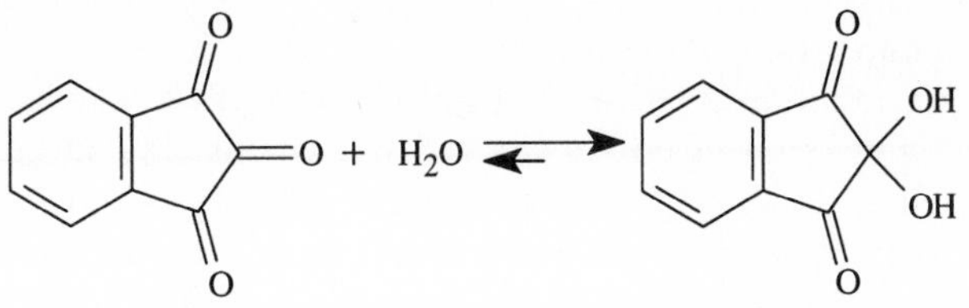

ninhydrin

The ninhydrin reaction with α-amino acids is a combination of simpler reactions that you have already studied. The first step is the formation of an imine (Sec. 19.11A on text p. 904). In this reaction, the amino group of the amino acid reacts with the central carbonyl group of the tricarbonyl form of ninhydrin. (Why is the central carbonyl group more reactive than the other carbonyl groups? See Sec. 19.7B in the text.)

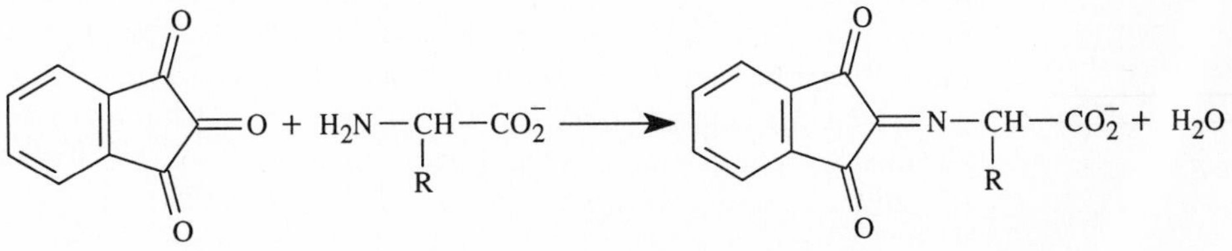

an imine

This imine is converted into a different imine by loss of $CO_2$ (decarboxylation), and the new imine hydrolyzes to an aldehyde and an amine. (Recall that imine formation is reversible.) Note that the aldehyde bears the side-chain R of the original amino acid.

The resulting amine then forms Ruhemann's purple, the final product, by reacting with a second molecule of ninhydrin in another imine-forming reaction.

Ruhemann's purple

Simple primary amines also react with ninhydrin to give Ruhemann's purple. The only alteration in the mechanism is the loss of a proton instead of decarboxylation in the first equation at the top of this page.

## 26.4    Amino Acid Analysis

The ion-exchange resin used in the separation in Fig. 26.3 contains negatively-charged groups much like the resin shown in Eq. 26.10 on text p. 1276. Notice that, as expected, the amino acids with negatively charged side chains (Asp and Glu) emerge from the column earlier, and those with positively charged side chains (Lys and Arg) emerge from the column later. To a first approximation, all the neutral amino acids would be expected to emerge from the column at the same time. Separation of the amino acids occurs because this is *not* the case!

Each neutral amino acid has a different affinity for the cation-exchange resin. For example, consider the fact that Ser emerges from the column much earlier than Phe. This can be understood from the structure of the cation-exchange resin. Although this resin contains charged $-SO_3^-$ groups, it also contains many substituted phenyl rings and an alkanelike backbone (see Eq. 26.10 on text p. 1276). Phe emerges later because the aromatic side chain of Phe interacts favorably with the hydrocarbon portion of the resin, and thus "sticks" to the resin longer; this interaction is essentially the same sort of attractive van der Waals interaction that is responsible for the solubility of hydrocarbons in other hydrocarbons. Ser emerges early because its —OH side-chain forms attractive hydrogen bonds with the solvent (an aqueous buffer) used to wash the column, and Ser cannot form such attractive interactions with the groups on the column itself. In other words, the behavior of amino acids in ion-exchange chromatography is similar conceptually to the solubility behavior of organic compounds. (Recall the "like-dissolves-like" rule of solubility solubility; Sec. 8.4B).

## 26.5    Solid-Phase Peptide Synthesis

The reactions of solid-phase peptide synthesis must be carried out in solvents that *solvate* the Merrifield resin, that is, solvents that interact well with the groups of the resin. If these groups are not solvated, reagents in solution cannot penetrate the resin and no reaction takes place. Because the polystyrene-based Merrifield resin is essentially a polymeric ethylbenzene, it is mostly a hydrocarbon. Yet many of the reactions of solid-phase peptide synthesis involve ionic compounds and intermediates. Water and alcohols do not work as solvents in solid-phase peptide synthesis because they do not solvate the resin. Dimethylformamide, *N*-methylpyrrolidone, and methylene chloride have been found empirically to solvate the Merrifield resin adequately and to dissolve the reagents used in solid-phase peptide synthesis. In fact, when such solvents are added to a Merrifield resin, a visually perceptible swelling of the resin bed occurs; this is caused by incorporation of the solvent into the resin.

The use of a cesium salt in Eq. 26.26 on text p. 1293 is related to the requirement for aprotic solvents. In DMF, ionic compounds tend to form ion pairs and other aggregates, and carboxylate salts have metal-oxygen bonds with a significant amount of covalent character. Yet the alkylation shown in this equation is fastest when the carboxylate oxygen has as much *ionic character* as possible, because a carboxylate oxygen with a full negative charge is more nucleophilic and therefore reacts much more rapidly in $S_N2$ reactions than an ion-paired (or covalent) carboxylate oxygen. Because cesium is the most electropositive of the readily available alkali metals, its carboxylate salts have more ionic character than lithium, sodium, or potassium salts and hence are most reactive in $S_N2$ reactions.

## ✓26.6 An All-D Enzyme

Although all naturally occurring enzymes are proteins that consist entirely of L-amino acids, a protein consisting entirely of D-amino acids is conceivable. What would be predicted for the catalytic activity of such a protein? It should catalyze only the reactions of the enantiomers of the substrates of its all-L counterpart, but on those enantiomeric substrates, its catalytic activity should be identical to that of the all-L enzyme.

In 1993 this idea was tested for the very first time by the synthesis of a small protein from human immunodeficiency virus (HIV), the virus that causes AIDS. The viral protein as it occurs in nature consists of only L-amino acids, and a research group at the Scripps Research Institute in La Jolla, California, had chemically synthesized this protein using solid-phase methods. The same group prepared the enantiomeric protein from D-amino acids and found that this enzyme has no catalytic activity on substrates of the L-enzyme, but has identical activity on *enantiomeric* substrates. This finding, of course, was completely expected. However, what made the work worth doing is that the expectation stems from the fundamental principles of stereochemistry, heretofore untested with enzymes. Had the expectation *not* been realized, these principles would have been called into question.

## P ◆ Solutions

### Solutions to In-Text Problems

**26.1**  (a)  This peptide is drawn in the form that exists at neutral pH.

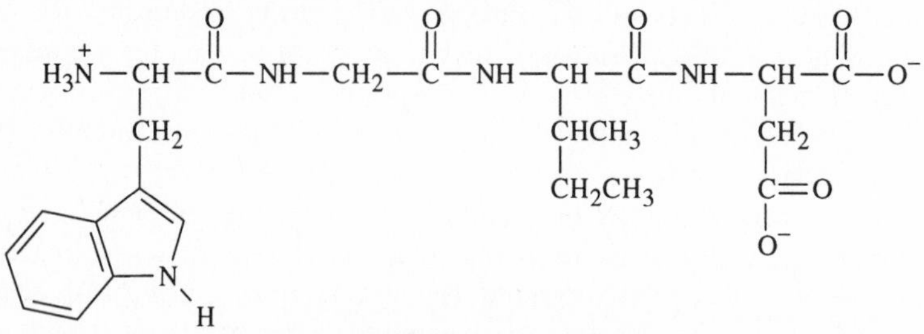

tryptophylglycylisoleucylaspartic acid
(Trp-Gly-Ile-Asp, or W-G-I-D)

**26.2**  His-Ile-Tyr-Met-Ser (histidylisoleucyltyrosylmethionylserine, or H-I-Y-M-S)

**26.3**  (a)  A Fischer projection of L-isoleucine:

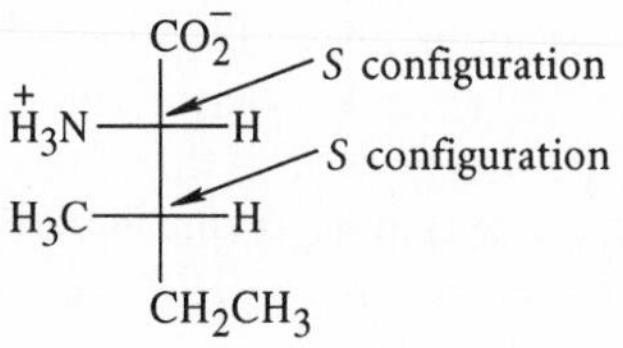

**26.4**  (a)  The $\alpha$-carbon of L-cysteine has the *R* configuration.

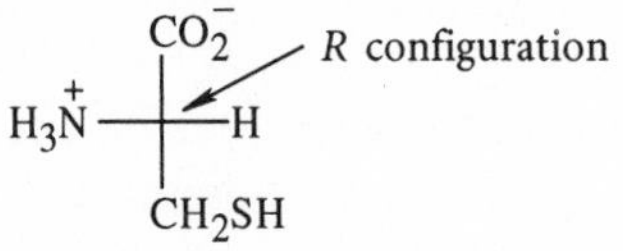

L-cysteine

**26.5**  (a)  In the major neutral form of A-K-V-E-M (Ala-Lys-Val-Glu-Met), all amino groups are protonated and all carboxy groups are ionized.

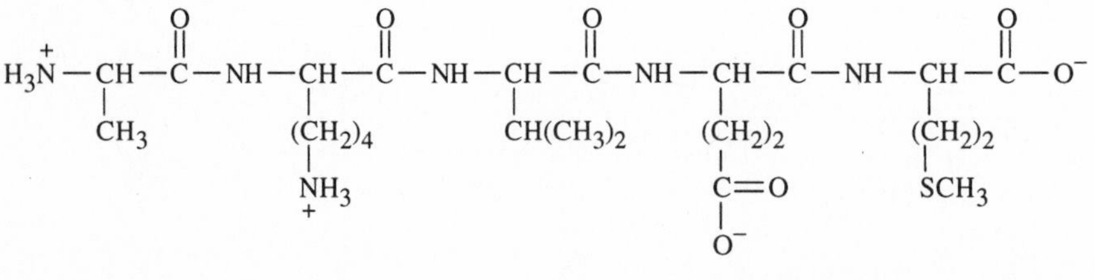

**26.6**  (a)  The amino group of tyrosine can be protonated, and both the carboxy group and the phenolic O—H group can be ionized.

(b)  At pH 6, the net charge on tyrosine is zero. A pH value of 6 is below the $pK_a$ of the conjugate acid of the amino group, and the amino group is therefore protonated; a pH value of 6 is above the $pK_a$ of the carboxy group, which is therefore ionized; and a pH value of 6 is below the $pK_a$ of the phenolic O—H group, which is therefore un-ionized.

(c) The structure of tyrosine in aqueous solution at pH 6:

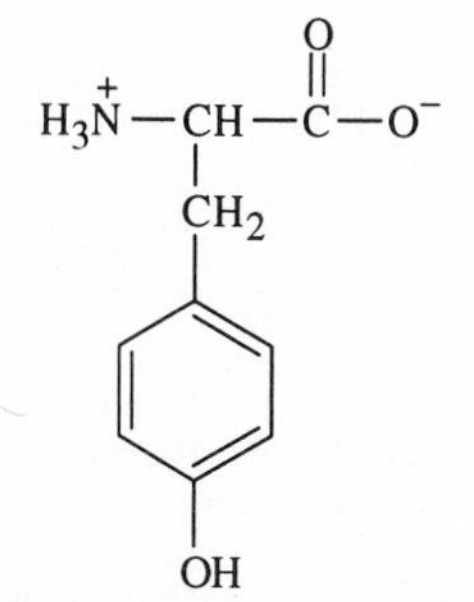

**26.8** The general rule of thumb is that if the peptide contains more acidic than basic groups it is an acidic peptide; if it contains more basic than acidic groups it is a basic peptide; and if the number of acidic and basic groups are equal, the peptide is neutral. (An acidic group is a group that is in its conjugate-base form at neutral pH, such as a carboxy group; a basic group is a group that is in its conjugate-acid form at neutral pH, such as an amino group.)

(a) The peptide is neutral because it contains one basic group (the terminal amino group, which is protonated at pH 6) and one acidic group (the terminal carboxy group, which is ionized at pH 6). Its net charge at pH 6 is 0.

(c) Acetylation eliminates the basicity of the terminal amino group. The peptide contains two basic groups (the Arg residues) and two acidic groups (the Asp residue and the terminal carboxy group). Thus, the peptide is neutral and its net charge is 0.

**26.9** Use Eqs. 26.5 on text p. 1272. Multiplication of the two equations gives

$$\frac{K_{a1}K_{a2}}{[H_3O^+]^2} = \frac{[N][B]}{[A][N]}$$

When the pH is equal to the isoelectric point, by definition $[A] = [B]$. Substituting this equality, and calling the hydronium-ion concentration at this pH $[H_3O^+]_i$, the equation above becomes

$$\frac{K_{a1}K_{a2}}{[H_3O^+]_i^2} = 1$$

or

$$K_{a1}K_{a2} = [H_3O^+]_i^2$$

Taking negative logarithms of both sides and letting $-\log[H_3O^+]_i = pI$,

$$pK_{a1} + pK_{a2} = 2pI$$

Division by 2 gives Eq. 26.7.

**26.10** (a) A resin containing a cationic group will serve as an anion exchanger. A quaternary ammonium salt is a suitable cationic group:

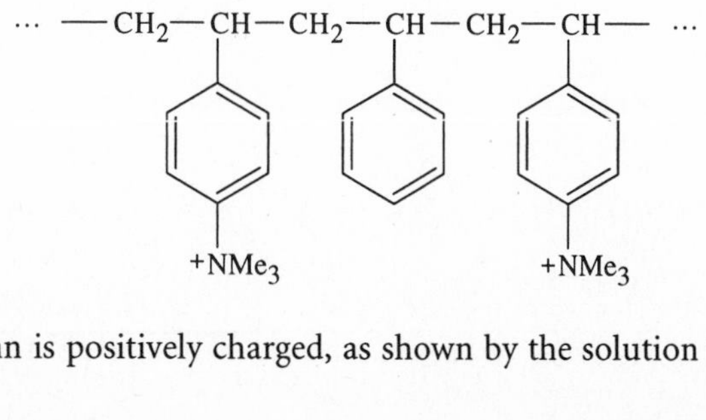

(b) The ion-exchange column is positively charged, as shown by the solution to part (a). At pH 6, the peptide A-V-G has a net charge of zero, is not retained by the column, and will emerge first; the peptide D-N-N-G has a net charge of −1, is retained by the column and will emerge next; and the peptide D-E-E-G has a charge of −3, is most strongly retained by the column, and will emerge last.

26.12  (a)  Alkylation of ammonia by α-bromophenylacetic acid will work, particularly since the alkyl halide is benzylic.

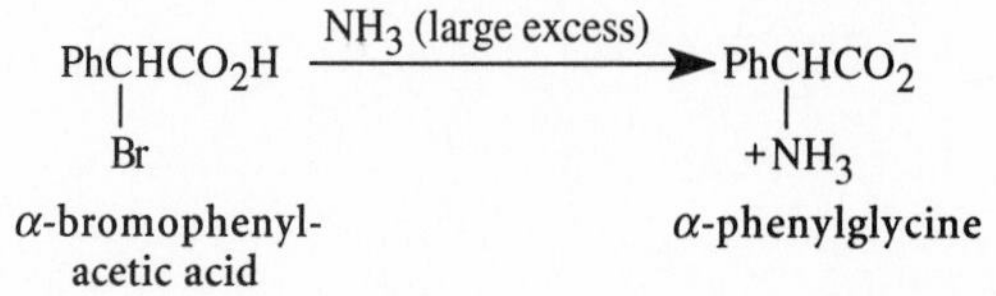

$$PhCHCO_2H \xrightarrow{\text{NH}_3 \text{ (large excess)}} PhCHCO_2^-$$

with Br below the left structure and $^+NH_3$ below the right structure

α-bromophenyl-
acetic acid              α-phenylglycine

The acetamidomalonate method would not work because it would require alkylation of a malonate anion by bromobenzene. Bromobenzene does not undergo $S_N2$ reactions. The Strecker synthesis would work.

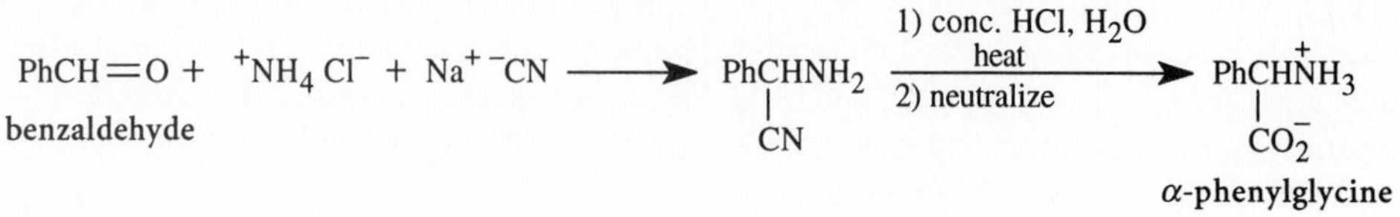

$$PhCH{=}O + {}^+NH_4\,Cl^- + Na^+\,{}^-CN \longrightarrow PhCHNH_2 \xrightarrow[\text{2) neutralize}]{\substack{\text{1) conc. HCl, H}_2\text{O} \\ \text{heat}}} PhCH\overset{+}{N}H_3$$

benzaldehyde                  with CN below middle; with $CO_2^-$ below right

α-phenylglycine

26.13  (a)

$$(CH_3)_2CHCH_2CHNH{-}\overset{\overset{O}{\|}}{\underset{\overset{\|}{O}}{S}}{-}\bigcirc{-}CH_3$$     *N*-tosylleucine

with $CO_2H$ below CHNH

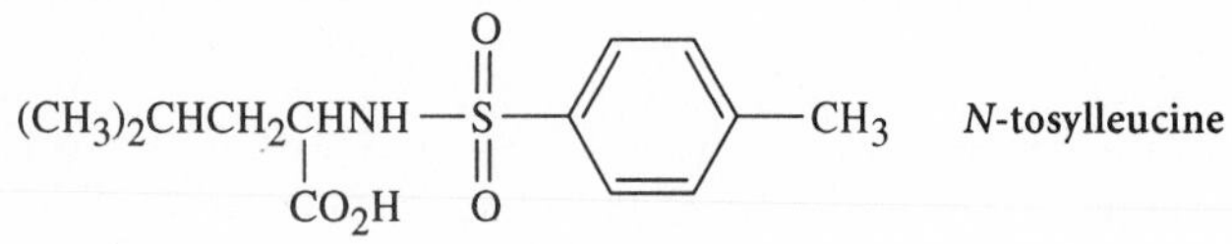

26.14  Polyglycine results from aminolysis of the ester group on one molecule by the amine on another:

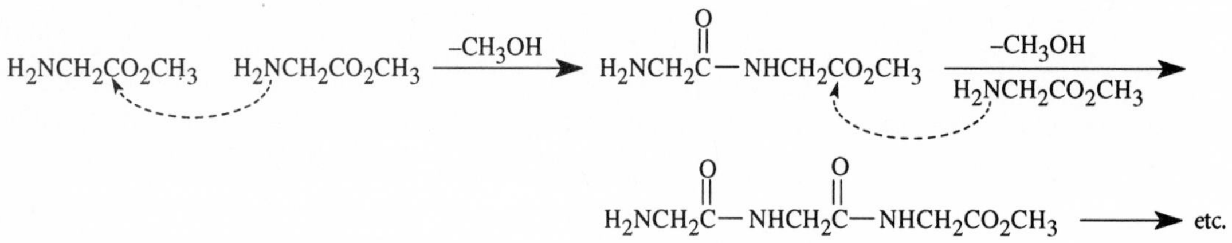

$$H_2NCH_2CO_2CH_3 \quad H_2NCH_2CO_2CH_3 \xrightarrow{-CH_3OH} H_2NCH_2\overset{\overset{O}{\|}}{C}{-}NHCH_2CO_2CH_3 \xrightarrow[H_2NCH_2CO_2CH_3]{-CH_3OH}$$

$$H_2NCH_2\overset{\overset{O}{\|}}{C}{-}NHCH_2\overset{\overset{O}{\|}}{C}{-}NHCH_2CO_2CH_3 \longrightarrow \text{etc.}$$

In this reaction the amino group of glycine acts as a nucleophile. In acidic solution the amino group of glycine is protonated and cannot act as a nucleophile; hence, the reaction does not occur.

26.15  (a)  The amide group of Asn hydrolyzes to give ammonium ion and Asp. Consequently, Asn appears in amino acid analysis as Asp.

26.16  (a)  The amino group serves as a nucleophile in an addition to the isothiocyanate.

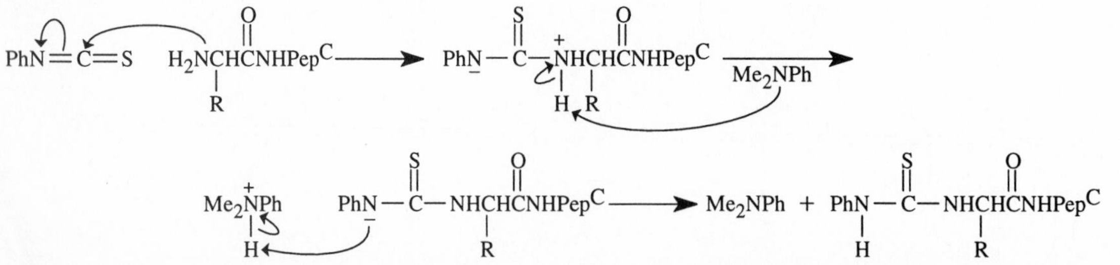

$$PhN{=}C{=}S \quad H_2N\overset{\overset{O}{\|}}{C}HCNHPep^C \longrightarrow PhN{-}\overset{\overset{S}{\|}}{C}{\overset{+}{\underset{H,R}{N}}}HCHCNHPep^C \xrightarrow{Me_2NPh}$$

with R below the left CHC

$$Me_2\overset{+}{N}Ph \quad PhN{-}\overset{\overset{S}{\|}}{C}{-}NHCHCNHPep^C \longrightarrow Me_2NPh + PhN{-}\overset{\overset{S}{\|}}{C}{-}NHCHCNHPep^C$$

with H and R labels as shown

(c)  Water serves as a nucleophile in opening the thiazolinone, and nitrogen of the resulting thiourea serves as the nucleophile in closing the ring to the PTH. The mechanism below begins with the carbonyl-protonated thiazolinone.

protonated thiazolinone

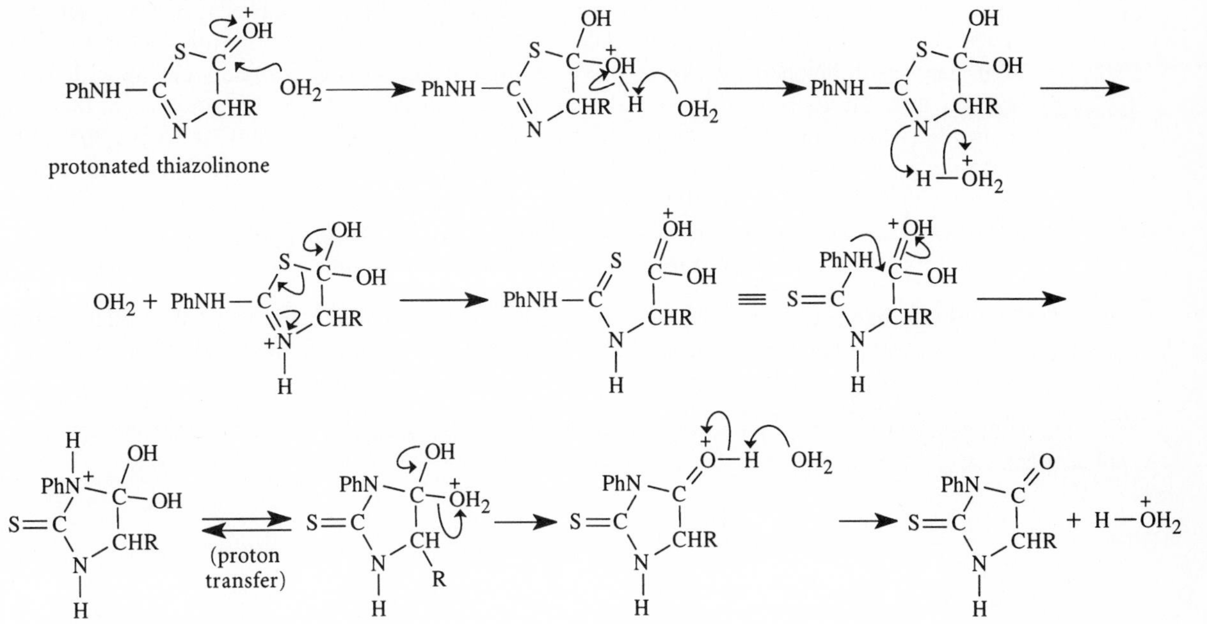

**26.17** As Eq. 26.21a on text p. 1284 as well as its mechanism in the solution to Problem 26.16(a) show, the Edman degradation depends on the presence of a free terminal amino group in the peptide. Because this group is blocked as an amide in acetylated peptides, such peptides cannot undergo the Edman degradation. (This is one reason that the development of a carboxy-terminal degradation is an important topic of current research.)

**26.18** (a) Because cyanogen bromide is an acid halide of $HC \equiv N$, it reacts with amines in the same way that other acid halides react with amines:

$$—(CH_2)_4NH_2 + Br—C \equiv N \longrightarrow —(CH_2)_4NH—C \equiv N + HBr$$

lysine side chain

(b) The acidic conditions of the reaction prevent the reaction. Under such conditions the amino group of lysine is protonated and therefore cannot act as a nucleophile.

**26.19** The peptide $C$ is derived from the amino end of $Q$ because it contains Leu, the amino-terminal residue of $Q$ itself. Peptide $D$ must be at the carboxy end of $Q$ because it contains no Lys or Arg, and therefore must have resulted from cleavage at its amino-terminal residue only. The order of $A$ and $B$, however, must be established by other data; the required order is provided by cleavage with chymotrypsin. Peptide $E$ shows that the Lys residue in $Q$ is followed by a Gly residue; since the other Gly residue follows a Pro (peptide $B$), this establishes that peptide $A$ follows peptide $C$ in the sequence of $Q$. The Pro-Gly-Arg-Ser sequence in peptide $F$ confirms that peptide $D$ follows peptide $B$ in the sequence of $Q$. The final sequence of $Q$, then, is $C$-$A$-$B$-$D$, or

Peptide $Q$ = Leu-Lys-Gly-Arg-Ile-Trp-Phe-Pro-Gly-Arg-Ser-Glu-Ile

**26.21** If the average yield of each step is $Y$, then the yield of the first step (assuming it is average) is $Y$; the yield of the second is $Y \cdot Y$; and that of the $n$th step is $Y^n$. Hence,

$$Y^{369} = 0.17$$
$$369 \log Y = \log 0.17 = -0.7696$$
$$\log Y = -0.0021, \text{ or } Y = 0.995$$

Therefore, the average yield of each step is 99.5%!

 The high yield of each step demonstrates two points about peptide synthesis. First, it has been developed into a remarkably efficient process. The second point, however, is that even with yields in excess of 99%, the overall yields of large proteins prepared by this method will be extremely small. Furthermore, these will be contaminated by large numbers of impurities that will be difficult to separate from the desired material. Fortunately, genetic-engineering methods (which you will study if you take biochemistry) allow scientists to prepare pure naturally occurring proteins in large amounts. At the present time, the chemical synthesis of peptides is most useful for the preparation of peptides containing about 2–50 residues, although a few noteworthy successes have been achieved in the synthesis of larger peptides and some proteins. (See, for example, Study Guide Link 26.6 on p. 729 of this manual.)

26.23 (a) Because lysine contains two amino groups, both must be protected in order to prevent nucleophilic side reactions, such as the reaction of the amino group of one lysine molecule with the DCC-activated carboxy group of another.

(b) The trifluoroacetic acid step removes both Boc protecting groups. Hence, both amino groups react in subsequent acylation reactions. (The Lys residue is drawn in more structural detail for clarity.)

$$\underset{\underset{\underset{NH_2}{|}}{\underset{(CH_2)_4}{|}}{Gly-NHCH\overset{\overset{O}{||}}{C}-Ala}} \qquad \underset{\underset{\underset{NH-Gly}{|}}{\underset{(CH_2)_4}{|}}{H_2NCH\overset{\overset{O}{||}}{C}-Ala}} \qquad \underset{\underset{\underset{NH-Gly}{|}}{\underset{(CH_2)_4}{|}}{Gly-NHCH\overset{\overset{O}{||}}{C}-Ala}}$$

26.24 Because the four chains of hemoglobin are held together by the same noncovalent forces that account for the tertiary structures of the individual chains, the subunits of hemoglobin would dissociate and would unfold into two random-coil $\alpha$-chains and two random-coil $\beta$-chains.

## Solutions to Additional Problems

26.25 Products from the reactions of valine:

(a)

$$\underset{\underset{CH(CH_3)_2}{|}}{H_3\overset{+}{N}CHCO_2Et} \ HSO_4^-$$

(b)

$$\underset{\underset{CH(CH_3)_2}{|}}{Ph\overset{\overset{O}{||}}{C}NHCH\overset{\overset{O}{||}}{C}O^-} \ Et_3\overset{+}{N}H$$

(c)

$$\underset{\underset{CH(CH_3)_2}{|}}{H_3\overset{+}{N}CHCO_2H} \ Cl^-$$

(d)

$$\underset{\underset{CH(CH_3)_2}{|}}{H_2NCHCO_2^-} \ Na^+$$

The reaction in (e) is a variation of the first reaction of the Strecker synthesis in which an amine—in this case, the amino acid valine—rather than ammonia is the source of nitrogen. (See Eq. 26.13 on text p. 1279.)

(e)

$$\underset{\underset{C\equiv N \ CH(CH_3)_2}{|\quad\quad|}}{PhCHNHCHCO_2^-} \ Na^+$$

(f)

$$\underset{\underset{CH(CH_3)_2}{|}}{(CH_3)_3CO\overset{\overset{O}{||}}{C}NHCH\overset{\overset{O}{||}}{C}O^-} \ Et_3\overset{+}{N}H$$

(g)

$$\underset{\underset{CH(CH_3)_2}{|}}{(CH_3)_3CO\overset{\overset{O}{||}}{C}NHCH\overset{\overset{O}{||}}{C}NHCH_2\overset{\overset{O}{||}}{C}OC(CH_3)_3}$$

(h)

$$\underset{\underset{CH(CH_3)_2}{|}}{H_3\overset{+}{N}CH\overset{\overset{O}{||}}{C}NHCH_2\overset{\overset{O}{||}}{C}O^-}$$

(i)

$$\underset{\underset{CH(CH_3)_2}{|}}{Cl^- \ H_3\overset{+}{N}CH\overset{\overset{O}{||}}{C}OH} \ + \ Cl^- \ H_3\overset{+}{N}CH_2\overset{\overset{O}{||}}{C}OH$$

26.26 (a) Aspartic acid. It has the lowest isoelectric point.

(c) Isoleucine and threonine. Each has two asymmetric carbon stereocenters.

(e) Asparagine and glutamine. Their side-chain amides are hydrolyzed to give aspartic acid and glutamic acid, respectively.

26.27 (a) Because the amino group of leucine is not protected, a certain amount of attachment to the resin will occur by attack of the $\alpha$-amino group on the chloromethyl groups of the resin. Subsequent peptide synthesis would then not proceed as planned.

26.28 Since lysozyme has many more basic residues (Lys and Arg) than acidic residues (Asp and Glu), lysozyme is expected to be (and is) a basic protein. Its isoelectric point should be $>>6$. (In fact, its isoelectric point is about 12.)

26.30 The amino-terminal residue is lysine. Both the side-chain amino group and the $\alpha$-amino group react with phenyl isothiocyanate (the Edman reagent). Thus, the side-chain amino group in the resulting PTH derivative (shown in the problem) is present as a thiourea derivative.

26.31 The amino-terminal residue of the peptide $P$ is valine. Because dansyl-valine is obtained, the $\alpha$-amino group of valine must have been free in the peptide, and hence, valine must have been the amino-terminal residue. The sequence of the other residues cannot be determined from the data given.

26.32 Number the residues of $Q$ from the amino terminus as 1,2,3, ...  First, the Edman degradation of $Q$ shows that leucine is the amino-terminal residue. The formation of only dipeptides, including the dipeptide Leu-Val, shows that DPAP catalyzes the hydrolysis of peptide $Q$ at every even-numbered residue from the amino terminus. Thus, this enzyme also catalyzes the hydrolysis of peptide $R$ at every other residue — the *odd* residues in the numbering of $Q$:

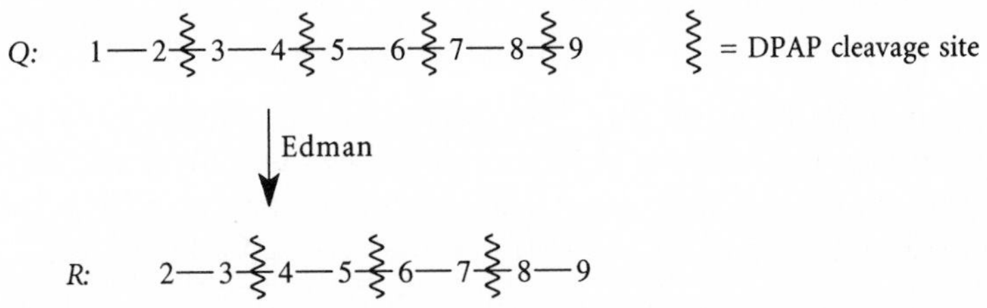

$$Q: \quad 1-2\ \ 3-4\ \ 5-6\ \ 7-8\ \ 9 \qquad \} = \text{DPAP cleavage site}$$

$$\Big\downarrow \text{Edman}$$

$$R: \quad 2-3\ \ 4-5\ \ 6-7\ \ 8-9$$

This analysis shows that Gly is residue 9, the carboxy-terminal residue of $Q$. To get the order of the dipeptides, work back and forth between the dipeptides derived from $Q$ and those derived from $R$. Since Gly has to be at the carboxy terminus of one of the dipeptides from $R$, and since the only dipeptide that meets this criterion is Ala-Gly, then Ala is the next residue in from the carboxy terminus. If Ala is in position 8, then position 7, from the $Q$ dipeptides, must be Gln; position 6, from the $R$ dipeptides, must be Asp; and so on. Alternatively, work from the amino terminus of $Q$: Leu-Val of $Q$ and Val-Arg of $R$ establish the sequence Leu-Val-Arg; Arg-Gly of $Q$ shows that the next residue is Gly; and so on. The final sequence of peptide $R$, then, is:

Leu-Val-Arg-Gly-Val-Asp-Gln-Ala-Gly

(Remember that the amino terminus by convention is on the left when peptides are written this way.) The ammonia arises from the hydrolysis of Gln to glutamic acid.

26.34 A molecular mass of about 1100 is consistent with the actual composition ($Ala_4,Arg_2,Gly_2,Ser_2$); amino acid analysis gives only the *relative* amounts of each amino acid. The absence of a reaction with the Edman reagent suggests that there is no terminal amino group. A *cyclic* peptide would give these results. Cleavage of a cyclic peptide at two arginine residues would give two peptides; but if the two peptides were identical, two equivalents of a single peptide would be obtained. The results are in accord with the following structure:

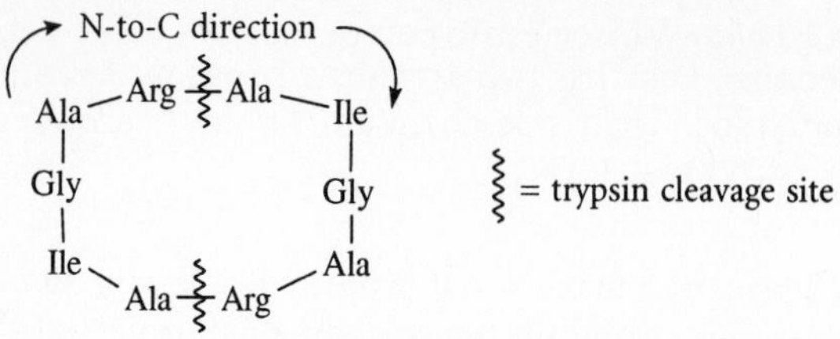

$\{$ = trypsin cleavage site

**26.36**   Amino acid $A$ is 2,4-diaminobutanoic acid. The key piece of structural data is the Hofmann rearrangement, which converts the side-chain amide of glutamine into an amino group. Compound $A$ is expected to be basic because it has two amino groups and only one carboxy group.

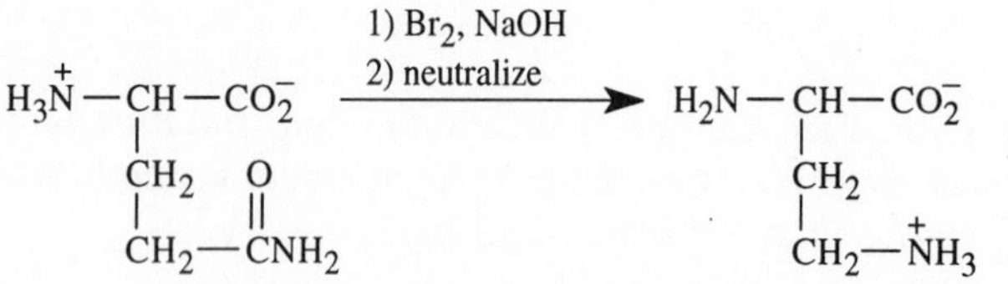

**26.38**   (a)   Reaction of chloromethylated polystyrene with trimethylamine would give the resin shown:

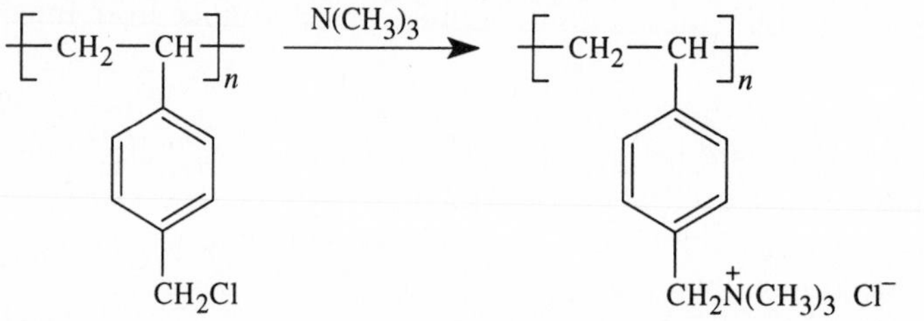

Notice that the alkyl halide is benzylic and is thus particularly reactive in $S_N2$ reactions.

(b)   This resin is an *anion-exchange resin*. (See the solution to Problem 26.10 on p. 731 of this manual.) At pH 6, Arg is positively charged and is therefore repelled by the column; it elutes earliest from the column. Leu has zero charge at pH 6; it elutes next. Glu is negatively charged at pH 6 and is therefore attracted to the positively-charged groups on the column; it elutes latest from the column.

**26.40**   (a)   Alanine (like the other amino acids) is a different compound in HCl, NaOH, and neutral $H_2O$ because its ionization state is different, and different compounds have different optical rotations. In acidic solution, the optical rotation is that of the acidic form; in base, the optical rotation is that of the basic form; and in water, the optical rotation is that of the neutral (zwitterion) form.

(b)   Lysine has two amino groups, and acetylation of each is possible.

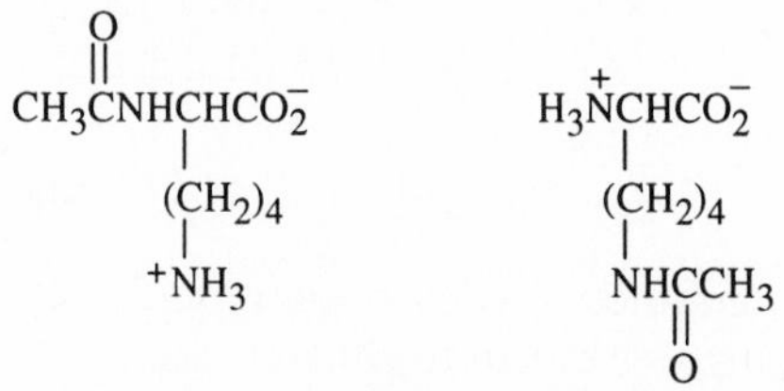

the two mono-*N*-acetylated lysine derivatives

(c)   The peptide is cleaved at the Arg-Ala bond at pH 8. However, in the presence of $8M$ urea, trypsin, like most enzymes, is denatured. Denatured enzymes are devoid of catalytic activity because their tertiary structures are disrupted. Thus, trypsin, when denatured, cannot catalyze cleavage of the peptide.

(d)   The 2-mercaptoethanol treatment ensures that disulfide bonds are reduced. Aziridine reacts much like an epoxide when the nitrogen is protonated. Aziridine is basic enough that a significant amount of it is protonated in aqueous solution. Reaction of protonated aziridine with the thiol group of a cysteine residue gives an amine:

$$—CH_2S—CH_2CH_2—NH_2 \underset{}{\overset{pH\ 8}{\rightleftarrows}} —CH_2S—CH_2CH_2—\overset{+}{N}H_3$$

The amino group of the modified residue is protonated at pH 8, the pH at which trypsin digestion is carried out. The side chain of the modified residue resembles the side chain of lysine in both length and charge, and trypsin hydrolyzes peptides at the modified residue much as it hydrolyzes peptides at lysine residues.

(e)  The two sulfoxides are diastereomers. They both have the same configuration at the $\alpha$-carbon, but differ in configuration at the sulfur of the sulfoxide group. This sulfur is an asymmetric atom and stereocenter because it has four different "groups" attached: the $CH_2$, the $CH_3$, the O, and the electron pair. Evidently, inversion at sulfur in a sulfoxide, unlike amine inversion, is very slow, because the individual diastereomers can be isolated.

(f)  Benzamidine contains a functional group that is very much like the guanidino group on the side chain of an arginine. Benzamidine, like an arginine residue in a peptide, is bound in the active site of trypsin. When the active site contains benzamidine, it cannot simultaneously bind a peptide. Hence, benzamidine blocks trypsin-catalyzed peptide hydrolysis. (Benzamidine is said to be a *competitive inhibitor* of trypsin catalysis because it competes with substrates for the trypsin active site.)

**26.41**  Figure 26.6 on text p. 1302 shows that the amino acid side chains extend outward from the periphery of the helix and are actually rather close to each other in space. The $pK_a$ of a protonated lysine side-chain amino group is about 10.5. At pH values below 10, the side-chain amino groups in polylysine are protonated; thus, these side chains are positively charged. The peptide avoids the $\alpha$-helical conformation at pH values below 10 to avoid the repulsive electrostatic interactions between the adjacent positively charged groups. At pH values above 11, the side-chain amino groups are unprotonated and hence uncharged. As a result, there are no charge-charge repulsions to destabilize the $\alpha$-helix.

The same principles operate in polyglutamic acid, but at the other end of the pH scale. In polyglutamic acid, the side chains are negatively charged at pH values above the $pK_a$ of the carboxy groups. Hence, the repulsions between negative charges cause the helix to be destabilized, and it unfolds at high pH. At low pH, the carboxy groups are un-ionized and hence uncharged. Consequently, the helix can form at low pH because there are no charge-charge repulsions to destabilize it.

**26.42**  (a)  This is a reductive amination of formaldehyde by the amino group of the lysine residue. Because excess formaldehyde is present, the amino group of lysine is methylated twice.

(c)  The thiol of cysteine serves as a nucleophile in a conjugate-addition reaction to maleimide. The pH must be high enough to form a small amount of the conjugate-base thiolate anion of cysteine, which is the actual nucleophilic species.

*(solution continues)*

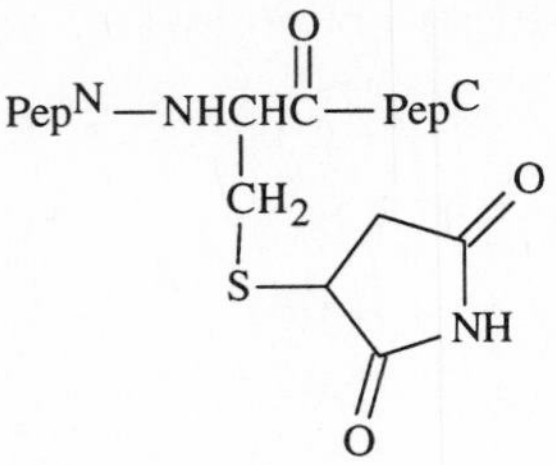

(e) The carbodiimide promotes a condensation between the side-chain carboxy group of aspartic acid and the amino group of the glycine ester.

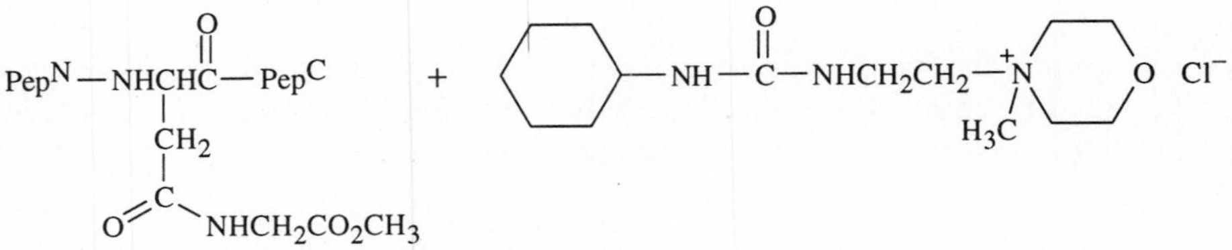

(f) The phenol ring of the tyrosine residue undergoes electrophilic substitution by the diazonium ion (see Sec. 23.4B on text p. 1143). Because the *para* position is blocked, substitution occurs at the position *ortho* to the hydroxy group.

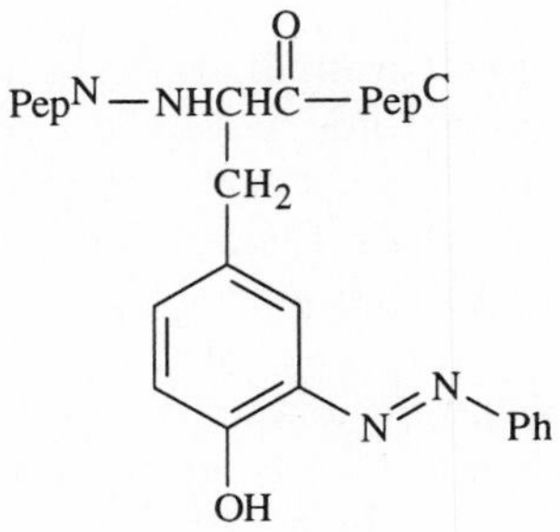

**26.43** (a)

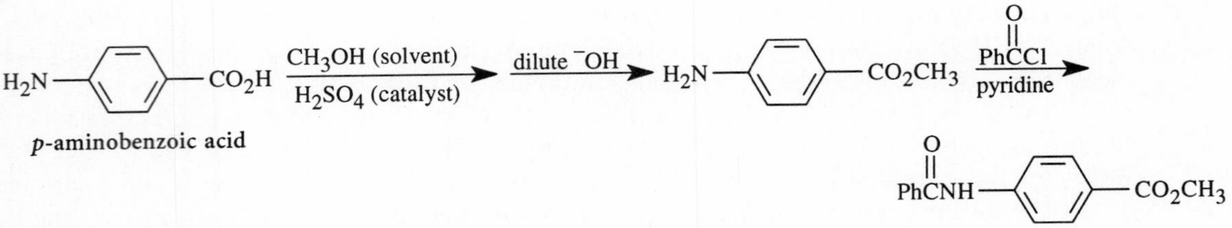

*p*-aminobenzoic acid

(b)

benzoic acid

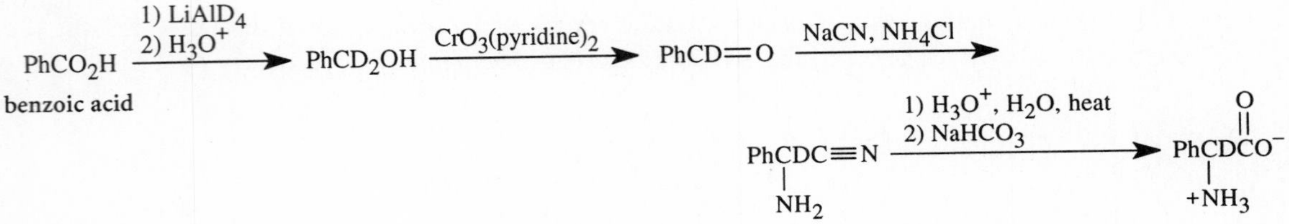

(c)

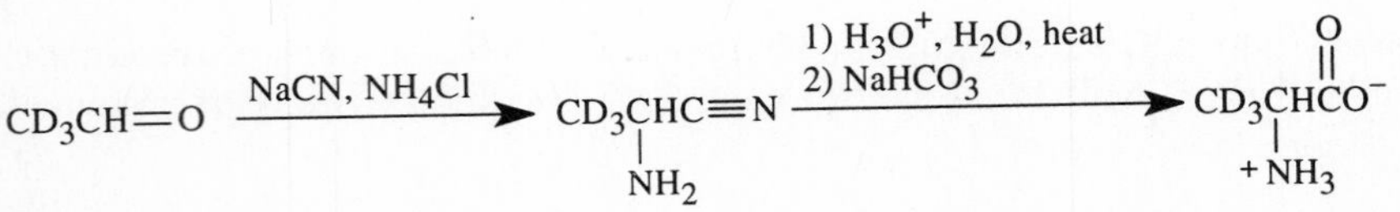

(d) First prepare Boc-proline from proline and Boc-alanine from alanine by the method shown in Eq. 26.25 on text p. 1292. Then form the cesium salt of Boc-proline with CsOH and attach it to the Merrifield resin as shown in Eq. 26.26 on text p. 1293 to give Boc-Pro—resin. Then continue as follows. (All amino acids and their derivatives have the L configuration, which is not indicated explicitly.)

Boc-Pro—resin —[1) $CF_3CO_2H$; 2) $Et_3N$]→ Pro—resin —[Boc-Ala, DCC]→ Boc-Ala—Pro—resin —[1) $CF_3CO_2H$; 2) $Et_3N$]→

Ala—Pro—resin —[di-Boc-lysine, DCC]→ di-Boc-Lys—Ala—Pro—resin —[1) liq. HF; 2) neutralize]→ Lys—Ala—Pro

(e)

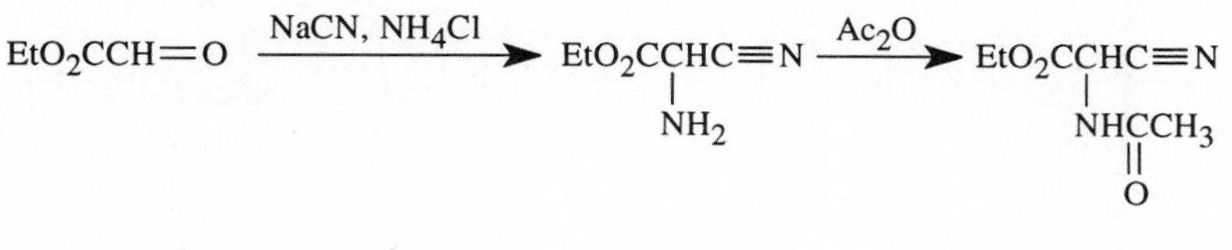


(f) First prepare 3-bromocyclopentene from cyclopentene:

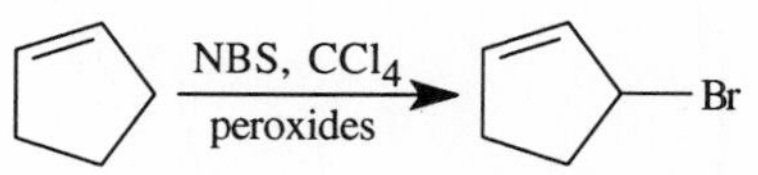


3-bromocyclopentene

Then use it in the following synthesis. Note that the $\alpha$-hydrogen of the product of (e) is about as acidic as an $\alpha$-hydrogen of diethyl malonate. (Why?) Hence, the techniques of the acetamidomalonate method can be used. Notice that 3-bromocyclopentene is an allylic halide and is therefore very reactive in $S_N2$ reactions. (See Sec. 17.4 on text p. 803.)

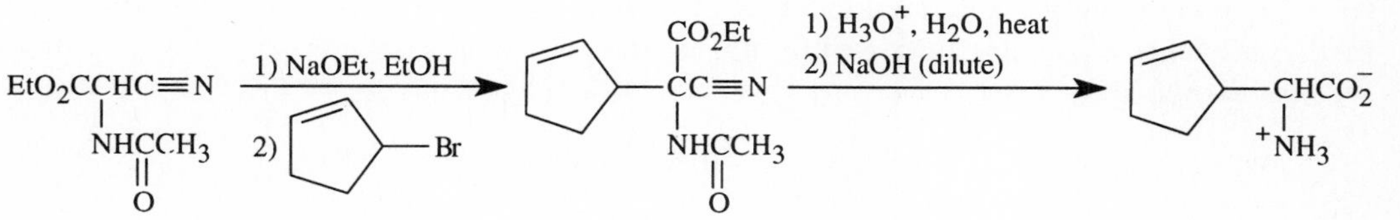


(g)

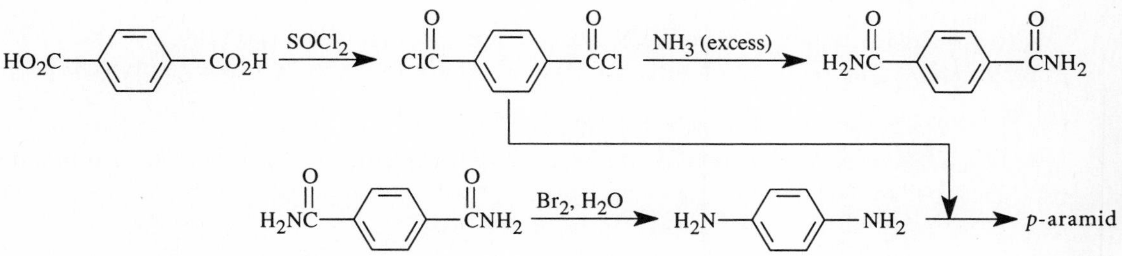


**26.45** In the first step, the amine adds to the Edman reagent; for the mechanism of this reaction, see the solution to Problem 26.16(a) on page 732 of this manual. The sulfur of the thiourea then serves as a nucleophile to close a six-membered ring and cleave the peptide. The mechanism below begins with the product of the reaction between the Edman reagent and the peptide, which is protonated on the carbonyl oxygen by the catalyzing acid.

*(solution continues)*

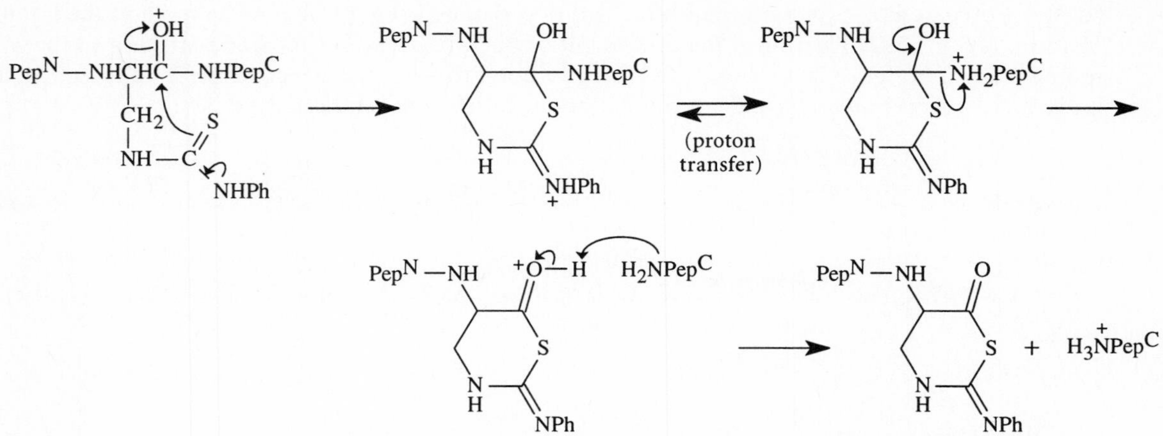

compound *X*

**26.47** (a)

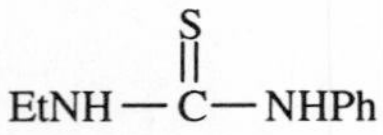

$$EtNH-\overset{\overset{\displaystyle S}{\|}}{C}-NHPh$$

(b) This is a modified Strecker synthesis in which methylamine is used in place of ammonia.

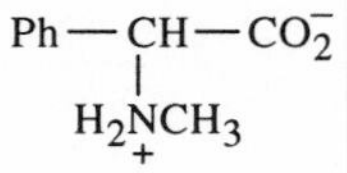

$$Ph-\underset{\underset{+}{\overset{|}{H_2NCH_3}}}{CH}-CO_2^-$$

(c) This is a reductive amination in which the amine is the $\alpha$-amino group of the amino acid.

$$(CH_3)_2CHCH_2NH\underset{\overset{|}{CH_3}}{CH}-CO_2^-$$

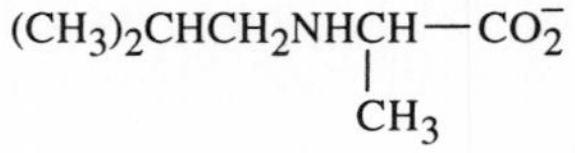

(d) The methyl ester is saponified much more rapidly that the *tert*-butyl ester for two reasons. First, the methyl branches of the *tert*-butyl group impede the approach of hydroxide to the carbonyl group. Second, the *tert*-butyl ester is also an amide, and amides hydrolyze much more slowly than esters.

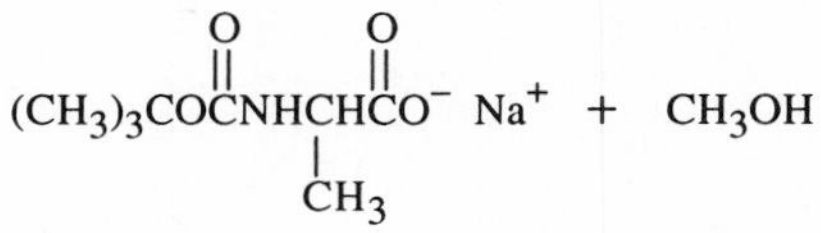

$$(CH_3)_3CO\overset{\overset{\displaystyle O}{\|}}{C}NH\underset{\overset{|}{CH_3}}{CH}\overset{\overset{\displaystyle O}{\|}}{C}O^-\ Na^+\ +\ CH_3OH$$

(e) This is a diazotization to form the *N*-diazo compound, that is, the acyl azide. (See Eq. 23.72 on text p. 1150.) [*Note:* The products of parts (e), (f), and (g) are summarized in the solution to part (g).]

(f) Heating the acyl azide gives a Curtius rearrangment to the isocyanate.

(g) The amino group of the amino acid ester adds to the isocyanate to give a urea. To summarize parts (e)–(g):

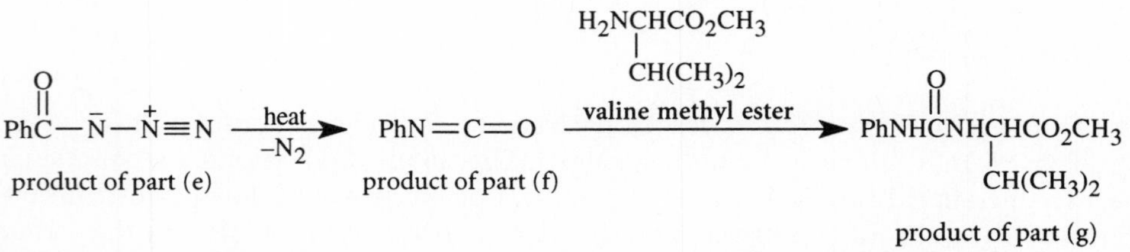

(h) This is a formylamidomalonate reaction, a variation of the acetamidomalonate synthesis.

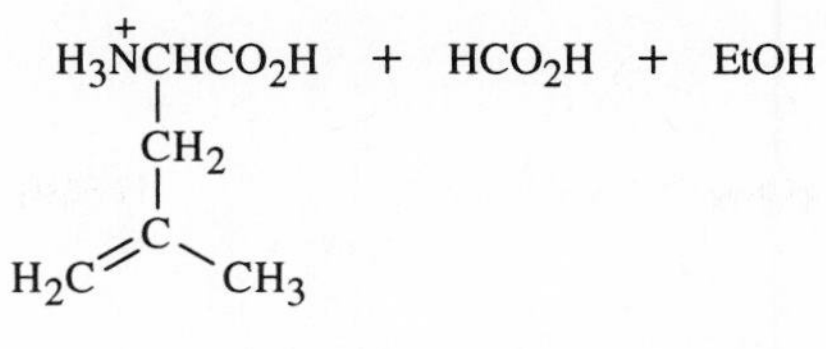

$$\overset{+}{H_3N}CHCO_2H \;+\; HCO_2H \;+\; EtOH$$

with the CH₂ and isopropenyl substituent structure:

$$\underset{\substack{| \\ CH_2 \\ | \\ H_2C\!\overset{C}{=}\!\!\!\overset{|}{\phantom{C}}\!CH_3}}{}$$

(i)  Hydrazine (H₂N—NH₂) displaces the ethoxy group of the ester to form a hydrazide *A*; diazotization forms the acyl azide *B*; heating this in ethanol gives a Curtius rearrangement in which the isocyanate is trapped by reaction with the solvent to give the ethyl carbamate *C*; and acidic hydrolysis generates the amino-protonated amino acid *D*.

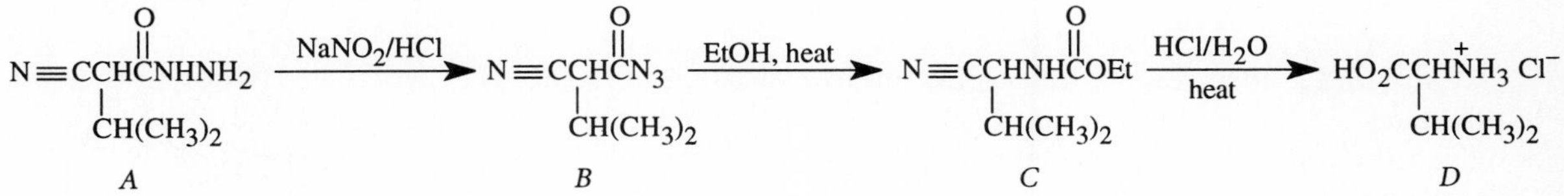

$$N\!\equiv\!CCHCNHNH_2 \xrightarrow{\text{NaNO}_2/\text{HCl}} N\!\equiv\!CCHCN_3 \xrightarrow{\text{EtOH, heat}} N\!\equiv\!CCHNHCOEt \xrightarrow[\text{heat}]{\text{HCl/H}_2\text{O}} HO_2CCHNH_3\ Cl^-$$

$$\quad A \qquad\qquad\qquad B \qquad\qquad\qquad C \qquad\qquad\qquad D$$

**26.49**  (a)  The first step of the mechanism is formation of an imine. (This is discussed in Sec. 19.11A on text pp. 904–905.) Reaction of this imine with the conjugate base of the thiol gives an addition product which is eventually transformed into the product. The following mechanism begins with the imine.

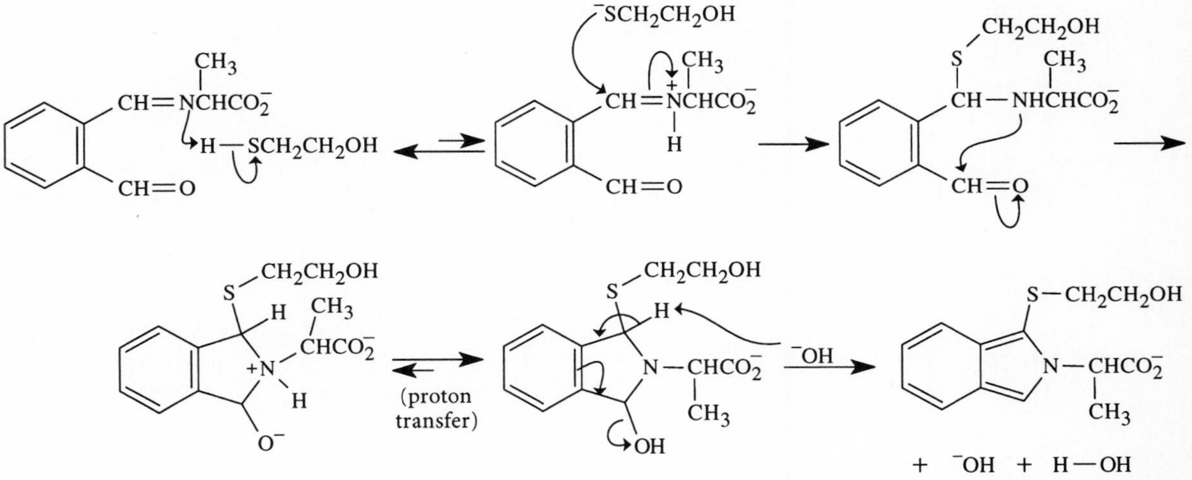

(b)  In the first part of the mechanism, ⁻SH displaces chloride in an S_N2 reaction. Recall that $\alpha$-halo carbonyl compounds are particularly reactive in S_N2 reactions (Sec. 22.3D on text p. 1053). Ammonia then reacts to forms an enamine. (Enamine formation is discussed in Sec. 19.11B on text pp. 907–908.) Although imine formation is favored, imines and enamines are in equilibrium just as aldehydes and enols are in equilibrium. Addition of both the —SH group and the —NH₂ group of this enamine to acetone (in a reaction much like acetal formation) and proton transfers give the first product *A*.

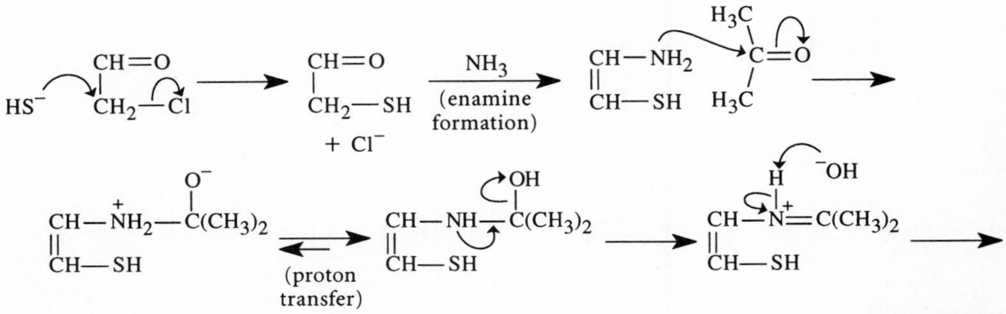

*(equation continues)*

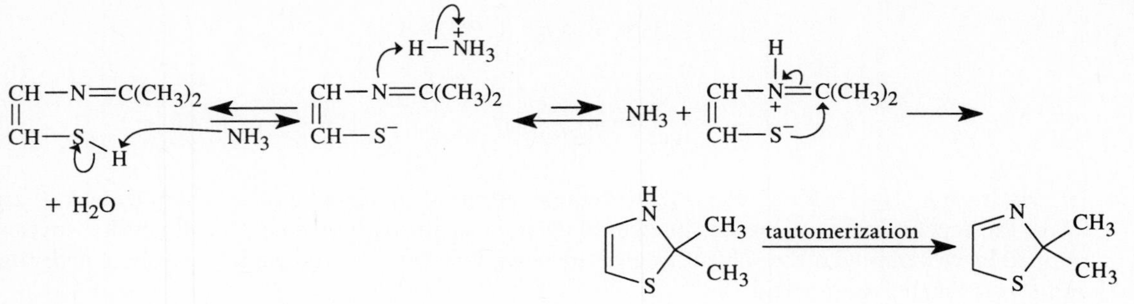

Compound *A* then reacts with cyanide ion in a variation of the Strecker synthesis. Hydrolysis in acid liberates acetone and the conjugate acid *B* of the α-cyano amine. The cyano group hydrolyzes in acid to give the conjugate acid of cysteine. (The mechanism of nitrile hydrolysis is given in Eqs. 21.20–21.21 on text pp. 994–995.)

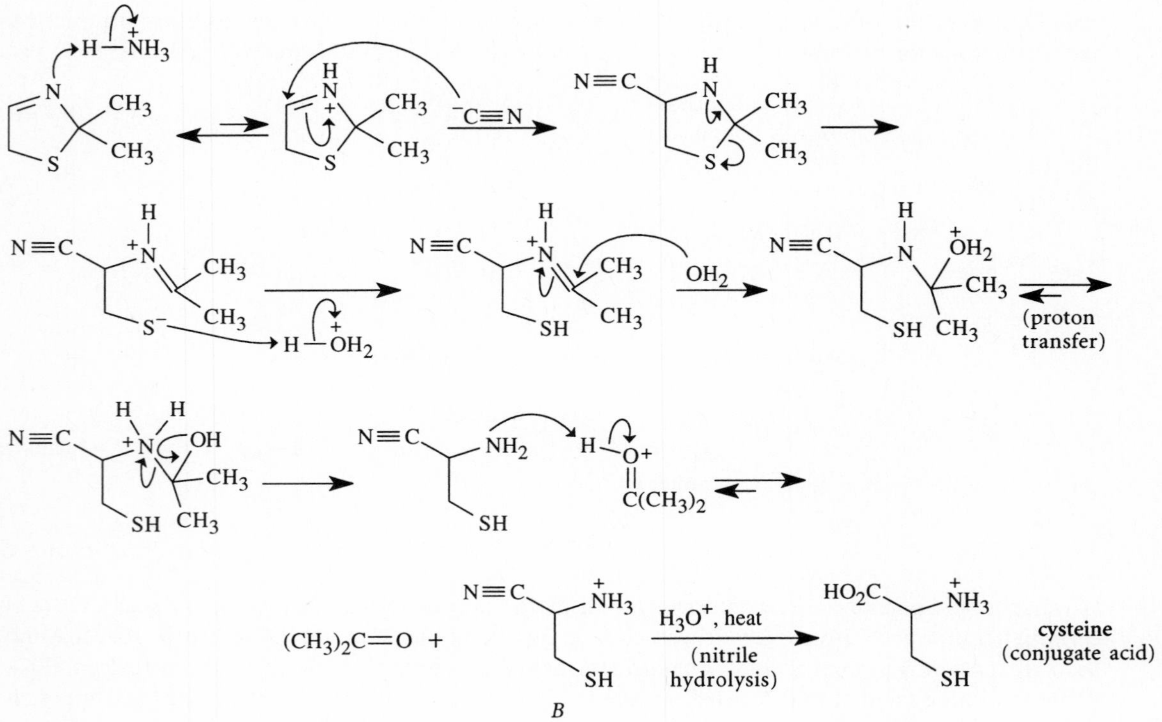

(c) Formaldehyde and the amino group of tryptophan react to give the conjugate acid of an imine, which acts as an electrophile in an intramolecular electrophilic aromatic substitution reaction to give the product. The mechanism below begins with the conjugate acid of the imine.

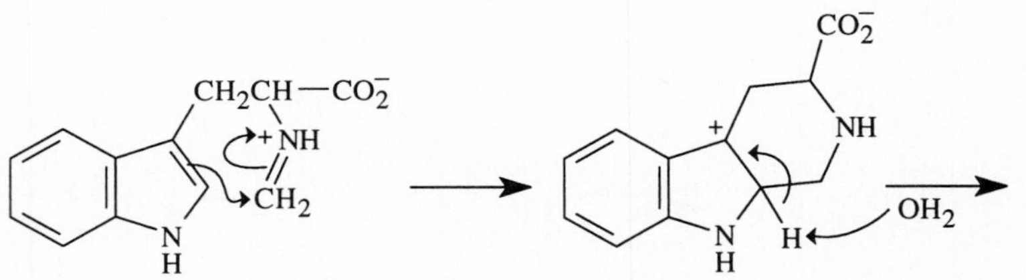

**(d)** This reaction is a Curtius rearrangement (with concomitant loss of $N_2$) to give an isocyanate, which reacts intramolecularly with the side-chain hydroxy group. The mechanism of the Curtius rearrangement is given in Eq. 23.76 on text p. 1151. The mechanism below begins with the isocyanate.

**26.50** **(a)** In the $Z$ configuration, the two large groups $Pep^N$ and the $N$-alkyl group $Pep^C$ are *anti*, and thus van der Waals repulsions between these groups cannot occur. In the the $E$ configuration. these two groups are close enough that van der Waals repulsions can result. These raise the energy of the peptide; hence, the $Z$ configuration is energetically preferred.

The word "configuration" rigorously should be used only when rotation about the appropriate bond is so slow that it practically speaking does not occur. For example, rotation about a carbon-carbon double bond is so slow that it for all practical purposes does not occur; consequently, we speak of $E$ and $Z$ *configurations* for alkenes rather than $E$ and $Z$ *conformations*. Because rotation rates about the carbonyl-nitrogen bond of amides are generally rapid, it is usually customary to speak of $E$ and $Z$ *conformations* about this bond. However, in large peptides and proteins, such rotations do not occur in most cases because, if they did, they would cause a large disruption in the remainder of the protein structure, which is also fixed by large numbers of interactions between other residues. Hence, we have used the word *configuration* in this problem to reflect the rigidity of most amino acid residues in *proteins* about this bond.

**(b)** The proline nitrogen bears two $N$-alkyl groups, whereas the peptide-bond nitrogens of other residues have only one. Thus, in either the $E$ or the $Z$ configuration, an alkyl group is *syn* to the $Pep^N$ group. Although the group that is *syn* to $Pep^N$ in the $E$ configuration is larger, there is a much smaller difference in energy between $E$ and $Z$ configurations for proline residues than there is for other residues. Hence, $E$ configurations in peptide bonds are most often found at proline.

proline residue
in an $E$ configuration

proline residue
in a $Z$ configuration

**26.51** Peptide *I* results from intramolecular nucleophilic attack of the conjugate-base anion of the neighboring amide bond. This occurs mostly at Asn-Gly because glycine has no carbon side chain; in residues other than glycine, the carbon side chains can cause van der Waals repulsions in the ring-closure step.

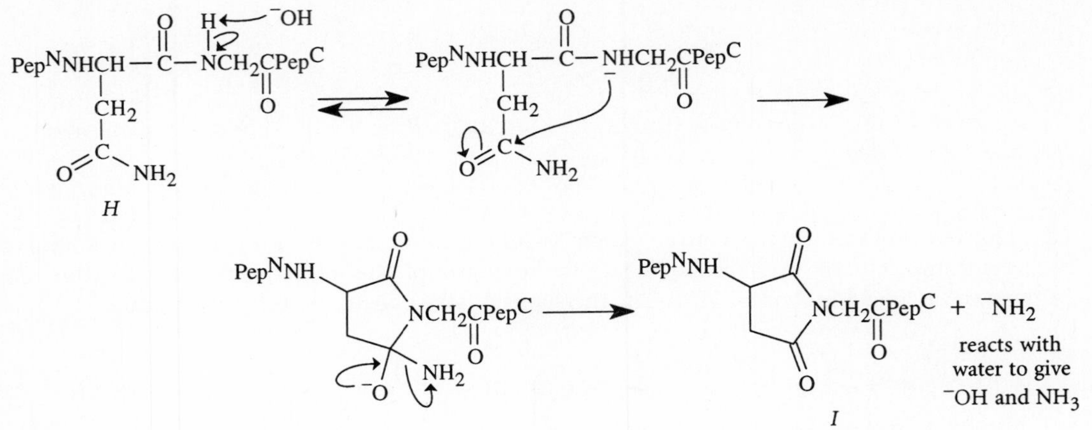

Derivative *I* is an imide, which is the nitrogen analog of an anhydride. Hydroxide can attack either carbonyl carbon of the imide ring to open the ring, thus generating either peptide *J* or peptide *K*. The mechanism for the formation of peptide *K* is as follows:

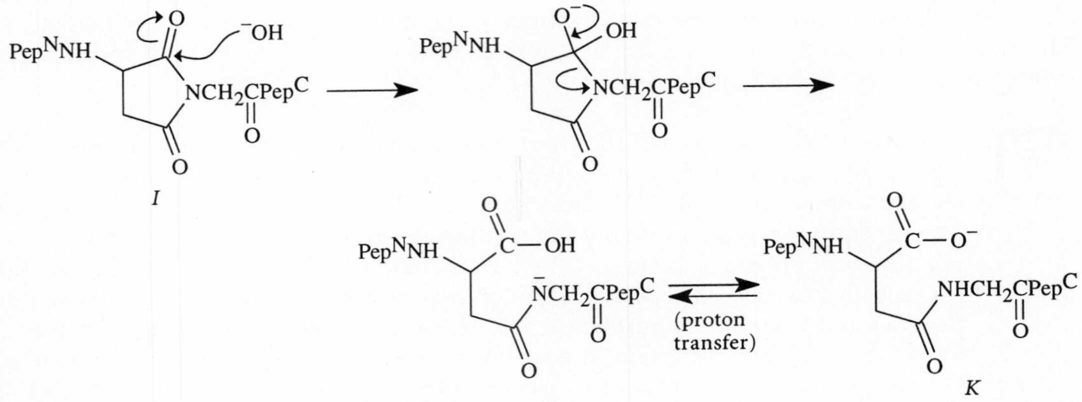

The structure of peptide *J* is shown below; you should show the mechanism for its formation.

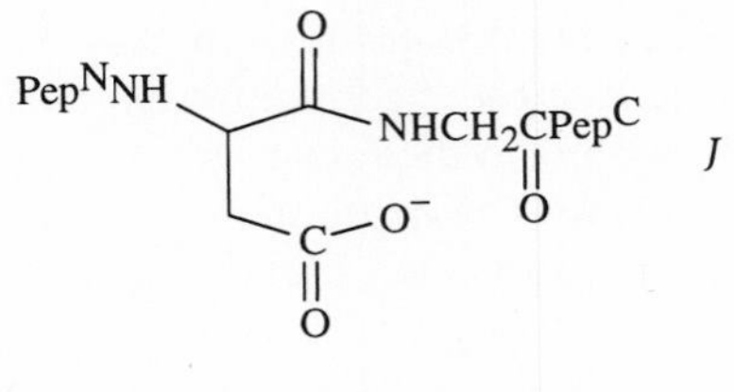

26.52 (a) The two monomethyl esters result from ester formation at each of the two carboxy groups.

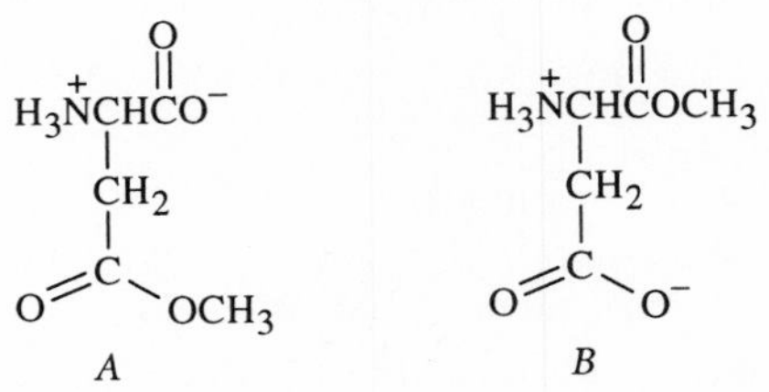

(b) and (c)    According to Table 26.1, the p$K_a$ values of the two carboxy groups in aspartic acid are about 1.9 (for the $\alpha$-carboxy group) and 3.6 (for the $\beta$-carboxy group). (How do we know which is which?) Assume that the p$K_a$ values for the respective esters are similar. At pH 3.0, the carboxy group of compound *A* is largely ionized; at the same pH, the carboxy group of *B* is mostly un-ionized. Hence, at pH 3.0, compound *B* carries a positive charge, whereas compound *A* is neutral. Therefore, compound *A* will be retained by an anion-exchange column (which contains positively-charged groups) and will be eluted more slowly than compound *B*. At pH 7, however, both compounds are negatively charged. Since they have the same charge, they are retained equally by an anion-exchange column and are therefore not separated.

   Early printings of the text contain an error in Problem 26.53. The sentence on the second line should read, "Treatment of *A* with the amino acid L-alanine . . ."

# 27

# Carbohydrates and Nucleic Acids

## Terms

# Concepts

......................................................................................................................

## I. Introduction to Carbohydrates

### A. CLASSIFICATION AND PROPERTIES OF CARBOHYDRATES

1. Carbohydrates are defined as aldehydes and ketones containing a number of hydroxy groups on an unbranched carbon chain (sugars), as well as their chemical derivatives.
   a. Most of the common sugars have formulas that fit a "hydrate of carbon" pattern, that is, a formula of the form $C_n(H_2O)_m$.
2. Carbohydrates can be classified using a variety of systems:
   a. by the type of carbonyl group in the carbohydrate:
      i. A carbohydrate with an aldehyde carbonyl group is called an aldose.
      ii. A carbohydrate with a ketone carbonyl group is called a ketose.
   b. by the number of carbon atoms:
      i. A six-carbon carbohydrate is called a hexose.
      ii. A five-carbon carbohydrate is called a pentose.
   c. by a combination of the two systems cited in (a) and (b) above:
      i. A six-carbon carbohydrate containing an aldehyde carbonyl group is called an aldohexose.
      ii. A five-carbon carbohydrate containing a ketone carbonyl group is called a ketopentose or pentulose.
      iii. A ketose can be indicated with the suffix *ulose*; thus, a six-carbon ketose is also termed a hexulose.

CH₂—CH—CH—CH—CH—C—H with OH groups

an aldose
a hexose
an aldohexose

CH₂—CH—CH—C—CH₂ with OH groups

a ketose
a pentose
a ketopentose or pentulose

   d. by their hydrolysis to simpler carbohydrates:
      i. Monosaccharides cannot be converted into simpler carbohydrates.
      ii. Disaccharides are converted into two monosaccharides.
      iii. Trisaccharides are converted into three monosaccharides.
      iv. Oligosaccharides are converted into several monosaccharides.
      v. Polysaccharides are converted into a large number of monosaccharides.
3. Carbohydrates are very soluble in water because of their many hydroxy groups but virtually insoluble in nonpolar solvents.

## II. Structure of Monosaccharides, Disaccharides, and Polysaccharides

### A. STEREOCHEMISTRY AND CONFIGURATION

1. The aldopentoses have three asymmetric carbons and $2^3$ or eight possible stereoisomers and can be divided into two enantiomeric sets of four diastereomers.
2. The aldohexoses have four asymmetric carbons and $2^4$ or sixteen possible stereoisomers and can be divided into two enantiomeric sets of eight diastereomers.
3. Each diastereomer is a different carbohydrate with different properties, known by a different name.
4. Although the *R,S* system could be used to describe the configuration of the asymmetric carbon atoms of a carbohydrate, it is more convenient to use the D,L-system, in which the configuration of a carbohydrate enantiomer is specified by applying the following conventions:
   a. The 2*R* enantiomer of the naturally occurring aldotriose glyceraldehyde is arbitrarily said to have the D configuration; its enantiomer, the 2*S* enantiomer, is then said to have the L configuration.

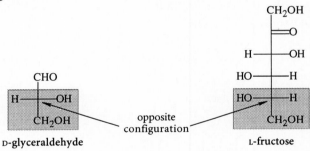

a D carbohydrate     an L carbohydrate

b. Other aldoses or ketoses are written in a Fischer projection with their carbon atoms in a straight vertical line, and the carbons are numbered consecutively as they would be in systematic nomenclature, so that the carbonyl carbon receives the lowest number.

$$\underset{\text{OH}}{\text{CH}_2}-\underset{\text{OH}}{\text{CH}}-\underset{\text{OH}}{\text{CH}}-\underset{\text{O}}{\overset{\text{O}}{\text{C}}}-\underset{\text{OH}}{\text{CH}_2} = \begin{array}{c}{}^{1}\text{CH}_2\text{OH} \\ 2=\!\!=\!\!\text{O} \\ \text{H}-3-\text{OH} \\ \text{H}-4-\text{OH} \\ {}^{5}\text{CH}_2\text{OH}\end{array}$$

c. The asymmetric carbon of highest number is designated as a reference carbon.
  i. If this carbon has the H, OH, and CH$_2$OH groups in the same relative configuration as the same three groups of D-glyceraldehyde, the carbohydrate is said to have the D configuration.
  ii. If this carbon has the same configuration as L-glyceraldehyde, the carbohydrate is said to have the L configuration.

D-glyceraldehyde    opposite configuration    L-fructose

5. There is no general correspondence between configuration and the sign of the optical rotation; there is also no simple relationship between the D,L system and the *R,S* system.
  a. The *R,S* system is used to specify the configuration of each asymmetric carbon atom in a molecule.
  b. The D,L system specifies a particular enantiomer of a molecule that might contain many asymmetric carbons.
6. A few of the aldoses and ketoses are particularly important, and their structures should be learned.

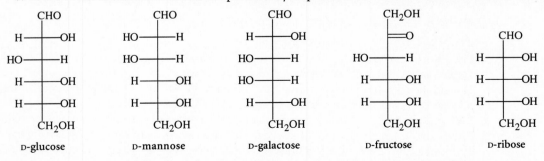

D-glucose    D-mannose    D-galactose    D-fructose    D-ribose

7. Compounds that differ in configuration at only one of several asymmetric carbons are called epimers. For example, mannose and galactose are epimers of glucose.

## B. CYCLIC STRUCTURES OF THE MONOSACCHARIDES

1. Aldoses and ketoses exist predominantly as cyclic hemiacetals.

2. In many carbohydrates both five-membered and six-membered cyclic hemiacetals are possible, depending on which hydroxy group undergoes cyclization.
    a. A five-membered cyclic acetal form of a carbohydrate is called a furanose.
    b. A six-membered cyclic acetal form of a carbohydrate is called a pyranose.

D-fructofuranose            D-glucopyranose

   c. The aldohexoses and aldopentoses exist predominantly as pyranoses, but the furanose forms of some carbohydrates are important.
3. A name such as glucose is used when referring to any or all forms of the carbohydrate.
4. To name a cyclic hemiacetal form of a carbohydrate, start with a prefix derived from the name of the carbohydrate followed by a suffix that indicates the type of hemiacetal ring.
    i. *furanose* for a five-membered ring
    ii. *pyranose* for a six-membered ring
5. A widely used convention for representing the cyclic forms of sugars is the Haworth projection.
    a. In this convention, the cyclic form of a sugar is represented as a planar ring perpendicular to the page.
    b. The shaded bonds are in front of the page, and the other bonds are in back.
    c. The substituents are indicated with up or down bonds.

D-fructofuranose            D-glucopyranose

## C. ANOMERS AND MUTAROTATION

1. The furanose or pyranose form of a carbohydrate has one more asymmetric carbon (carbon-1) than the open-chain form.
    a. When two cyclic forms of a carbohydrate differ in configuration only at the hemiacetal carbon (carbon-1 of an aldose), they are said to be anomers.
    b. The hemiacetal or acetal carbon of a carbohydrate is called the anomeric carbon.
2. The ring in a Fischer projection of these cyclic compounds is represented as a long bond.
3. Anomers are named with the Greek letters $\alpha$ and $\beta$.
    a. In the $\alpha$-anomer, the hemiacetal —OH group is on the same side of the Fisher projection as the oxygen at the configurational carbon.
    b. In the $\beta$-anomer, the hemiacetal —OH group is on the side of the Fisher projection opposite the oxygen at the configurational carbon.

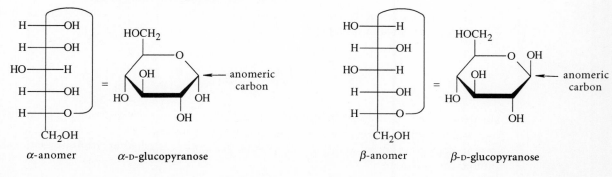

$\alpha$-anomer    $\alpha$-D-glucopyranose        $\beta$-anomer    $\beta$-D-glucopyranose

4. When a pure anomer of a carbohydrate is dissolved in aqueous solution, its optical rotation changes with time; this change, called mutarotation, is catalyzed by both acid or base, and it results from the formation of an equilibrium mixture of both the $\alpha$- and $\beta$-anomers.
5. The mechanism of mutarotation begins as the reverse of hemiacetal formation.
   a. A 180° rotation about the bond to the carbonyl group permits attack of the hydroxy group on the opposite face of the carbonyl carbon.
   b. Hemiacetal formation then gives the other anomer.

6. Some general conclusions:
   a. Most aldohexoses exist primarily as pyranoses, although a few have substantial amounts of furanose forms.
   b. There are relatively small amounts of noncyclic carbonyl forms of most monosaccharides.
   c. Mixtures of $\alpha$- and $\beta$-anomers are usually found, although the exact amounts of each vary from case to case.

D. CONFORMATIONAL REPRESENTATIONS OF PYRANOSES
   1. The six-membered ring of a pyranose exists in two chair conformations related by the chair-chair interconversion.
   2. To go from a Fisher projection to a chair conformation:
      a. Using an allowed manipulation of Fisher projections, first redraw the molecule in an equivalent Fisher projection in which the ring oxygen is in a down position.
      b. Draw a Haworth projection by turning the plane of the ring 90° so that the anomeric carbon is on the right and the ring oxygen is in the rear.
         i. The groups in the up positions are those that are on the left in the Fischer projection.
         ii. The groups in the down positions are those that are on the right in the Fischer projection.
      c. Draw either one of the two chair conformations in which the anomeric carbon and the ring oxygen are in the same relative positions as they are in the Haworth projection above.
      d. Place the up and down groups in axial or equatorial positions, as appropriate.

   f. If the configuration of the anomeric carbon is uncertain, or if there is a mixture of anomers, the bond is represented by a squiggly line.

   3. The five-membered rings of furanoses are nonplanar, but they are close enough to planarity that Haworth formulas are good approximations to their actual structures.
   4. In some cases it is simpler to derive a cyclic structure of one carbohydrate from its relationship to another carbohydrate.

## E. GLYCOSIDES

1. Most monosaccharides react with alcohols under acidic conditions to yield cyclic acetals called glycosides, which are named as derivatives of the parent carbohydrate.
   a. The term pyranoside indicates that the glycoside ring is a six-membered ring.
   b. The term furanoside indicates that the glycoside ring is a five-membered ring.

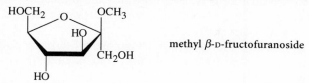

methyl *β*-D-fructofuranoside

2. Glycoside formation is catalyzed by acid.
3. Because glycosides are acetals, they are stable to base, but are hydrolyzed in dilute aqueous acid back to their parent carbohydrates.
4. Many compounds occur naturally as glycosides; the glycoside of a natural product can be hydrolyzed to its component alcohol or phenol and carbohydrate.

## F. STRUCTURES OF DISACCHARIDES AND POLYSACCHARIDES

1. Disaccharides consist of two monosaccharides connected by a glycosidic linkage.
   a. (+)-Lactose (milk sugar) is a disaccharide in which galactose is linked by a *β*-glycosidic bond to the oxygen at carbon-4 of glucose.
   b. (+)-Sucrose (table sugar), a disaccharide in which glucose is linked by an *α*-glycosidic bond to the oxygen at carbon-2 of fructose, is a nonreducing sugar.

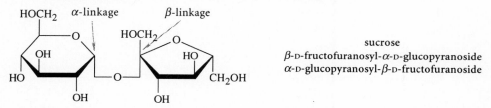

sucrose
*β*-D-fructofuranosyl-*α*-D-glucopyranoside
*α*-D-glucopyranosyl-*β*-D-fructofuranoside

2. Any number of monosaccharide residues can be linked together with glycosidic bonds to form chains; long chains of connected monosaccharide residues are called polysaccharides.
   a. Cellulose is a polymer of D-glucopyranose residues connected by *β*-1,4-glycosidic linkages.

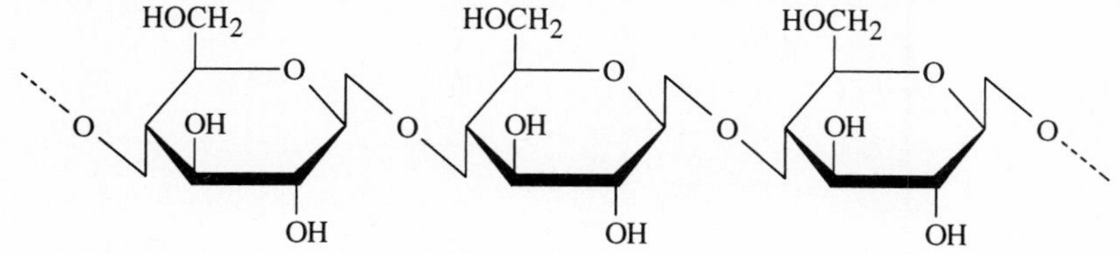

cellulose

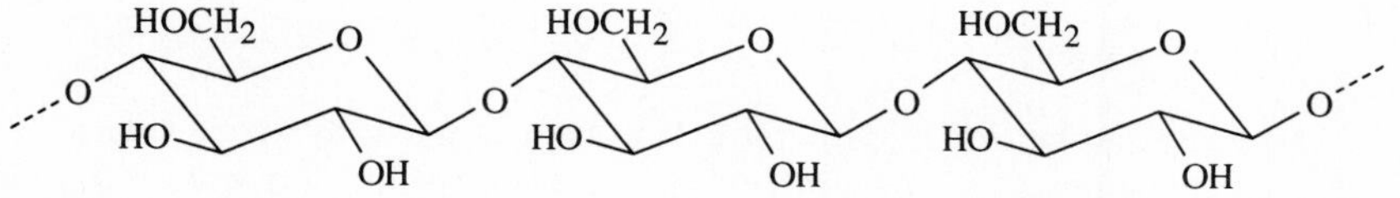

cellulose

   b. Starch is a polymer of D-glucopyranose. Starch consists of two components:
      i. Amylose has glucose residues connected by *α*-1,4-glycosidic linkages.
      ii. Amylopectin has glucose residues connected by *α*-1,4-glycosidic and *α*-1,6-glycosidic linkages.
   c. Chitin is a polysaccharide of *N*-acetyl-D-glucosamine connected by *β*-1,4-glycosidic linkages.

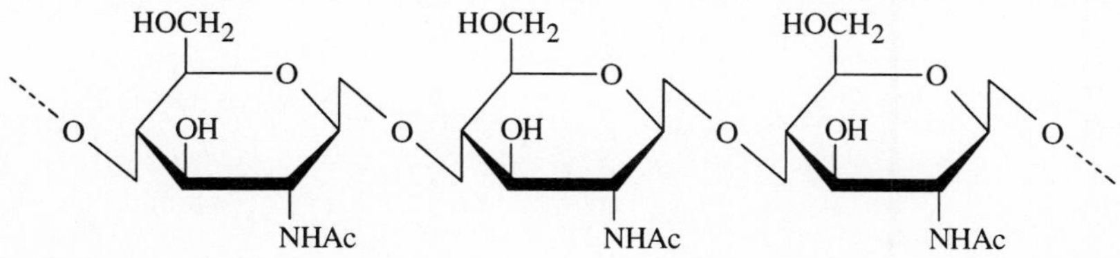

chitin

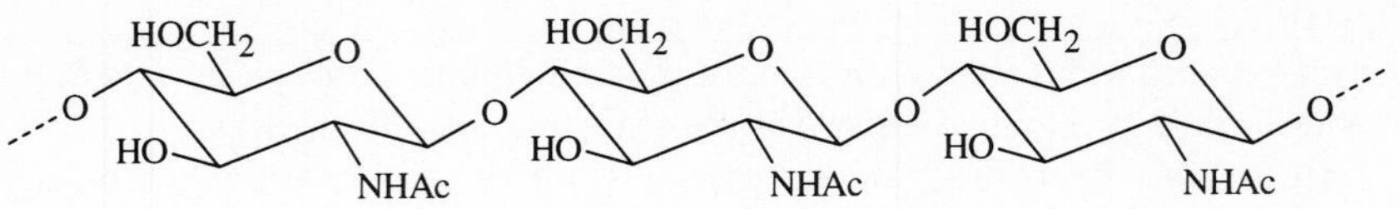

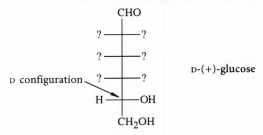

chitin

3. Polysaccharides are mostly long chains with some branches.
   a. There are no highly cross-linked, three-dimensional networks.
   b. Cyclic oligosaccharides are known.
4. The linkages between monosaccharide units are in every case glycosidic linkages; thus, all polysaccharides can be converted into their component monosaccharides by acid-catalyzed hydrolysis.
5. A given polysaccharide incorporates only one stereochemical type of glycoside linkage. Thus, the glycoside linkages in cellulose are all β; those in starch are all α.

## III. *Proof of Glucose Stereochemistry*

### A. THE FISCHER PROOF

1. Fischer arbitrarily assumed that carbon-5 (the configurational carbon in the D,L system) of (+)-glucose has the —OH on the right in the standard Fischer projection.

CHO
? ┼ ?
? ┼ ?
D configuration ↘ ? ┼ ?
H ┼ OH
CH₂OH

D-(+)-glucose

2. The subsequent logic involved can be summarized in four steps:
   a. (−)-Arabinose is converted into both (+)-glucose and (+)-mannose by a Kiliani-Fischer synthesis. Therefore:
      i. (+)-Glucose and (+)-mannose are epimeric at carbon-2.
      ii. (−)-Arabinose has the same configuration at carbons-2, 3, and 4 as that of (+)-glucose and (+)-mannose at carbons-3, 4, and 5, respectively.

D-(−)-arabinose

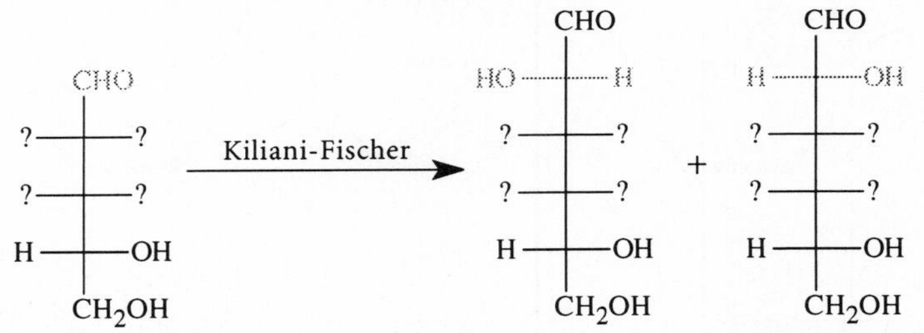

CHO
? ┼ ?
? ┼ ?
H ┼ OH
CH₂OH

→ Kiliani-Fischer →

CHO
HO ┼ H
? ┼ ?
? ┼ ?
H ┼ OH
CH₂OH

+

CHO
H ┼ OH
? ┼ ?
? ┼ ?
H ┼ OH
CH₂OH

D-(+)-glucose
and
D-(+)-mannose

   b. (−)-Arabinose can be oxidized by dilute $HNO_3$ to an optically active aldaric acid. Therefore:
      i. The —OH group at carbon-2 of arabinose must be on the left; otherwise, arabinose would be *meso*, regardless of the configuration of the —OH group at carbon-3.
      ii. The —OH group at carbon-3 of (+)-glucose must be on the left.

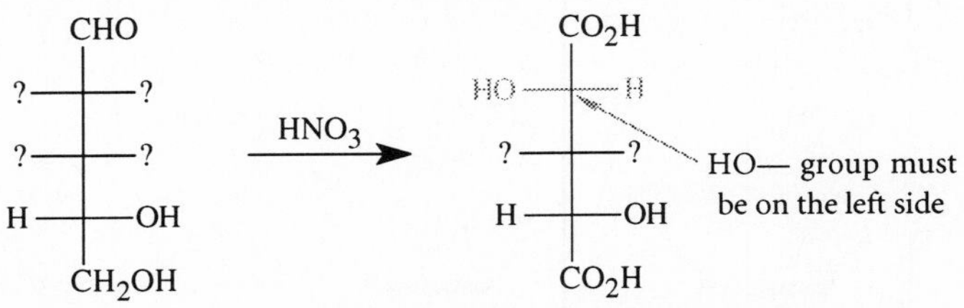

CHO
? ┼ ?
? ┼ ?
H ┼ OH
CH₂OH

D-(−)-arabinose

→ $HNO_3$ →

CO₂H
HO ┼ H
? ┼ ?
H ┼ OH
CO₂H

HO— group must be on the left side

optically active aldaric acid

c. Oxidation of both (+)-glucose and (+)-mannose with $HNO_3$ give optically active aldaric acids. Therefore:

   *i.* The —OH group at carbon-4 is on the right in both (+)-glucose and (+)-mannose.

   *ii.* The configuration at carbon-4 of (+)-glucose and (+)-mannose is the same as that at carbon-3 of (−)-arabinose.

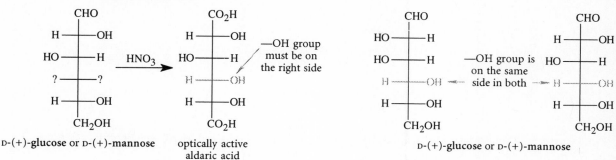

D-(+)-**glucose** or D-(+)-**mannose**     optically active                      D-(+)-**glucose** or D-(+)-**mannose**
                                 aldaric acid

   *iii.* The structure of D-(−)-arabinose is thus determined.

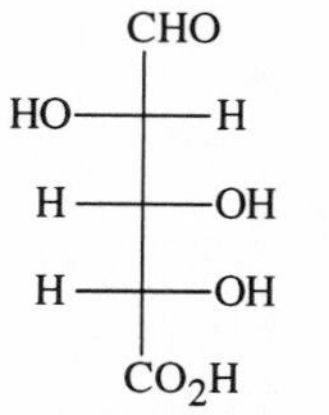

$$\begin{array}{c} \text{CHO} \\ \text{HO} \longrightarrow \text{H} \\ \text{H} \longrightarrow \text{OH} \\ \text{H} \longrightarrow \text{OH} \\ \text{CO}_2\text{H} \end{array} \qquad \text{D-(−)-arabinose}$$

d. (+)-Gulose can be oxidized with $HNO_3$ to the same aldaric acid as (+)-glucose.

   *i.* The structure with the —OH group at carbon-2 on the right forms an aldaric acid upon oxidation with $HNO_3$ that is identical to the aldaric acid formed by the oxidation of (+)-gulose with $HNO_3$; this must be (+)-glucose.

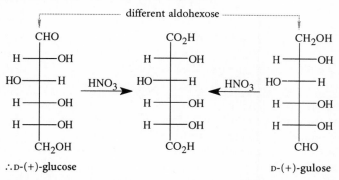

∴ D-(+)-**glucose**                                       D-(+)-**gulose**

   *ii.* The structure with the —OH group at carbon-2 on the left forms a unique aldaric acid upon oxidation with $HNO_3$; this must be (+)-mannose.

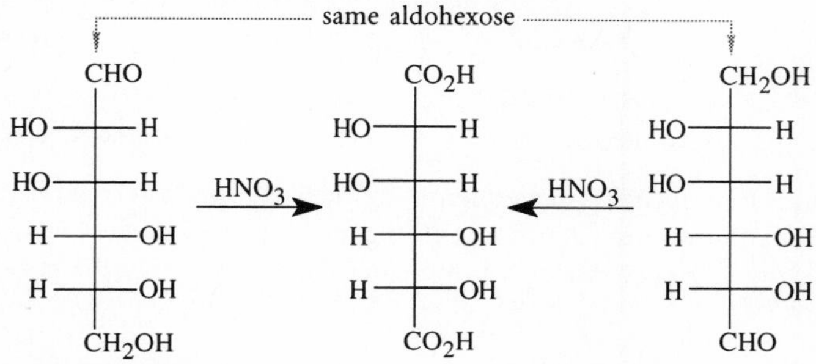

same aldohexose

$$\begin{array}{ccc} \text{CHO} & \text{CO}_2\text{H} & \text{CH}_2\text{OH} \\ \text{HO}\!-\!\text{H} & \text{HO}\!-\!\text{H} & \text{HO}\!-\!\text{H} \\ \text{HO}\!-\!\text{H} & \text{HO}\!-\!\text{H} & \text{HO}\!-\!\text{H} \\ \text{H}\!-\!\text{OH} & \text{H}\!-\!\text{OH} & \text{H}\!-\!\text{OH} \\ \text{H}\!-\!\text{OH} & \text{H}\!-\!\text{OH} & \text{H}\!-\!\text{OH} \\ \text{CH}_2\text{OH} & \text{CO}_2\text{H} & \text{CHO} \end{array}$$

∴ D-(+)-**mannose**

B.  ABSOLUTE CONFIGURATION OF GLUCOSE
1.  Two cycles of the Ruff degradation convert (+)-glucose into (–)-erythrose.
    a.  D-Glyceraldehyde, in turn, is related to (–)-erythrose by a Kiliani-Fischer synthesis.

CHO
H——OH
HO——H
H——OH   **2 Ruff** →   CHO
H——OH        H——OH
CH₂OH         H——OH
                          CO₂H

D-(+)-glucose            D-(–)-erythrose

   b.  (+)-Glucose, (–)-erythrose,(–)-threose, and (+)-glyceraldehyde are all of the same
       stereochemical series—the D series.

CHO                    CHO          CHO
H——OH   **Kiliani-**   H——OH   HO——H
        **Fischer** →   H——OH  +  H——OH
CO₂H                   CO₂H         CO₂H

D-(+)-glyceraldehyde   D-(–)-erythrose   D-(–)-threose

2.  Oxidation of D-(–)-threose with dilute $HNO_3$ gives D-(–)-tartaric acid; (+)-tartaric acid was shown
    by X-ray crystallography to possess the L configuration; hence, (+)-glucose has the D configuration.

CHO                        CO₂H
HO——H   **$HNO_3$** →   HO——H
H——OH                      H——OH
CO₂H                       CO₂H

D-(–)-threose              D-(–)-tartaric acid

## IV.  *Nucleosides, Nucleotides, and Nucleic Acids*

A.  NOMENCLATURE OF NUCLEOSIDES, NUCLEOTIDES, AND NUCLEIC ACIDS
1.  A β-glycoside of a heterocyclic nitrogen base is called a nucleoside; the base and the sugar ring
    systems are numbered separately; primes (′) are used to refer to the sugar carbon atoms.
    a.  A ribonucleoside is derived from D-ribose.
    b.  A deoxyribonucleoside is derived from D-2-deoxyribose; 2-deoxyribose lacks the —OH group at
        carbon-2 of ribose.

a ribonucleoside                 a deoxyribonucleoside

2.  The bases that occur most frequently in nucleosides are derived from two heterocyclic ring systems:
    a.  Three pyrimidines occur most commonly and are attached to the sugar at the *N*-1 position.
    b.  Two purines occur most commonly and are attached to the sugar at the *N*-9 position.

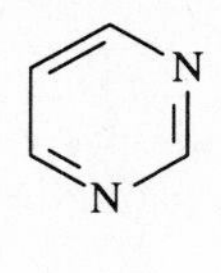

pyrimidine 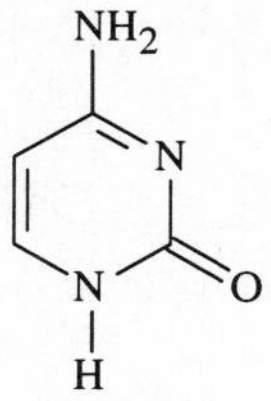       cytosine (C) 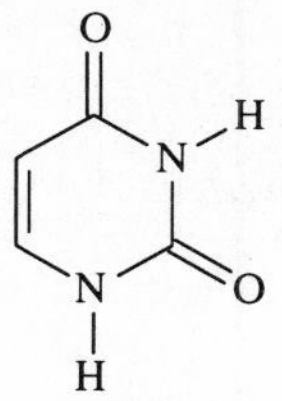       uracil (U) 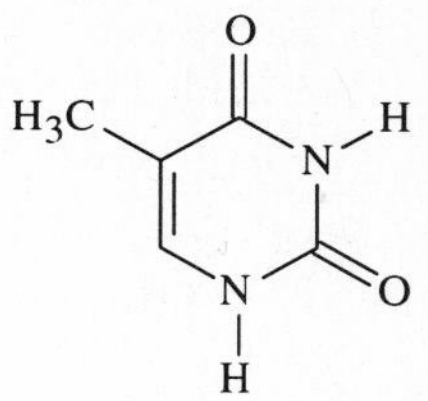       thymine (T)

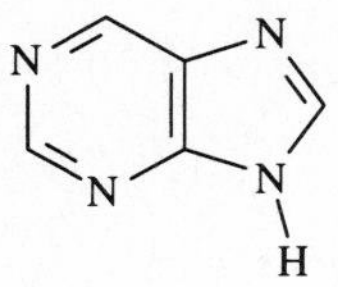

purine 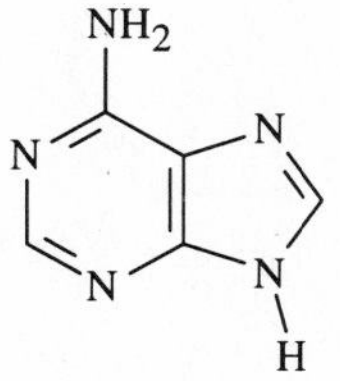       adenine (A) 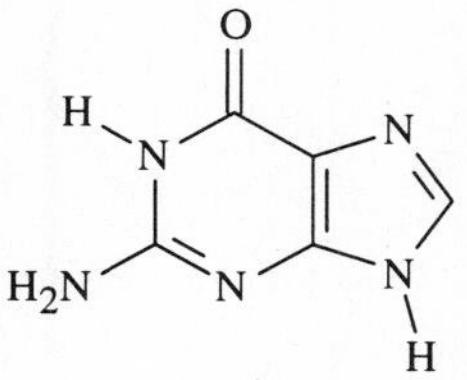       guanine (G)

3. The 5′—OH group of the ribose in a nucleoside is often found esterified to a phosphate group; a 5′-phosphorylated nucleoside is called a nucleotide.
   a. A ribonucleotide is derived from D-ribose.
   b. A deoxyribonucleotide is derived from 2′-deoxyribose.

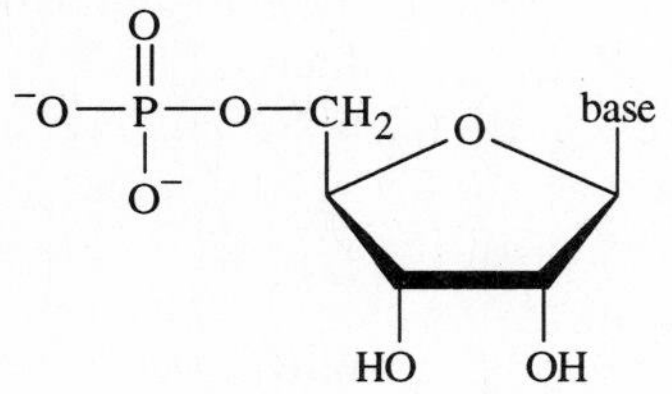

a ribonucleotide 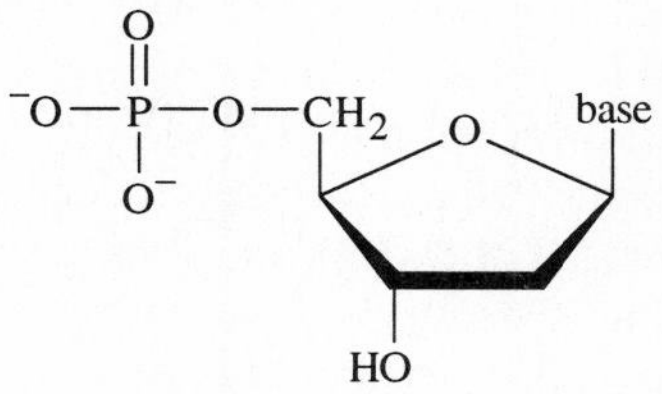       a deoxyribonucleotide

4. Some nucleotides contain a single phosphate group; others contain two or three phosphate groups condensed in phosphoric anhydride linkages.
5. The nomenclature and abbreviations of the five common bases and their corresponding nucleosides and nucleotides are summarized in Table 27.3, text page 1369; the corresponding 2′-deoxy derivatives are named by appending the prefix *2′-deoxy* (or *deoxy*) to the names of the corresponding ribose derivatives, or by appending a *d* prefix to the abbreviation.

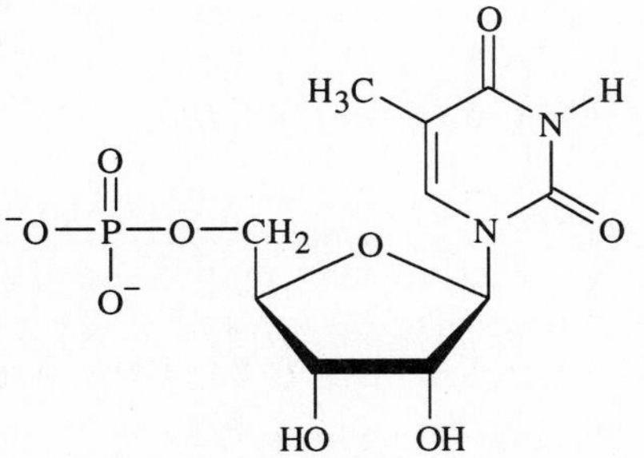

thymidylic acid
or thymidine monophosphate (TMP)       deoxyguanosine diphosphate (*d*GDP)

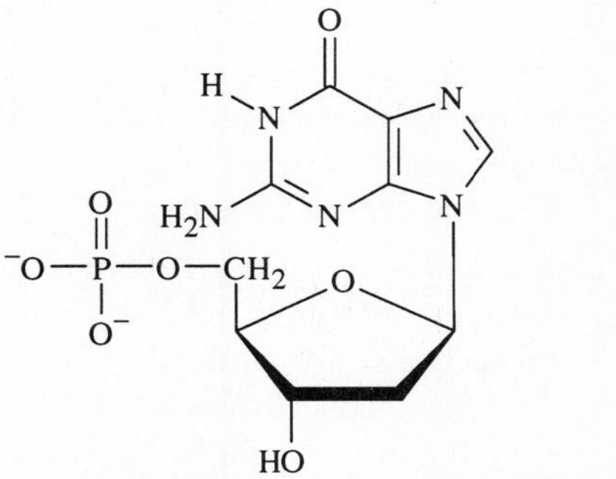

6. One of the most ubiquitous nucleotides is ATP (adenosine triphosphate), which serves as the fundamental energy source for the living cell.
7. Nucleic acids are of two general types:
   a. Deoxyribonucleic acid (DNA) is a principal component of the cell nucleus and is the storehouse of genetic information in the cell.

b. Ribonucleic acid (RNA) serves various roles in translating and processing the information encoded in the structure of DNA.

## B. STRUCTURE OF DNA AND RNA

1. Deoxyribonucleic acid (DNA) is a polymer of deoxyribonucleotides.
   a. The individual nucleotide residues are connected by a phosphate group that is esterified both to the 3′—OH group of one ribose and the 5′—OH of another.

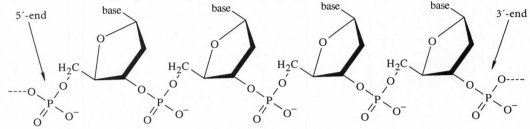

a deoxyribonucleic acid

   b. DNA incorporates adenine, thymine, guanine, and cytosine as the bases; each residue in a polynucleotide is distinguished by the identity of its base.
   c. The ratios of adenine to thymine, and guanosine to cytosine, in DNA are both 1.0; these observations are called Chargaff's rules.
   d. A typical strand of DNA, which carries genetic information, might be thousands of nucleotides long and is replicated, or copied, during cellular reproduction.
2. Ribonucleic acid (RNA) polymers are formally much like DNA polymers, except that ribose is the sugar; RNA incorporates essentially the same bases as DNA, except that uracil occurs in RNA instead of thymine, and some rare bases are found in certain types of RNA.
3. The Watson-Crick structure of DNA has the following important features:
   a. The structure contains two right-handed helical polynucleotide chains that run in opposite directions, coiled around a common axis; the structure is therefore that of a double helix.
   b. The sugars and phosphates, which are rich in —OH groups and charges, are on the outside of the helix.
   c. The chains are held together by hydrogen bonds between bases, which are on the inside of the double helix.
      i. Adenine (A) in one chain always hydrogen-bonds to thymine (T) in the other.
      ii. Guanine (G) in one chain always hydrogen-bonds to cytosine (C) in the other.
   d. The planes of the complementary base pairs are stacked, one on top of the other, and are perpendicular to the axis of the helix.
   e. There is no restriction on the sequence of bases in a polynucleotide; however, the sequence of one polynucleotide strand in the double helix is complementary to that of the other strand.
4. The proper sequence of each new DNA strand during cellular reproduction is assured by hydrogen-bonding complementary.

## C. DNA, RNA, AND THE GENETIC CODE

1. A strand of DNA is the code for the synthesis of a complementary strand of RNA; this RNA is called messenger RNA (mRNA), and the process by which it is assembled is called transcription.
   a. The sequence of the mRNA transcript is complementary to one DNA strand of the gene and runs in the opposite direction to that of its parent DNA.
   b. The mRNA sequence is used by the cell as the code for the synthesis of a specific protein from its component amino acids; this process is called translation.
      i. Each successive three-residue sequence of mRNA is translated as a specific amino acid in the sequence of a protein according to the genetic code given in Table 27.4, text page 1377; some amino acids have multiple codes.
      ii. The precise sequence of bases in DNA (by way of its complementary mRNA transcription product) codes for the successive amino acids of a protein.

c. There is a specific start signal (either of the nucleotide sequences AUG or GUG) at the appropriate point in the mRNA; specific stop signals (UAA, UGA, or UAG) cause protein synthesis to be terminated.

d. It is possible for the change of only one base in the DNA of an organism to cause the change of an amino acid in the corresponding protein.

2. There are many different types of RNA besides messenger RNA, each with a specific function in the cell.

### D. DNA MODIFICATION AND CHEMICAL CARCINOGENESIS

1. There is strong circumstantial evidence that chemical damage to DNA can interfere with its hydrogen-bonding complementary and can trigger the state of uncontrolled cell division known as cancer.

a. One type of chemical damage is caused by alkylating agents.

b. Another type of chemical damage is caused by ultraviolet light. In adjacent positions on a strand of DNA, ultraviolet light promotes the [2 + 2] cycloaddition of the two pyrimidines. People who lack the enzymes to restore the original DNA structure die at an early age.

# Reactions

## I. Reactions of Carbohydrates

### A. BASE-CATALYZED ISOMERIZATION OF CARBOHYDRATES

1. In base, aldoses and ketoses rapidly equilibrate to mixtures of other aldose and ketoses; this transformation is an example of the Lobry de Bruyn-Alberda van Ekenstein reaction.

a. An aldose can ionize to give a small amount of its enolate ion in base.

b. Protonation of this enolate ion at one face gives back the aldose; protonation at the other face gives an epimer.

$$\begin{array}{ccc}
\text{D-(+)-glucose} & \text{enolate ion} & \text{D-(+)-mannose}
\end{array}$$

c. The enolate ion can also protonate on the oxygen to give a new enol, called an enediol, which contains a hydroxy group at both ends of the double bond.

enolate ion of
D-(+)-glucose or D-(+)-mannose          an enediol          D-(+)-fructose

2. Several transformations of this type are important in metabolism.

### B.   ETHER AND ESTER DERIVATIVES OF CARBOHYDRATES

1. In the presence of concentrated base, carbohydrates are converted into ethers by reactive alkylating agents (Williamson ether synthesis).

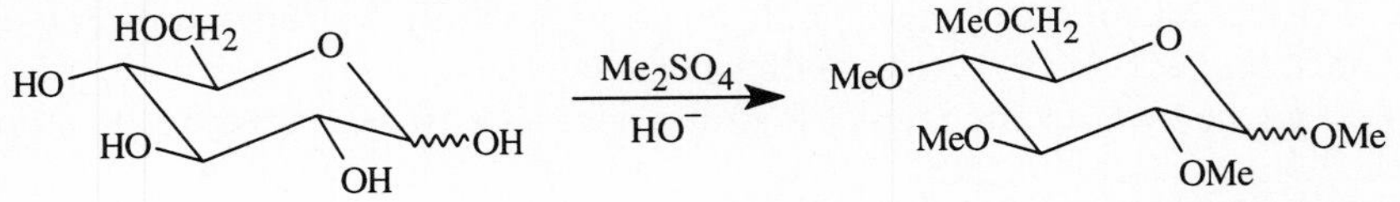

methyl 2,3,4,6-tetra-*O*-methyl-D-glucopyranoside

2. The alkoxy group at the anomeric carbon is different from other alkoxy groups in the alkylated carbohydrate.
   a. The alkoxy group at the anomeric carbon is part of a glycosidic linkage.
   b. Since it is an acetal, it can be hydrolyzed in aqueous acid under mild conditions; the other alkoxy groups are ordinary ethers and do not hydrolyze under these conditions.

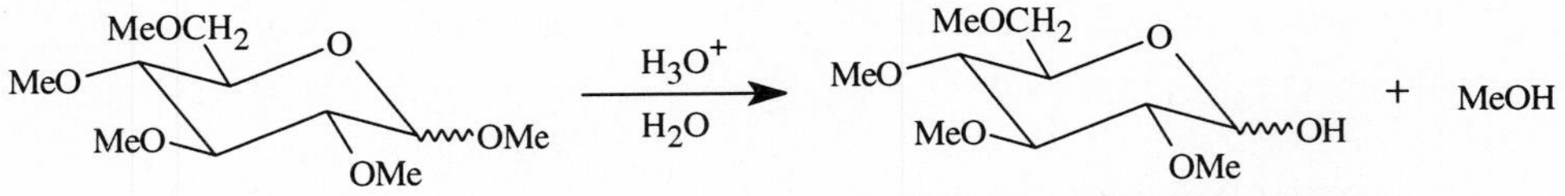

2,3,4,6-tetra-*O*-methyl-D-glucopyranoside

3. The hydroxy groups of carbohydrates can be esterified; the resulting esters can be saponified in base or can be removed by transesterification with an alkoxide.

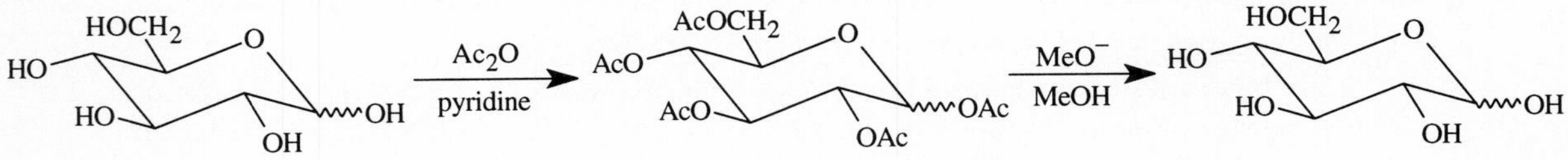

4. Ethers and esters are used as protecting groups in reactions involving carbohydrates; furthermore, they have broader solubility characteristics and greater volatility than the carbohydrates themselves.

### C.   OXIDATION AND REDUCTION REACTIONS OF CARBOHYDRATES

1. Treatment of an aldose with bromine water oxidizes aldoses to aldonic acids; this reaction is a useful test for aldoses. (Carbohydrates that can be oxidized by bromine water are called reducing sugars.)
   a. Aldonic acids exist in acidic solution as five-membered lactones called aldonolactones.

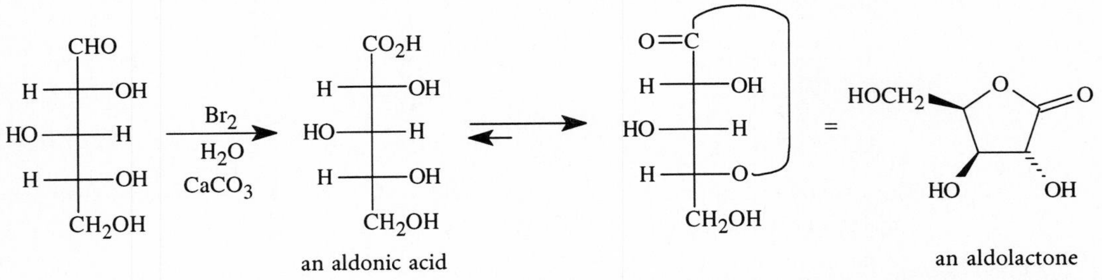

an aldonic acid                an aldolactone

   b. Glycosides are not oxidized by bromine water because the aldehyde carbonyl group is protected as an acetal. (Carbohydrates that cannot be oxidized by bromine water are called nonreducing sugars.)
2. Aldoses can also be oxidized with other reagents; the alkaline conditions of Tollens' test, however, also promote the equilibration of ketoses and aldoses; thus, ketoses also give a positive Tollens' test.
3. Both ends of an aldose are oxidized to carboxylic acid groups by dilute $HNO_3$, but the secondary alcohol groups are not affected; the oxidation product is an aldaric acid.
   a. Aldaric acids in acidic solution form lactones.

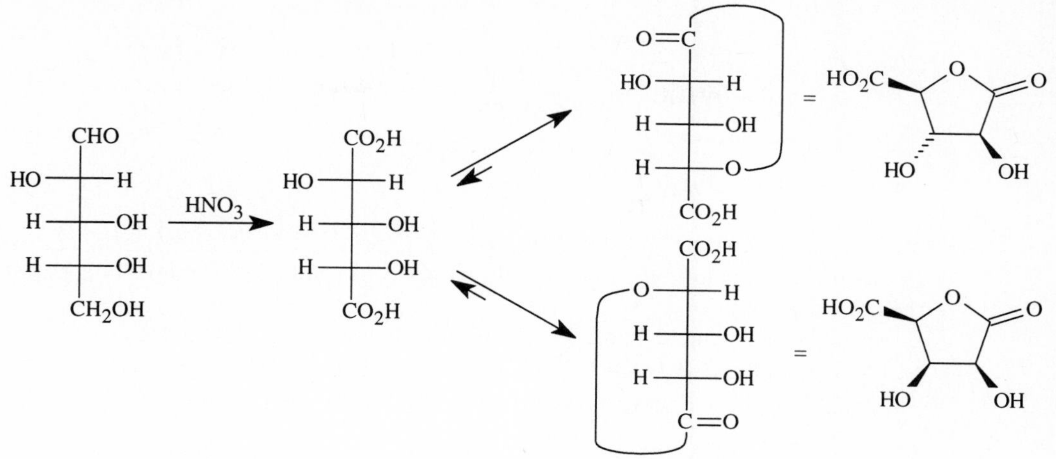

b. Two different five-membered lactones are possible from a hexose or pentose, depending on which carboxylic acid group undergoes lactonization.

c. Under certain conditions, some aldaric acids can be isolated as dilactones, in which both carboxylic acid groups are lactonized. (An example can be found on text p. 1351.)

4. Many carbohydrates contain vicinal glycol units and are oxidized by periodic acid.

a. $\alpha$-Hydroxy aldehydes are oxidized to formic acid and another aldehyde with one fewer carbon.

b. $\alpha$-Hydroxymethyl ketones are oxidized to formaldehyde and a carboxylic acid.

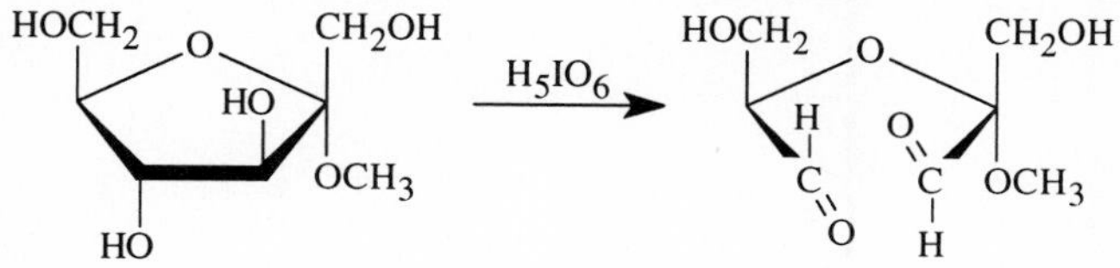

c. Because it is possible to determine accurately both the amount of periodic acid consumed and the amount of formic acid produced, periodic acid oxidation can be used to differentiate between pyranose and furanose structures for saccharide derivatives.

5. Aldoses and ketoses undergo many of the usual carbonyl reductions; an aldose is reduced to a primary alcohol known as an alditol.

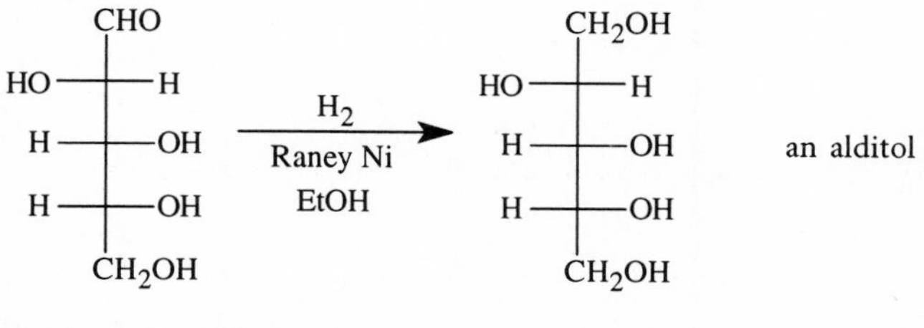

## II. Synthesis of Sugars from other Sugars

### A. KILIANI-FISCHER SYNTHESIS; INCREASING THE LENGTH OF THE ALDOSE CHAIN

1. Addition of hydrogen cyanide to aldoses gives cyanohydrins; because the cyanohydrin product of such a reaction has an additional asymmetric carbon, it is formed as a mixture of two epimers.

a. These epimers are diastereomers and are typically formed in different amounts.

b. The mixture of cyanohydrins can be converted by catalytic hydrogenation into a mixture of aldoses, which can be separated.

c. The hydrogenation reaction involves reduction of the nitrile to the imine (or a cyclic carbinolamine derivative), which, under the aqueous reaction conditions, hydrolyzes to the aldose and ammonium ion.

2. The cyanohydrin formation-reduction sequence converts an aldose into two epimeric aldoses with one additional carbon; this process is known as the Kiliani-Fischer synthesis.

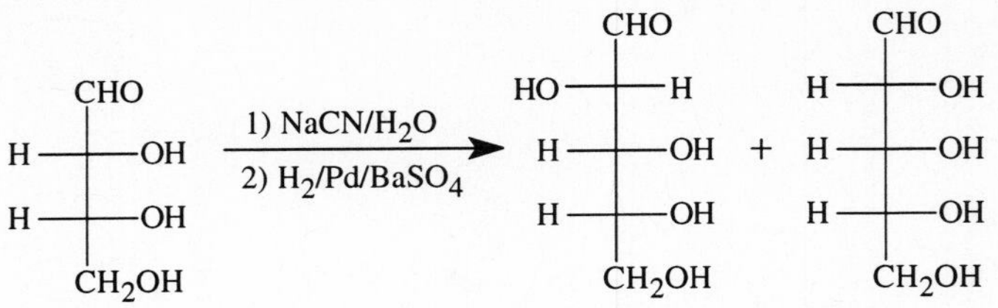

$$
\begin{array}{c}
\text{CHO} \\
\text{H}\!-\!\!\!-\!\text{OH} \\
\text{H}\!-\!\!\!-\!\text{OH} \\
\text{CH}_2\text{OH}
\end{array}
\quad
\xrightarrow[\text{2) H}_2\text{/Pd/BaSO}_4]{\text{1) NaCN/H}_2\text{O}}
\quad
\begin{array}{c}
\text{CHO} \\
\text{HO}\!-\!\!\!-\!\text{H} \\
\text{H}\!-\!\!\!-\!\text{OH} \\
\text{H}\!-\!\!\!-\!\text{OH} \\
\text{CH}_2\text{OH}
\end{array}
\;+\;
\begin{array}{c}
\text{CHO} \\
\text{H}\!-\!\!\!-\!\text{OH} \\
\text{H}\!-\!\!\!-\!\text{OH} \\
\text{H}\!-\!\!\!-\!\text{OH} \\
\text{CH}_2\text{OH}
\end{array}
$$

## B.  RUFF DEGRADATION; DECREASING THE LENGTH OF THE ALDOSE CHAIN

1. In the Ruff degradation, the calcium salt of an aldonic acid is oxidized with hydrogen peroxide in the presence of $Fe^{3+}$.
   a. An aldose is degraded to another aldose with one fewer carbon atom.
   b. The stereochemistry of the remaining groups remains the same.

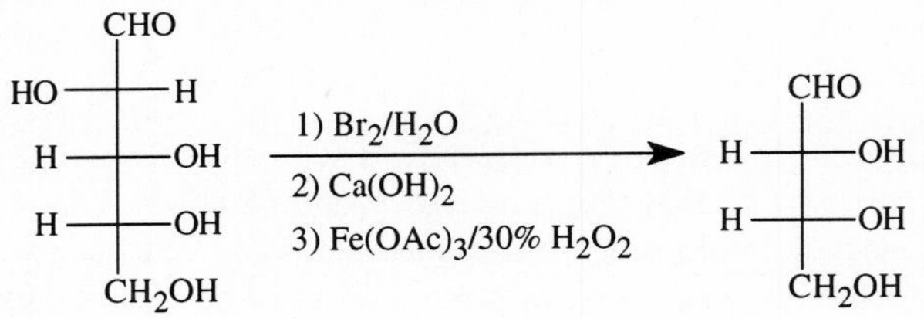

$$
\begin{array}{c}
\text{CHO} \\
\text{HO}\!-\!\!\!-\!\text{H} \\
\text{H}\!-\!\!\!-\!\text{OH} \\
\text{H}\!-\!\!\!-\!\text{OH} \\
\text{CH}_2\text{OH}
\end{array}
\quad
\xrightarrow[\substack{\text{2) Ca(OH)}_2 \\ \text{3) Fe(OAc)}_3\text{/30\% H}_2\text{O}_2}]{\text{1) Br}_2\text{/H}_2\text{O}}
\quad
\begin{array}{c}
\text{CHO} \\
\text{H}\!-\!\!\!-\!\text{OH} \\
\text{H}\!-\!\!\!-\!\text{OH} \\
\text{CH}_2\text{OH}
\end{array}
$$

# Study Guide Links

## 27.1   Nomenclature of Anomers

Long before the actual structures of the anomeric forms of many carbohydrates were known, the $\alpha$-anomer of a D-carbohydrate was simply defined as "the more dextrorotatory of the two anomers." This convention, proposed by C. S. Hudson (1881–1952), an American carbohydrate chemist, is cumbersome because, in order to apply it, one has to know the optical rotations of the two anomers. When the structures of several anomers became known, it was found that in most cases the more dextrorotatory anomer of each D-carbohydrate has the configuration shown in the text for the $\alpha$-anomer. Hudson's definition in terms of a physical property—optical rotation—was subsequently replaced by the structural definition used in the text.

## ✓27.2   Acid Catalysis of Carbohydrate Reactions

Acid catalysis of glycoside formation involves protonation of the hydroxy group at the anomeric carbon; acid catalysis of mutarotation (Eqs. 27.8 on text p. 1338 and Problem 27.6 on text p. 1340) involves protonation of the ring oxygen. You may be asking why these specific oxygens are protonated—why not others? The answer is that *any one* of the oxygens can indeed be protonated to small extents in acidic solution. However, *only* protonation at the ring oxygen leads to mutarotation, and *only* protonation of the —OH group at the anomeric carbon leads to glycoside formation. In other words, the various protonated forms are in rapid equilibrium and are present whether they are shown explicitly or not. Only certain of the protonated forms can react further, and these are the ones shown in the equations.

## ✓27.3   Configurations of Aldaric Acids

Consideration of the stereochemistry of aldaric acids reveals a subtle yet important aspect of the D,L configurational system. Certain aldaric acids can be derived from either a D or an L carbohydrate.

D-glucaric acid (from D-glucose) or L-gularic acid (from L-gulose)
(the same compound)

In this case, whether the aldaric acid is classified as D or L depends on which carbon is specified as carbon-1. Because the two ends of the molecule are *constitutionally equivalent* (see text p. 476), the choice is completely arbitrary! The names D-glucaric acid and L-gularic acid are both correct names for this aldaric acid. The two names reflect the fact that it can be formed by nitric acid oxidation of either D-glucose or L-gulose.

This situation arises because the —OH groups on the endmost asymmetric carbons are on the same side of the Fischer projection. However, when these —OH groups are on opposite sides, the configuration is unambiguous. For example, D-tartaric acid also has constitutionally equivalent ends, but it has the D configuration no matter how it is turned.

D-tartaric acid

As you might expect, a similar situation arises in other derivatives with constitutionally equivalent ends, such as alditols.

This issue is explored in Problem 27.18 of the text.

## 27.4    More on the Fischer Proof

The Fischer proof of glucose stereochemistry has a number of interesting facets that are not discussed in the text. One of the most important experimental aspects of carbohydrate chemistry is that many carbohydrates are notoriously difficult to crystallize; they are frequently isolated as "syrups" and "gums." Because chemists in Fischer's day did not have available the sophisticated purification methods available today, they had to rely almost exclusively on crystallization as a purification method. Thus, because carbohydrates were difficult to crystallize, they were very difficult to purify. Few chemists were interested in working on problems fraught with so many experimental difficulties; consequently, little progress had been made in the field of carbohydrate chemistry when Fischer began his work. In fact, Fischer himself, in some of his correspondence, deplored the dreadfully slow pace of carbohydrate research in his own laboratory brought about by the experimental difficulties of handling carbohydrates.

Two solutions to the problem of carbohydrate crystallization emerged from Fischer's long-standing interest in phenylhydrazine, $PhNHNH_2$. He had already shown that this compound could be used to prepare indoles (Fischer indole synthesis; Sec. 24.4A on text p. 1190). When he allowed phenylhydrazine to react with aldohexoses, he found that very interesting compounds called *osazones* were obtained.

This reaction is very reminiscent of phenylhydrazone formation (Table 19.3 on text p. 906). However, an unusual aspect of this reaction is that one equivalent of phenylhydrazine serves as an oxidizing agent, with the net result that a second substituted imine group is introduced. (For the mechanism of this reaction, see Loudon, *Organic Chemistry*, first edition, Addison-Wesley Publishing Co., 1984, pp. 1417–1418.) Osazones proved to be much more easily crystallized, and thus more easily characterized, than aldoses themselves. Furthermore, they served to confirm some of Fischer's stereochemical deductions. Notice that *neither carbon-1 nor carbon-2* are asymmetric in the osazone. Consequently, isomeric aldoses that are epimeric only at carbon-2 give the same osazone. Fischer found that (+)-glucose and (+)-mannose give *different* phenylhydrazones (formed under milder conditions with phenylhydrazine), but give the same osazone. This fact, along with their simultaneous preparation in the Kiliani-Fischer synthesis from (−)-arabinose, showed that they are epimers at carbon-2.

Phenylhydrazine figured prominently in Fischer's research in a second way when Fischer found that the aminolysis reactions of aldonic acids (undoubtedly reacting as their lactones) with phenylhydrazine give phenylhydrazides, which, unlike the parent acids, are beautifully crystalline compounds.

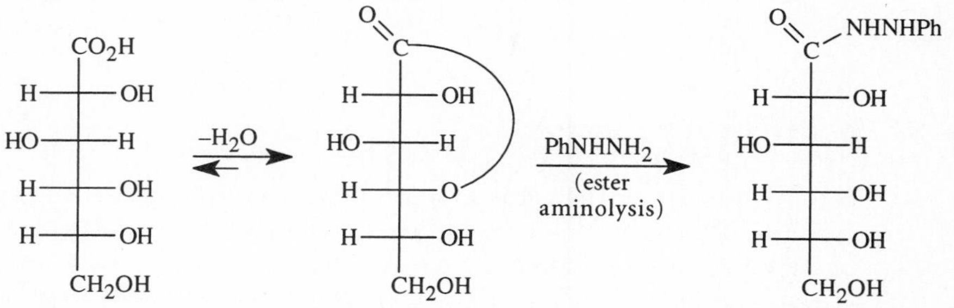

D-gluconic acid     D-γ-gluconolactone     D-gluconic acid phenylhydrazide

One other interesting aspect of the Fischer proof has to do with the last step, which was made possible by a remarkable bit of serendipity. It is noted in the text that Fischer had prepared L-(+)-gulose in the course of his research; recall that it was the oxidation of D-glucose and L-gulose to the same aldaric acid that completed the Fischer proof. Equation 27.27 on text p. 1351 shows that D-glucaric acid can exist as a mixture of two different lactones. Fortuitously, Fischer's procedure for isolating the lactones gave mainly the 1,4-lactone. His synthesis of L-gulose began with this lactone.

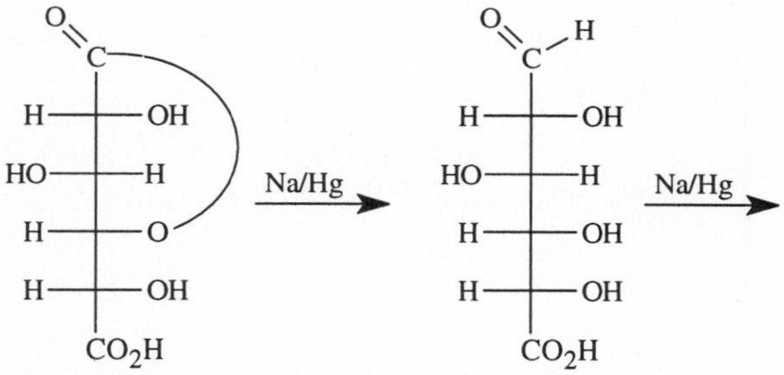

D-glucaric acid-1,4-lactone     D-glucuronic acid

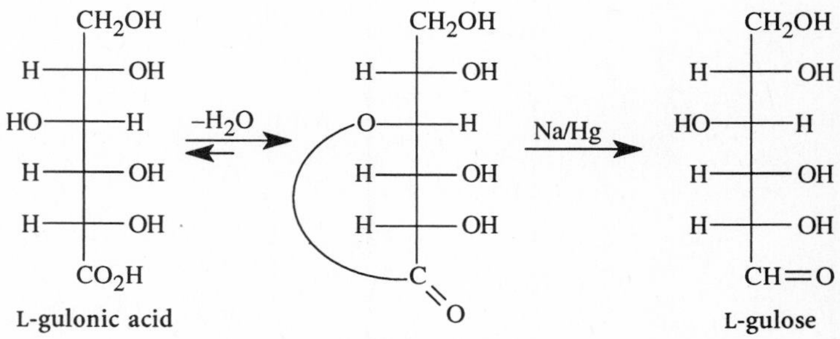

L-gulonic acid     L-gulonolactone     L-gulose

"Na/Hg" is sodium amalgam, a solution of sodium in mercury, which was widely used as a reducing agent until lithium aluminum hydride and sodium borohydride were developed. The sodium-amalgam reductions shown above can be carried out in stages by controlling the conditions of each step. Recall that D-glucose is readily oxidized to D-glucaric acid-1,4-lactone with dilute nitric acid. Consequently, beginning with D-glucose, the overall sequence of glucose oxidation followed by the sodium amalgam reductions effects the *net interchange* of the CH=O and CH$_2$OH groups, and thus gives L-gulose from D-glucose. The serendipitous element was this: had Fischer isolated the 3,6-lactone of D-glucaric acid, it would have given back D-glucose when taken through this sequence of reactions, and the last step of the proof would not have been possible.

An intriguing account of Fischer's life and work, including details of the Fischer proof, was written by F. W. Lichtenthaler and published in *Angewandte Chemie International Edition in English,* vol. 31, No. 12, pp. 1541–1596 (1992). This journal is available in most chemistry libraries.

## ✓27.5    DNA Transcription

The process of DNA transcription presented in the text is simplified. The DNA transcript in higher organisms contains long sections of "intervening RNA," or *introns,* that are not part of the final mRNA product; these are excised out of the transcript, and the resulting shortened strand of RNA undergoes certain chemical modifications to produce mRNA. (You can find a description of this process in any modern biology or biochemistry text.) However, the presence of introns does not alter the fundamental idea, namely, that information flows from DNA to RNA to protein.

## Solutions

## Solutions to In-Text Problems

**27.1** (a) Transform the given Fischer projection into a standard form in which all of the backbone carbons are in a vertical line. This will show that the —OH group on carbon-5 is on the right. Thus, this aldose has the D-configuration.

CH₃——H  ⇒  H——OH
     OH              CH₃

carbon-5        OH on the right;
                ∴ D-configuration

(b) An analysis of the carbon-3 configuration of D-glucose shows that carbon-3 has the *S* configuration. Consequently, the L-enantiomer of glucose has the *R* configuration at carbon-3.

Remember that the configuration of an asymmetric carbon in a Fischer projection can be analyzed directly. To review, see Study Guide Link 6.7 on p. 120, Vol. 1, of the Study Guide & Solutions Manual.

**27.3** Put all the sugars in the standard Fischer projection to see the relative location of corresponding groups:

|       (a)       |       (b)       |       (c)       |
|-----------------|-----------------|-----------------|
| CH=O            | CH=O            | CH=O            |
| H——OH           | H——OH           | HO——H           |
| HO——H           | HO——H           | H——OH           |
| H——OH           | H——OH           | HO——H           |
| HO——H           | H——OH           | HO——H           |
| CH₂OH           | CH₂OH           | CH₂OH           |

Compounds (b) and (c) are enantiomers. Compound (a) is an epimer of (b) and a diastereomer of both (b) and (c). (Remember that an epimer is a particular type of diastereomer.)

**27.4** (a) A Fischer projection for D-fructose is given on text page 1331.

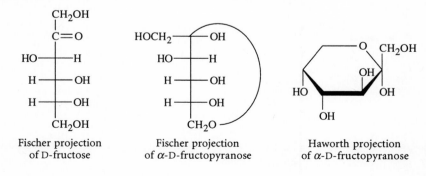

Fischer projection      Fischer projection          Haworth projection
of D-fructose          of α-D-fructopyranose       of α-D-fructopyranose

*(solution continues)*

chair conformations
of α-D-fructopyranose

(c) Start with the structure of D-xylose in Figure 27.1 and perform the usual manipulations. An easier solution is to recognize that D-xylose is epimeric to D-ribose at carbon-3. Thus, the structures for β-D-ribofuranose at the top of text p. 1336 can be used if the configuration of carbon-3 is inverted.

Fischer projection
of β-D-xylofuranose

Haworth projection
of β-D-xylofuranose

(e) Draw the enantiomers of the corresponding structures for α-D-glucopyranose, which are shown on text pp. 1334–1335.

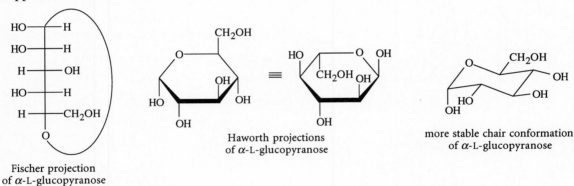

Fischer projection
of α-L-glucopyranose

Haworth projections
of α-L-glucopyranose

more stable chair conformation
of α-L-glucopyranose

**27.5**   (a) Perform the reverse of the analysis in Study Problem 27.2 on text p. 1334–1335 to reveal that this is α-D-ribofuranose.

Haworth projection
of furanose form

Fischer projection
of furanose form

Fischer projection
of aldehyde form

**D-ribose**

**27.6**   The mechanism below begins with the α-anomer protonated on the ring oxygen and ends with the β-anomer, also protonated on the ring oxygen.

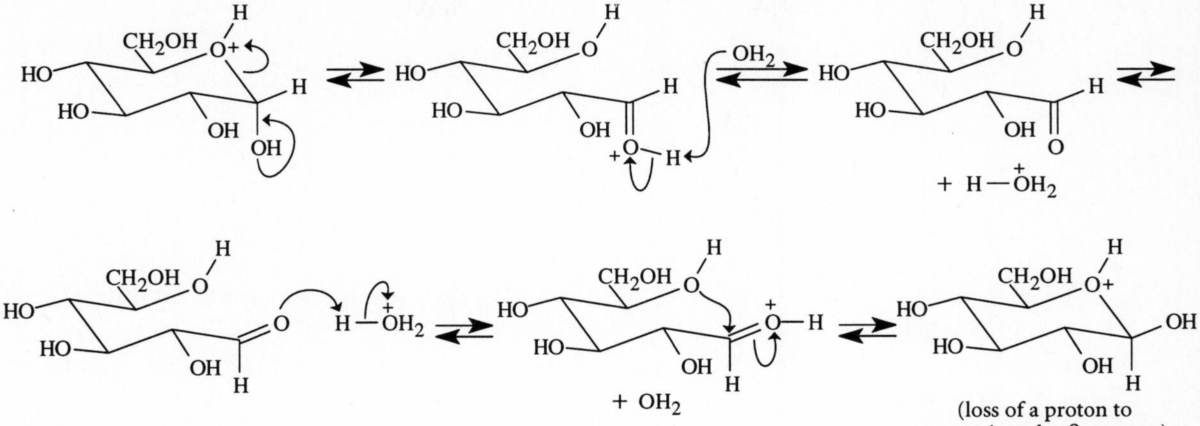

**27.8** The β-pyranose forms of D-glucose and D-talose are as follows:

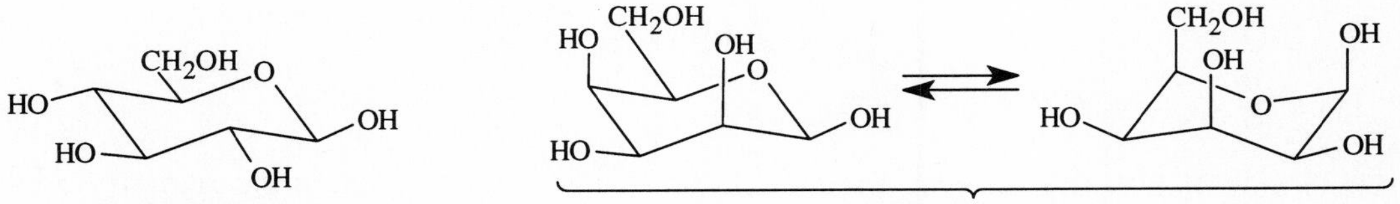

β-D-glucopyranose

β-D-talopyranose

Both chair forms of the β-D-talopyranose molecule have a significantly greater number of 1,3-diaxial interactions than β-D-glucopyranose. These unfavorable interactions in β-D-talopyranose raise its energy and thus lower the relative amounts of these forms at equilibrium. Hence, talose contains a lower percentage of the β-pyranose form (and a lower percentage of the α-pyranose form for the same reason) and a higher percentage of the furanose and aldehyde forms at equilibrium.

**27.10** Each form of the sugar contributes its own optical rotation in proportion to the amount that is present. Let $N_i$ = the fraction of form $i$.

$$N_{\text{total}} = 52.7° = N_\alpha(112.0°) + N_\beta(18.7°)$$

Since $N_\alpha + N_\beta = 1$, substitute $N_\beta = (1 - N_\alpha)$ and obtain

$$52.7° = N_\alpha(112.0°) + (1 - N_\alpha)(18.7°)$$
$$N_\alpha(112.0° - 18.7°) = 52.7° - 18.7° = 34.0°$$
$$N_\alpha = \frac{34.0°}{93.3°} = 0.36$$
$$N_\beta = 1 - N_\alpha = 0.64$$

Table 27.1 on text p. 1340 confirms that there is 36% of the α-form and 64% of the β-form of D-glucopyranose at equilibrium.

**27.11** (a) D-Galactose would be transformed into the aldohexose that is epimeric at carbon-2, namely, D-talose. The ketose formed would be D-tagatose which has the following structure:

*(solution continues)*

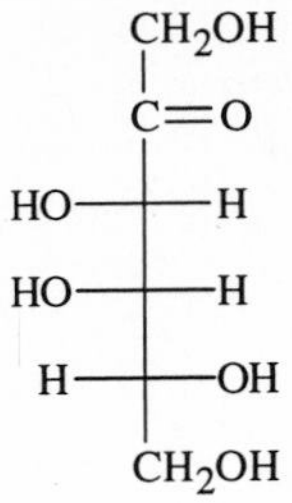

$$
\begin{array}{c}
CH_2OH \\
| \\
C{=}O \\
HO{-}\!\!\!-\!\!\!-H \\
HO{-}\!\!\!-\!\!\!-H \\
H{-}\!\!\!-\!\!\!-OH \\
| \\
CH_2OH
\end{array}
\qquad \text{D-tagatose}
$$

**27.12**  (a)  The compound is the $\beta$-*p*-nitrophenyl glycoside of D-galactopyranose; its name is *p*-nitrophenyl $\beta$-D-galactopyranoside.

(b)  This compound is hydrolyzed in aqueous acid to *p*-nitrophenol and a (mutarotated) mixture of $\alpha$- and $\beta$-D-galactopyranose.

**27.14**  (a)  Start with the structure of $\beta$-D-fructofuranose on text p. 1339 and use it to draw the $\beta$-glycoside of methanol.

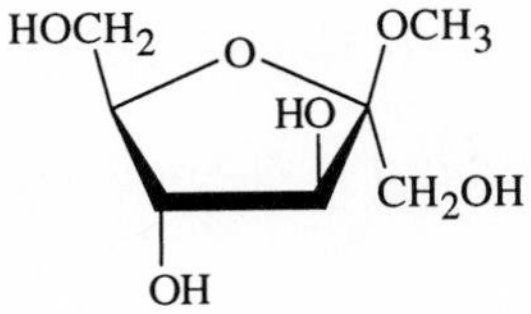

methyl $\beta$-D-fructofuranoside

**27.15**  The protonation of an oxygen of the acetal group (the glycosidic oxygen in the example below) and loss of an alcohol (methanol in the example below) gives a *resonance-stabilized* carbocation intermediate. A similar mechanism applied to ordinary ethers gives carbocations that are not resonance-stabilized. Hammond's postulate implies that the hydrolysis reaction which involves the more stable carbocation intermediate should occur more rapidly.

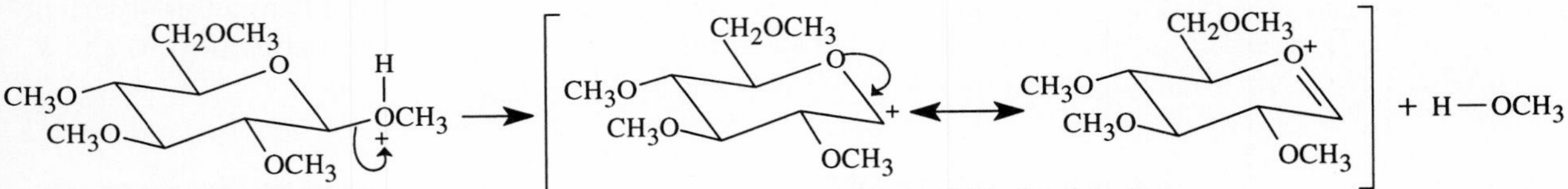

glycoside protonated
on the C-1 oxygen

a resonance stabilized carbocation

**27.16**  (a)  Prepare the 1-*O*-ethyl derivative and use the Williamson synthesis to introduce the methyl groups.

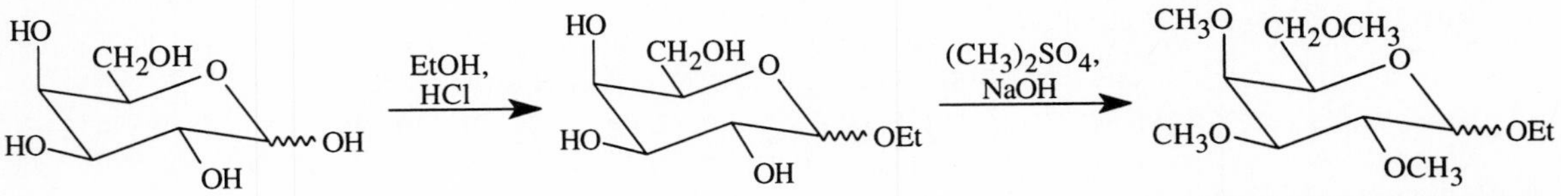

ethyl 2,3,4,6-tetra-*O*-methyl-
D-galactopyranoside

**27.17**  (a)  The structure of D-galacturonic acid:

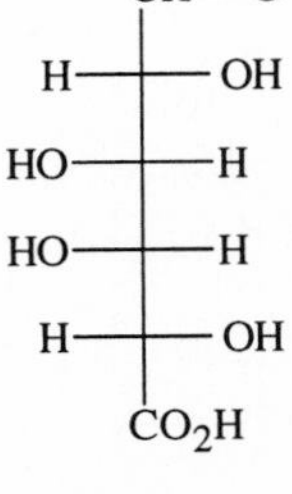

$$
\begin{array}{c}
CH{=}O \\
H{-}\!\!\!-\!\!\!-OH \\
HO{-}\!\!\!-\!\!\!-H \\
HO{-}\!\!\!-\!\!\!-H \\
H{-}\!\!\!-\!\!\!-OH \\
| \\
CO_2H
\end{array}
\qquad \text{D-galacturonic acid}
$$

**27.18**  The structures of D-glucaric and L-gularic acids are as follows. Turning either structure 180° in the plane of the page shows that the two structures are identical. (This identity was a key element in the Fischer proof of glucose stereochemistry, which is discussed in text Sec. 27.9A.)

D-glucaric acid
(derived from D-glucose)

L-gularic acid
(derived from L-gulose)

**27.19**  (a)  The structure of the aldaric acid derived from oxidation of D-galactose, and its 1,4-lactone:

D-galactaric acid

D-galactaric acid 1,4-lactone

**27.20**  (a)  As is the case with carbohydrates, the primary alcohol is selectively oxidized by $HNO_3$. (Note: Early printings of the text show a boxed "D27.35" symbol. This is a compositor error and should be deleted.)

**27.21**  As Eq. 27.29a on text p. 1352 shows, periodate oxidation of a methyl pyranoside in the D-series gives a product containing carbons 1, 2, 4, 5, and 6 of the pyranoside. The secondary alcohols at carbons 2 and 4 are converted into aldehyde groups in all cases. In the oxidation products, only the carbons corresponding to carbons 1 and 5 in the pyranoside starting material are asymmetric. Since the configuration of carbon-5 determines whether the pyranoside has the D or the L configuration, and the configuration of carbon-1 determines whether the pyranoside is $\alpha$ or $\beta$, it follows that all methyl $\alpha$-D-pyranosides give the same oxidation product with periodate.

**27.23**  The aldopentose *A* is D-ribose, and the aldohexose *B* is D-allose. The other hexose formed in the Kiliani-Fischer synthesis is D-altrose. (See Fig. 27.1 on text page 1329 for the structures of these aldoses.) Although D-xylose would also be oxidized to an optically inactive aldaric acid, its Kiliani-Fischer products, D-gulose and D-idose, would both be oxidized to optically active aldaric acids.

**27.24**  The aldopentose *A* is D-lyxose. The aldohexoses *B* and *C* are D-galactose and D-talose, respectively. (See Fig. 27.1 on text page 1329 for the structures of these aldoses.) Notice that the aldaric acid formed by oxidation of both "ends" of D-galactose is a *meso*-compound and is therefore achiral and thus optically inactive. Although D-arabinose would also be oxidized to an optically inactive aldaric acid, its Kiliani-Fischer products, D-gluose and D-mannose, would both be oxidized to optically active aldaric acids.

**27.26** Oxidize both (with dilute nitric acid) to their respective aldaric acids (which in this case are stereoisomeric tartaric acids). The compound that gives the optically active tartaric acid is D-threose; the compound that gives *meso*-tartaric acid (which is achiral and therefore optically inactive) is D-erythrose.

**27.28** (a) Aqueous HCl (1 *M*) would bring about the hydrolysis of lactobionic acid into one equivalent each of D-galactose and D-gluconic acid, which, under the acidic conditions, exists primarily as D-γ-gluconolactone (structure in Eq. 27.25 on text p. 1350).

(b) Dimethyl sulfate in the presence of NaOH methylates all hydroxy groups.

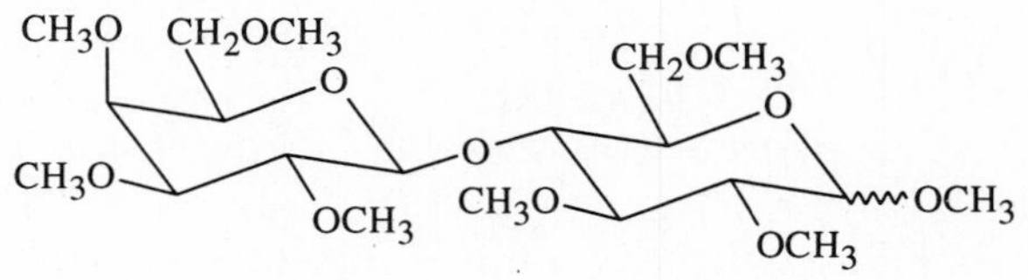

(c) Hydrolysis of all acetal and hemiacetal groups takes place.

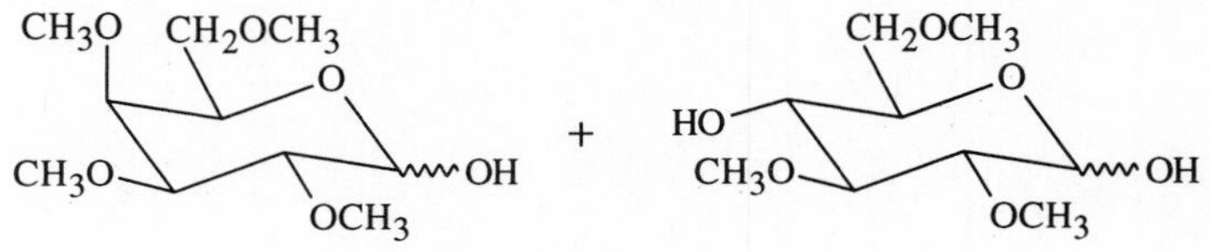

**27.30** The products would be a small amount of 2,3,4,6-tetra-*O*-methyl-D-glucopyranose (*A*, from the residue at the nonreducing end of the polymer) and mostly 2,3,6-tri-*O*-methyl-D-glucopyranose (*B*):

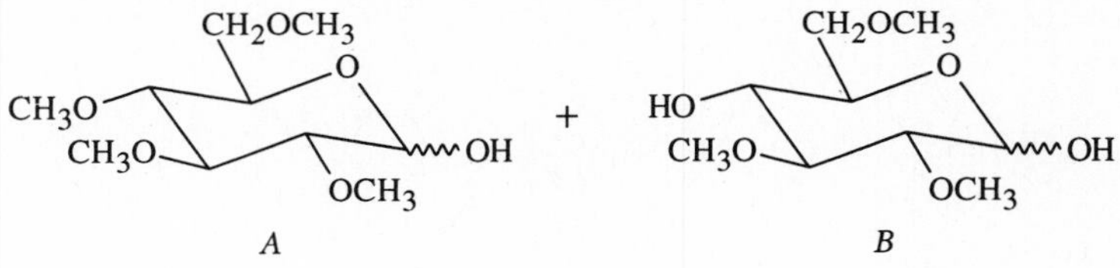

**27.31** (a)

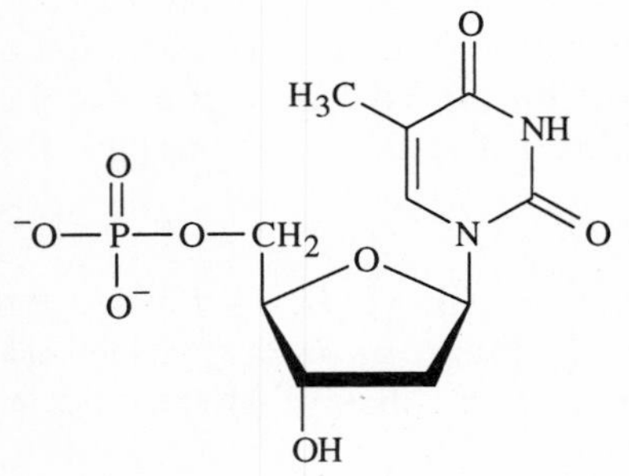

deoxythymidine monophosphate (*d*TMP)

**27.32** The structure of a tetranucleotide (a four-residue segment of RNA) with the sequence A-U-C-G is shown in Fig. SG27.1 on the following page.

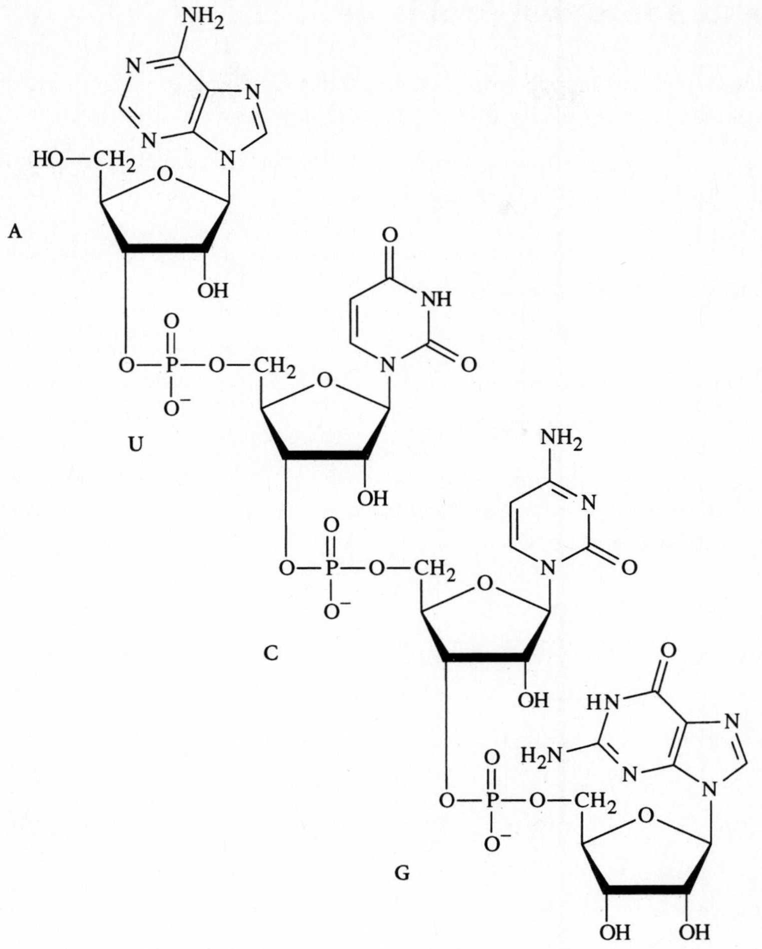

**Figure SG27.1**  *The structure of a tetranucleotide A-U-C-G to accompany the solution to Problem 27.32.*

27.34  (a)  The structure of an *O4*-methylated thymine residue in DNA is as follows. (For the numbering of purine and pyrimidine rings see text p. 1368.)

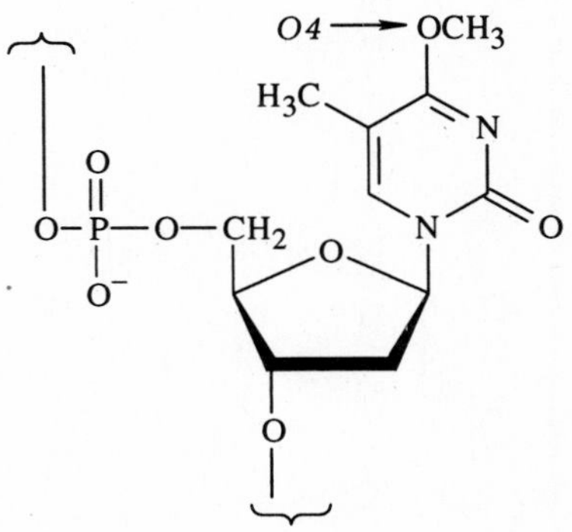

## Solutions to Additional Problems

· · · · · · · · · · · · · · · · · · · · · · · · · · · · · · · · · · · · · · · · · · · · · · · · · · · · · · · · · · · · · · · · · · · · · · · · · · · · · · · · ·

27.35  Note that carboxylic acids derived from carbohydrates usually exist as lactones, and aldoses exist primarily as cyclic hemiacetals. For simplicity they are represented below in their noncyclic forms.

(a)                                                          (b)  No reaction except mutarotation

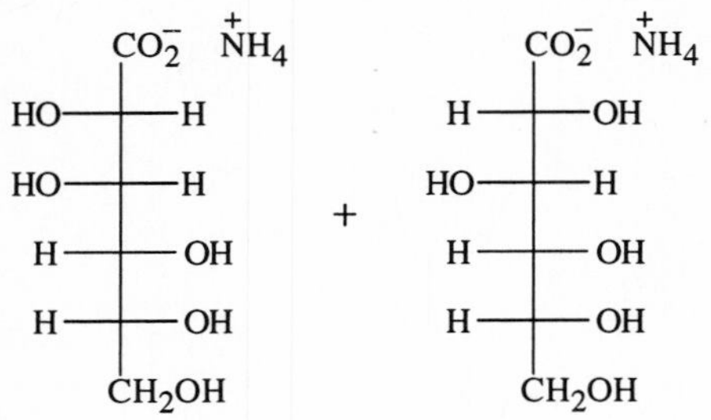

Note that in part (a), the basic conditions under which the Tollens' reagent is used (aqueous ammonium hydroxide) promote the de Bruyn–van Ekenstein rearrangement (see text p. 1341 and p. 1350). Thus, carbon-2 of D-mannose is epimerized.

(c)                                                                         (d)

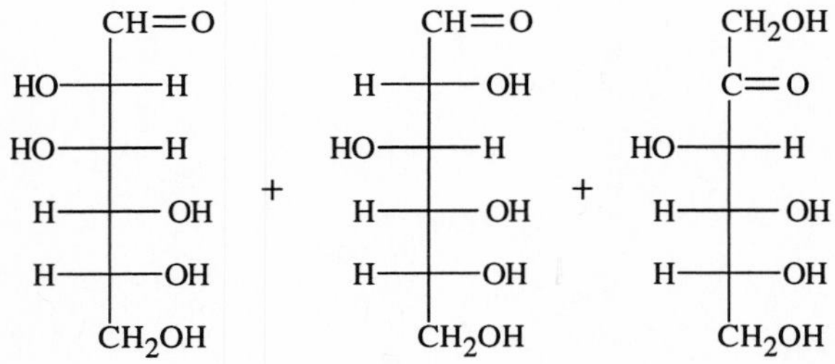

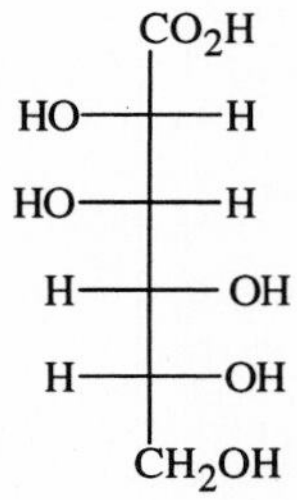

(e) 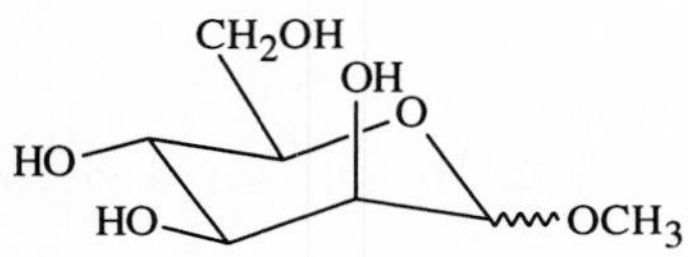                                                          (f)

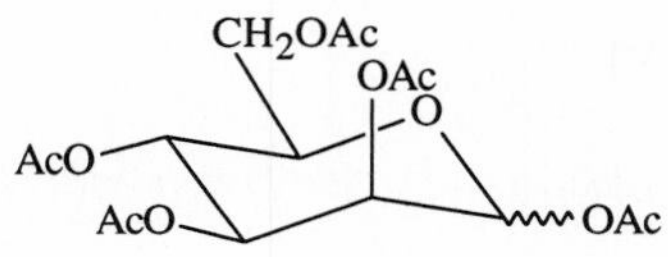

(g) 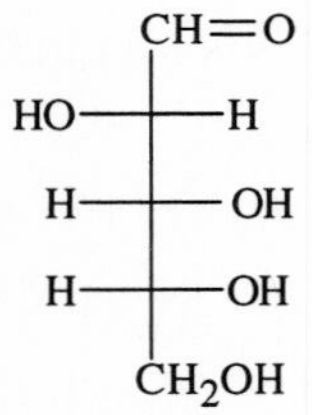                                          (h)

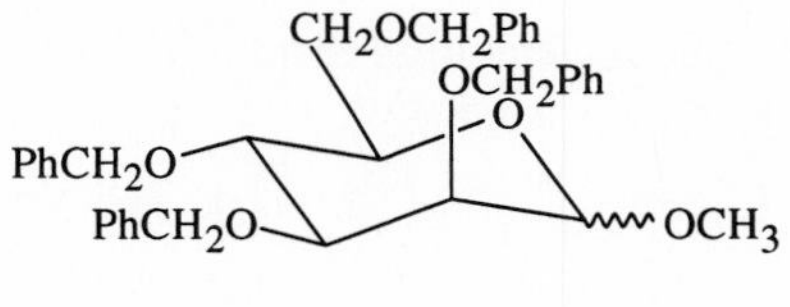

27.37  (a)                                                    (c)

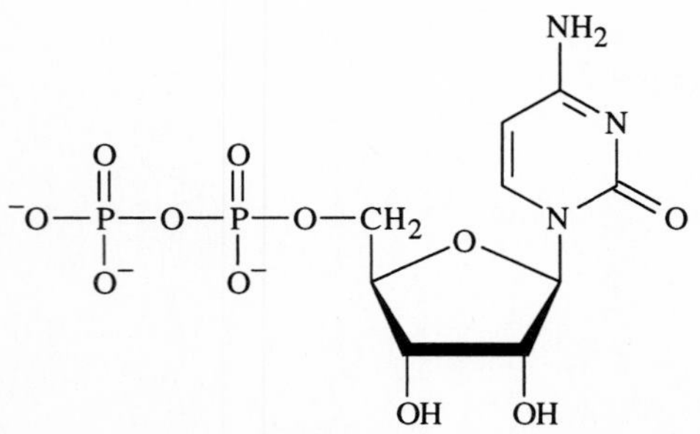

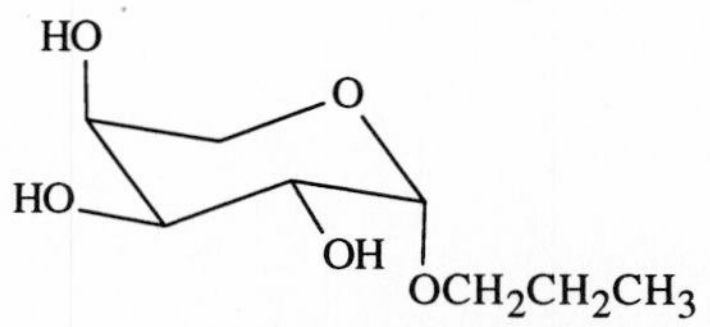

propyl β-L-arabinopyranoside

CDP

Remember that there are several valid ways to draw the chair conformations of pyranoses. Because the text probably has made you accustomed to seeing β-anomers with equatorial groups at carbon-1, you may be thinking that the chair conformation in the solution to part (c) shows the α-anomer. However, this *is* a correct representation of the β-anomer. Whether the β-anomer has an equatorial or axial group at the anomeric carbon depends, of course, on which chair conformation you choose to draw. The following equally valid representations of a different chair conformation of the same compound do have the group at the anomeric carbon in an equatorial position.

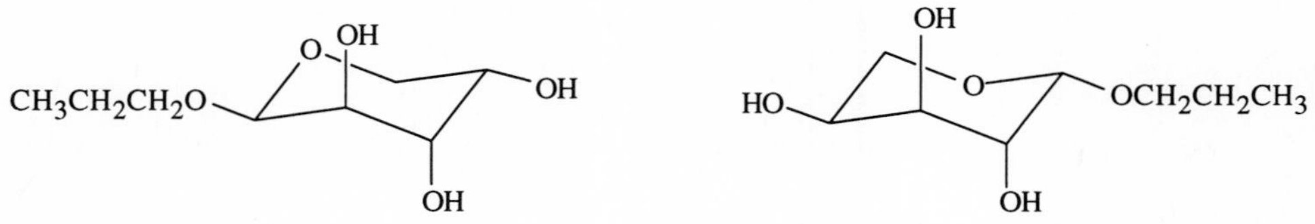

**27.38** (a) Turning this structure over 180° shows that it is the nonsuperimposable mirror image of β-D-galacto-pyranose; therefore it is β-L-galactopyranose.

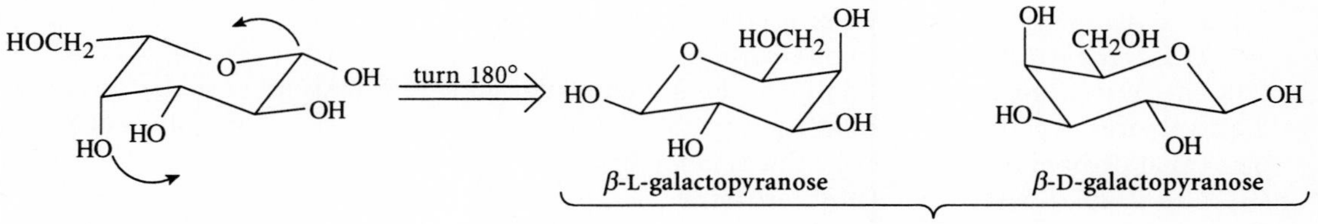

**β-L-galactopyranose**       **β-D-galactopyranose**

enantiomers

(c) First, notice that carbon-5 of this furanose has the *R* configuration. Since the D-hexoses have the *R* configuration at carbon-5 (the configurational carbon), and since the hemiacetal carbon bears a hydrogen, this must be a D-aldohexose. This can be converted into a Fischer projection as follows:

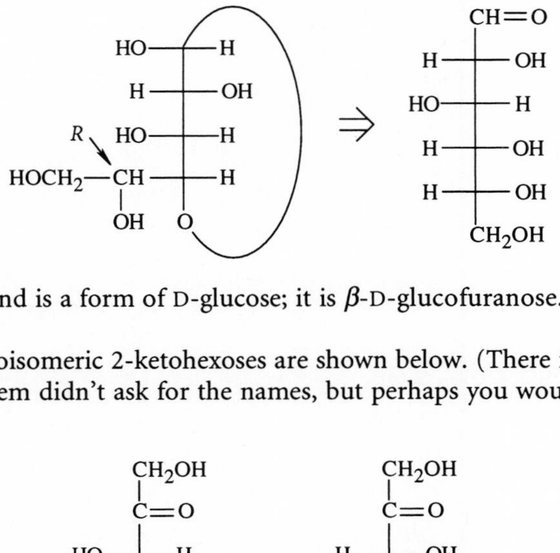

This shows that this compound is a form of D-glucose; it is β-D-glucofuranose.

**27.39** (a) The D-series of the stereoisomeric 2-ketohexoses are shown below. (There is, of course, an enantiomeric L-series.) The problem didn't ask for the names, but perhaps you would like to know them anyway.

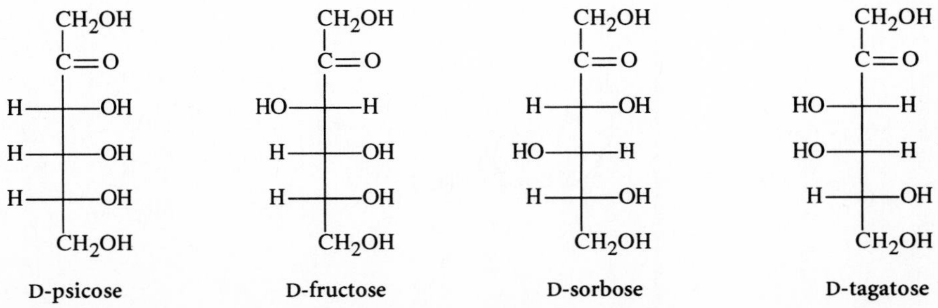

D-psicose       D-fructose       D-sorbose       D-tagatose

(c) The structure of $\alpha$-D-galactofuranose is derived from the Fischer projection as follows:

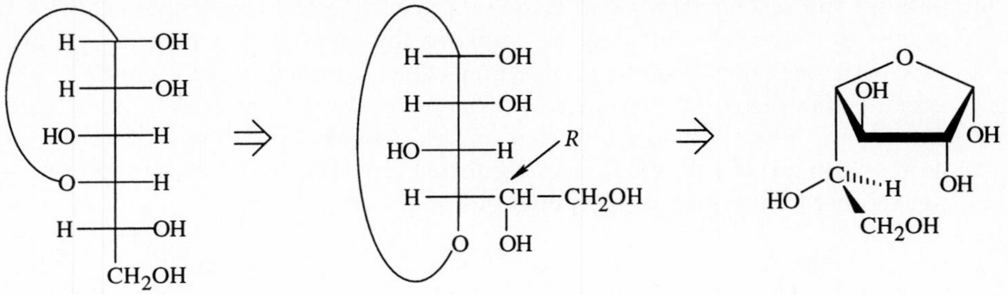

**27.40** (a) Diastereomers, epimers, and anomers. Remember that epimers and anomers are special types of diastereomers.

(c) Enantiomers

(e) Constitutional isomers

**27.41** (a) L-Sorbose is a hexose.

(c) L-Sorbose is not a glycoside because it is not involved in acetal formation with an alcohol or a phenol.

(e) This is another valid Fischer projection of L-sorbose.

(g) This is the $\alpha$-furanose form of L-sorbose.

(i) The relationship of this furanose to the one shown in part (g) is best discerned by turning either of the structures over 180°. This operation reveals that this structure differs in configuration from the one in part (g) at carbons 3 and 4. Hence, this is not a form of L-sorbose.

(k) This is the $\beta$-pyranose form of L-sorbose.

**27.42** (a) Raffinose is a nonreducing sugar because it has no hemiacetal groups; the saccharide residues at both ends are involved in acetal linkages.

(b) The glycoside linkages in raffinose:

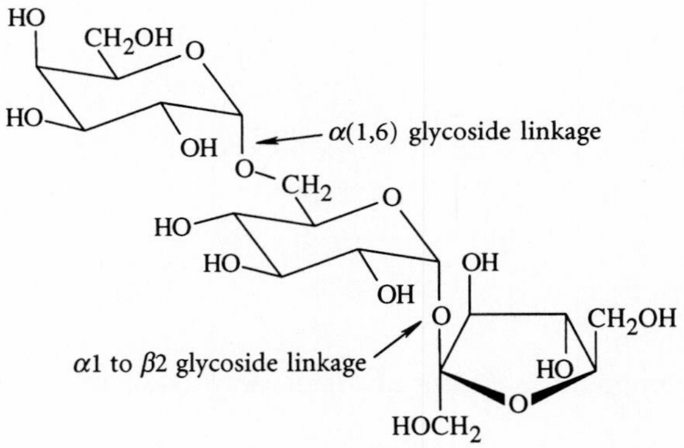

The type of linkage in the furanose residue can be determined systematically by recasting it in a different Haworth projection and using this to form the standard Fischer projection. (The glycosidic linkage is indicated as RO— .)

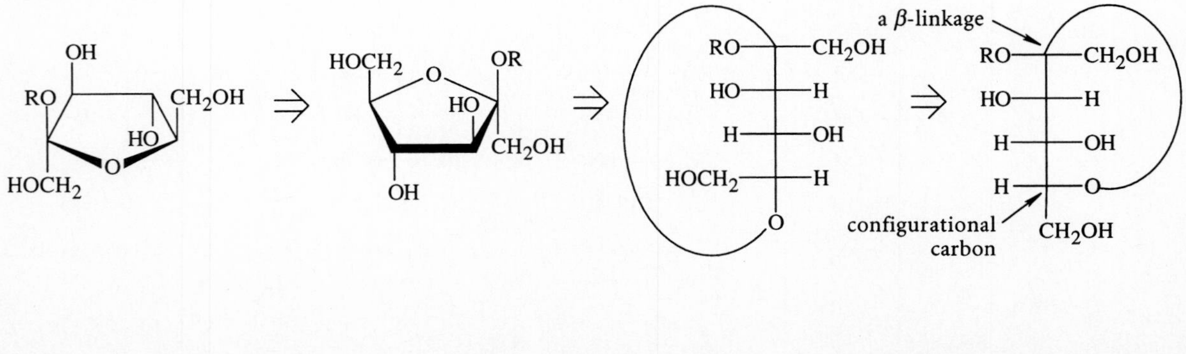

The glycoside linkage at the furanose residue is $\beta$ because the —OR group is on the opposite side of the Fischer projection from the oxygen at carbon-5, the configurational carbon.

 In early printings of the text, the —OH group at carbon-4 in the middle saccharide residue is missing in the structure of raffinose.

(c) Raffinose is hydrolyzed into equal amounts of D-galactose, D-glucose, and D-fructose.

(d) The products of methylation followed by hydrolysis are the following; each is a mixture of anomers.

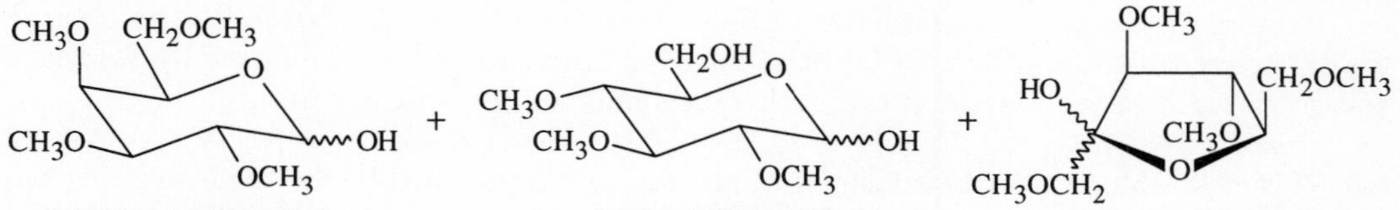

2,3,4,6-tetra-*O*-methyl-
D-galactopyranose
+
2,3,4,-tri-*O*-methyl-
D-glucopyranose
+
1,3,4,6-tetra-*O*-methyl-
D-fructofuranose

**27.44** (a) Just as the acetal units in carbohydrates undergo *hydrolysis* in water, they also undergo *methanolysis* in methanol. The result is a mixture of the methyl glycosides along with the by-product phenol.

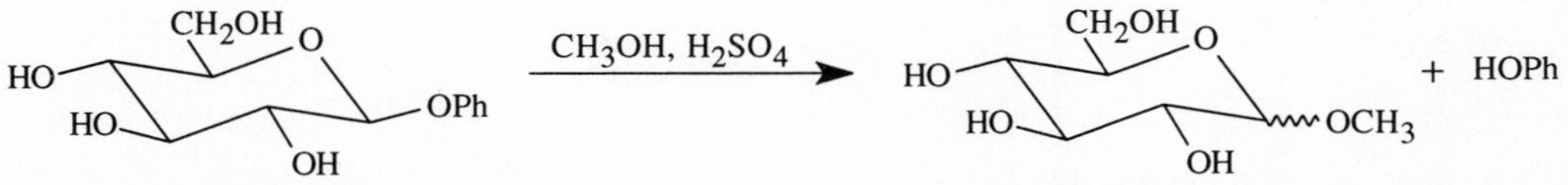

phenyl $\beta$-D-glucopyranoside $\xrightarrow{\text{CH}_3\text{OH, H}_2\text{SO}_4}$ methyl D-glucopyranoside (mixture of $\alpha$ and $\beta$) + HOPh

(c) Osmium tetroxide forms *cis*-1,2-cyclohexanediol; periodic acid cleaves this diol to form the dialdehyde 1,6-hexanedial, $O{=}CHCH_2CH_2CH_2CH_2CH{=}O$; and this dialdehyde is reduced by sodium borohydride to 1,6-hexanediol, $HOCH_2CH_2CH_2CH_2CH_2CH_2OH$.

(e) All of the free hydroxy groups in sucrose are methylated. (For the structure of (+)-sucrose see text p. 1363.)

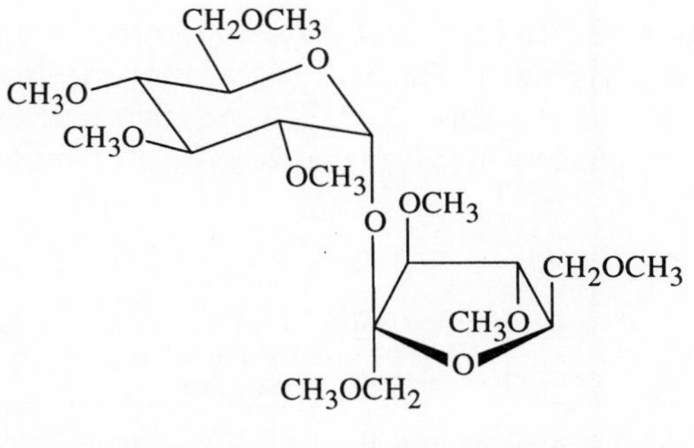

(g) This reaction is not unlike the one in part (a), in which one acetal is converted into another in methanol. Acid-catalyzed loss of methanol is followed by cyclization to give a mixture of the two methyl pyranosides. (Note that $CH_3\overset{+}{O}H_2\ Cl^-$, the conjugate acid that results from protonation of the solvent by HCl, is the effective acid catalyst.)

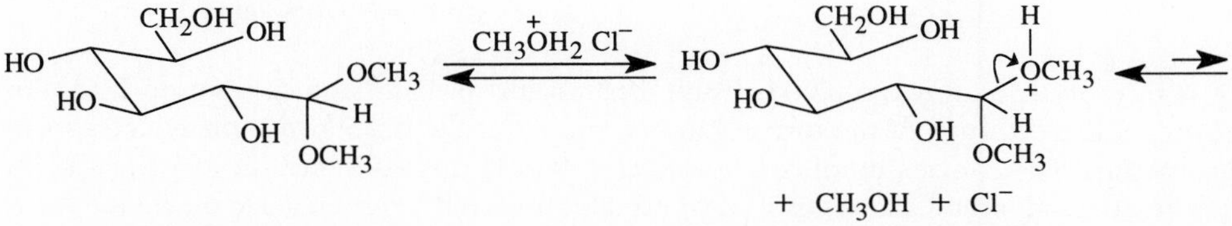

$\xrightarrow{CH_3\overset{+}{O}H_2\ Cl^-}$

+ $CH_3OH$ + $Cl^-$

*(equation continues)*

CH₂OH structures (reaction scheme at top of page showing mechanism with methyl glycoside formation):

$$CH_2OH \quad OH \quad \cdots \quad HO \quad HO \quad OH \quad OCH_3 \quad H \quad + \; HOCH_3 \longrightarrow$$

$$CH_2OH \quad O \quad HO \quad HO \quad OH \quad OCH_3 \quad H \; HOCH_3 \rightleftharpoons$$

$$CH_2OH \quad O \quad HO \quad HO \quad OH \quad OCH_3 \quad H \quad + \; H_2\overset{+}{O}CH_3$$

**27.45** (a) Because the osazones from the two carbohydrates are the same, and given that both are aldohexoses, the only stereochemical difference between them—a difference that is obliterated by osazone formation—is at carbon-2. Consequently, the two carbohydrates must be epimeric at carbon-2.

**27.46** (a) This is an isotopically labeled analog of glucose, and is prepared by the Kiliani-Fischer synthesis shown in Eqs. 27.32 and 27.33 on text page 1354, except that radioactive sodium cyanide (Na¹⁴CN) is used instead of ordinary NaCN. In this synthesis, ¹⁴C-D-mannose will be a by-product.

(c) The synthesis of compound (c) requires removal of an *unlabeled* carbon from D-arabinose, replacing it with a labeled carbon, and then completion of an ordinary Kiliani-Fischer synthesis to give labeled mannose. (* = ¹⁴C)

Fischer projection series:

CH=O / HO—H / H—OH / H—OH / CH₂OH  →(Ruff degradation)→  CH=O / H—OH / H—OH / CH₂OH  →(Kiliani-Fischer synthesis with *NaCN)→  *CH=O / HO—H / H—OH / H—OH / CH₂OH  →(Kiliani-Fischer synthesis)→  CH=O / HO—*—H / HO—H / H—OH / H—OH / CH₂OH

D-Ribose-1-¹⁴C is a by-product of the first Kiliani-Fischer synthesis, and D-glucose-2-¹⁴C is a by-product of the second Kiliani-Fischer synthesis.

**27.47** The results of the oxidation of *A* indicate that compound *A* is not a methyl pyranoside or furanoside, that is, the methyl ether is not at carbon-1. The Ruff degradation yields another reducing sugar. This means that the methyl ether cannot be at carbon-2. Only if the ether is at carbon-3 can one of the Kiliani-Fischer products be oxidized to an optically inactive (that is, a *meso*) compound:

Fischer projection series:

CO₂H / H—OH / CH₃O—H / H—OH / H—OH / CH₂OH (*B*)  ←(Br₂ / H₂O)—  CH=O / H—OH / CH₃O—H / H—OH / H—OH / CH₂OH (*A*)  →(Kiliani-Fischer synthesis)→  CH=O / H—OH / H—OH / CH₃O—H / H—OH / H—OH / CH₂OH  +  CH=O / HO—H / H—OH / CH₃O—H / H—OH / H—OH / CH₂OH

HNO₃ oxidation gives a *meso*-compound

**27.49** Compound *A* is the corresponding oxime (see Table 19.3 on text p. 906), and compound *B* results from acetylation of the hydroxy groups of the oxime. Compound *C* results from a sodium acetate-promoted β-elimination reaction. The conversion of compound *C* to D-arabinose is the result of a transesterification reaction, which removes all acetyl groups as methyl acetate, followed by the reverse of cyanohydrin formation.

*Identity of compounds A and B:*

*The mechanism for the conversion of compound B into compound C:*

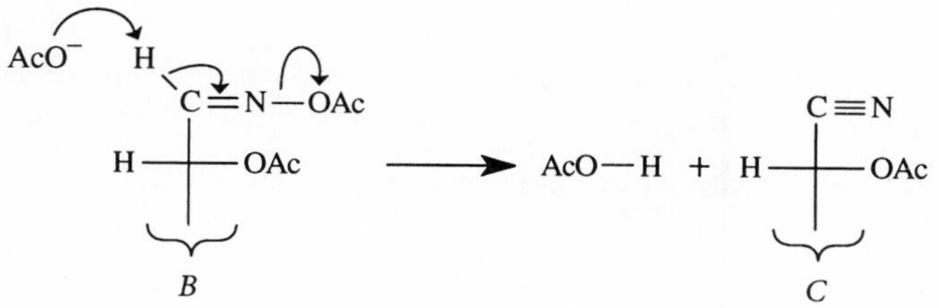

*The mechanism for the conversion of compound C into* D-*arabinose:*

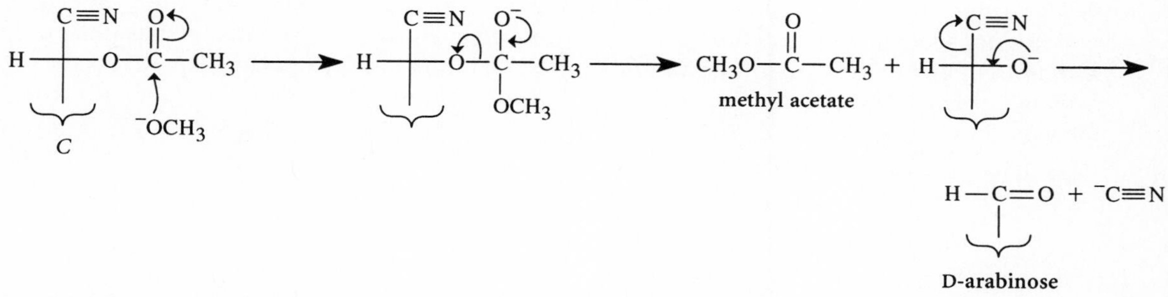

Note that the other acetate groups are removed in transesterification reactions, the mechanisms of which are identical to the first two steps of the foregoing mechanism.

**27.50** Compound *A* is the lactone of the corresponding aldonic acid, and compound *B* is the amide that results from lactone ammonolysis. Chlorine and NaOH bring about a Hofmann rearrangement of amide *B* to give *C*, a carbinolamine, which, under the aqueous reaction conditions, breaks down spontaneously to ammonia and an aldehyde. (See Eq. 19.56a on text p. 1105.) This aldehyde is the aldose that has one fewer carbon atom than the starting aldose. (This is illustrated below with D-glucose, but any aldose whose aldonic acid can form a lactone would undergo the same chemistry.)

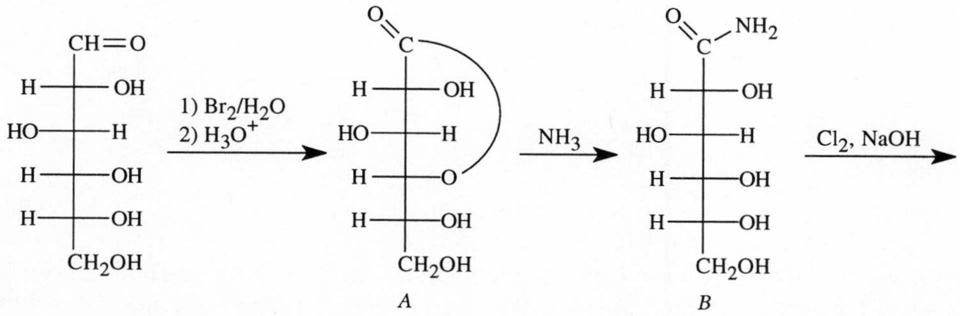

*(equation continues)*

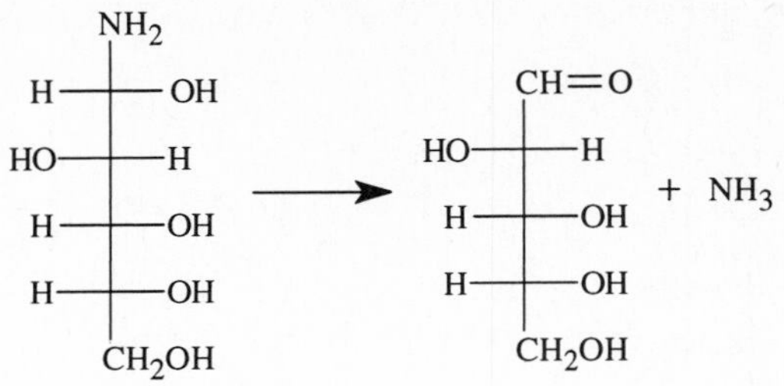

*C*
(a carbinolamine)

**27.51** The product *X* might be reasonably expected to result from treatment of methyl $\alpha$-L-rhamnopyranoside with periodic acid:

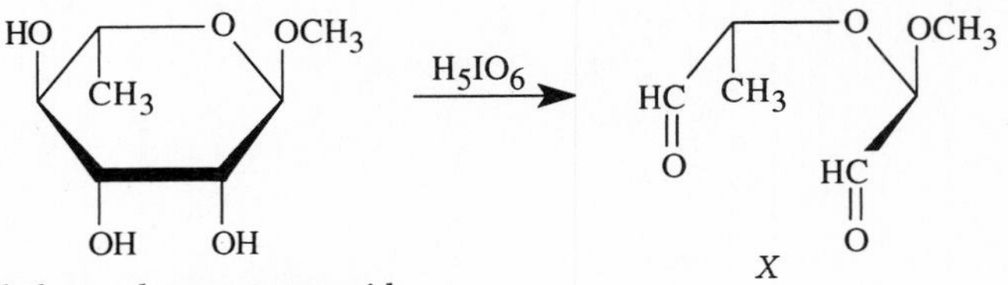

methyl $\alpha$-L-rhamnopyranoside

The formula of *X*, $C_6H_{10}O_4$, is short of the formula of actual product *A* by the elements of $H_2O$. Nevertheless, reduction of this product by $NaBH_4$ would give compound *C*. Evidently, compound *A* has two —OH groups, because methylation introduces two carbon atoms. Compound *A* results, first, from hydration of one of the carbonyl groups of *X*; an —OH group of the hydrate then attacks the other carbonyl carbon to give a "double hemiacetal" *A*. The structure of *A* and its methylation product *B* are as follows:

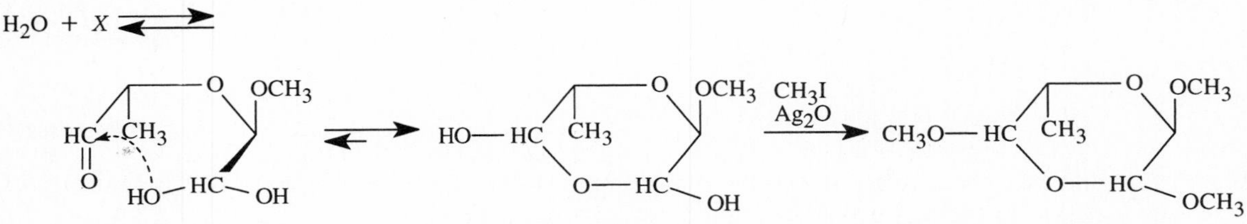

$H_2O + X \rightleftharpoons$

a hydrate of *X*          *A*          *B*

(Compound *B* is undoubtedly a mixture of stereoisomers.) Because compound *A* is a hemiacetal, it shows no carbonyl absorption. However, it is in equilibrium with a small amount of the dialdehyde *X*, which is reduced by $NaBH_4$ to give compound *C*. The reduction pulls the equilibrium between *A* and *X* toward *X* until all of this compound has been reduced.

**27.53** (a) The pairing of an imine tautomer of C with A:

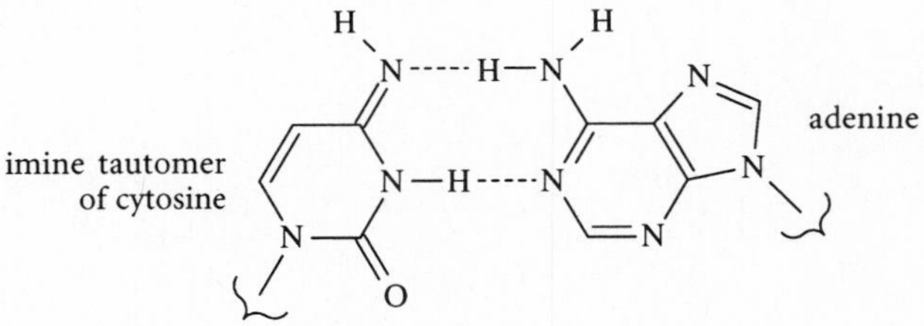

imine tautomer
of cytosine

adenine

**27.54** (a) Figure 27.6 on text p. 1374 shows that a G-C pair has three hydrogen bonds, but an A-T pair has only two. The hydrogen bonds hold the strands of the double helix together. Since the melting temperature is a measure of the forces holding the strands of the double helix together, the higher melting temperature of the polyG-polyC double helix is accounted for by the greater number of hydrogen bonds per residue.

27.55 The hydrolysis of maltose by $\alpha$-amylase shows that maltose contains two glucose residues connected by an $\alpha$-glycosidic bond. Because maltose is a reducing sugar, a free hemiacetal group must be at carbon-1 of one of the glucose units. The 2,3,4,6-tetra-O-methyl-D-glucose must arise from the glucose residue at the nonreducing end. The question is which oxygen in the glucose residue at the reducing end is involved in the glycosidic linkage; the other product of methylation-hydrolysis provides evidence on this point. Since the oxygens at carbons 4 and 5 are not methylated in this product, one of these oxygens is involved in pyranoside or furanoside ring formation, and the other is involved in the glycosidic linkage. Two structures satisfy these requirements:

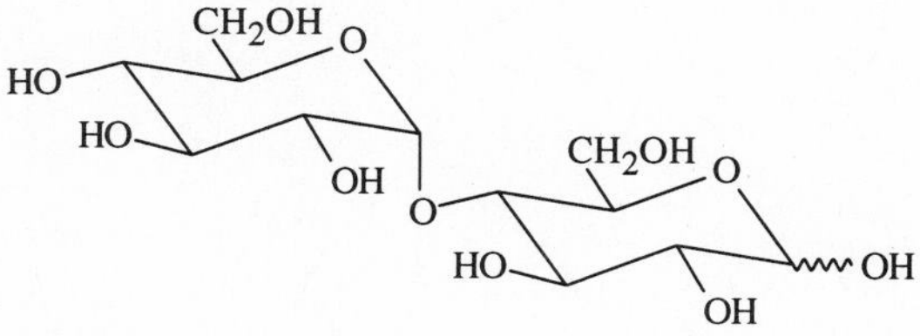

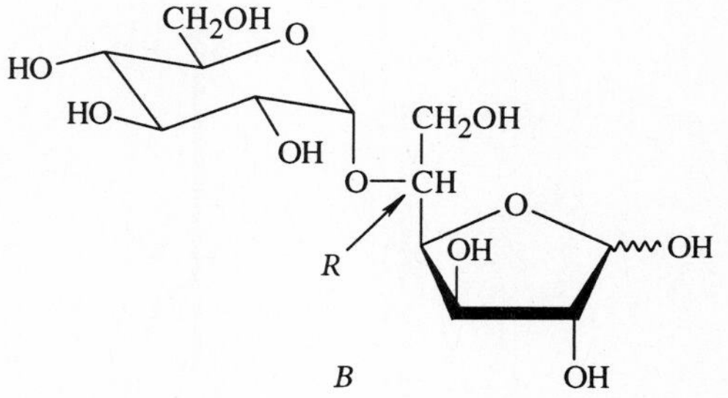

Methylation of maltobionic acid occurs under basic conditions; under these conditions any lactone present would be saponified. Hence, the additional oxygen that is methylated under these conditions must be the oxygen that was *within* the glycoside ring at the reducing end of maltose. In other words, oxidation and saponification expose this oxygen as an —OH group and thus make it available for methylation. Since the oxygen at carbon-5 is methylated, it must have been the oxygen within the saccharide ring. Since the oxygen at carbon-4 is *not* methylated, it must have been the oxygen involved in the glycosidic bond. Hence, structure A is the correct one for maltose. To summarize the chemistry using the correct structure:

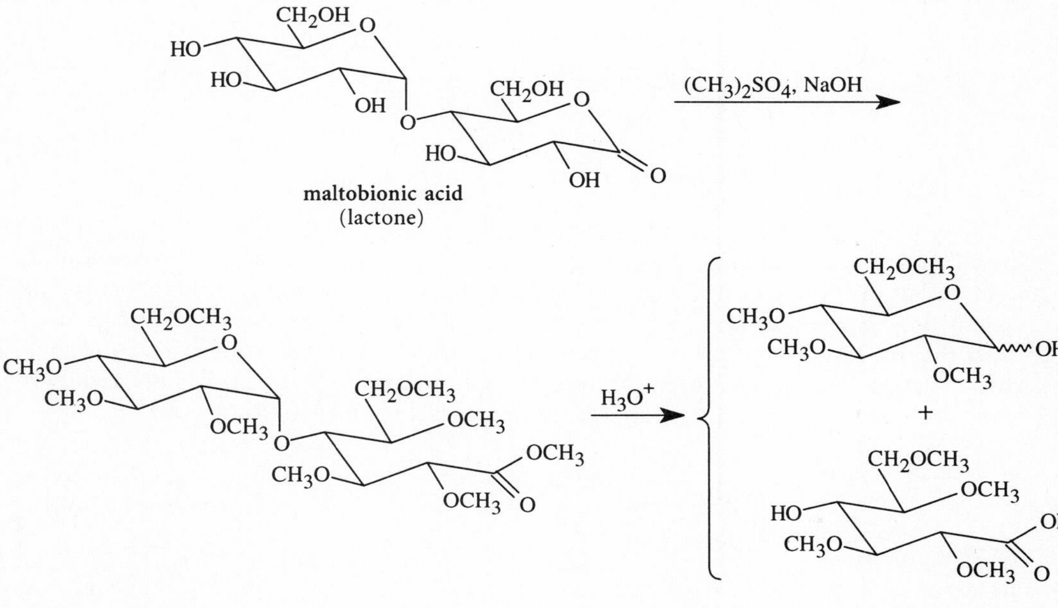

27.57 (a) The ribonucleotide sequence is translated from the 5'-end according to the genetic code in Table 27.4 on text p. 1377:

A-U-G-A-A-A-C-A-A-G-A-U-U-U-U-U-A-U-U-G-G-G-G-G

Met — Lys — Gln — Asp — Phe — Tyr — Trp — Gly

(c)  The codon at positions 16, 17, and 18 is U-A-U; the indicated mutation would change this to U-A-A. Because this is a "stop" codon, the resulting peptide would be Met-Lys-Gln-Asp-Phe.

27.58  The mechanism below begins with the imine, which is formed by reaction of the aldehyde group of D-glucose with the amine. (The mechanism of imine formation is discussed on text p. 905.) The *p*-methylphenyl group of the amine is abbreviated R, and only the part of D-glucose that is involved in the reaction is shown explicitly.

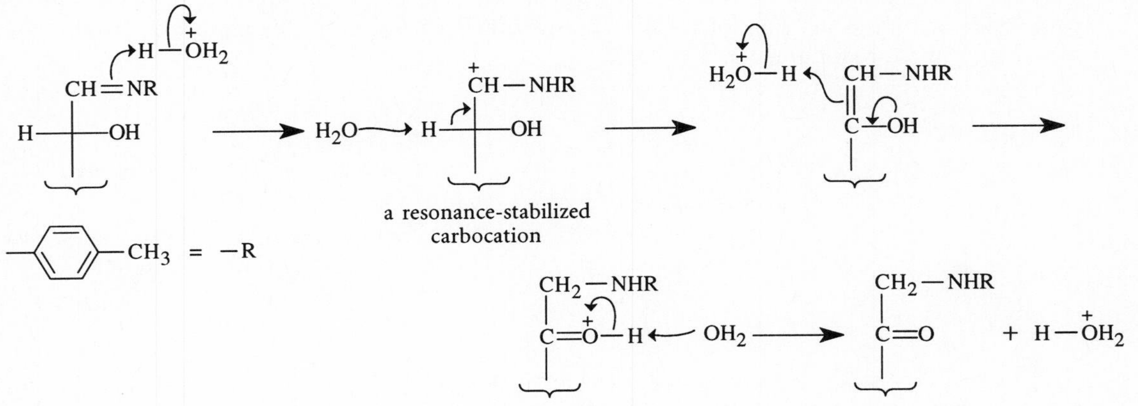

27.60  (a)  The acidic hydrogen is the one that ionizes to give a conjugate-base anion in which the negative charge is delocalized into the carbonyl group. This anion is stabilized by the electron-withdrawing polar effects of nearby oxygens and by the same type of resonance interactions that are present in a carboxylate ion. Consequently, ascorbic acid is about as acidic as a carboxylic acid.

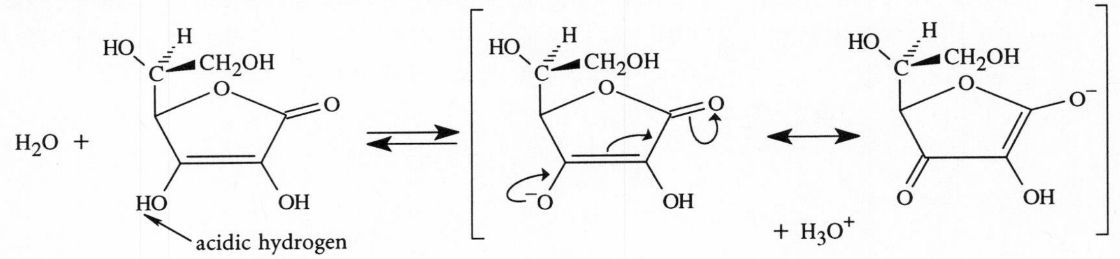

(b)  The reaction sequence with the missing structures added begins with compound *A*. The fact that carbon-2 of L-sorbitol (or carbon-5 of D-glucitol) becomes the acetal carbon as a result of the reaction labeled "*C*" in the problem shows that the —OH group on this carbon is the one that is oxidized; this carbon is therefore at the ketone level of oxidation. KMnO$_4$ oxidizes the —CH$_2$OH group to a carboxylate; because the other —OH groups are protected as acetals, they are stable under the basic conditions of permanganate oxidation. Hydrolysis of the oxidation product in acid removes the acetal protecting groups, protonates the carboxy group, and promotes lactone formation.

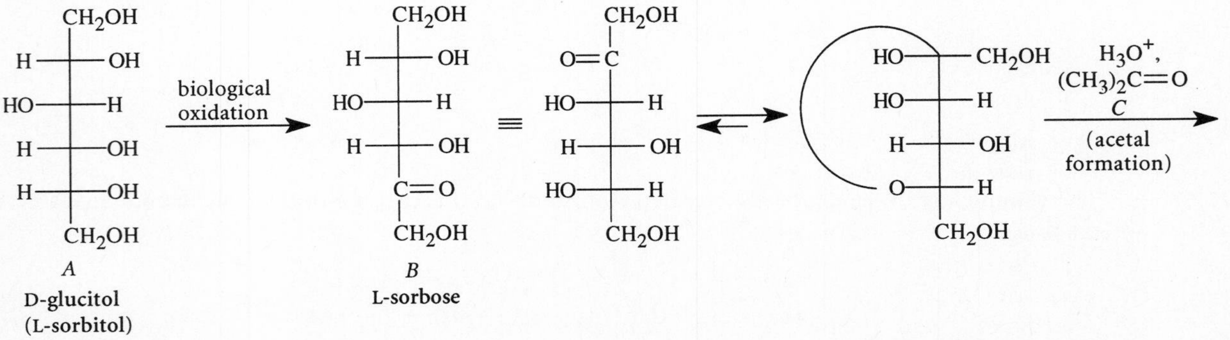

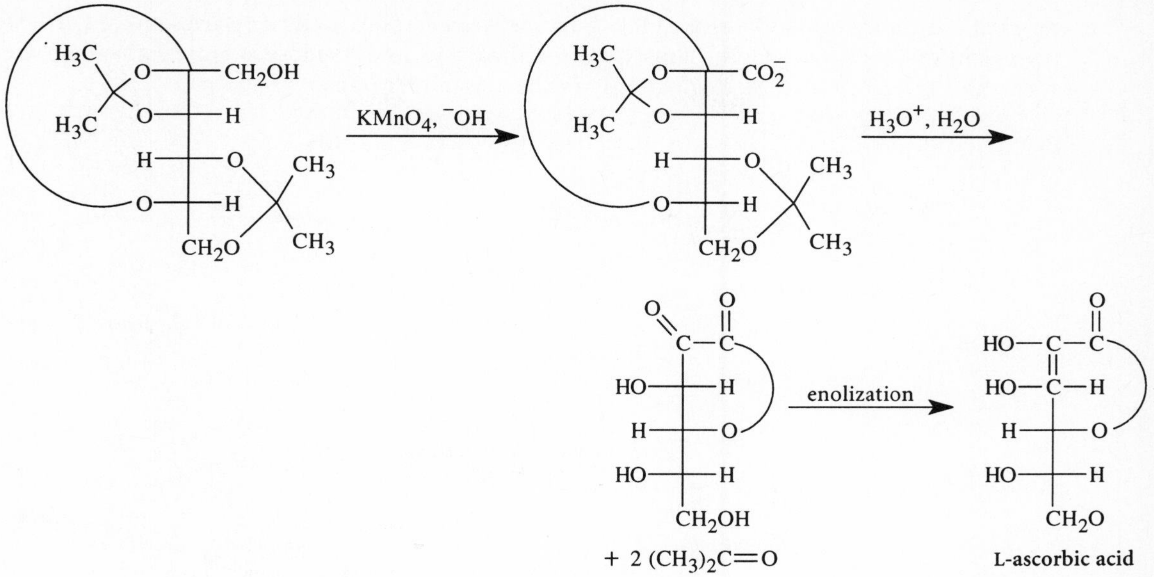

Can you think of one or more reasons why ascorbic acid should exist as an enediol rather than as an $\alpha$-keto lactone? (See Sec. 22.2 on text p. 1044 and the discussion of carbonyl dipoles on text p. 843.)

**27.61** (a) Periodic acid cleaves the *cis*-2′,3′-diol group at the 3′-end of RNA to form a dialdehyde. (Note that only the 3′ residue of RNA contains a 2′,3′-diol.) The presence of an aldehyde carbonyl group makes the $\alpha$-hydrogen at carbon-4′ more acidic. Removal of this hydrogen sparks an E2 reaction in which the internucleotide bond is cleaved.

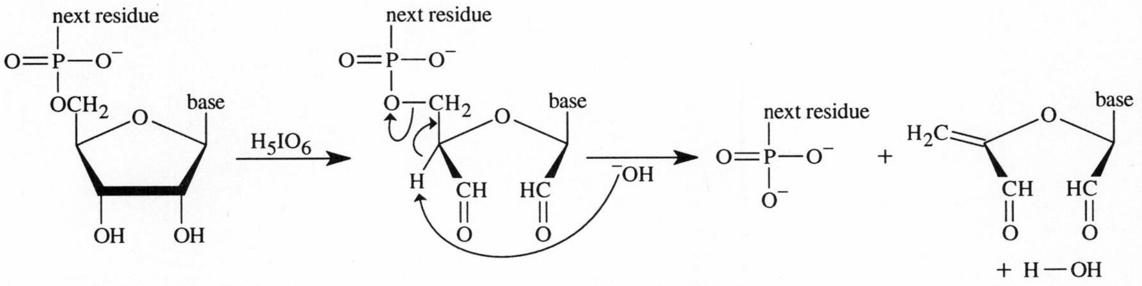

(b) DNA cannot undergo this reaction because a *cis*-2′,3′-diol group is needed for the periodate cleavage reaction. DNA, of course, has no 2′-hydroxy group at the 3′ end, and thus, no 2′,3′-diol.

**27.63** (a) The anhydro form of D-idose has the following chair conformation:

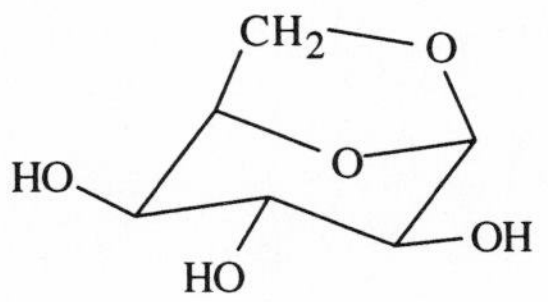

1,6-anhydro-D-idopyranose

(b) The issue is the relative stabilities of the anhydro and ordinary pyranose forms of D-idose and D-glucose. In the chair conformation of 1,6-anhydro-D-glucopyranose, all of the hydroxy groups are axial. The many resulting 1,3-diaxial interactions destabilize this form of glucose. In the more stable chair form of D-glucopyranose, in contrast, all of the hydroxy groups are equatorial. Thus, the pyranose form of D-glucose is much more stable than the 1,6-anhydro form, and D-glucose exists primarily as an ordinary pyranose. In 1,6-anhydro-D-idopyranose, all of the hydroxy groups are equatorial. In the more stable of the chair conformations of the ordinary pyranose form of D-idose, the —CH$_2$OH group is forced into an axial position. Formation of the ether bond between carbons 1 and 6 relieves one set of 1,3-diaxial interactions in this form of D-idose. Thus, the 1,6-anhydro form of D-idose is more stable than the pyranose form, and therefore D-idose exists mostly in the 1,6-anhydro form.

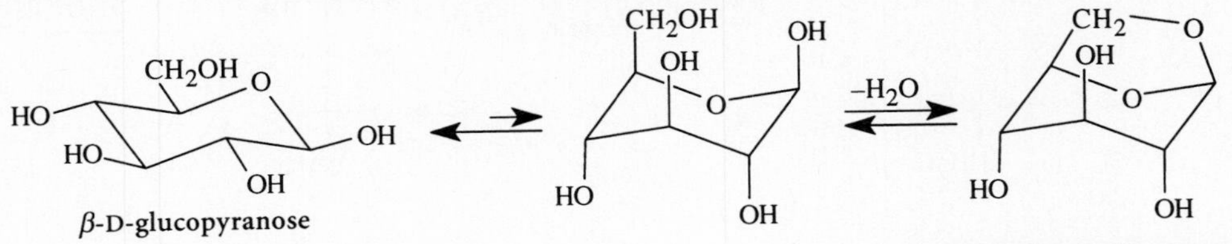

β-D-glucopyranose    1,6-anhydro-D-glucopyranose

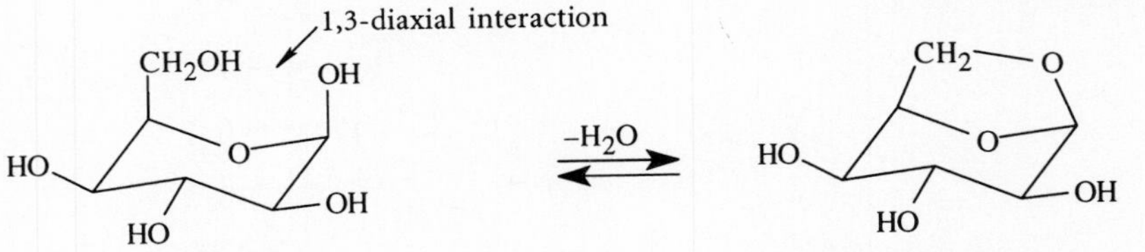

β-D-idopyranose    1,6-anhydro-D-idopyranose

1,3-diaxial interaction